Advanced Research in Flexible Risers and Pipelines

Advanced Research in Flexible Risers and Pipelines

Guest Editors

Yong Bai
Weidong Ruan

MDPI

Basel • Beijing • Wuhan • Barcelona • Belgrade • Novi Sad • Cluj • Manchester

Guest Editors

Yong Bai
College of Civil Engineering
and Architecture
Zhejiang University
Hangzhou
China

Weidong Ruan
College of Civil Engineering
Zhejiang University of
Technology
Hangzhou
China

Editorial Office
MDPI AG
Grosspeteranlage 5
4052 Basel, Switzerland

This is a reprint of the Special Issue, published open access by the journal *Journal of Marine Science and Engineering* (ISSN 2077-1312), freely accessible at: https://www.mdpi.com/journal/jmse/special_issues/4KD1P5CY67.

For citation purposes, cite each article independently as indicated on the article page online and as indicated below:

Lastname, A.A.; Lastname, B.B. Article Title. *Journal Name* **Year**, *Volume Number*, Page Range.

ISBN 978-3-7258-7174-2 (Hbk)
ISBN 978-3-7258-7175-9 (PDF)
https://doi.org/10.3390/books978-3-7258-7175-9

Contents

About the Editors

Yong Bai

Yong Bai is affiliated with the College of Civil Engineering and Architecture, Zhejiang University. His main research interests lie in structural engineering, ocean engineering, offshore structures, subsea pipelines and risers, structural safety and risk assessment, and interdisciplinary research integrating ocean engineering with intelligent manufacturing, information technology, automation, big data, and artificial intelligence. He has presided over a number of major research and engineering projects, covering ship structures, offshore platform structural design, subsea engineering, structural stability, dynamic positioning, risk analysis, and other related fields. His research achievements include proposing innovative design theory for the buckling behavior and ultimate bearing capacity of deep-water subsea pipelines, which has broken through traditional design approaches, upgraded relevant industrial design standards, and been widely applied in engineering practice, with an internationally leading level of performance. He has also established close cooperative relationships with government agencies, enterprises, and renowned universities and research institutions both at home and abroad, and is committed to introducing cutting-edge technologies to promote technological innovation and research breakthroughs in related fields both internationally and in China.

Weidong Ruan

Weidong Ruan is affiliated with the College of Civil Engineering, Zhejiang University of Technology. His main research interests lie in the dynamic response of marine risers and dynamic cables, the local response and cross-section design of composite flexible pipes, and the mechanical behavior and high-precision numerical simulation of deep-water slender marine structures—fields that are crucial to advancing the development of offshore energy exploitation and marine engineering industries. He has presided over a number of key research and engineering projects, covering the dynamic response mechanisms of slender marine structures, multi-scale numerical simulation of marine risers, local deformation and buckling failure of subsea pipelines, and other frontier and critical fields in marine structural engineering. His research achievements include systematically revealing the intrinsic dynamic response mechanisms of deep-water slender structures, and developing an innovative multi-scale coupled numerical algorithm for slender risers matching beam solid elements based on the multi-point constraint method. This groundbreaking algorithm has broken through the limitations of traditional numerical simulation technologies, achieving high-precision and efficient simulation of local deformation characteristics in hazardous areas of deep-water risers, and clearly clarifying the plastic development process, local stress distribution law, and buckling failure mechanism of hazardous pipeline sections. These achievements have contributed to a progressive understanding of the safety and risk control of deep-water marine structures, providing a technical basis for the design and management of related engineering projects, as well as laying a foundation for ongoing technological advancement and innovation in the field of marine engineering.

Journal of
*Marine Science
and Engineering*

MDPI

Editorial

Advanced Research in Flexible Risers and Pipelines

Yong Bai [1] and Weidong Ruan [2,*]

1 College of Civil Engineering and Architecture, Zhejiang University, Hangzhou 310058, China; baiyong@zju.edu.cn
2 College of Civil Engineering, Zhejiang University of Technology, Hangzhou 310014, China
* Correspondence: wdruan@zju.edu.cn

1. Introduction

Over the past few decades, with the continuous expansion of offshore oil and gas resource development into deepwater and ultra-deepwater environments, slender structures such as flexible risers, subsea pipelines, mooring cables, and underwater cables have served as core links connecting offshore platforms, subsea production systems, and onshore terminals. They have been widely applied in the fluid transportation and energy transmission between various offshore production units, becoming indispensable key equipment for deep-sea resource development [1]. Benefiting from prominent characteristics such as convenient installation, excellent fatigue resistance, and high adaptability, such slender structures have demonstrated significant advantages in both technical feasibility and economic efficiency and are regarded as efficient solutions to promote the advancement of marine resource development towards the far-reaching sea [2].

However, multiple loads in the marine environment, such as strong currents, massive waves, complex terrain, corrosive media, extreme temperature and pressure, and internal flow coupling effects, combined with extreme operating conditions including typhoons, internal waves, and subsea geological hazards, pose severe challenges to the mechanical response characteristics, structural integrity, and long-term service safety of slender structures such as flexible risers and subsea pipelines [3–6]. This series of technical challenges not only imposes higher requirements on structural design theories, numerical simulation methods, and engineering test technologies but also provides important opportunities for technological innovation and disciplinary development in the field of marine engineering. They continuously drive researchers and engineers to conduct systematic exploration and breakthrough research in aspects such as material modification, structural configuration optimization, intelligent monitoring and operation, and maintenance technologies.

At present, research on slender structures such as flexible risers and subsea pipelines has formed a complete system covering theoretical research, numerical simulation, and model testing, achieving systematic progress. In terms of theoretical research, the Morison equation is generally adopted to calculate the hydrodynamic loads on slender structures, and improved finite element methods are combined with S-N curve approaches to conduct in-depth analysis on the global dynamic response and fatigue damage of marine slender structures under the combined action of wind, waves, and currents [7–11]. In the research on local failure mechanisms, based on elastoplastic mechanics and shell theory with nonlinear buckling theory as the core analytical framework, energy principles are widely applied to establish governing equations, thereby revealing the local buckling failure mechanisms of flexible risers, subsea pipelines, and other structures [12,13].

Numerical simulation technology is characterized by equal emphasis on global and local analysis. At the level of global dynamic response analysis, professional marine engineering software such as OrcaFlex (version 11.6b), relying on its powerful nonlinear time-domain finite element analysis capabilities, is widely used to simulate the global dynamic response of structures in complex marine environments [14–16] and can achieve coupling and integration with hydrodynamic calculation software such as ANSYS AQWA (version 2025 R2) and MOSES (version 2025). At the level of local refined analysis, general finite element software such as ABAQUS (version 2025) and ANSYS (version 2025 R2) plays a key role. They are mainly employed to construct local refined models of multi-layer structures such as unbonded flexible risers/pipes so as to explore the stress distribution, contact behavior, and local failure modes (e.g., critical collapse pressure, tensile stiffness, minimum bending radius, etc.) of each layer under various load conditions [17].

Despite significant advancements in theoretical research and numerical simulation, both often involve numerous assumptions, and the reliability of their results still needs verification. Therefore, model testing remains an important approach for the marine engineering community to obtain accurate results. However, offshore platforms and their riser/mooring systems typically have large spatial scales. Even when models are constructed using conventional scale ratios (e.g., 1/50 to 1/100), their sizes often exceed the carrying capacity of existing test facilities (such as the water depth and flow velocity conditions of test tanks), and the test costs are high. This makes it difficult to carry out full-model tests on the global dynamic response of flexible risers/moorings. Therefore, existing model tests focus more on the research of local phenomena, such as local buckling failure [18] and vortex-induced vibration (VIV) response [19,20]. Such tests involve relatively small spatial scales and are more feasible to implement in laboratories.

This Special Issue, entitled "Advanced Research in Flexible Risers and Pipelines", aims to promote international academic development and technical exchange in the field of marine slender structures such as flexible risers, subsea pipelines, mooring cables, and underwater cables. The Special Issue collects cutting-edge research achievements and practical reviews from experts in different countries and regions around the world, and presents an in-depth discussion on the key scientific issues and engineering challenges currently existing in this field. Despite significant progress in research, such marine slender structures still face multiple challenges in practical applications: first, the high computational cost of numerical models makes it difficult to meet the needs of real-time monitoring; second, the scarcity of experimental data, especially the lack of full-scale test data; third, the insufficient simulation of multi-physics coupling mechanisms (such as thermo-fluid–structure coupling), which affects the accurate prediction of the global structural behavior; and fourth, the long-term impact mechanisms caused by material aging, fatigue damage, and climate change all urgently need in-depth research. Future research directions should include the development of efficient reduced-order models and AI-aided tools, strengthening long-term on-site monitoring, exploring the structural response in extreme environments (such as the Arctic region), integrating sustainable materials (such as environmentally friendly polymers), and promoting the application of digital twin technology for full-life cycle management.

2. Published Papers

Wu et al. (Contribution 1) systematically investigated the cyclic tensile shakedown behavior of flexible risers by incorporating armor wire winding-induced residual stresses into an ABAQUS numerical model. Simulations accurately characterized the spatially inhomogeneous residual stress distribution across wire cross-sections. The team quantified the influence of initial residual stresses on the riser's cyclic tensile response. It reveals that within the lower load range (0–16.1 kN), the armor wire cross-section tends to attain a shakedown state with increasing loading cycles; in contrast, within the higher range

https://doi.org/10.3390/jmse14050428

(0–30.2 kN), the cross-sectional strain accumulates progressively per loading–unloading cycle, creating a ratcheting effect. This highlights the necessity of incorporating such shakedown behavior into the riser design criteria.

Saneian et al. (Contribution 2) developed and compared three ABAQUS-based finite element models (a simplified two-dimensional (2D) axisymmetric model, a fully detailed three-dimensional (3D) helical model, and an efficient periodic boundary condition (PBC) model) to address internal pressure-induced bursting of flexible risers' pressure armor layer. The results demonstrate that the PBC model, which enforces structural periodicity to simulate an infinitely long pressure armor layer, achieves an optimal trade-off between predictive accuracy and computational efficiency—yielding burst pressure predictions within 1% of analytical solutions while reducing computational expense by more than 15-fold compared to a full 3D helical model. By effectively eliminating end effects, the PBC model can deliver high-fidelity stress distributions that closely reflect real structural behavior. This study further underscores the importance of capturing helix-induced stiffness and contact interactions, as evidenced by the significant underestimation of collapse pressure in the oversimplified 2D axisymmetric model.

To resolve traditional numerical model inaccuracy in predicting the bending stiffness of flexible pipes, Costa et al. (Contribution 3) developed a finite-difference model integrated with a return-mapping algorithm to study the response of a 2.5″ flexible pipe under combined axisymmetric/cyclic bending loads, validated against non-degraded samples and samples degraded for 18 months (flooded with CO_2-saturated water at 1 bar and 40 °C). The model aligned well with experiments, capturing hysteretic behavior driven by initial adhesion (0.9 MPa) and a friction coefficient (0.05). The model reveals that higher adhesion and friction increase bending forces until the no-slip condition occurs, while energy dissipation peaks at partial slip; degraded samples retain an unchanged global bending response but require updated fatigue curves. The 2D axisymmetric model's limitations in ignoring interlayer interactions are highlighted, underscoring the crucial role of interlayer mechanics in flexible pipe performance.

Liu et al. (Contribution 4) proposed a comprehensive analytical–numerical framework to clarify the coupled failure mechanism of pressure- and tensile-armor layers in unbonded flexible risers. They established a tension–pressure coupled failure boundary for ultimate-limit-state design with theoretical verification. The research shows that pre-applied internal pressure barely affects axial stiffness, but initial axial tension notably boosts burst resistance (35% higher failure pressure at 500 kN tensile load). During tensile failure, both inner and outer tensile-armor layers assume primary load-bearing responsibility; however, the internal tensile-armor layer experiences higher stresses compared to the external counterpart. Under combined loading, the pressure-armor layer carries most of the load, which first undergoes a stress drop then re-loads until burst; after burst, the inner tensile-armor layer immediately takes over as the primary pressure-bearing component. This work provides a novel tension–pressure coupled failure boundary for ultimate-limit-state design, enabling more accurate prediction of failure thresholds and contributing to safer, more cost-effective deepwater riser engineering.

You et al. (Contribution 5) explored the impacts of internal fluid conditions (closed or constant pressure) and fluid types (incompressible or compressible) on the axial stiffness of fiber-reinforced flexible pipes via numerical simulation and experimental validation. Pioneering the fluid cavity method, they developed a fluid–structure coupling model to overcome the limitations of the traditional uniformly distributed load (UDL) assumption and adopted a hyperelastic constitutive model for an accurate description of rubber's nonlinear behavior. With only a 1.3% discrepancy in experimental validation, the UDL assumption is applicable to gas-filled pipes (both in the closed and constant pressure

https://doi.org/10.3390/jmse14050428

states) and liquid-filled pipes in the constant pressure state; however, the incompressibility of the filled liquid significantly enhances the pipe's axial stiffness, rendering the UDL approximation invalid for liquid-filled pipes in the closed state.

To address the limitations of iterative simulation/experiential tuning-based conventional methods, Fu et al. (Contribution 6) proposed an empirical model to predict optimal valve opening for offshore riser severe slugging elimination. Its core innovation integrates two empirical closures (valve resistance factor equivalence between risers and simple vertical pipes under slugging elimination; control target quantification as dual-frequency fluctuations) into a framework linking measurable parameters (e.g., pressure) to valve characteristics (e.g., flow coefficient). Using a non-slip gas fraction model and real valve data, it achieved high predictive accuracy (average absolute error of +0.01%), significantly outperforming industry-standard tools like OLGA software (+4.91% before tuning and +0.08% after tuning). Validated via laboratory experiments and an S-shaped riser field case (deviation < ±2%), it enables valve selection/operational optimization in the design stage without complex transient simulations.

Unlike most prior studies focusing on small-diameter pipes (<12 inches) or simplified boundary conditions, Wang et al. (Contribution 7) investigated the mechanical properties of 24-inch X65+Alloy625 metallurgically clad pipes via combined experimental and numerical approaches. This study adopted an integrated multi-scale, multi-physics methodology, correlating macroscopic mechanical properties with microstructural characteristics and quantifying the influence of initial ovality and residual stresses on bending performance. Digital image correlation (DIC) was employed to measure full-field strain, accurately capturing the complete sequential failure process—from uniform deformation between the base and clad layers to interfacial debonding (at an average strain threshold of 34.17%), followed by clad layer fracture and ultimately base layer fracture. The proposed finite element model incorporated measured initial ovality (0.407%) and manufacturing-induced residual stresses (54.58 MPa), ensuring simulation alignment with engineering reality.

Traditional methods predict the free-span fatigue damage of submarine pipelines assuming a constant span length, yet in practice, span length and position undergo spatiotemporal variations induced by local scour and sand wave migration. Jiang et al. (Contribution 8) proposed a probabilistic approach to account for such variations and predict the pipeline free-span fatigue life. A validated VIV numerical model based on Euler–Bernoulli beam theory and nonlinear seabed–pipeline contact was developed. The approach's core lies in establishing a mathematical transformation between span length and fatigue life: span length uncertainty from local scour is characterized via truncated Gaussian, Rayleigh, and uniform distributions to derive the probability density function of fatigue life. The findings demonstrate that compared with the fatigue life of a fixed span, the proposed approach leads to an increase in the fatigue life by about ten times.

To investigate the overbending-induced unlocking of the pressure armor layer, Wei et al. (Contribution 9) developed a three-layer finite element model implemented with uniform bending via pre-defined displacement and rotation loads and boundary constraints. This model accurately reproduced a true Zeta cross-sectional shape and helical structure of the pressure armor wire. In addition, by simplifying the complex carcass layer into an analytical cylindrical surface and adopting sophisticated meshing and coupling constraint strategies, it achieved an effective simulation of the complex contact nonlinearity unlocking problem. They pointed out that the thickness and internal diameter of the armor wire profile may be crucial variables for the analytical model of unlocking curvature radius, and internal pressurization reduces the difficulty of unlocking to a limited extent.

Vieira and de Sousa (Contribution 10) developed 2D and 3D finite element permeation models in ANSYS (Release 2023 R1) to address oversimplified pipeline geometry

https://doi.org/10.3390/jmse14050428

in existing permeation models. These models use fugacity (instead of concentration) to describe mass transfer, incorporate radial temperature gradients, and adopt a decoupled analysis approach for heat and mass transfer. Compared to traditional concentration-based models, fugacity can more accurately characterize non-ideal gas behavior, preserve continuity at material interfaces, and eliminate computational complexity from concentration discontinuities. Results show that under dry annulus conditions, annulus fugacity distribution is uniform, and 2D/3D model results are highly consistent (complex geometric effects are insignificant). In flooded annulus conditions, water causes uneven fugacity distribution; notably, the 3D model captures the critical influence of the metallic armor's helical geometry—an effect the 2D axisymmetric model misses—leading to substantial discrepancies between the two models.

Malta et al. (Contribution 11) proposed a decoupled numerical approach to analyze the mechanical response of multi-layer flexible pipelines buried in clay under large lateral soil displacement. Traditional methods, limited by computational complexity, over-simplify pipelines (as rigid sections) or soil (as elastic springs), resulting in insufficient calculation accuracy. This proposed approach decomposes the pipe–soil interaction analysis into two core steps: first, using the coupled Eulerian–Lagrangian (CEL) ABAQUS (version 2016) model to simulate the soil large deformation and obtain the soil pressure distribution on the pipe section resulting from the pipeline's interaction with the seabed; second, importing the pressure data into an ANSYS-built refined pipeline model to quantify cross-sectional deformation and local stress. Additionally, parametric studies were conducted to explore the influence of soil strength, the pipeline's internal diameter, and burial depth on the pipeline failure risk. A collapse analysis was performed based on the deformed cross-section to determine the critical loads at the maximum operational depth. The study demonstrates that soil strength and internal diameter critically govern pipe failure and the risk of collapse.

Through a systematic review, Schnepf and Gudmestad (Contribution 12) systematically expounded the core elements in the configuration design of dynamic inter-array power cables for floating offshore wind turbines (FOWTs). This study pointed out that the primary objective of the design is to achieve optimal performance and cost minimization by optimizing power cable selection, overall configuration design, and the integration of specialized components and ancillaries on the premise of ensuring safety and reliability. The authors emphasized that design constraints (such as tension/compression limits, minimum bend radius, and required minimum fatigue life) and environmental conditions (especially wind, waves, currents, and the impacts of marine growth and climate change) are the key drivers of the optimization process, while local specific conditions determine which constraints take the highest priority.

A comprehensive review by Qi et al. (Contribution 13) provides valuable insights into cables and towed objects within ocean engineering towing systems. These towing operations involve a complex multi-body system comprising tugboats, cables, and towed objects that are subjected to environmental loads such as wind, waves, and currents. Current studies primarily employ numerical simulations (e.g., OrcaFlex software), theoretical models, and experimental methods (both reduced-scale and full-scale) to analyze motion responses, yet they often overlook the integration of mechanical response and structural strength with actual towing conditions. The authors identified key gaps, such as insufficient attention to cable behavior during initial towing stages, and proposed future directions, including mechanical response during the initial stage of towing, multi-tug towing experiments, composite fiber cable studies using finite element methods, and the structural optimization of towing-related components to enhance safety and efficiency.

Ruan et al. (Contribution 14) presented a systematic review of the latest research progress on lazy wave risers (LWRs), focusing on key issues including their structural

https://doi.org/10.3390/jmse14050428

characteristics, hydrodynamic loads, global responses, fatigue damage assessment, and structural optimization. This study elaborated on the applications of numerical simulation (e.g., CFD and finite element analysis), theoretical models (e.g., nonlinear large deformation beam theory), and experimental methods in the research on LWR global responses. It emphasized the importance of multi-physics coupling effects such as vortex-induced vibration (VIV) and wave–current interaction and identified deficiencies in current research regarding model accuracy, experimental validation, and the integration of multiple environmental factors. Finally, the authors called for the use of advanced algorithms (e.g., genetic algorithms) for fatigue life assessment and structural optimization to achieve the safe and efficient application of LWRs in deepwater oil and gas development, providing directional guidance for future research.

Funding: This research was funded by the National Natural Science Foundation of China, grant numbers 52571314 and 51909236.

Conflicts of Interest: The authors declare no conflicts of interest.

List of Contributions:

1. Wu, S.; Liu, J.; Shang, E.; Yue, X.; Shen, Z. Study on Cyclic Tensile Shakedown Behaviour of Flexible Risers Considering Winding Process. *J. Mar. Sci. Eng.* **2025**, *13*, 2186. https://doi.org/10.3390/jmse13112186.
2. Saneian, M.; Zhang, Y.; Fang, P.; Bai, Y. Numerical Investigation of Burst Capacity in Pressure Armour Layer of Flexible Risers. *J. Mar. Sci. Eng.* **2025**, *13*, 2182. https://doi.org/10.3390/jmse13112182.
3. Costa, B.d.P.M.d.; Sousa, J.R.M.d.; Roitman, N.; Radke, E. Theoretical and Experimental Analysis of the Global Response of a Flexible Pipe Under Combined Axisymmetric and Bending Loads. *J. Mar. Sci. Eng.* **2025**, *13*, 2180. https://doi.org/10.3390/jmse13112180.
4. Liu, Y.; Wu, Q.; He, J.; Liu, Q.; Li, M.; Wang, G. Coupled Burst and Fracture Failure Characteristics of Unbonded Flexible Riser Under Internal Pressure and Axial Tension. *J. Mar. Sci. Eng.* **2025**, *13*, 1866. https://doi.org/10.3390/jmse13101866.
5. You, J.; Zhao, Y.; Zhang, B. The Effect of In-Pipe Fluid States and Types on Axial Stiffness Characteristics of Fiber-Reinforced Flexible Pipes. *J. Mar. Sci. Eng.* **2025**, *13*, 1069. https://doi.org/10.3390/jmse13061069.
6. Fu, J.; Wu, Q.; Sun, J.; Wang, H.; Zou, S. The Prediction of the Valve Opening Required for Slugging Control in Offshore Pipeline Risers Based on Empirical Closures and Valve Characteristics. *J. Mar. Sci. Eng.* **2025**, *13*, 981. https://doi.org/10.3390/jmse13050981.
7. Wang, F.; Gao, Y.; Hu, Z.; Yang, S.; Cui, Z.; Fu, R.; Yuan, L. Experimental and Numerical Investigation on Mechanical Properties of Large-Diameter Metallurgically Clad Pipes. *J. Mar. Sci. Eng.* **2025**, *13*, 880. https://doi.org/10.3390/jmse13050880.
8. Jiang, D.; Huang, X.; Zhao, D.; Yang, H.; Tang, G. Fatigue Life Prediction of Submarine Pipelines with Varying Span Length and Position. *J. Mar. Sci. Eng.* **2025**, *13*, 763. https://doi.org/10.3390/jmse13040763.
9. Wei, R.; Li, X.; Vaz, M.A.; Custódio, A.B. Finite Element Modelling of Pressure Armour Unlocking Failure in Unbonded Flexible Risers. *J. Mar. Sci. Eng.* **2025**, *13*, 411. https://doi.org/10.3390/jmse13030411.
10. Vieira, J.M.B.; de Sousa, J.R.M. Numerical Permeation Models to Predict the Annulus Composition of Flexible Pipes. *J. Mar. Sci. Eng.* **2024**, *12*, 2294. https://doi.org/10.3390/jmse12122294.
11. Malta, E.R.; Dong, X.; Shiri, H. Decoupled Analysis of a Multi-Layer Flexible Pipeline Buried in Clay Subjected to Large Lateral Soil Displacement. *J. Mar. Sci. Eng.* **2024**, *12*, 1238. https://doi.org/10.3390/jmse12071238.
12. Schnepf, A.; Gudmestad, O.T. Key Parameters for Design Analysis and Optimization of Dynamic Inter-Array Power Cable Configurations in Floating Offshore Wind Farms. *J. Mar. Sci. Eng.* **2025**, *13*, 875. https://doi.org/10.3390/jmse13050875.

https://doi.org/10.3390/jmse14050428

13. Qi, K.; Zhang, J.; Zhang, L.; Zhang, J.; Gan, R. A Review of Research Progress on Cables and Towed Objects of the Ocean Engineering Towing System. *J. Mar. Sci. Eng.* **2025**, *13*, 257. https://doi.org/10.3390/jmse13020257.
14. Ruan, W.; Zhou, C.; Yang, H.; Wang, Z.; Sun, B.; Bai, Y. A Review on the Recent Process of Lazy Wave Risers. *J. Mar. Sci. Eng.* **2024**, *12*, 2000. https://doi.org/10.3390/jmse12112000.

References

1. Amaechi, C.V.; Reda, A.; Ja'e, I.A.; Wang, C.; An, C. Guidelines on Composite Flexible Risers: Monitoring Techniques and Design Approaches. *Energies* **2022**, *15*, 4982. [CrossRef]
2. Ruan, W.; Nie, Q.; Lu, Y.; Chen, M.; Liu, D.; Sun, B. A global-local approach for dynamic stress evaluation of lazy wave flexible risers subjected to random wave and vessel motion. *Ships Offshore Struct.* **2023**, *18*, 810–826. [CrossRef]
3. Jhingran, V.; Zhang, H.; Lie, H.; Braaten, H.; Vandiver, K. Buoyancy spacing implications for fatigue damage due to vortex-induced vibrations on a steel lazy wave riser (SLWR). In Proceedings of the Offshore Technology Conference, Houston, TX, USA, 30 April–3 May 2012; Paper No. OTC-23672-MS. [CrossRef]
4. Xu, P.; Zheng, J.; Lai, X.; Du, Z. Pipe-soil interaction behaviors of deepwater J-lay pipeline on sloping seabed. *Appl. Ocean Res.* **2023**, *141*, 103806. [CrossRef]
5. Chen, Y.; Zhang, H.; Zhang, J.; Liu, X.; Li, X.; Zhou, J. Residual bending capacity for pipelines with corrosion defects. *J. Loss Prev. Process Ind.* **2014**, *32*, 33–42. [CrossRef]
6. Drumond, G.P.; Pasqualino, I.P.; Pinheiro, B.C.; Estefen, S.F. Pipelines, risers and umbilicals failures: A literature review. *Ocean Eng.* **2018**, *148*, 412–425. [CrossRef]
7. Chen, X.; Zhang, J.; Ma, W. On dynamic coupling effects between a spar and its mooring lines. *Ocean Eng.* **2001**, *28*, 863–887. [CrossRef]
8. Bai, X.; Huang, W.; Vaz, M.A.; Yang, C.; Duan, M. Riser-soil interaction model effects on the dynamic behavior of a steel catenary riser. *Mar. Struct.* **2015**, *41*, 53–76. [CrossRef]
9. Kim, S.; Kim, M.H. Dynamic behaviors of conventional SCR and lazy-wave SCR for FPSOs in deepwater. *Ocean Eng.* **2015**, *106*, 396–414. [CrossRef]
10. Jeong, H.; Jang, B.S.; Kim, J.D.; Park, G.; Choi, J.; Lee, D. A study on effects of slug flow on dynamic response and fatigue damage of risers. *Ocean Eng.* **2020**, *217*, 107965. [CrossRef]
11. Ruan, W.; Shi, J.; Sun, B.; Qi, K. Study on fatigue damage optimization mechanism of deepwater lazy wave risers based on multiple waveform serial arrangement. *Ocean Eng.* **2021**, *228*, 108926. [CrossRef]
12. Gong, S.F.; Ni, X.Y.; Yuan, L.; Jin, W.L. Buckling response of offshore pipelines under combined tension and bending. *Struct. Eng. Mech.* **2012**, *41*, 627–636. [CrossRef]
13. Bai, Y.; Ruan, W.D.; Cheng, P.; Yu, B.; Xu, W. Buckling of reinforced thermoplastic pipe (RTP) under combined bending and tension. *Ships Offshore Struct.* **2014**, *9*, 525–539. [CrossRef]
14. Jia, J.; Gu, J.; Huang, J.; Gao, L.; Chen, L.; Wang, S. Numerical simulation of loop formation in catenary risers on nonlinear uneven seabed. *Ocean Eng.* **2023**, *270*, 113615. [CrossRef]
15. Amaechi, C.V.; Butler, H.O.; Beddu, S.B.; Syamsir, A.; Ja'e, I.A.; Reda, A.; Ju, X. Parametric analysis on the global design of flexible riser under different environmental conditions using OrcaFlex. *PLoS ONE* **2024**, *19*, e0310360. [CrossRef] [PubMed]
16. Zhao, H.; Xu, N.; Li, Y.; Wang, Y. The effect of new steep wave riser configurations on the stability of the suspension cluster manifold based on Orcaflex. *Ships Offshore Struct.* **2022**, *17*, 1132–1141. [CrossRef]
17. Ruan, W.; Zhou, C.; Qiu, E.; Zheng, X.; Shang, Z.; Fang, P.; Bai, Y. Structural Response Research for a Submarine Power Cable with Corrosion-Damaged Tensile Armor Layers Under Pure Tension. *J. Mar. Sci. Eng.* **2025**, *13*, 2026. [CrossRef]
18. Yuan, L.; He, J.; Wang, C.; Ren, D. Experimental and numerical study on the indentation response of mechanically lined pipe. *Int. J. Press. Vessel. Pip.* **2025**, *214*, 105433. [CrossRef]
19. Ren, H.; Fu, S.; Zhang, M.; Ren, H. Hydrodynamic coefficients for an oscillating prototype riser with surface roughness ratio of 1×10^{-5}. *Mar. Struct.* **2025**, *107*, 103991. [CrossRef]
20. Li, X.; Peng, Y.; Liu, N.; Sun, P. Experimental and numerical study on vibration suppression mechanism of vortex induced vibration in riser. *Ocean Eng.* **2025**, *329*, 121098. [CrossRef]

https://doi.org/10.3390/jmse14050428

Journal of
*Marine Science
and Engineering*

MDPI

Article

Study on Cyclic Tensile Shakedown Behaviour of Flexible Risers Considering Winding Process

Shanghua Wu [1], Junyu Liu [1], Ersu Shang [2], Xiufeng Yue [3,*] and Zhuoyuan Shen [1]

[1] School of Chemical Engineering, Ocean and Life Sciences, Dalian University of Technology, Panjin 124221, China; shanghuawu@dlut.edu.cn (S.W.); ljyrxy@163.com (J.L.); 13363088075@163.com (Z.S.)
[2] VH-Marinetech Co., Ltd., Shenyang 110023, China; shangersu@vh-marinetech.com
[3] Faculty of Economics and Management, Dalian University of Technology, Dalian 116024, China
* Correspondence: yuexiufeng@dlut.edu.cn

Abstract: Flexible risers are subjected to significant tensile loads during manufacturing, installation, and in-service phases, and they experience multiple cyclic tensile loads throughout their entire service life. Whether the armour wires can achieve shakedown under cyclic tensile loads remains an open question to be investigated. In this study, first, the winding process of the tensile armour layers was explored, and the residual stress distribution in the cross-section of the armour wires after the winding process was obtained. Subsequently, a numerical simulation model of the flexible riser that considers residual stress was established based on the ABAQUS 2021 software to study the shakedown behaviour of the flexible riser under cyclic tensile loads. The results show that, during the initial loading–unloading process of the example pipe, the stress in the armour wire cross-section undergoes obvious redistribution. When cyclic loading is applied with a tensile force range of 0–16.1 kN, the armour wire cross-section tends to reach a shakedown state as the number of loading cycles increases. However, when cyclic loading is applied with a tensile force range of 0–30.2 kN, the strain of the armour wire cross-section gradually increases with each loading–unloading cycle, thus exhibiting a ratcheting effect. The cyclic tensile shakedown prediction model proposed in this study can provide a reference for the design of armour layers in deepwater flexible risers.

Keywords: flexible riser; armour wire; shakedown behavior; residual stress; ratcheting effect

1. Introduction

Flexible risers are critical equipment connecting subsea wellheads to upper floating structures. During their in-service operation, they bear tensile loads induced by self-weight, which increase as the application water depth increases. In marine engineering applications, deepwater flexible risers are subjected to complex dynamic loads over long-term service. Due to the uncertainty of the marine environment, the periodic motion of waves and currents causes the risers to undergo reciprocating tensile and bending deformations; cyclic tensile loads, in particular, are among the key factors affecting the risers' service performance. Additionally, during the installation and maintenance of risers, operations such as traction and dragging also impose cyclic tensile loads on the structure. These cyclic tensile loads under actual working conditions pose severe challenges to the mechanical properties of the armour wire layers. Figure 1 shows the typical structure of a flexible riser, where the tensile armour layer bears the majority of the pipeline's tensile load. A typical deepwater flexible riser adopts a multi-layer composite structure design, with each layer working synergistically to meet multiple functional requirements including sealing, pressure resistance, tensile resistance, wear resistance, and corrosion resistance. Among these layers, the tensile armour layer is a core load-bearing component specifically designed to bear most of the axial tensile load of the pipeline. It is usually composed of multiple layers of high-strength steel wires wound helically at a specific angle, and its mechanical properties directly determine the overall tensile strength and fatigue life of the riser. The winding process of tensile armour wires itself has a significant impact on their final performance. During this process, the steel wires are wound onto a moving mandrel via a high-speed rotating pay-off reel. While being wound, the steel wires are simultaneously subjected to tensile, bending, and torsional loads. A key characteristic is that after winding is completed, clamps and pipe end joints are usually installed directly without an unloading process being performed. This causes the internal forces exerted on the steel wire cross-section during winding—especially the stresses induced by tension and torsion—to be partially retained inside the wires, thereby forming pre-tightening forces from the processing stage. This residual stress state, combined with complex external cyclic tensile loads during service, profoundly affects the long-term mechanical behaviour and tensile failure mode of the armour layer and even the entire flexible riser. The tensile fracture failure process is shown in Figure 2.

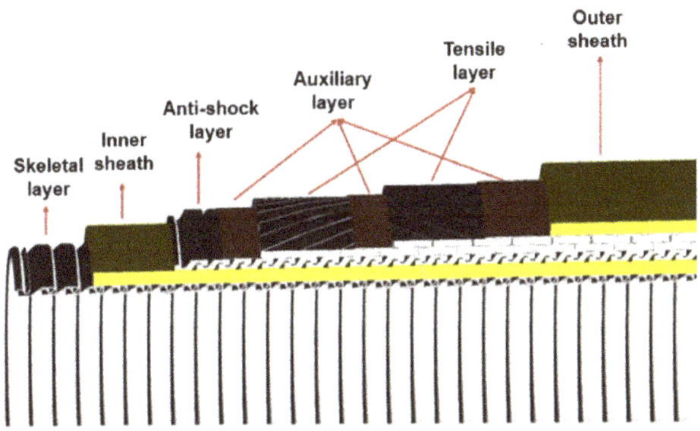

Figure 1. Schematic diagram of typical flexible riser structure.

Figure 2. Tensile failure of deepwater flexible riser.

In the research on the tensile bearing capacity of flexible risers, Knapp [1–3] conducted an in-depth analysis of the stress–displacement relationship of helically wound armoured cables under tensile and torsional loads. The study considered the material nonlinearity and geometric nonlinearity under large deformation conditions and provided a nonlinear solution based on numerical methods. Meanwhile, through a linear simplification algorithm, the stiffness matrix of helically wound armoured cables under combined tensile and torsional loads was derived, and the torsional equilibrium condition was established accordingly. Lanteigne et al. [4] further considered the coupling effect of helically wound armoured cables under tensile, bending, and torsional loads and proposed a stiffness matrix expression for the combined mechanical behaviour. LeClair et al. [5] put forward a method to predict the response of a single helically wound structure under tensile, bending, torsional loads, and frictional forces. Jolicoeur et al. [6] systematically summarized the existing prediction models for the mechanical behaviour of helically wound steel wires under axisymmetric loads, organized these models into a standardized form, obtained the results of each theoretical model through numerical methods, and compared them with experimental results. Witz et al. [7] derived the governing equations for flexible structures such as flexible pipes, umbilicals, and marine cables, and solved the load–displacement relationship under combined tensile and torsional conditions. It is worth noting that there is a difference in stress analysis between flexible pipes and cables—the stress analysis of cables does not involve internal pressure loads. Custodio et al. [8] proposed the response analysis equations and their solution methods for umbilicals and unbonded flexible pipes under tensile and torsional loads combined with internal and external pressures. Ramos et al. [9] studied the tensile-torsional performance of unbonded flexible pipes under internal and external pressure loads, proposed a linear solution for the stress and strain of each layer of the pipe, and verified the calculation results of the theoretical method through numerical and experimental means. Yue, Tang, and other researchers [10], respectively, proposed analytical models to predict the axial tensile stiffness of flexible pipes and umbilicals based on their radial contraction deformation and verified the correctness of the models by combining them with pipe tensile experiments. Xiang et al. [11] developed a new model to describe the response characteristics of multi-strand steel wire ropes under axial tension and axial torque. Francesco et al. [12] proposed a new analytical model to study the elastoplastic behaviour of metal helical wires under axial torsional loads. However, all of the above studies on tensile bearing capacity focus on single tension, and none involve cyclic tension, which is related to the issue of shakedown behaviour.

In the field of shakedown theory and experimental research, many scholars have carried out in-depth mechanical analyses. The German scholar Bleich [13] first proposed the concept of adaptability of truss structures under cyclic loads, and then Dr. Melan [14] extended it to three-dimensional ideally elastoplastic structures. The classical lower bound theorem of shakedown proposed by Melan [15] and the kinematic shakedown theorem proposed by Koiter [16] jointly laid the foundation of the shakedown theory. Since then, scholars have continuously promoted the development and improvement of the shakedown theory: Ponter and Karadeniz [17] proposed the calculation criteria for plastic shakedown; Polizzotto [18] expanded the application field of plastic shakedown analysis and improved the concept of plastic shakedown. After that, Prager [19] first proposed a simple linear kinematic hardening model. This model has the advantages of high calculation efficiency and only requiring one plastic parameter (C), but its main drawback is that it cannot predict the ratcheting effect and can only exhibit the shakedown phenomenon. Subsequently, Besseling [20] proposed a multi-linear kinematic hardening model. By dividing multiple linear segments, this model can more accurately reflect the hardening rate of materials at different stages, thereby better simulating the behaviour of actual materials. When the number of linear segments is sufficient, the model can simulate a smooth uniaxial stress–strain relationship. Similarly to the BKH model, this model has the advantages of easy parameter determination and wide application, but its disadvantages are that it cannot predict uniaxial ratcheting effects, and the predicted values of multiaxial ratcheting effects are relatively low. Armstrong and Frederic [21] proposed the most well-known nonlinear kinematic hardening model to date—the Armstrong–Frederic model. This model introduces a dynamic recovery term into the linear kinematic hardening rate, endowing it with the instantaneous memory effect of strain paths and nonlinear characteristics. Under uniaxial asymmetric cyclic conditions, the dynamic recovery term in the model can cause cumulative plastic deformation (i.e., ratcheting strain) in the material or structure during the loading and unloading processes, marking a major breakthrough in cyclic plastic constitutive models. However, this model still has the defect of being unable to accurately predict the ratcheting effect. Chaboche [22] and other researchers proposed a superimposed nonlinear kinematic hardening model, which superimposes multiple AF models and is also known as the CH3 model. The CH3 model usually divides the uniaxial tensile curve into three stages: the initial yielding stage with high plasticity; the transient nonlinear stage; and the constant modulus stage with a large strain range. Nevertheless, this model still produces a constant ratcheting strain rate (which indicates that the model simplification leads to distortion, failing to reflect the dynamic change in the material's ratcheting strain rate and resulting in non-convergent simulation, thus requiring experimental data to correct parameters). On this basis, Chaboche [23] superimposed a fourth hardening rate with the concept of a threshold onto the CH3 model, forming the CH4 model. Han Ke [24] summarized the research progress on the ratcheting effect of pressurized pipelines in recent years and determined the methods for defining the ratcheting boundary of pressurized pipelines.

In addition, the residual stress of the pipe cross-section has a significant impact on shakedown behaviour, and the processing process cannot be ignored. Fernando [25] measured the residual stress in pressure/tensile armour wires using neutron diffraction. The evolution of residual stress in tensile wires at various stages of the steel pipe manufacturing process was studied using contour methods and diffraction methods. Research was also conducted on the manufacturing process of the structural layers of flexible pipes and the prediction methods of residual stress. Saevik [26] proposed a dual-strategy method based on the two-dimensional curved beam theory to calculate the residual and transverse stress effects of the pressure armour in flexible pipes. Fernando [27] established a finite element structural model to study the detailed local and residual stress changes during the forming process, as well as stress relaxation after the Factory Acceptance Test (FAT). Tang [28] proposed a three-dimensional finite element model to study the distribution and variation in residual stress during the forming process of the flexible pipe body layer in detail. However, the research on the generation mechanism of residual stress and its impact on the mechanical properties of armour wires is still insufficient. Lu, Wu, and other researchers [29] established a mechanical analysis model for the winding process of armour wires in flexible pipes and quantified the influence of winding tension on the internal force of the steel wire cross-section after winding under elastic conditions using analytical methods. Pan [30] deeply studies the research status of marine flexible risers in fatigue analysis, comprehensively reviews the literature in related fields, and discusses the fatigue problems that may be encountered in the use of marine flexible risers in the marine environment. Bao [31] presents a theoretical analysis of the stress and failure properties of carbon fibre reinforced plastics in marine flexible risers under combined pure torsion and thermomechanical loading.

To summarize, the current research on the shakedown behaviour of flexible risers after processing does not consider the influence of residual stress from processing. However, the residual stress of the pipe cross-section has a significant impact on shakedown behaviour, and the processing process cannot be ignored. Therefore, this study investigates the shakedown behaviour while considering the residual stress from processing.

2. Simulation Model of the Winding Process of Tensile Armour Layers

2.1. Establishment of the Numerical Model for the Winding Process of Armour Wires

In this study, a mechanical model for steel wire winding was first established, with a selected winding tension of 1780 N. The detailed parameters are shown in Table 1 below.

Table 1. Armour wire structural parameters.

Width	Thickness	Winding Radius	Winding Angle	Elastic Modulus	Yield Strength	Wire Merging Die Radius
12 mm	4 mm	127 mm	30°	2×10^5 MPa	800 Mpa	132 mm

In this model, both the deformation of the steel wire during the winding process of armour wires and the contact between the steel wire and the cylinder surface are nonlinear problems. To conduct parameter research and optimization more efficiently and improve the efficiency of engineering design and analysis, a numerical simulation model for the winding process of flexible risers was established based on ABAQUS to quantify the tension, bending moment, and torque of the steel wire cross-section after winding under different winding tensions. Figure 3 shows the winding process of the flexible riser. The main equipment for the winding process of flexible risers includes a wire merging die, a steel wire reel, a pay-off reel, steel wires, and the flexible riser. In the actual steel wire winding processing, the flexible riser moves horizontally while the wire merging die remains stationary. At the same time, the steel wire reel rotates around its central axis to wind around the flexible riser. Due to symmetry, under ideal conditions, all armour wires experience the same stress state during the winding process and exhibit the same residual stress distribution after winding. Therefore, in the numerical analysis of this paper, only the residual stress of a single steel wire after winding around the central cylinder is analyzed.

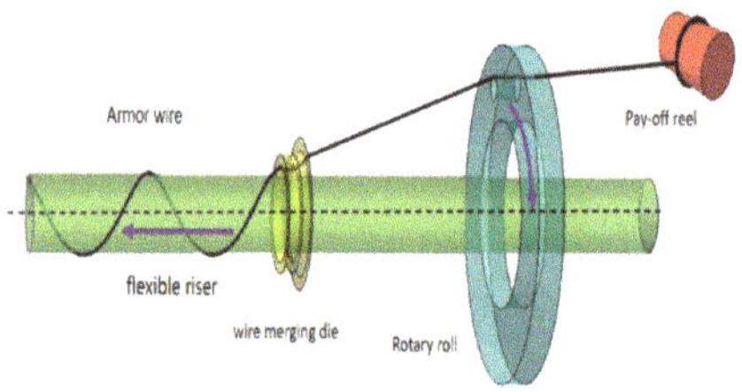

Figure 3. Winding process for armour wires of flexible risers.

A solid model is usually used to simulate the steel wire, and its cross-section is divided into six elements, as shown in Figure 4 below. The deformation of the wire merging die and the steel wire reel is negligible, so an analytical rigid body approach is used to establish the model. The steel wire winding process usually involves large deformations and nonlinear behaviours. To ensure the accuracy of calculations, a dynamic implicit analysis step is selected in this study. A reference point (RP) is set at the contact end of the steel wire and the flexible riser. Meanwhile, a reference point (RP-1) is set on the central axis at one end of the cylinder. The steel wire and the cylinder are constrained by motion coupling through the reference points. According to the relativity of motion, the cylinder can be rotated instead of the steel wire reel to simulate the 30° winding process. To ensure the stability of the steel wire winding process, the cylinder is constrained to have zero degrees of freedom except for translation and rotation in the Y-direction. The wire merging die is fixed, and a second reference point (RP-1) is set at the other end of the steel wire. A local coordinate system (Datum csys-1) is established along the direction of the steel wire, and the winding tension load is applied along the axial direction of the steel wire. The simulation model of a single steel wire winding is shown in Figure 5 below.

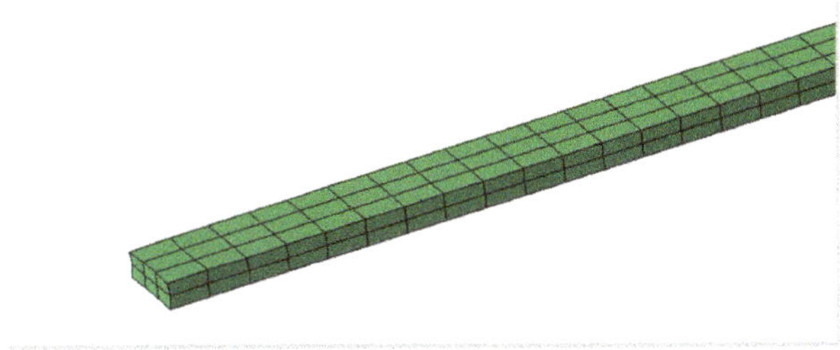

Figure 4. Mesh generation of armour wire cross-section.

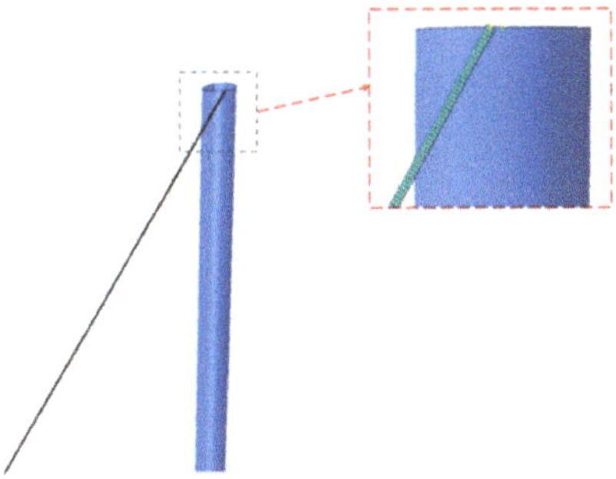

Figure 5. Numerical model for winding process of armour wire.

2.2. Analysis of Residual Stress in Steel Wire Cross-Section After Winding

To accurately study the distribution characteristics and uniformity of residual stress along the length of the steel wire after the winding process, the middle section of the steel wire—far from the end constraint effect—was selected as the representative analysis area, as shown in Figure 6. This area is intended to avoid the atypical stress state near the starting and ending ends of winding, which is caused by boundary constraints, sudden changes in contact status, and end effects. Thus, it ensures that the analyzed residual stress field mainly reflects the core influence of the winding process. The simulation calculation adopts a field variable output request, focusing on recording the total stress and plastic strain to comprehensively evaluate the model response and obtain key mechanical information. The stress–strain diagram of the calculation results of the steel wire winding model is shown in Figure 6 below. Five representative cross-sections were selected at equal intervals along the length of the middle section for analysis, and the positions of the selected cross-sections are shown in Figure 7 below. Among them, the stress-time curve of Cross-section 1 is shown in Figure 8; the stress distribution diagram is shown in Figure 9; the stress-time results of all five cross-sections are shown in Figure 10; and the stress comparison diagram of the five cross-sections of the steel wire after winding is shown in Figure 11.

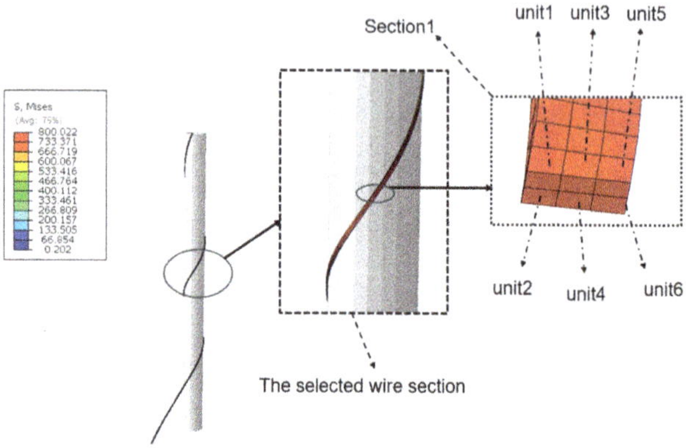

Figure 6. Residual stress distribution of armour wire cross-section after winding.

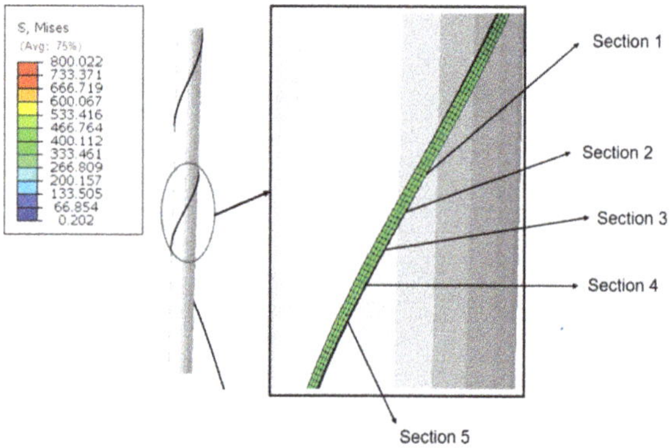

Figure 7. Stress contour of armour wire after winding.

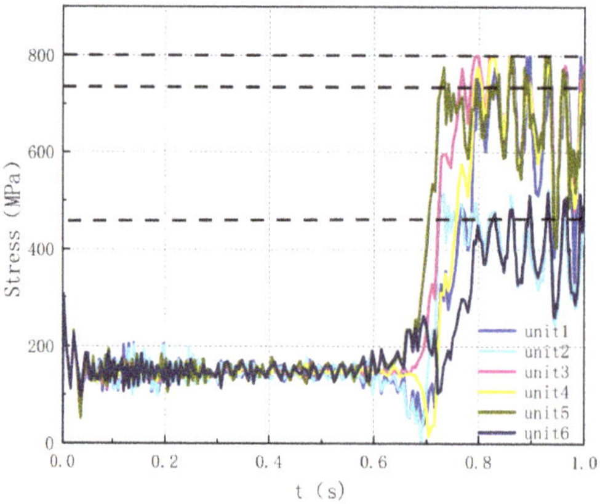

Figure 8. Variation in mises stress with time in the armour wire winding process.

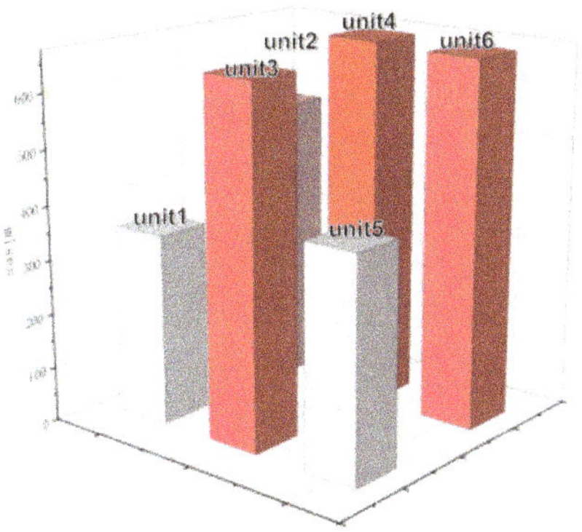

Figure 9. Mises stress distribution of armour wire cross-section after winding.

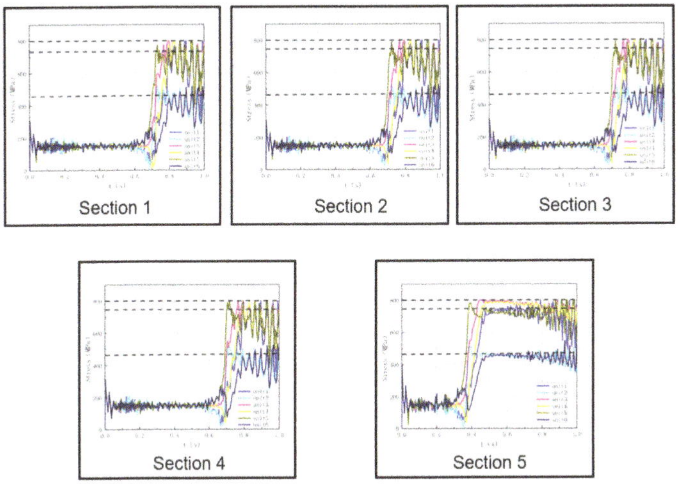

Figure 10. Variation in mises stress with time in the different cross-sections of the armour wire after winding.

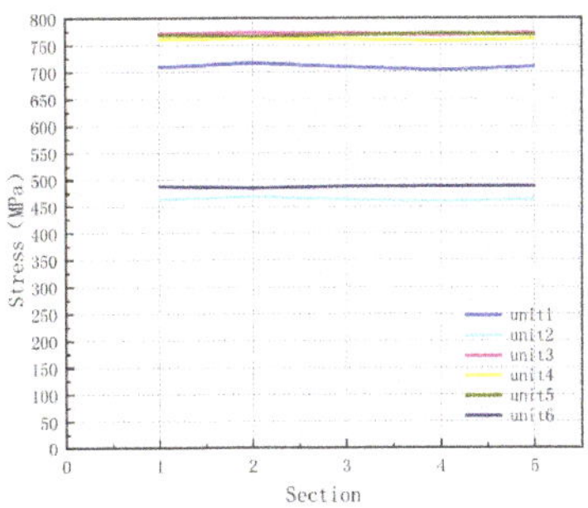

Figure 11. Comparison of mises stress of different cross-sections of armour wire after winding.

From Figure 9 above, it can be concluded that the residual stress distribution across the cross-section exhibits obvious spatial inhomogeneity. Among the six elements of the cross-section, the maximum residual stress is concentrated in the inner element with the largest bending curvature and the two adjacent middle elements. Their mises stress values are significantly higher than those in other regions, reaching the yield strength, which indicates that these regions have entered the plastic yielding stage during the winding process, while the other three elements show no plastic strain and remain in the elastic deformation state. This phenomenon confirms that the peak residual stress highly coincides with the plastic deformation zone, and the plastic deformation during the bending process is the direct cause of high residual stress. Moreover, the inner element with the largest bending curvature first undergoes plastic deformation due to bearing the maximum tensile strain, and irreversible residual tensile stress is generated after unloading. The adjacent middle elements also reach the yield state and contribute to high residual stress under the combined effect of stress transfer from the plastic zone and their own bending. In contrast, the outer elements remain in the elastic state due to small bending strains, resulting in low residual stress levels. This distribution law confirms the bending-dominated mechanism of the winding process—the peak residual stress and plastic strain zone strictly correspond to the region with the largest geometric curvature.

From Figures 10 and 11, it can be observed that, although there is a stress gradient within the cross-section, there is a high degree of consistency along the axial direction of the steel wire. The mises stress fields of the five cross-sections are basically consistent in axial distribution, and there is no significant difference in the stress values of the elements at the corresponding positions. The residual stress values along the axial path are continuous and stable without local fluctuations. This indicates that, after avoiding the end effect, the winding process forms a stable and reproducible residual stress field in the middle section of the steel wire.

To verify the accuracy of the numerical model for winding processing, the residual stress results from numerical simulation are compared with the theoretical formulas in Reference [29]. Reference [29] provides the relationship between the winding tension and the internal force of the steel wire cross-section after winding:

$$F_A = \frac{1}{\sqrt{\cos^2 \alpha + \frac{\sin^2 \alpha}{\cos^2 \theta}}} \cdot F \tag{1}$$

The winding tension in the numerical model is F = 1780 N. The internal force of the armour wire cross-section can be calculated using this theoretical method: $F_A = 1735$ N. The output of the simulation model is shown in Figure 12. Numerical model calculation: Total cross-sectional area of steel strand: S_total = 12 mm × 4 mm = 48 mm^2. The cross-section is divided into six units, each with an area of S = 48/6 = 8 mm^2. Axial stresses in the six units are 200 MPa, 210 MPa, −220 MPa, 15 MPa, 12 MPa, and 3 MPa. Total axial force calculation: F = 8 × (200 + 210 − 220 + 15 + 12 +3) = 1760 N. The result closely matches the theoretical value, demonstrating the accuracy of the finite element simulation model and the rationality of the winding process analysis.

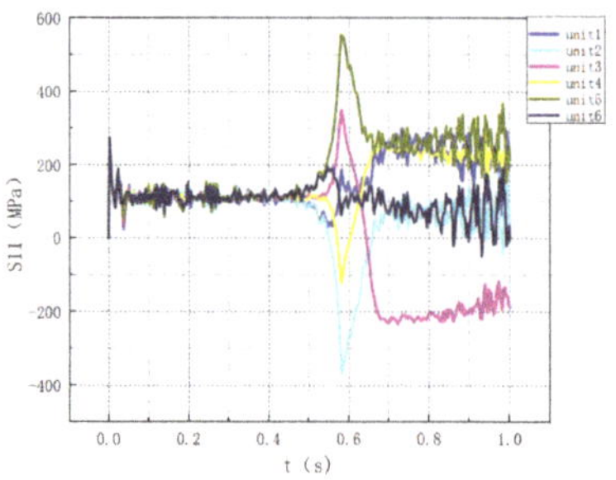

Figure 12. Axial stress of six elements in the armour wire section.

3. Tensile Loading–Unloading Model of Armour Wires with Processing-Induced Residual Stress

3.1. Model Establishment

In the simulation of the cyclic tensile behaviour of steel wires after the winding process, to accurately reflect the actual state of the steel wires before service, the final state obtained from the numerical simulation of the winding process was used as the initial condition and imported into the cyclic tension model in this study. Using the predefined field function of ABAQUS, the result file at the end of the winding simulation was mapped to the steel wire component of the cyclic tension model. This process accurately imported the residual stress field distribution across the steel wire cross-section after winding, the geometric deformation state caused by winding, and the key material hardening history. Mesh sensitivity and convergence analyses were conducted to ensure that the mapped residual stress field retained its spatial distribution and magnitude without significant numerical diffusion. The model after the successful import of the residual stress field is shown in Figure 13, and the stress distribution diagram after import is shown in Figure 14. This import method maximizes the physical authenticity of the starting point of the cyclic tension simulation, avoiding significant errors caused by starting calculations from a state of no residual stress, no geometric deformation, and no hardening. It is crucial for accurately studying the mechanical response of the steel wire under cyclic tensile loads. For the load application mechanism, displacement loading was used to directly control the boundary motion, avoiding the risk of "infinite deformation" in force loading. The loading history was precisely defined through a piecewise linear amplitude curve to simulate the actual tension-unloading process. The mandrel and the tail end of the steel wire winding were set with fully fixed constraints. The tail end of the wire is coupled to a reference point (RP) on the mandrel with a constraint, and a fixed constraint is applied (all degrees of freedom are restricted). At the starting end (loading end) of the steel wire winding, it was coupled with another reference point (RP-1) on the mandrel for motion. The axial cyclic tensile displacement load was directly applied to RP-1. To simulate the dominant axial tension behaviour, only the translational degree of freedom along the axial direction of the steel wire was retained in the model, while the other degrees of freedom were constrained to prevent unintended rigid body motion or bending. The element type selected for the steel wire was C3D8R, mainly because the C3D8R element achieves a good balance between calculation accuracy and efficiency, making it suitable for large-model calculations. The reduced integration technology makes it insensitive to shear locking when simulating bending and contact problems, which is crucial for problems involving bending deformation and complex contact during the winding process and can provide more accurate stress results, as well as effectively handle material nonlinearity. This element type is generally compatible with the residual stress field and deformation state imported from the winding stage. In terms of the contact algorithm, a dynamic implicit analysis step was adopted. The contact behaviour between the steel wire and the mandrel is usually highly nonlinear. The implicit dynamic analysis is generally more robust than static analysis in handling the convergence of such complex, state-changing contacts. During the cyclic loading process, especially during yielding, reverse yielding, or sudden changes in contact status, local convergence difficulties may occur. The implicit dynamic analysis step helps overcome these local instabilities by introducing inertial effects, improving the robustness of the overall solution. The cyclic tensile analysis employed an elastic-perfectly plastic material model to determine the shakedown limit load based on the initial yield criterion.

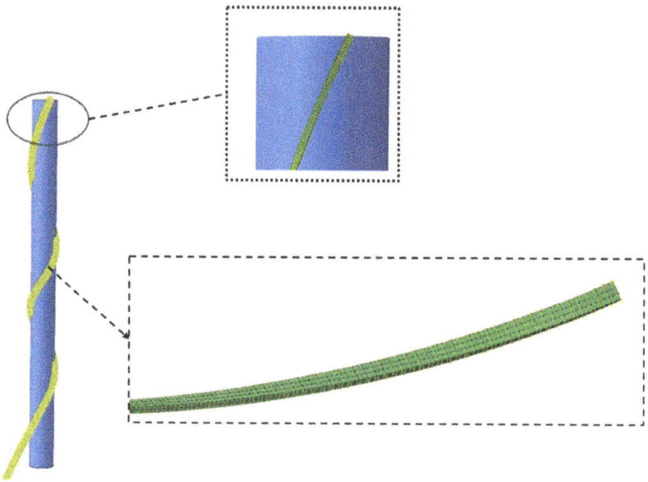

Figure 13. Numerical model for cyclic tensile loading of armour wire.

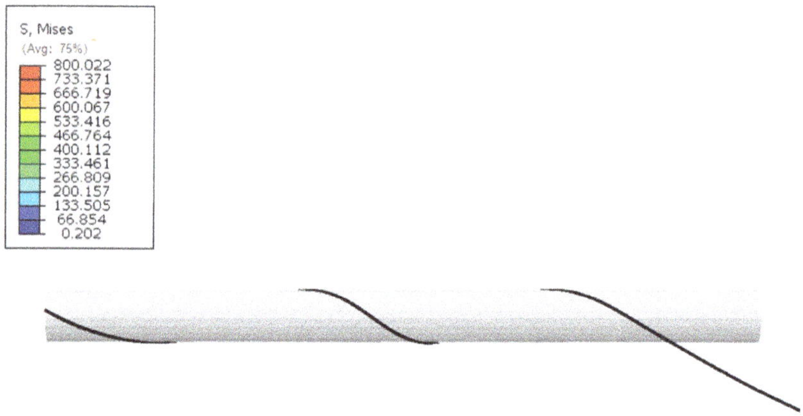

Figure 14. Stress contour of armour wire after residual stress import.

3.2. Discussion on the Residual Stress Distribution in Steel Wire Cross-Section After Loading and Unloading

To systematically study the evolution behaviour of residual stress in the wound steel wire under service loads, based on the initial residual stress field, this study imported the stress field as the initial condition and applied cyclic tension-unloading loads to the steel wire. The stress distribution diagram of Cross-section 1 is shown in Figure 15; the simulation calculation continued the field variable output request, and a displacement of 0–16.1 kN was applied. The stress–strain diagrams of Cross-section 1 of the steel wire under different unloading times are shown in Figure 16 below; the strain-time curves of Cross-section 1 with and without residual strain output from the model are shown in Figures 17 and 18 below.

As shown in Figure 18, the structure exhibits complete stress recovery upon unloading in the absence of residual stress, confirming its ideal elastic shakedown behaviour. In contrast, Figure 17 demonstrates that residual stress fundamentally undermines this self-restoring capacity, leaving the structure with persistent non-zero strain and locking it into a preloaded state. These findings confirm that residual stress not only compromises structural stability but also elevates the risk of progressive failure under cyclic loading.

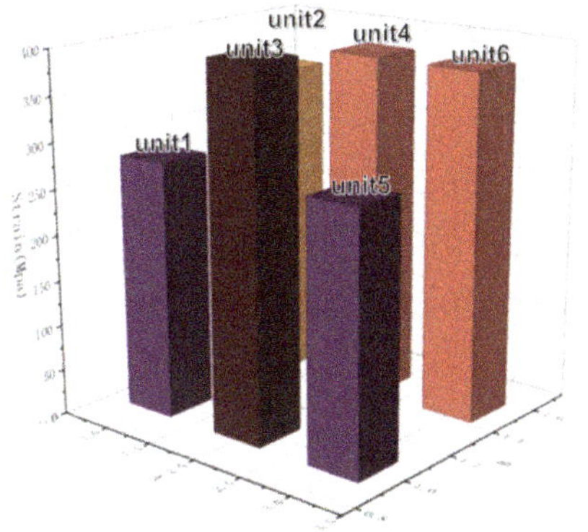

Figure 15. Mises stress distribution of cross-section after tensile loading and unloading.

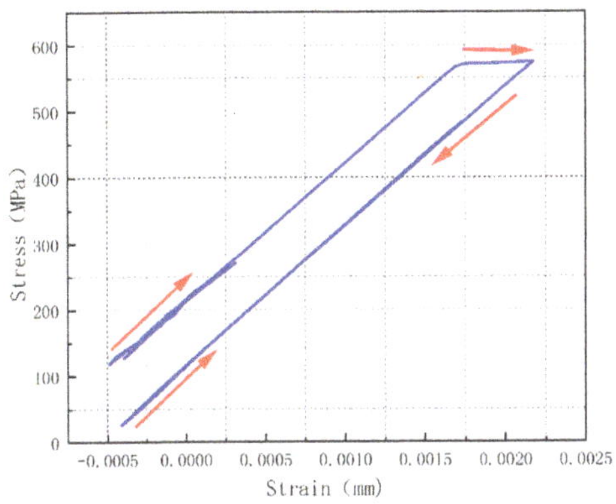

Figure 16. Stress–strain diagram of cross-section of armour wire.

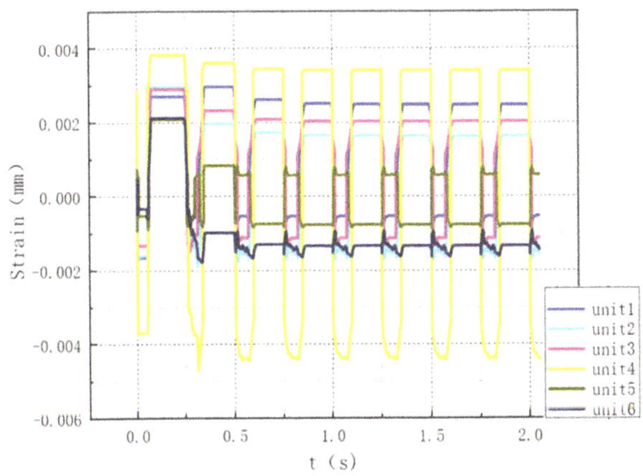

Figure 17. Variation in strain with time in cross-section with residual stress during loading–unloading process.

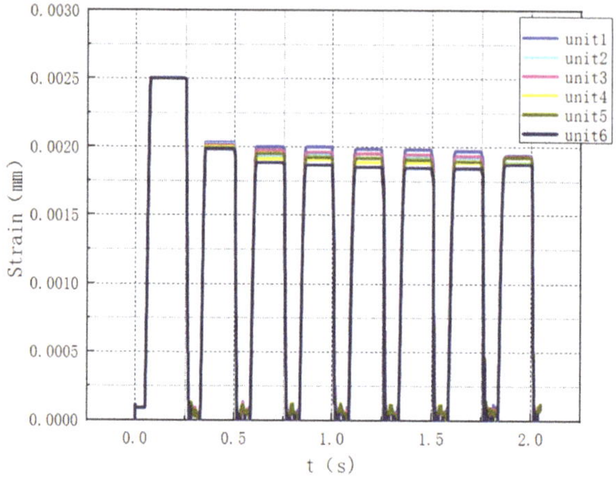

Figure 18. Variation in strain with time in cross-section without residual stress during loading–unloading process.

4. Study on Cyclic Tension Shakedown Behaviour

The stress–strain diagram of the calculation results of the cyclic tension model of steel wires with residual stress is shown in Figure 19 below, which is consistent with the model without residual stress and is divided into three parts. To improve the accuracy of the output curve, a field variable output request was adopted in this model calculation, and the output variables included stress and strain. The comprehensive recording of these output variables facilitates the comprehensive evaluation of the model, thereby enabling a more comprehensive and accurate analysis of the system response. The transition between shakedown and ratcheting was identified based on the convergence or divergence of plastic strain accumulation over successive cycles. Shakedown criterion: Stability refers to the condition where a structure, after undergoing a finite number of cyclic loading cycles, no longer accumulates plastic deformation but only exhibits elastic deformation. Melan's static stability criterion: A structure is considered stable if there exists a time-independent static stress field such that the deviation between the actual stress at every point in the structure and this static stress field remains within the material's elastic limit. Ratcheting judgement criterion refers to the 'cyclic accumulation characteristic of plastic strain' in structures under asymmetric cyclic loading. The strain accumulation criterion states the following: If plastic strain increases monotonically with each cycle (without a convergence tendency), the structure exhibits a ratchet effect. The shakedown state was quantified by the decay of the plastic strain increment per cycle to below 1×10^{-5}, while ratcheting was characterized by a persistent strain increment exceeding this threshold.

When a displacement of 0–16.1 kN was applied, Cross-section 1 of the steel wire was selected, and the plastic strain of the steel wire was obtained, as shown in Figure 20 below.

As shown in Figure 20, the structure attains an ideal cyclic shakedown state, with its maximum deformation capacity being consistently maintained under the prescribed loading conditions. This behaviour reflects a fully stabilized structural response. The uniform peak strain observed in each cycle serves as a clear indicator of cyclic stability, demonstrating that the structure has developed a repeatable hysteresis loop after initial conditioning. These results confirm the structure's ability to sustain cyclic loading without progressive strain accumulation or stiffness degradation.

When a displacement of 0–30.2 kN was applied, Cross-section 1 of the steel wire was selected, and the plastic strain diagram of a wire unit is shown in Figure 21 below.

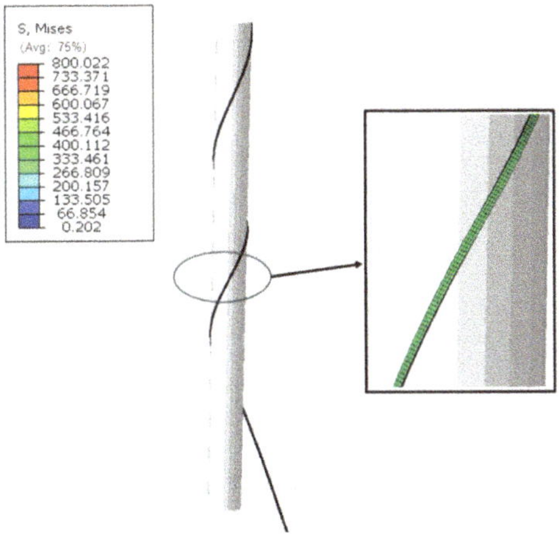

Figure 19. Mises stress diagram of numerical model.

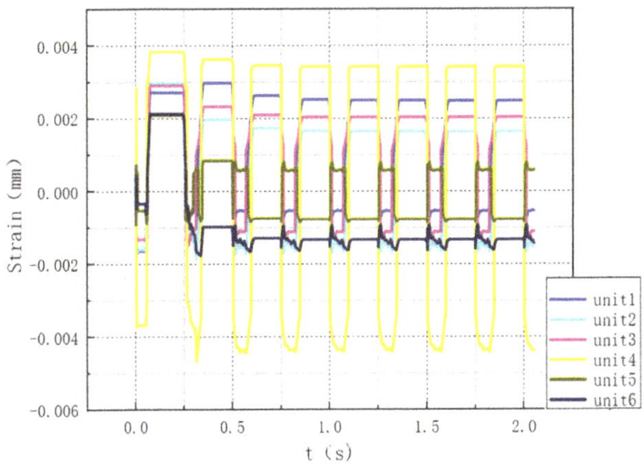

Figure 20. Plastic strain of armour wire cross-section during 0–16.1 kN loading–unloading process.

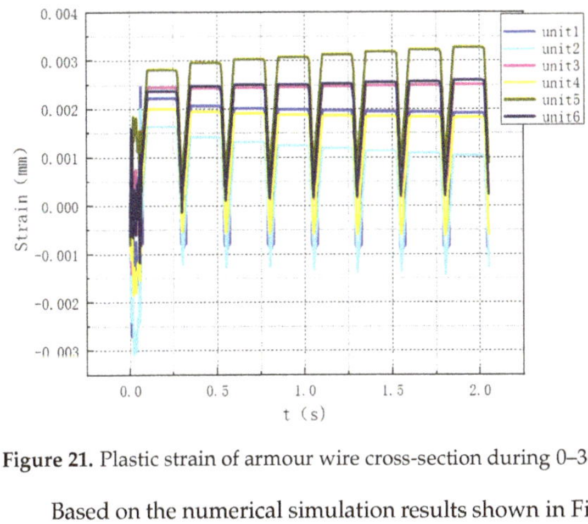

Figure 21. Plastic strain of armour wire cross-section during 0–30.2 kN loading–unloading process.

Based on the numerical simulation results shown in Figure 21, the strain accumulation exhibits a characteristic stair-step growth pattern with each loading cycle, demonstrating typical ratcheting behaviour. After several loading–unloading cycles, the cumulative strain reaches the maximum allowable strain limit of the material, eventually leading to the tensile fracture failure of the armour wire.

It is worth noting that, in the tensile resistance design of flexible risers, the design is usually based on the ultimate tensile breaking force, and the estimation method of the ultimate tensile breaking force is shown in the following:

$$F = \sigma A \cos \alpha \tag{2}$$

The ultimate tensile breaking fore of one armour wire can be calculated from Table 1, which is 33.3 kN. Through Figure 21, we discovered that repeated tensile forces induce a ratchet effect in riser armour plates. In this case, the critical load for the ratchet effect is 30.1 kN, which is lower than the ultimate tensile strength. Since flexible risers are subjected to repeated tensile loads during processing, installation, and operational conditions, it is essential to apply the proposed method in this study to analyze the stability of riser pipes under cyclic tensile loads in practical engineering applications.

The currently established numerical model does not account for the effects of temperature, corrosion, and varying friction coefficients on shakedown behaviour. Moving forward, research that incorporates these factors and conducts relevant experiments would be of great value for advancing this investigation further.

5. Conclusions

In this study, a numerical simulation model for the winding process of armour wires in flexible risers was established, and the stress distribution in the cross-section of the armour wires after processing was quantified. Furthermore, residual stress was incorporated into the analysis, and a simulation methodology for investigating the shakedown behaviour of flexible risers during the cyclic tensile load was developed. The key findings are as follows:

The residual stress of armour wires after the winding process exhibits inhomogeneity in the cross-section, with the maximum residual stress concentrated in the inner elements. During the initial loading–unloading process of the example pipe, the stress in the armour wire cross-section undergoes obvious redistribution.

When cyclic loading is applied with a tensile force of 0–16.1 kN, the armour wire cross-section tends to reach a shakedown state as the number of loading cycles increases.

When cyclic loading is applied with a tensile force of 0–30.1 kN, the strain of the armour wire cross-section gradually increases with each loading–unloading cycle, exhibiting a ratcheting effect.

When a flexible riser is subjected to cyclic tensile loads, the riser may fail due to the ratcheting effect occurring inside the armour wires, even if the tensile load is less than the ultimate tensile breaking force of the pipe. Therefore, in the design of flexible risers, the shakedown behaviour of the riser under cyclic tensile loads should be considered.

Author Contributions: Conceptualization, S.W.; Software, S.W.; Validation, E.S.; Formal analysis, X.Y.; Investigation, J.L.; Writing—original draft, X.Y.; Writing—review & editing, Z.S. All authors have read and agreed to the published version of the manuscript.

Funding: This research was funded by National Natural Science Foundation of China grant number (52201395).

Data Availability Statement: Should you need to access the data for research verification or related purposes, please feel free to contact the corresponding author directly.

Conflicts of Interest: Author Ersu Shang was employed by the company VH-Marinetech Co., Ltd. The remaining authors declare that the research was conducted in the absence of any commercial or financial relationships that could be construed as a potential conflict of interest.

References

1. Knapp, R. Nonlinear analysis of a helically armoured cable with nonuniform mechanical properties in tension and torsion. In Proceedings of the Ocean 75 Conference, San Diego, CA, USA, 22–25 September 1975; IEEE: New York, NY, USA, 1975; pp. 155–164.
2. Knapp, R.H. Derivation of a new stiffness matrix for helically armoured cables considering tension and torsion. *Int. J. Numer. Methods Eng.* **1979**, *14*, 515–529. [CrossRef]
3. Knapp, R.H. Torque and stress balanced design of helically armoured cables. *J. Eng. Ind.* **1981**, *103*, 61–66. [CrossRef]
4. Lanteigne, J. Theoretical estimation of the response of helically armoured cables to tension, torsion, and bending. *J. Appl. Mech.* **1985**, *52*, 423–432. [CrossRef]
5. LeClair, R.A.; Costello, G.A. Axial, bending and torsional loading of a strand with friction. *J. Offshore Mech. Arct. Eng.* **1988**, *110*, 38–42. [CrossRef]
6. Jolicoeur, C.; Cardou, A. A numerical comparison of current mathematical models of twisted wire cables under axisymmetric loads. *J. Energy Resour. Technol.* **1991**, *113*, 241–249. [CrossRef]
7. Witz, J.A.; Tan, Z. On the axial-torsional structural behaviour of flexible pipes, umbilicals and marine cables. *Mar. Struct.* **1992**, *5*, 205–227. [CrossRef]
8. Custódio, A.; Vaz, M. A nonlinear formulation for the axisymmetric response of umbilical cables and flexible pipes. *Appl. Ocean. Res.* **2002**, *24*, 21–29. [CrossRef]
9. Ramos, R., Jr.; Pesce, C.P. A consistent analytical model to predict the structural behavior of flexible risers subjected to combined loads. *J. Offshore Mech. Arct. Eng.* **2004**, *126*, 141–146. [CrossRef]
10. Yue, Q.; Lu, Q.; Yan, J.; Zheng, J.; Palmer, A. Tension behavior prediction of flexible pipelines in shallow water. *Ocean. Eng.* **2013**, *58*, 201–207. [CrossRef]
11. Xiang, L.; Wang, H.; Chen, Y.; Guan, Y.; Wang, Y.; Dai, L. Modeling of multi-strand wire ropes subjected to axial tension and torsion loads. *Int. J. Solids Struct.* **2015**, *58*, 233–246. [CrossRef]
12. Foti, F.; di Roseto, A.D.L. Analytical and finite element modelling of the elastic–plastic behaviour of metallic strands under axial–torsional loads. *Int. J. Mech. Sci.* **2016**, *115*, 202–214. [CrossRef]
13. Blkich, H. Über die Bemessung statisch unbestimmter Stahlwerke unter Berücksichtigung des elastischplastischen Berhaltens des Baustoffes. *Bauingenieur* **1932**, *13*, 261–267.
14. Melan, E. *Theorie Statisch Unbestimmter Systeme aus Ideal-Plastischem Baustoff*; Hölder-Pichler-Tempsky in Komm: Prambachkirchen, Austria, 1936.
15. Melan, E. Zur plastizität des räumlichen kontinuums. *Ing.-Arch.* **1938**, *9*, 116–126. [CrossRef]
16. Koiter, W.T. A new general theorem on shakedown of elastic-plastic structures. *Proc. Koninkl. Ned. Akad. Wet. B* **1956**, *59*, 24–34.
17. Ponter, A.R.S.; Karadeniz, S. An extended shakedown theory for structures that suffer cyclic thermal loading, Part 1: Theory. *J. Appl. Mech.* **1985**, *52*, 877–882. [CrossRef]
18. Polizzotto, C. On the conditions to prevent plastic shakedown of structures: Part I—Theory. *J. Appl. Mech.* **1993**, *60*, 15–19. [CrossRef]
19. Prager, W. A new method of analyzing stresses and strains in work-hardening plastic solids. *J. Appl. Mech.* **1956**, *23*, 493–496. [CrossRef]
20. Besseling, J.F. A theory of elastic, plastic, and creep deformations of an initially isotropic material showing anisotropic strain-hardening, creep recovery, and secondary creep. *J. Appl. Mech.* **1958**, *25*, 529–536. [CrossRef]
21. Armstrong, P.J.; Frederick, C.O. *A Mathematical Representation of the Multiaxial Bauschinger Effect*; Nuclear Laboratories: Berkeley, CA, USA, 1966; Volume 731.
22. Chaboche, J.L. Time-independent constitutive theories for cyclic plasticity. *Int. J. Plast.* **1986**, *2*, 149–188. [CrossRef]
23. Chaboche, J.L. On some modifications of kinematic hardening to improve the description of ratchetting effects. *Int. J. Plast.* **1991**, *7*, 661–678. [CrossRef]
24. Han, K.; Chen, Z.-F.; Wang, W.; Shi, L.; Lu, K.-Q. A short review of ratcheting effect in pressurized piping. *Int. J. Press. Vessel. Pip.* **2023**, *205*, 105009. [CrossRef]
25. Fernando, U.S.; Davidson, M.; Simpson, C.; Pirling, T.; Yan, K.; Callaghan, M.D.; Roy, M.; Francis, J.A.; Withers, P.J. Measurement of residual stress shakedown in pressure/tensile armour wires of flexible pipes by neutron diffraction. In Proceedings of the ASME 2015 34th International Conference on Ocean, Offshore and Arctic Engineering, St. John's, NL, Canada, 31 May–5 June 2015; American Society of Mechanical Engineers: New York, NY, USA, 2015.

26. Sævik, S.; Gray, L.; Phan, A. A method for calculating residual and transverse stress effects in flexible pipe pressure spirals. In Proceedings of the 20th International Conference on Offshore Mechanics and Arctic Engineering (OMAE 2001 CD), Rio de Janeiro, Brazil, 3–8 June 2001; pp. 3–8.
27. Fernando, U.S.; Tan, Z.; Sheldrake, T.; Clements, R. The stress analysis and residual stress evaluation of pressure armour layers in flexible pipes using 3D finite element models. In Proceedings of the International Conference on Offshore Mechanics and Arctic Engineering, Vancouver, BC, Canada, 20–25 June 2004; Volume 37459, pp. 57–65.
28. Tang, M.; Yan, J.; Chen, J.; Yang, Z.; Yue, Q. Residual Stress Evaluation and Buckling Analysis of Carcass Layers in Flexible Pipes Using 3D Finite Element Model. In Proceedings of the ASME 2015 34th International Conference on Ocean, Offshore and Arctic Engineering, St. John's, NL, Canada, 31 May–5 June 2015; American Society of Mechanical Engineers: New York, NY, USA, 2015.
29. Lu, Q.; Wu, S.; Wang, D.; Yang, Z.; Yin, Y.; Hu, H.; Yan, J.; Yue, Q. Study on mechanical behavior of tensile armour wires of marine flexible pipes and cables during winding process. In Proceedings of the International Conference on Offshore Mechanics and Arctic Engineering, Glasgow, UK, 9–14 June 2019; American Society of Mechanical Engineers: New York, NY, USA, 2019; Volume 58806.
30. Pan, Y.; Zheng, L.; Yang, S.; Wang, Y. A review of fatigue analysis of marine flexible risers. *Geoenergy Sci. Eng.* **2024**, *246*, 213555. [CrossRef]
31. Bao, R.; Liu, J.; Xiao, Z. Stress and failure analysis of composite flexible riser subjected to torsion and thermomechanical loading. In Proceedings of the International Ocean and Polar Engineering Conference, Rhodes, Greece, 16–21 June 2024; ISOPE: Mountain View, CA, USA, 2024.

Journal of
Marine Science
and Engineering

MDPI

Article

Numerical Investigation of Burst Capacity in Pressure Armour Layer of Flexible Risers

Mohsen Saneian [1], Yuteng Zhang [1,*], Pan Fang [1,2] and Yong Bai [1,3]

1 College of Civil Engineering and Architecture, Zhejiang University, Hangzhou 310058, China
2 Department of Materials, Textiles and Chemical Engineering (MaTCh), Ghent University, Technologiepark-Zwijnaarde 46/70, B-9052 Zwijnaarde, Belgium
3 OPR Offshore Engineering Equipment Co., Ltd., Ningbo 315000, China
* Correspondence: yutengzhang@zju.edu.cn

Abstract: Unbonded flexible risers are key components in offshore oil and gas production, enabling safe hydrocarbon transport under extreme deepwater conditions. The pressure armour (PA) layer is the primary metallic layer, which is designed to resist internal pressure; however, its complex helical geometry poses significant modelling challenges. This study presents a comprehensive numerical and analytical investigation of the burst behaviour of the PA layer under internal pressure. Three FE modelling strategies are developed and compared with theoretical predictions: a simplified axisymmetric 2D model, a fully detailed 3D helical model, and an efficient periodic boundary condition (PBC) model. The proposed PBC-based approach offers a practical balance between accuracy and efficiency, providing an effective tool for parametric analysis and design optimisation of flexible risers.

Keywords: flexible pipe; pressure armour; periodic boundary condition; burst pressure

1. Introduction

Pipelines serve as vital infrastructure in offshore oil and gas developments, enabling the reliable transportation of hydrocarbons from subsea wells to processing facilities [1,2]. With the global expansion of deep and ultra-deepwater exploration, the demand for pipelines that can withstand harsh environments and heavy loads has greatly increased [3,4]. Among these, unbonded flexible risers have emerged as a critical technology due to their superior mechanical performance, geometric adaptability, and ease of installation in harsh offshore environments [5]. Unlike rigid pipelines, flexible risers are able to accommodate vessel motions, seabed irregularities, and thermal or pressure variations without significant structural damage, making them particularly suitable for deep-water field development [6]. Their internal architecture, schematically illustrated in Figure 1, is composed of metallic and polymeric layers. The metallic constituents primarily include the carcass, pressure armour, and tensile armour, while the polymeric components consist of the inner and outer sheaths together with anti-wear and high-strength anti-birdcage tapes.

The hydrocarbon stream flows inside the carcass, which is enclosed by an inner polymer sheath that acts as a seal and prevents fluid from entering the space where the pressure and tensile armours are located [7,8]. Since the polymer sheath provides only sealing and does not contribute significantly to load-bearing, the entire internal pressure is transferred to the metallic pressure armour [9]. This makes the pressure armour the primary structural layer resisting burst failure, which is one of the most critical failure modes in flexible risers [10]. Failure of this layer can lead to catastrophic consequences such as hydrocarbon leakage, interruption of production, severe environmental damage, and high replacement and repair costs. For this reason, accurate prediction of burst resistance is of paramount importance in order to ensure safe and efficient offshore operations [11].

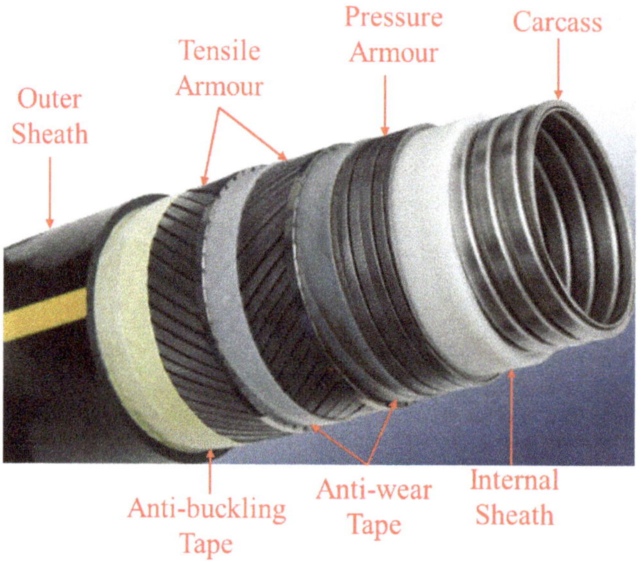

Figure 1. Typical unbounded flexible pipe.

Early research efforts have attempted to predict the collapse and burst behaviour of risers through analytical formulations. Bournazel and Feret [12] introduced a simplified equilibrium approach to approximate internal pressure loading and to provide design guidelines. Oliviera et al. [13] extended this work by incorporating axial–circumferential interaction and structural deformation, thereby improving accuracy. Colquhoun et al. [14] adapted classical thin-walled cylinder theory with correction factors to capture the unique geometry of Z-shaped pressure armour profiles. Fernando et al. [15] later compared simplified 2D axisymmetric models with more detailed 3D representations that incorporated manufacturing-induced deformations, showing the limitations of purely analytical formulations. Zhu and Leis [16] proposed the average shear stress yield (ASSY) theory, positioned between the Tresca and von Mises criteria, noting that pipeline failures usually occur between these bounds. While von Mises is often adopted in commercial finite element codes as a conservative assumption, its direct application to the highly complex geometry of pressure armour layers remains questionable.

Building on these foundations, several authors proposed nonlinear models that consider both geometric and material nonlinearity. Neto et al. [17,18] introduced linear and nonlinear modelling approaches for the pressure armour, accounting for hardening behaviour, radial displacements, and thickness variations under internal pressure. More recently, Cuamatzi-Melendez et al. [19] investigated burst capacity by modelling only the inner sheath together with the pressure armour, validating their results against analytical solutions. O'Halloran et al. [20] explored the influence of internal pressure on fatigue performance, showing that it significantly reduces the predicted number of cycles to crack initiation in the pressure armour. Liu et al. [21] developed a finite element model of a 2.5-inch unbonded riser, combining stress distribution and deformation analyses with yield/ultimate stress theory to predict burst failure. Wei et al. [22] further investigated the unlocking mechanism of the pressure armour layer under internal pressure using

3D nonlinear FEM, showing that increasing internal pressure can promote interlock loss, leading to unlocking and eventual birdcage failure.

Although these contributions have advanced the understanding of flexible riser mechanics, significant challenges remain. Experimental testing of risers under burst conditions is extremely costly and time-consuming, and requires specialised facilities. Consequently, most available research relies on numerical simulations. However, fully detailed 3D FE simulations that explicitly capture interwire contact, friction, and geometric imperfections are computationally expensive and impractical for routine design or large parametric studies. As a result, researchers often use simplified equivalent models to reduce computational costs, but such approaches frequently compromise accuracy by neglecting key physical effects. Santos and Pesce [23] proposed a space-filling mesh scheme to represent voids between wires, validated against analytical and experimental data. While this method reduced simulation time, it did not adequately address the end-cap effect under internal pressure, leaving a gap in predictive reliability.

These limitations show the need for efficient and accurate modelling methods that can predict the burst pressure of the pressure armour layer without the high computational cost of fully detailed 3D models. In the present work, three FE modelling approaches are explored, and the results are compared with the theoretical equations. First, a 2D axisymmetric model is developed to capture the global stress distribution of the pipe under internal pressure in a highly efficient manner. Second, a fully detailed 3D model is constructed to reproduce the complete helical geometry of the pressure armour and provide a benchmark for accuracy. Finally, a simplified finite element framework based on periodic boundary conditions (PBCs) is proposed to represent an infinitely long pipe under internal pressure. The PBC approach is shown to combine the accuracy of the full model with the efficiency of the axisymmetric model, making it the most suitable choice for parametric studies and design optimisation.

By combining computational efficiency with physical accuracy, this study aims to provide a practical methodology for assessing burst failure in flexible risers, contributing to safer and more cost-effective offshore pipeline design.

2. Analytical Model

For a thick cylindrical layer with inner radius a, outer radius b, internal pressure p_i and external pressure p_o, the axial (longitudinal) stress, due to Timoshenko and Goodier [24], is

$$\sigma_z = \frac{p_i a^2 - p_o b^2}{b^2 - a^2} \tag{1}$$

which is constant through the thickness. In this work, we adopt the open–ends boundary condition ($\sigma_z = 0$).

The Lamé radial and circumferential (hoop) stresses at radius $r \in [a, b]$ [25] are

$$\sigma_r(r) = \frac{p_i a^2 - p_o b^2}{b^2 - a^2} - \frac{a^2 b^2 (p_i - p_o)}{(b^2 - a^2)\, r^2} \qquad \sigma_\theta(r) = \frac{p_i a^2 - p_o b^2}{b^2 - a^2} + \frac{a^2 b^2 (p_i - p_o)}{(b^2 - a^2)\, r^2} \tag{2}$$

For the common special case when there is no external pressure ($p_o = 0$)

$$\sigma_r(r) = \frac{p_i a^2}{b^2 - a^2}\left(1 - \frac{b^2}{r^2}\right) \qquad \sigma_\theta(r) = \frac{p_i a^2}{b^2 - a^2}\left(1 + \frac{b^2}{r^2}\right) \tag{3}$$

and, at the inner surface $r = a$

$$\sigma_r(a) = -\, p_i \qquad \sigma_\theta(a) = p_i\, \frac{a^2 + b^2}{b^2 - a^2} \tag{4}$$

The von Mises equivalent stress is

$$\sigma_{\mathrm{vM}}(r) = \sqrt{\frac{(\sigma_\theta - \sigma_r)^2 + (\sigma_r - \sigma_z)^2 + (\sigma_z - \sigma_\theta)^2}{2}} \tag{5}$$

which, at $r = a$ using (4), becomes a linear function of pressure

$$\sigma_{\mathrm{vM}}(a) \;=\; S_{\mathrm{vM}}\, p_i \qquad S_{\mathrm{vM}} \;=\; \sqrt{K^2 + K + 1} \qquad K \;\equiv\; \frac{a^2 + b^2}{b^2 - a^2} > 1 \tag{6}$$

Given elastic data pairs $(p_i, \sigma_{\mathrm{vM},i})$, a through-origin least-squares estimate of the slope is

$$\widehat{S}_{\mathrm{vM}} \;=\; \frac{\sum_i p_i\, \sigma_{\mathrm{vM},i}}{\sum_i p_i^2} \tag{7}$$

where, for the open–ends (plane–stress) case to be physically admissible with $b > a$, one must have $\widehat{S}_{\mathrm{vM}} \geq \sqrt{3}$.

To relate the outer radius b to measurable section properties, Martins et al. [26] introduced an equivalent plate thickness based on bending–stiffness equivalence.

$$t_{eq} \;=\; \sqrt[3]{\frac{12\,(1+\psi)\, I_{G\,\mathrm{min}}}{\ell}} \tag{8}$$

where ℓ is the effective strip width used in the plate equivalence, $I_{G\,\mathrm{min}}$ is the minimum principal second moment of area about the centroidal axes, and ψ is a geometric factor defined here as

$$\psi \;=\; \frac{x}{s} \tag{9}$$

where x and s denote the characteristic axial repeat length of the interlocked profile and its axial pitch, respectively. On the other hand, the thickness links directly to the cylinder radii via

$$t \;=\; b - a \tag{10}$$

Substituting (10) into (6) yields a closed-form σ_{vM}–p relation entirely in terms of $(a, \psi, \ell, I_{G\,\mathrm{min}})$ The quantity $I_{G\,\mathrm{min}}$ is obtained from the eigenvalues of the symmetric inertia matrix.

$$\mathbf{I} = \begin{bmatrix} I_{11} & I_{12} & I_{13} \\ I_{21} & I_{22} & I_{23} \\ I_{31} & I_{32} & I_{33} \end{bmatrix} \qquad I_{G\,\mathrm{min}} \;=\; \min\{\lambda_j : \lambda_j > 0\} \tag{11}$$

where $\{\lambda_j\}$ are the eigenvalues of $\mathbf{I}$. Finally, if a measured elastic slope S_{vM} is available (from (7)), one may invert (6) to infer the geometry: first solve

$$K = \frac{-1 + \sqrt{4S_{\mathrm{vM}}^2 - 3}}{2} \qquad (K > 1) \tag{12}$$

then recover

$$b \;=\; a\sqrt{\frac{K+1}{K-1}} \tag{13}$$

All formulas are dimensionally consistent.

3. Numerical Model

3.1. The Basics

The material data for the polyethylene (PE) layer were obtained from uniaxial tensile tests reported by Zhang et al. [4]. The corresponding true stress–strain curve, shown in Figure 2a, was directly implemented into the finite element model to capture both the elastic and plastic responses of the polymer material. For the pressure armour (PA) layer, the material was selected as stainless steel 316L [27]. The Young's modulus was determined based on tensile tests performed by Krishna et al. [28], while, according to the findings of Yan et al. [29], the cold-worked and annealed 316L stainless steel exhibits a yield strength of approximately 900 MPa. No strain-hardening term was introduced ($K = 0$); therefore, the PA material was idealised as an elastic–perfectly plastic material, as shown in Figure 2b, to simplify the numerical implementation while accurately representing the yielding behaviour of metallic layers under high internal pressure [30]. This assumption also ensures consistency between the numerical and analytical predictions while preserving the load-carrying capacity of the PA layer. The complete set of elastic properties for both PE and PA layers is summarised in Table 1.

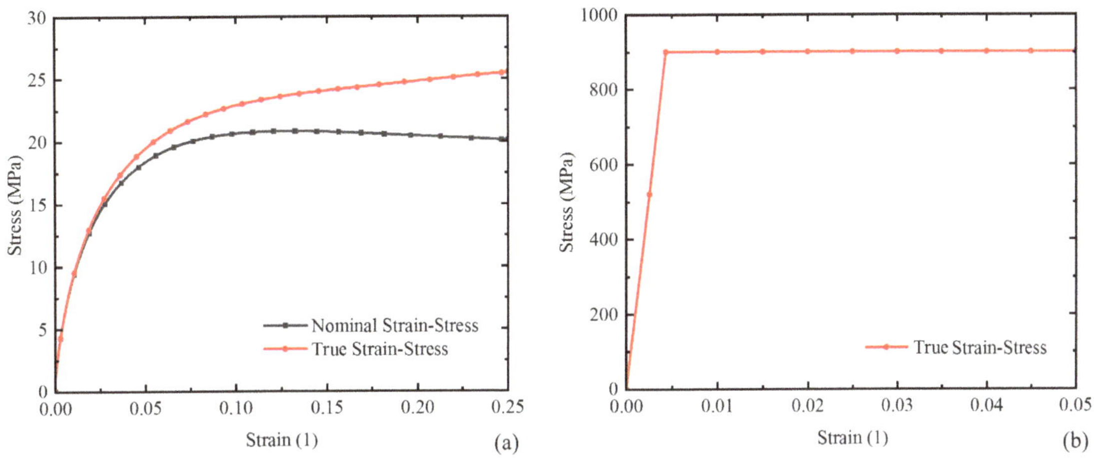

Figure 2. Stress-strain curves (**a**) HDPE and (**b**) pressure armour steel.

Table 1. Properties of the test specimen.

Parameter	Pressure Armour	HDPE
Poisson's ratio (1)	0.29	0.45
Young's Modulus (MPa)	208,160	1134
Yield strength (MPa)	900	-

The pressure armour profile adopted in this study was selected in accordance with API 17B [31], which provides industry guidelines for flexible riser design. The detailed cross-sectional geometry of the Z-shaped armour is shown in Figure 3a. All critical dimensions and curvature radii are specified, highlighting the interlocking mechanism that ensures structural integrity under high internal pressure. The interlocking geometry of the PA layer and its corresponding pitch length are illustrated in Figure 3b. The thickness of the PE layer is considered to be 8 mm for the FE models. All finite element models are executed using ABAQUS/Standard 2024 [32], employing the Static General step with NLGEOM enabled.

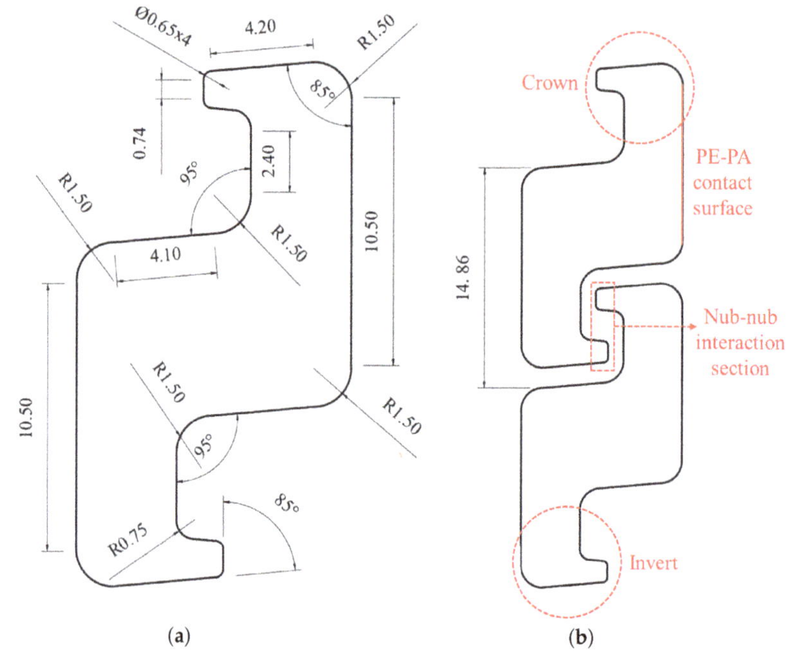

Figure 3. Pressure armour geometry: (**a**) cross-section with dimensions (mm) and curvature radii (mm), with angles between contour lines indicated in degrees (°), showing the Z-shaped interlocking profile; (**b**) longitudinal section with pitch length (mm).

3.2. Axisymmetric Model

The pressure armour layer presents significant modelling challenges due to its pronounced geometric nonlinearity, which arises from the radial displacements of the Z-shaped profiles, the contact interactions between successive profiles, and the helical winding pitch. However, since the winding angle is close to 90°, this feature can be reasonably neglected, allowing the assumption of axisymmetry about the pipe's longitudinal axis. Under this simplification, the loading conditions are considered uniform along the circumference of the pipe, enabling the use of a two-dimensional axisymmetric representation. In this model, the polymer sheath and the pressure armour were included, both modelled in a planar domain. To capture uniform stress distribution along the Z-profiles, the model incorporated three complete Z-profiles together with two half-profiles, following the approach of Cuamatzi-Melendez et al. [19], as illustrated in Figure 4.

Internal pressure (IP) was applied along the inner surface of the polymer sheath. To reproduce the stress field within the Z-profile closures, boundary conditions were carefully defined. Y-symmetry was imposed on the PE layer boundaries to maintain circumferential continuity. A tie-constraint strategy was then employed to ensure consistent deformation across the PA cross-section. Two separate Tie constraints, as illustrated in Figure 4 (Pair 1 in green and Pair 2 in yellow), were introduced. The first ties the two vertical end surfaces of the PA profile, ensuring that the entire section deforms as a continuous strip without introducing artificial gaps or mismatched motion at the boundaries. The second ties the two interlocking nub surfaces, enforcing compatible displacement and rotation between the mating regions of the Z-shaped closure. In both cases, ABAQUS's Tie option fully constrains all translational and rotational degrees of freedom of the secondary surface to follow the main surface, thereby preventing unintended relative sliding or separation in the 2D axisymmetric formulation. Both the PE and PA were meshed using CAX4R 4-node bilinear axisymmetric elements with reduced integration and hourglass control. Figure 5 shows the PA cross-section in its actual global coordinate orientation used in the FE model, together with its mesh details.

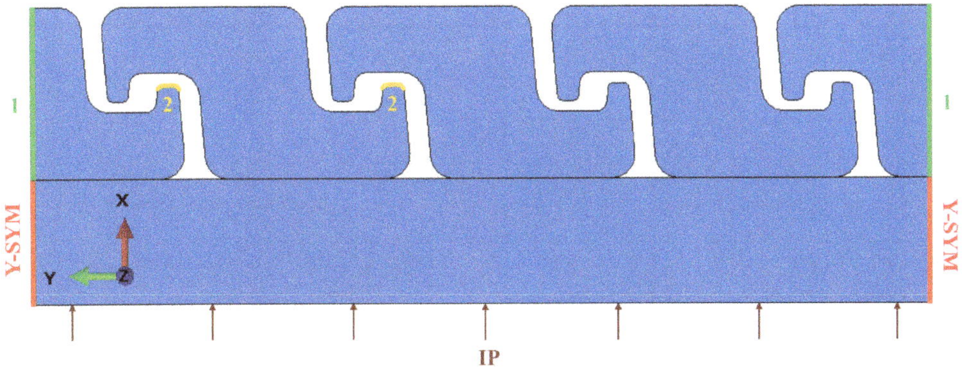

Figure 4. Axisymmetric model of the PE and PA layers.

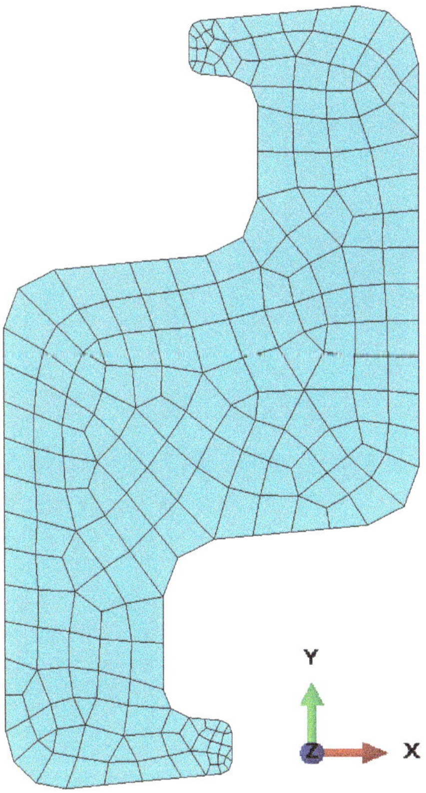

Figure 5. Finite element mesh of the PA cross-section.

3.3. Full 3D Model

A 3D FE model of the interlocked pressure armour (PA) layer, including its pitch and lay angle, was established using the commercial software ABAQUS version 2024. To simulate load transfer, a polymer layer was added to the inside of the PA layer. The inner diameter of the pipe is 154 mm (6 inch). To minimise the influence of end effects on the results, several models with varying lengths were tested, and the model with a length approximately twice the outer diameter of the pipe was ultimately selected for analysis. Solid elements were used, and the material properties were applied based on the results of the previous section. The analysis was carried out as a geometrically nonlinear problem using the static general step. The layers were meshed with C3D8R elements (8-node linear brick, reduced integration, hourglass control), a standard element formulation in ABAQUS known for balancing accuracy and computational efficiency in both linear and nonlinear problems [33]. The general resulting mesh configuration of the pipe is depicted in Figure 6.

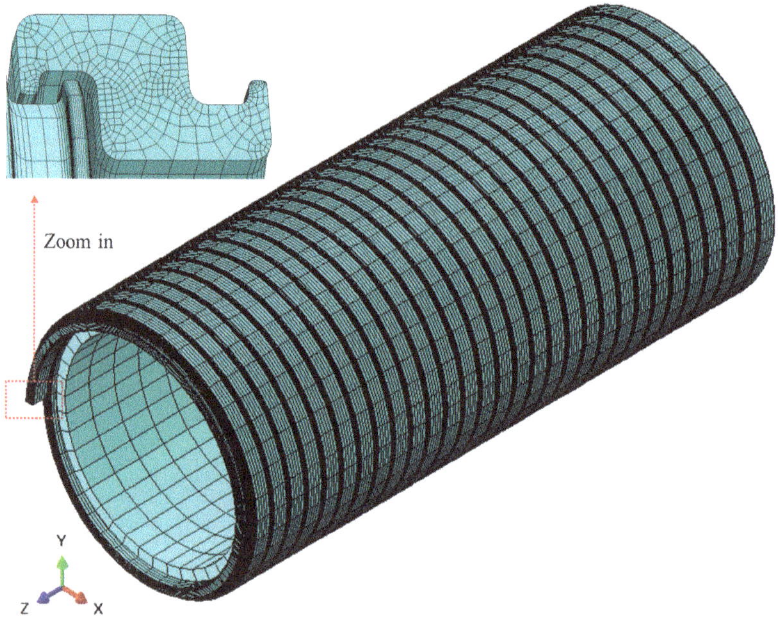

Figure 6. Finite element mesh of the pipe.

Boundary conditions were applied to both ends of the pipe model, where all translational degrees of freedom were restricted, and internal pressure was applied to the inner surface of the PE layer, as illustrated in Figure 7. To prevent interpenetration between surfaces, general contact was defined across the entire model, with an additional explicit contact pair defined between the outer surface of the PE and the inner surface of the PA. The contact was modelled with hard pressure-overclosure behaviour in the normal direction and a penalty formulation in the tangential direction, allowing separation after contact. Based on experimental findings, the friction coefficient between the PA and PE was set to 0.15 [34], and between the armours is considered to be 0.1 [35].

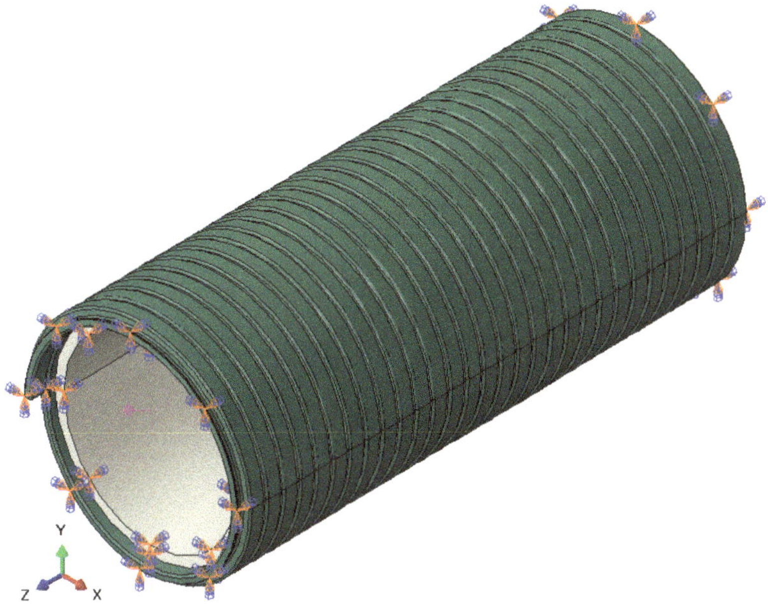

Figure 7. Applied loads and boundary conditions.

3.4. PBC Model

Because full-length internal pressure simulations of flexible pipes are computationally prohibitive due to their complex multilayered structure, researchers have proposed efficient numerical strategies that exploit the periodic geometry of helical components, using periodic boundary conditions to represent infinitely long structures while minimising boundary effects. The concept of periodic boundary conditions (PBC) originates from multiscale analysis, where homogenization theory, often employing asymptotic expansions, has been applied to bridge the microscale and macroscale behaviours in slender beam-like structures, such as cables, flexible pipes, and helical ropes [36]. In the PBC model, the length is reduced as:

$$l = k\frac{p_i}{m_i} \tag{14}$$

where $k \in \mathbb{N}$, p_i is the pitch length, m_i is the number of helixes, and i denotes the sequence of the current layer.

A periodic boundary condition between two surfaces can be implemented in Abaqus by imposing an appropriate constraint between each corresponding pair of their nodes [37]. A single strip of the interlocked pressure armour is divided into two parts and modelled with contact interactions, as illustrated in Figure 8, which is sufficient to represent the behaviour of the entire pressure armour layer. In addition, the internal polymer layer is modelled inside the PA layer to transmit the applied internal pressure, while an outer sheath is introduced around the PA layer to replicate the functional role of the external sheath in a real flexible pipe and to restrain excessive radial expansion. Periodic boundary conditions were applied to both ends of the PA and PE layers to represent the repeating helical architecture of the flexible pipe. Specifically, the nodal displacements on one end of the model were coupled with those on the opposite end along the longitudinal direction, ensuring that deformation and stress fields were continuous and repeatable across the boundaries. This approach allows the model to simulate an infinitely long structure using only one representative segment, thereby eliminating artificial boundary effects and significantly reducing computational cost [38]. Moreover, to prevent rigid body motion and ensure numerical stability, both ends of the outer sheath were fully constrained in all degrees of freedom (DOFs).

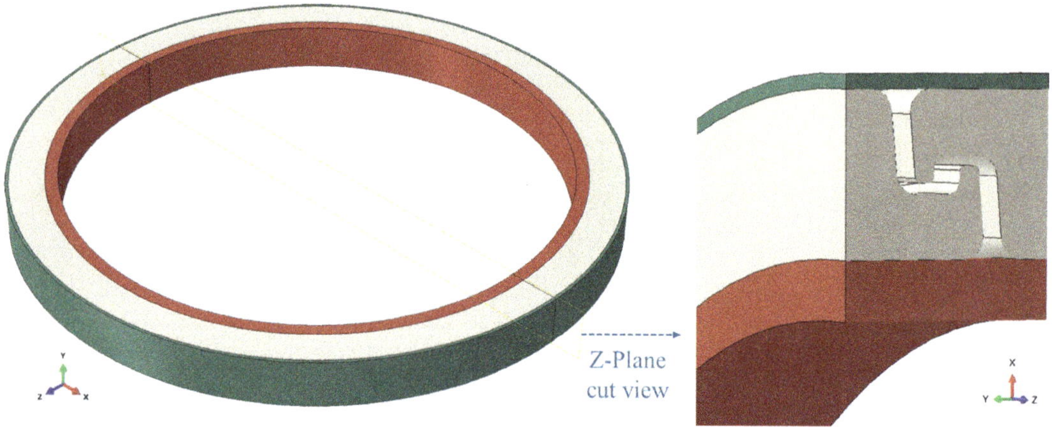

Figure 8. The FE geometry of the PBC model.

To ensure reproducibility, the explicit multi-point constraint (MPC) equations are used in the PBC model. Each node on the $+Z$ boundary is paired with its corresponding node on the $-Z$ boundary such that

$$\mathbf{u}^{+}(i) - \mathbf{u}^{-}(i) = \mathbf{R}\,\Delta\theta \tag{15}$$

where $\mathbf{R}$ is the rotation matrix corresponding to one helical pitch and $\Delta\theta = 2\pi/m$ is the circumferential advance of a single strip. This explicit node-pair formulation ensures that the stress, strain, and displacement fields are periodic, allowing for the unambiguous reproduction of the model.

The Static General step in ABAQUS was chosen, as it efficiently handles nonlinearities such as contact, plasticity, and large deformation, making it suitable for flexible pipe simulations. In this study, the step time used an initial increment size of 0.1, a minimum allowable increment of 10^{-5}, and the default force residual tolerance of 10^{-3}, providing stable convergence throughout the nonlinear loading process. The internal pressure was applied uniformly to the inner surface of the PE layer. Moreover, three surface-to-surface contact interactions were established to capture the interfacial behaviour between layers:

(1) Between the PE and PA layers, a rough tangential formulation was applied with hard normal contact and separation allowed after contact. This choice ensures that the PE acts purely as a load transmitter without allowing unrealistic sliding at the interface, while still permitting detachment under high internal pressure [39].

(2) Between the PA strips, a penalty tangential formulation was used with a friction coefficient of 0.1, while normal behaviour was defined as hard contact with separation allowed after contact. The penalty formulation with a calibrated friction coefficient reflects experimental observations of limited slip between PA strips and prevents artificial locking, consistent with the experimental results [34].

(3) Between the PA and the outer sheath, a rough tangential formulation was employed with hard normal contact, but separation after contact was not permitted. This constraint was necessary to mimic the physical role of the outer sheath in resisting radial expansion and providing continuous confinement to the PA layer during pressurisation.

The PBC formulation represents an equivalent lay angle of 90°, consistent with published descriptions of pressure armour winding geometry [40]. Moreover, under pure internal pressure, shear transfer across interfaces is minimal and frictional effects on burst behaviour become secondary [41].

All layers were meshed using C3D8R elements, an 8-node linear brick element with reduced integration and hourglass control, which is widely recommended in ABAQUS for contact-dominated problems. This choice ensures reliable accuracy for nonlinear analyses while maintaining computational efficiency. The mesh details are shown in Figure 9.

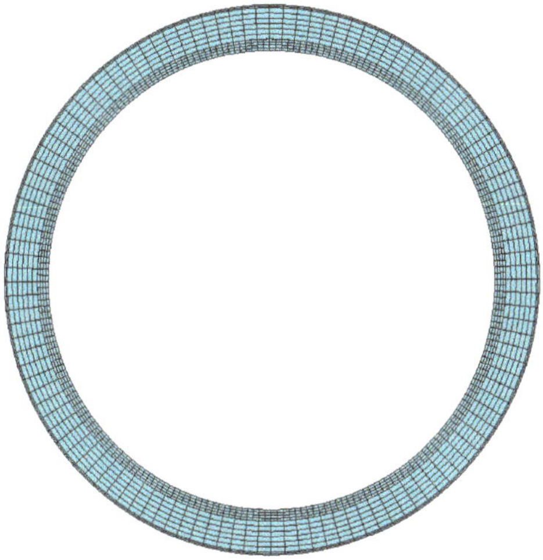

Figure 9. The mesh schematic of the PBC model.

A mesh sensitivity analysis was performed using six different mesh densities to identify a suitable element size for the model. The predicted critical pressure stabilised once the mesh was refined below approximately 12–15 mm, and further refinement produced changes of less than 1%. Meshes coarser than this range (corresponding to fewer than about 28,800 elements in the PE layer, 21,600 in the PA layer, and 960 in the outer sheath) underpredicted the critical pressure by several percent due to insufficient resolution of the localised plastic deformation in the PA strips. In contrast, meshes with element sizes between 8 mm and 20 mm exhibited nearly identical pressure–response behaviour, indicating that the solution had reached mesh independence within this range. Based on this observation, an element size of 10 mm was selected as the optimal compromise between computational efficiency and numerical accuracy.

3.5. Results

Figure 10 presents the von Mises stress distribution obtained from the axisymmetric FE model under internal pressure. The stress field exhibits a fully periodic pattern along the circumferential direction, consistent with the repeated Z-shaped geometry of the pressure armour. Local stress concentrations develop at the crown regions of the Z-lock, where geometric curvature induces bending effects and governs the load transfer mechanism. Throughout loading, the polymer liner sustains comparatively low stress levels, confirming that the PA layer carries the dominant portion of the hoop stresses while the PE layer primarily functions as a stabilising internal sheath that transmits internal pressure to the metallic reinforcement. A small displacement in the X-direction is visible on the right boundary of the model; this results from the reduced stiffness of the half-modelled PA segment but is negligible and does not influence the global stress response or the predicted collapse pressure.

Figure 10. Von Mises stress distribution in the axisymmetric FE model under internal pressure (unit: MPa).

Figure 11 presents the von Mises stress distribution from the full 3D FE model of the pipe section under internal pressure. High stress zones (in red) appear primarily at the contact interfaces and curved regions of the metallic interlocks, indicating these areas experience the greatest plastic deformation. In contrast, the PE liner experiences substantially lower stress levels, remaining far below its yield strength and acting mainly as a pressure-transmitting medium. Along the PA tips, the interlocked geometry efficiently carries hoop stresses but also generates critical stress hotspots that may trigger plastic deformation and potential instability.

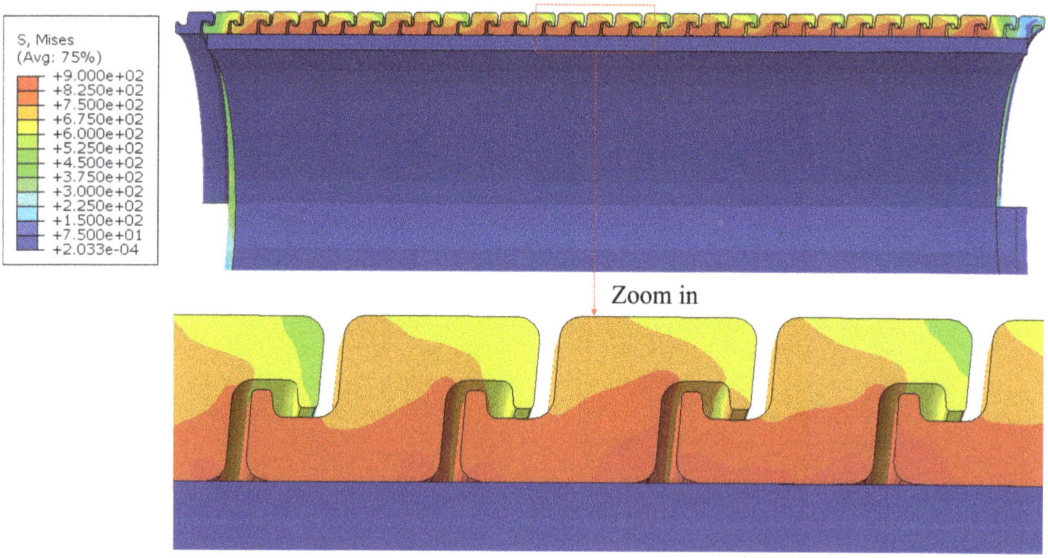

Figure 11. Von Mises stress distribution in the full 3D FE model under internal pressure (unit: MPa).

Figure 12 shows the evolution of the von Mises stress distribution in the PBC finite element model of the Z-shaped closing profile when subjected to increasing internal pressure levels of approximately 25 MPa, 50 MPa, and 75 MPa. At the lowest internal pressure (a), the von Mises stresses are concentrated mainly at the inner corner of the interlocking region, while the rest of the cross-section remains in a relatively low-stress state. As the pressure increases (b), the overall stress pattern remains similar; however, the polymer begins to penetrate further into the PA cross-section. At the failure stage (c), this polymer penetration becomes more pronounced, which significantly increases the local contact pressure and drives deformation in the weaker section of the PA. Consequently, the von Mises stress reaches its maximum values (red zone) within the inner fillet and the load-bearing section of the interlock, indicating the regions most prone to yielding or damage initiation under pressurisation. The overall stress pattern remains symmetric and continuous, confirming that the boundary conditions realistically capture the mechanical response of the periodic structure.

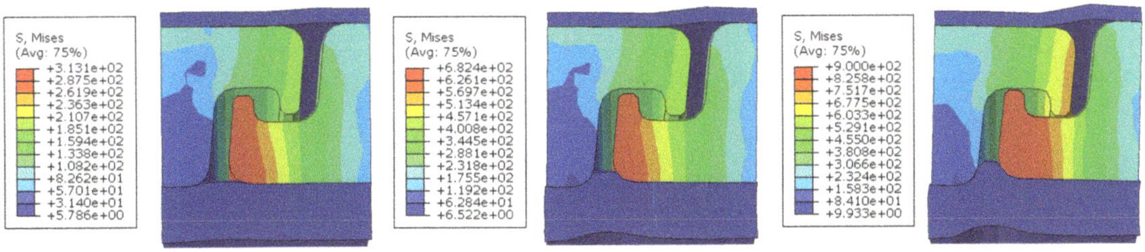

(a) (b) (c)

Figure 12. Von Mises stress distribution in the PBC FE model under internal pressures of (**a**) 25 MPa, (**b**) 50 MPa, and (**c**) 75 MPa.

It is important to note that the stress distribution obtained from the PBC model (Figure 12) differs from that of both the full 3D helix (Figure 11) and the 2D axisymmetric model (Figure 10), because the three formulations represent fundamentally different physical assumptions. The full 3D model captures the behaviour of a finite-length helical reinforcement, including end effects, helix-induced bending, and non-uniform contact along the pitch; as a result, it produces a Lamé-type stress field with maximum von Mises stresses concentrated along the inner surface of the PA strip. In contrast, the PBC model represents an infinitely long, periodically repeating unit cell in which periodic constraints suppress global hoop deformation and eliminate end effects. This forces the stress field to become fully periodic and governed almost entirely by the local geometry of the Z-shaped interlock. Consequently, the PBC model localises its peak stresses at the locking tip—exactly where curvature, contact, and geometric constraint dominate the load transfer—rather than along the inner surface as in the full 3D helix. This localisation behaviour is also more consistent with the 2D axisymmetric representation, where hoop deformation is constrained and the stress field is driven primarily by profile curvature rather than helix bending.

Figure 13 compares the hoop stress response of the analytical model with the three FE approaches: the axisymmetric 2D model, the full 3D model, and the PBC model. All models exhibit a linear increase in von Mises stress with internal pressure until reaching the plastic region of the PA material. The axisymmetric 2D model predicts a significantly lower collapse pressure than the other approaches because its formulation cannot reproduce the inherently three-dimensional behaviour of the pressure armour. Moreover, the 2D model neglects lay-angle-induced bending–torsion coupling, the discrete Z-lock contact mechanism, and the shear transfer between adjacent strips—mechanisms that collectively provide substantial stiffness under internal pressure. In contrast, the full 3D model captures these effects explicitly, including geometric stiffening from the helix angle and localised compaction at the interlocking region, which results in a slightly higher predicted capacity than the analytical solution. The PBC model agrees closely with the analytical response and remains slightly below the full 3D prediction, as the periodic formulation suppresses end effects and pitch-induced deformation.

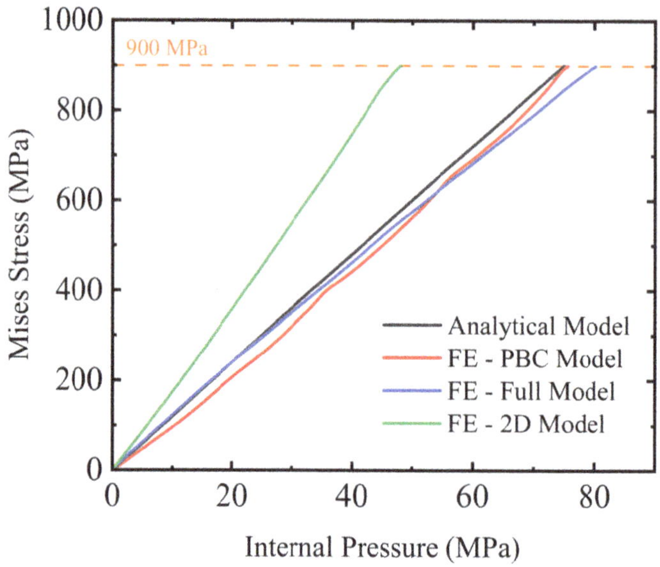

Figure 13. The comparison between FE models and theory.

The computational cost differs significantly among the FE methods, and all simulations were carried out on a workstation equipped with an Intel Core i9-14900KF processor (24 cores, 32 threads) and 64 GB RAM. The results, together with the corresponding critical internal pressure (IP) values predicted by each method, are summarised in Table 2. These results demonstrate that the PBC model yields a conservative yet sufficiently accurate prediction of collapse resistance while offering a dramatic reduction in computational cost. In contrast, the full 3D helical model provides the most realistic stress field but requires significantly longer runtimes, making it less practical for extensive design studies. The axisymmetric model is computationally much more efficient but underestimates the collapse pressure because it cannot capture helix-induced stiffness and three-dimensional contact behaviour. Taken together, the very close agreement in critical pressures between the PBC model and the analytical solution confirms the robustness, accuracy, and conservative reliability of the PBC finite element strategy.

Table 2. Critical internal pressure (IP) amounts of different models.

Models	Analytical	FE-2D	FE-Full	FE-PBC
Critical IP (MPa)	74.90	48.00	80.19	75.55
Computational cost (s)	/	40	16,408	1075

4. Discussion

To further examine the pipe response under increasing internal pressure, the radial displacement of the PA in the PBC model is shown in Figure 14. The distribution pattern shows that the crown of the PA cross-section exhibits the greatest displacement due to its reduced thickness along the radial axis, making it more flexible. In contrast, the invert region shows the smallest displacement because pressure is transferred indirectly through the PA body before reaching the invert. Much of the load is therefore absorbed along the PA thickness, reducing its radial deformation. Correspondingly, A line of nodes in the PE-PA contact surface was selected, where the largest radial expansion occurs. The displacement of these nodes was tracked and averaged to represent the overall radial deformation of the PA, with results plotted as a function of internal pressure in Figure 15.

Figure 14. The radial displacement distribution among the PA (unit: mm).

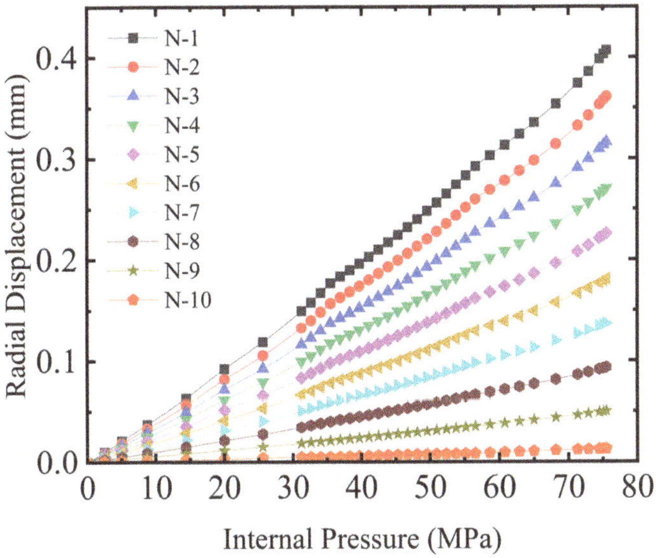

Figure 15. The internal pressure-radial displacement relationship.

The finite element analyses reveal that the model fails because the pressure armour reaches its material stress limit before any contact occurs in the nub–nub interaction section (Figures 10–12). This outcome indicates that there is sufficient clearance between neighbouring nubs for the selected material, meaning that failure is not governed by nub–nub interaction or crown–invert compression, which are often the dominant mechanisms in the PA configurations. Instead, the results suggest that the chosen cross-sectional design is particularly well suited for this material, as it prevents premature interaction-driven failure and allows the load-bearing capacity of the PA to be fully activated.

The contact pressure distribution along the PE–PA interface is shown in Figure 16. The evaluated nodes correspond to those used in Figure 14. The crown region exhibits the lowest contact pressure, confirming it as the most compliant and weakest section of the PA under internal pressure. The figure, obtained at the failure point, shows that the contact pressures at the interlocking tips exceed the pipe's burst pressure, indicating that sealing integrity is maintained and failure does not occur due to leakage. The evolution of contact pressure with internal pressure is presented in Figure 17. In all cases, the contact pressure at the PE–PA interface increases nonlinearly and remains higher than the applied internal fluid pressure throughout the loading range. This behaviour confirms that the interface stays in compression, preventing fluid penetration or the formation of leakage paths. Consequently, the overall structural failure is governed by yielding or plastic collapse of the metallic PA, rather than by loss of sealing or liner separation.

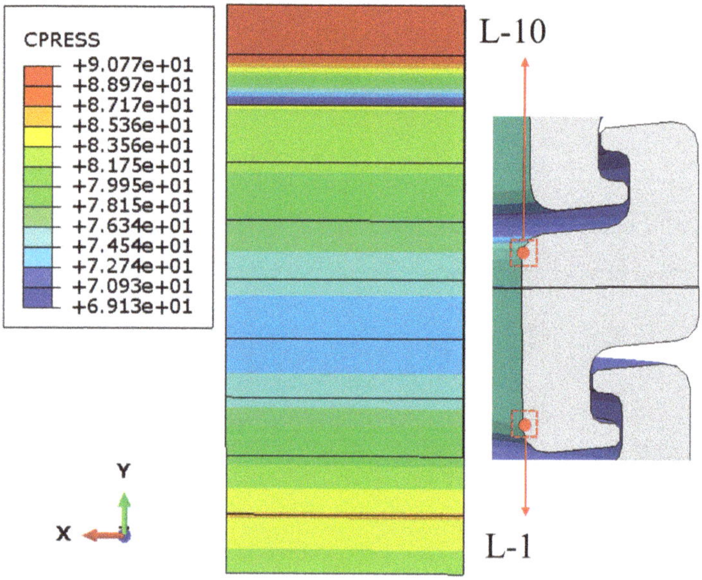

Figure 16. The contact pressure in the PE-PA contact surface (unit: MPa).

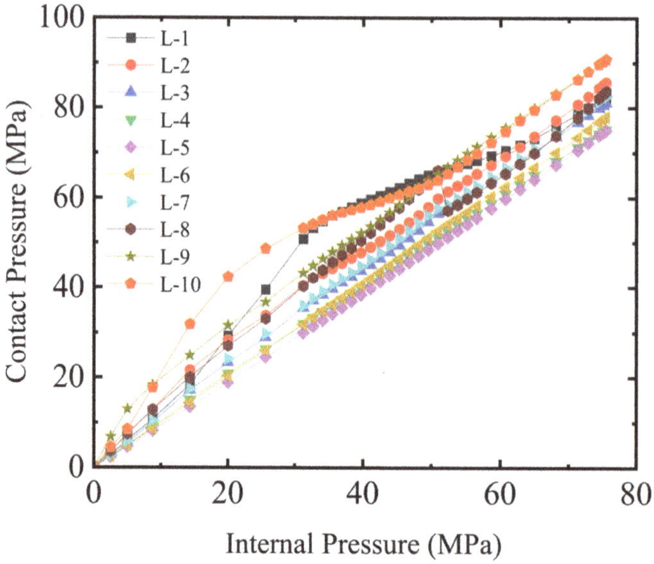

Figure 17. The internal pressure-contact pressure relationship.

5. Conclusions

This study presented a comprehensive investigation into the burst behaviour of the pressure armour (PA) layer under internal pressure. Three finite element (FE) approaches were developed and assessed: a simplified two-dimensional (2D) axisymmetric model, a fully detailed three-dimensional (3D) helical model, and an efficient periodic boundary condition (PBC) model. These were benchmarked against analytical solutions to evaluate their accuracy and computational performance.

- The axisymmetric 2D model, although extremely fast, significantly underestimated the collapse pressure because it cannot represent the inherently three-dimensional mechanics of the helical strip. The absence of lay-angle-induced bending–torsion coupling, discrete Z-lock contact, and shear transfer between neighbouring strips leads to an underprediction of global stiffness.
- The full 3D helical model captured all geometric and contact-related effects explicitly, including localised locking deformation and helix-induced stiffening. As a result, it predicted a slightly higher collapse pressure than the analytical reference and provided the most realistic stress field. However, its computational cost was prohibitively high for use in design optimisation or parametric studies.
- The PBC model closely reproduced the analytical hoop stress response while maintaining a slightly conservative prediction relative to the full 3D model. By eliminating end effects and enforcing periodicity, it retained the essential mechanics of the interlocking Z-shaped strip while achieving a reduction in runtime of more than an order of magnitude.

Across all FE approaches, the governing failure mechanism was consistent: collapse occurred when the material strength of the PA layer was exceeded. No premature nub–nub interaction or crown–invert compression was observed, confirming that the selected PA cross-section provides sufficient clearance to avoid interaction-driven failure. The polymer liner remained far below its strength limit in all simulations and primarily acted as a pressure-transmitting medium rather than as a structural load-bearing component.

Future work may focus on validating the numerical findings through dedicated burst tests on full-scale PA specimens, enabling direct experimental confirmation of the predicted collapse pressures.

Author Contributions: M.S.: software, data curation, writing—original draft preparation; Y.Z.: methodology, validation, writing—review and editing; P.F.: visualization; Y.B.: supervision. All authors have read and agreed to the published version of the manuscript.

Funding: This research received no external funding.

Data Availability Statement: The original contributions presented in this study are included in the article. Further inquiries can be directed to the corresponding author.

Conflicts of Interest: Author Yong Bai was employed by the company OPR Offshore Engineering Equipment Co., Ltd. The remaining authors declare that the research was conducted in the absence of any commercial or financial relationships that could be construed as a potential conflict of interest.

References

1. Saneian, M.; Han, P.; Jin, S.; Bai, Y. Fracture response of steel pipelines under combined tension and torsion. *Thin-Walled Struct.* **2020**, *154*, 106870. [CrossRef]
2. Saneian, M.; Han, P.; Jin, S.; Bai, Y. Fracture response of steel pipelines under combined tension and bending. *Thin-Walled Struct.* **2020**, *155*, 106987. [CrossRef]
3. Zhang, Y.; Liu, W.; Kuang, W.; Saneian, M.; Bai, Y. Mechanical behaviour analysis of steel strips reinforced thermoplastic pipe under torsion. *Ocean. Eng.* **2024**, *303*, 117655. [CrossRef]

4. Zhang, Y.; Liu, W.; Saneian, M.; Bai, Y. A comprehensive study on the failure modes of steel strips reinforced thermoplastic pipe under torsion. *Int. J. Press. Vessel. Pip.* **2025**, *216*, 105490. [CrossRef]
5. Bai, Y.; Zhang, Y.; Saneian, M.; Fang, P.; Quintero, J.; Ye, Q. Structural integrity of pipe-joint entirety under multi-axial loading: Comparative assessment of butt-joints and mid-joints. *Ocean. Eng.* **2025**, *342*, 122911. [CrossRef]
6. de Sousa, J.R.M.; Sagrilo, L.V.S.; Rocha, D.M. Semi-empirical equations to predict the bisymmetric dry and wet hydrostatic collapse strengths of flexible pipes. *Appl. Ocean. Res.* **2022**, *122*, 103129. [CrossRef]
7. Saneian, M.; Bai, Y. Assessment of the Carcass Role in Designing Deepwater Riser Connectors. In Proceedings of the International Conference on Offshore Mechanics and Arctic Engineering, Glasgow, UK, 9–14 June 2019; Volume 58806, p. V05AT04A033. [CrossRef]
8. Saneian, M.; Bai, Y. Evaluation of the Inner Liner in Unbounded Flexible Pipelines in the Sealing System of End-Fitting. In Proceedings of the International Conference on Offshore Mechanics and Arctic Engineering, Singapore, 9–14 June 2024; Volume 87806, p.V003T04A004. [CrossRef]
9. Zhang, Y.; Saneian, M.; Bai, Y.; Fang, P.; Cheng, P.; Zhao, L. Test-based equivalent-material method for collapse qualification of helically wound and layered cylindrical structures-with application to flexible-riser carcass. *Nat. Commun. Eng.* **2025**. [CrossRef]
10. Bai, Y. *Deepwater Flexible Risers and Pipelines*; John Wiley & Sons: Hoboken, NJ, USA, 2020.
11. Det Norske Veritas. *Project Recommended Practice: Composite Risers*; Technical Report; Report No. 2002-0067; Det Norske Veritas: Høvik, Norway, 2002.
12. Féret, J.J.; Bournazel, C.L. Calculation of Stresses and Slip in Structural Layers of Unbonded Flexible Pipes. *J. Offshore Mech. Arct. Eng.* **1987**, *109*, 263–269. [CrossRef]
13. De Oliveira, J.; Goto, Y.; Okamoto, T. Theoretical and methodological approaches to flexible pipe design and application. In Proceedings of the Annual Offshore Technology Conference, Houston, TX, USA, 6–9 May 1985; pp. 517–526. [CrossRef]
14. Colquhoun, R.; Hill, R.; Nielsen, R. Design and materials considerations for high pressure flexible flowlines. In Proceedings of the SUT Advances in Subsea Pipeline Engineering and Technology, Aberdeen, UK, 30–31 May 1990; pp. 145–178.
15. Fernando, U.S.; Tan, Z.; Sheldrake, T.; Clements, R. The stress analysis and residual stress evaluation of pressure armour layers in flexible pipes using 3D finite element models. In Proceedings of the International Conference on Offshore Mechanics and Arctic Engineering, Vancouver, BC, Canada, 20–25 June 2004; Volume 37459, pp. 57–65. [CrossRef]
16. Zhu, X.K.; Leis, B.N. Theoretical and Numerical Predictions of Burst Pressure of Pipelines. *J. Press. Vessel. Technol.* **2007**, *129*, 644–652. [CrossRef]
17. Gay Neto, A.; Martins, C.v.d.A.; Pesce, C.P.; Meirelles, C.O.C.; Malta, E.R.; Barbosa Neto, T.F.F.; Godinho, C.A.F. Burst prediction of flexible pipes. In Proceedings of the International Conference on Offshore Mechanics and Arctic Engineering, Shanghai, China, 6–11 June 2010; Volume 49132, pp. 511–520. [CrossRef]
18. Neto, A.G.; Martins, C.D.A.; Pesce, C.P.; Meirelles, O.C.; Malta, E.R. Prediction of burst in flexible pipes. *J. Offshore Mech. Arct. Eng.* **2013**, *135*, 011401. [CrossRef]
19. Cuamatzi-Melendez, R.; Castillo-Hernández, O.; Vázquez-Hernández, A.; Albiter, A.; Vaz, M. Finite element modeling of burst failure in unbonded flexible risers. *Eng. Struct.* **2015**, *87*, 58–69. [CrossRef]
20. O'halloran, S.; Connaire, A.; Harte, A.M.; Leen, S.B. Modelling of fretting in the pressure armour layer of flexible marine risers. *Tribol. Int.* **2016**, *100*, 306–316. [CrossRef]
21. Liu, X.; Qu, Z.; Liu, Y.; He, J.; Si, G.; Wang, S.; Liu, Q. Numerical Simulation of Burst Failure in 2.5-Inch Unbonded Flexible Riser Pressure Armor Layers. *Metals* **2024**, *14*, 762. [CrossRef]
22. Wei, R.; Li, X.; Vaz, M.A.; Custódio, A.B. Finite Element Modelling of Pressure Armour Unlocking Failure in Unbonded Flexible Risers. *J. Mar. Sci. Eng.* **2025**, *13*, 411. [CrossRef]
23. Santos, C.C.; Pesce, C.P. An alternative strategy for offshore flexible pipes finite element analysis. *Mar. Struct.* **2019**, *65*, 376–416. [CrossRef]
24. Timoshenko, S.P.; Goodier, J.N. *Theory of Elasticity*, 2nd ed.; McGraw–Hill: New York, NY, USA, 1951.
25. Lamé, G. *Leçons sur la Théorie Mathématique de L'élasticité des Corps Solides*; Bachelier: Paris, France, 1852.
26. Martins, C.v.A.; Pesce, C.P.; Aranha, J.A. Structural behavior of flexible pipe carcass during launching. In Proceedings of the International Conference on Offshore Mechanics and Arctic Engineering, Cancun, Mexico, 8–13 June 2003; Volume 36827, pp. 537–546. [CrossRef]
27. American Petroleum Institute. *API Specification 17J: Specification for Unbonded Flexible Pipe*; American Petroleum Institute: Washington, DC, USA, 2014.
28. Krishna, S.; Manna, I.; Glushych, V.; Weisheit, A.; Dutta Majumdar, J. Microstructural Evolution and Wear Resistance of Surface Alloyed AISI 316L Stainless Steel with Equi-atomic CoCrFeMnTi by Continuous Wave Diode Laser. *Opt. Laser Technol.* **2024**, *179*, 111345. [CrossRef]
29. Yan, F.; Liu, G.; Tao, N.; Lu, K. Strength and ductility of 316L austenitic stainless steel strengthened by nano-scale twin bundles. *Acta Mater.* **2012**, *60*, 1059–1071. [CrossRef]

30. Cornacchia, F.; Liu, T.; Bai, Y.; Fantuzzi, N. Tensile strength of the unbonded flexible pipes. *Compos. Struct.* **2019**, *218*, 142–151. [CrossRef]
31. API. *RP17B: Recommended Practice for Flexible Pipe*; American Petroleum Institute: Washington, DC, USA, 2008.
32. Dassault Systèmes. *ABAQUS/Standard User's Manual*; SIMULIA: Johnston, RI, USA, 2024.
33. Belytschko, T.; Bindeman, L.P. Assumed strain stabilization of the eight node hexahedral element. *Comput. Methods Appl. Mech. Eng.* **1993**, *105*, 225–260. [CrossRef]
34. Saevik, S.; Berge, S. Fatigue testing and theoretical studies of two 4 in flexible pipes. *Eng. Struct.* **1995**, *17*, 276–292. [CrossRef]
35. Dong, L.; Huang, Y.; Zhang, Q.; Liu, G. An analytical model to predict the bending behavior of unbonded flexible pipes. *J. Ship Res.* **2013**, *57*, 171–177. [CrossRef]
36. Smith, D.M.; Cunningham, L.S.; Chen, L. Efficient finite element modelling of helical strand cables utilising periodicity. *Int. J. Mech. Sci.* **2024**, *263*, 108792. [CrossRef]
37. Fang, P. Development of an Effective Modelling Method for the Local Mechanical Analysis of Submarine Power Cables. Ph.D. Thesis, Delft University of Technology, Delft, The Netherlands, 2024.
38. Li, S.; Sitnikova, E. An excursion into representative volume elements and unit cells. *Compr. Compos. Mater. II* **2017**, *1*, 451–489. [CrossRef]
39. Zhang, M.; Chen, X.; Fu, S.; Guo, Y.; Ma, L. Theoretical and numerical analysis of bending behavior of unbonded flexible risers. *Mar. Struct.* **2015**, *44*, 311–325. [CrossRef]
40. Bai, Y.; Bai, Q. *Subsea Pipelines and Risers*, 2nd ed.; Elsevier Science: Oxford, UK, 2005.
41. Provasi, R.; Toni, F.G.; de Arruda Martins, C. Friction coefficient influence in a flexible pipe: A macroelement model. *Ocean. Eng.* **2022**, *266*, 112719. [CrossRef]

Journal of
Marine Science and Engineering

MDPI

Article

Theoretical and Experimental Analysis of the Global Response of a Flexible Pipe Under Combined Axisymmetric and Bending Loads

Bernardo de P. M. da Costa [1,*], José Renato M. de Sousa [2], Ney Roitman [2] and Erik Radke [3]

[1] Petrobras, Petróleo Brasileiro S.A., Avenida República do Chile, 65, Rio de Janeiro 20031-170, RJ, Brazil
[2] Civil Engineering Department, COPPE/UFRJ—Post Graduate Institute, Federal University of Rio de Janeiro, Rio de Janeiro 21945-970, RJ, Brazil; jose.sousa@coc.ufrj.br (J.R.M.d.S.); roitman@coc.ufrj.br (N.R.)
[3] LAMEF, Physical Metallurgy Laboratory, Federal University of Rio Grande do Sul, Av. Bento Gonçalves, 9500, Porto Alegre 91504-000, RS, Brazil; erik.radke@ufrgs.br
* Correspondence: bernardocosta@petrobras.com.br

Abstract: The bending stiffness of flexible pipes is highly dependent on curvature, driven by the interaction between their structural layers—a behavior often misrepresented by traditional numerical models. To overcome this limitation, a finite-difference-based model was developed, integrating previously proposed formulations for monotonic bending and axisymmetric responses into a return-mapping algorithm to capture hysteretic behavior under cyclic loading. The model was calibrated against pure bending and pressurized tests, accounting for interlayer adhesion and friction, which govern stiffness variation, force levels, and energy dissipation. Results showed excellent agreement with experimental data across different load combinations, confirming the model's predictive capability. Parametric analyses revealed that higher adhesion and friction coefficients increase imposed forces until a no-slip condition is achieved, while energy dissipation follows a nonlinear dependence on interlayer friction, peaking at intermediate values and vanishing under no-slip conditions. Cyclic bending tests performed on degraded samples demonstrated that, despite wire deterioration, the global bending response remains essentially unchanged, reinforcing the stability of riser behavior over time. However, fatigue resistance must still be reassessed through updated S–N curves to account for material degradation. These findings underscore the crucial role of interlayer mechanics in determining the overall performance of flexible pipes and offer a validated framework for assessing fatigue and integrity.

Keywords: flexible pipe; finite difference method; global analysis; flexible pipe bending; cyclic bending test

1. Introduction

Unbonded flexible pipes are crucial in offshore systems, serving as conduits for oil production, water injection, gas transportation, and service lines. They offer key advantages, such as enabling early production systems, adapting to subsea layout changes, and reducing installation time compared to rigid pipelines. In a typical unbonded flexible pipe, as in Figure 1, mechanical resistance is mainly provided by three metallic layers: the carcass, the pressure armor, and the tensile armor layers. The carcass, when present, is the innermost layer, composed of profiled stainless steel strips wound at nearly 90° to withstand external pressure. The pressure armor layer resists internal pressure and typically consists of one or two Z-shaped carbon steel strips, laid at angles close to 90°. The tensile armor layers comprise high-alloy carbon steel wires with approximate rectangular cross-sections arranged in pairs (one or two pairs) and wound in opposing directions at angles between 20° and 55°. Their primary function is to resist tensile loads, torsion, and end cap effects. In contrast, internal and outer polymeric sheaths ensure tightness, while antiwear polymeric tapes reduce friction and wear between the metallic layers during the pipe operation. Furthermore, depending on design requirements, the structure may include insulation layers, a protective outer sheath, and high-strength tapes, as specified in [1].

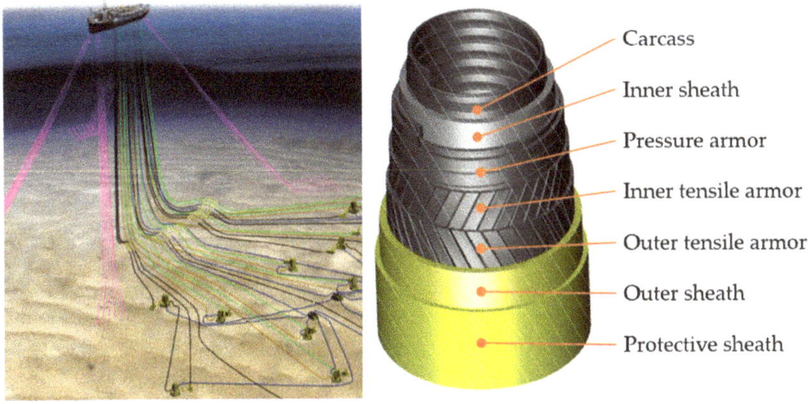

Figure 1. Subsea flexible pipe production system with a detail of the pipe's internal layers.

Brazil's flexible pipe industry benefits from strong manufacturing capabilities and a substantial fleet of installation vessels. However, flexible pipe technology faces challenges at extreme depths, in high-temperature environments, and in environments with elevated contaminant levels. These limitations may impact their effectiveness in offshore hydrocarbon exploration, particularly in Brazil's Pre-Salt region [2]. For instance, flexible risers (e.g., Figure 1) are used in dynamic applications, connecting production units to flowlines, seabed equipment, and other floating structures [1]. These structures must resist operational and environmental loads that subject them to cyclic tensile forces and bending moments, especially at the riser support, where high axial loads and curvatures co-occur. The intensity of these loads increases with depth and operating pressures, which have both steadily risen over recent decades. In 2000, the average operating depth was approximately 650 m, with design pressures of 225 bar. By 2020, these values had increased to 1450 m and 320 bar [3].

The design of flexible risers involves iterative local and global analyses [4]. Local analyses model each pipe layer to determine the mechanical properties of the pipe (axial, torsional, and bending stiffnesses), which are then used in global finite element (FE) models to evaluate the riser's response to environmental loads and vessel motions. In global models, while axial and torsional stiffnesses are generally assumed constant, accurately representing the nonlinear variation in bending stiffness with curvature remains a challenge. The design process concludes with refined local analyses to verify stresses, strains, and fatigue resistance under axisymmetric and bending loads. Hence, this process requires specific local models to account for the effects of axisymmetric and bending loads in the pipe structure, as well as global models that incorporate the nonlinear response of these structures.

The local mechanical response under axisymmetric loading has been extensively studied using analytical and numerical models. Early analytical formulations, which are based on equilibrium and compatibility equations with simplifying assumptions [5–7], evolved with the inclusion of the Clebsch–Kirchhoff theory for curved beams [8], polymeric sheath effects, and possible layer separation [9,10], which improved predictions of axial compression and torsion. Their predictive capability was validated by blind tests [11] and subsequent experimental studies [12,13]. Later works incorporated material nonlinearities [14,15], pressure and thermal effects [16,17] into the extensional–torsional response of the pipe, as well as structural defects [13,18]. Dedicated analytical models have also addressed failure mechanisms such as collapse, burst, and tensile armor buckling [19–23].

Numerical models, typically FE-based, require fewer assumptions but entail higher computational cost. They capture geometric imperfections, nonlinear interlayer contact (including friction and gap opening), and material nonlinearities, making them suitable for both operational and failure analyses. Simplified representations, such as equivalent layers, unit cell approaches, or macroelements [17,24–29], reduce computational effort while maintaining accuracy. These models have demonstrated good agreement with experimental data and analytical predictions [11–13,24,30]. For radial failure modes (collapse and burst), both equivalent and direct modeling approaches [31–35] are employed; direct models, although more costly, remain feasible due to the limited model length required.

While axisymmetric responses are well understood, the bending behavior of flexible pipes remains less clearly characterized. Analytical bending models typically focus on the moment–curvature relationship or stresses in tensile armor wires, whereas FE models offer greater versatility at a higher cost. A central challenge for both approaches is representing interlayer friction, which governs stiffness transitions from no-slip to full-slip states and produces a hysteretic response dependent on curvature and axial load.

Axisymmetric loads generate interlayer contact pressures and friction forces that restrict relative motion between layers when the pipe is bent. At low curvatures, this friction prevents tensile armor wire slippage, and the pipe exhibits its maximum (no-slip) bending stiffness (1), as shown in Figure 2. As curvature increases, friction is overcome and wires begin to slide (2), reducing stiffness until it stabilizes at the minimum (full-slip) value (3). Further bending may cause wire contact, increasing stiffness, while load reversal restores adhesion and the no-slip condition (4). This friction-controlled process produces a hysteretic response, whose accurate stiffness evaluation must also consider polymeric layers and axial loads.

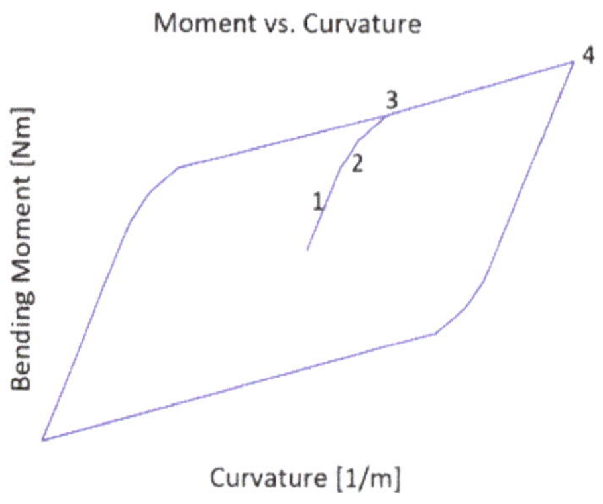

Figure 2. Moment vs. curvature in a typical flexible pipe.

Recent studies have deepened the understanding of flexible pipe bending by focusing on layer interaction and initial contact pressure. Dong et al. [36], building upon the work of Kebadze and Kraincanic [37], developed an analytical model that incorporates helical element bending and torsion, and validated it against Witz's [11] tests, which required an initial contact pressure of 1.5 MPa for convergence. Ye et al. [38] introduced a correction factor for antiwear tapes and found discrepancies in hysteresis at low pressures due to manufacturing-induced contact, which diminished with increasing internal pressure. Kim et al. [39] refined these models by including shear deformation in polymeric layers and curvature-dependent contact forces, achieving better agreement with FE simulations. Dai et al. [40] further explored friction modeling, showing that a smoothed Coulomb law best balanced accuracy and convergence, with friction coefficients requiring pressure-based calibration. Wang et al. [41] proposed a simplified 3D FE model using a double-helix beam contact, significantly reducing computational cost while preserving accuracy, calibrated through initial stresses and strains. At the wire scale, Fang et al. [42] demonstrated that analytical models perform well for round wires but deviate for rectangular ones. Finally, Tang et al. [43] conducted combined tension–bending tests on an 8-inch pipe, confirming good agreement between the experiments and analytical predictions based on tendon slip behavior.

Another key aspect concerns the local–global coupling in flexible pipe design. Global analyses of flexible pipes are typically performed using dedicated software, such as Orcaflex (version 11.5e) [44–46], Riflex (version V11.0) [47], or ANFLEX (Petrobras' in-house software) [48], which model long flexible risers as beam elements with equivalent mechanical properties. Although these tools can represent the nonlinear bending moment–curvature relationship, the underlying modeling frameworks are often not fully documented, which may hinder potential modifications or, for example, the implementation of similar formulations in more general FE packages such as ANSYS (version 2025 R2) or Abaqus (version 2025 FD03).

In summary, while the axisymmetric response of flexible pipes is well established, uncertainties persist regarding bending behavior, mainly due to interlayer friction. Global models still rely on specialized black-box software, and few experimental studies address cyclic bending or armor degradation issues, critical for long-term performance [2]. Therefore, this work proposes a theoretical approach to analyze the response of a 2.5″ flexible pipe subjected to combined axisymmetric and bending loads in a four-point bending test. The approach integrates a finite difference (FD) global model with a local model that captures the nonlinear variation in bending stiffness. Experimental tests were conducted to calibrate and verify the proposed model, including a second testing phase after controlled armor degradation.

Next, the 2.5″ pipe is described, followed by a detailed description of the experimental apparatus and test procedures. Subsequently, the proposed theoretical approach is introduced and applied to simulate the experimental tests described. The obtained results are compared to provide insights into the pipe's response to combined bending and axisymmetric loading for both non-degraded and degraded tensile armors. Finally, the study's main conclusions are established.

2. Experimental Tests

2.1. Flexible Pipe Characteristics

The experimental tests conducted in this study employed a sample of a 2.5″ flexible pipe (internal diameter of 63.5 mm), whose characteristics are presented in Table 1. The pipe operated as a flowline at a water depth close to 1000 m with an internal design pressure of 20.7 MPa.

Table 1. 2.5″ flexible pipe characteristics.

Layer (Material)	Properties
Inner carcass (Stainless 316 L)	Thickness = 3.50 mm, No. of tendons = 1, Lay angle = 87.7°, Interlocked profile, Area = 16.00 mm^2, Moment of inertia = 11.24 mm^4, Young modulus = 193 GPa, Poisson coefficient = 0.3
Internal plastic (Polyamide 12)	Thickness = 5.00 mm, Young modulus = 372 MPa, Poisson coefficient = 0.45
Pressure armor (Carbon steel)	Thickness = 6.35 mm, No. of tendons = 1, Lay angle = 87.9°, Z-profile, Area = 49.95 mm^2, Moment of inertia = 186.29 mm^4, Young modulus = 205 GPa, Poisson coefficient = 0.30
Antiwear layer (Polypropylene)	Thickness = 0.3 mm, Young modulus = 350 MPa, Poisson coefficient = 0.30
Inner tensile armor (Carbon steel)	Thickness = 2.01 mm, No. of wires = 43, Lay angle = 38°, Rectangular profile, Width = 5.00 mm, Young modulus = 205 GPa, Poisson coefficient = 0.30
Antiwear layer (Polypropylene)	Thickness = 0.6 mm, Young modulus = 350 MPa, Poisson coefficient = 0.30
Anti-buckling tape (Glass filament)	Thickness = 0.81 mm, Young modulus = 350 MPa, Poisson coefficient = 0.30
Antiwear layer (Polypropylene)	Thickness = 0.30 mm Young modulus = 350 MPa, Poisson coefficient = 0.30
Outer tensile armor (Carbon steel)	Thickness = 2.01 mm, No. of wires = 47, Lay angle = 36°, Rectangular profile, Width = 5.0 mm Young modulus = 205 GPa, Poisson coefficient = 0.30
Antiwear layer (Polypropylene)	Thickness = 0.30 mm, Young modulus = 350 MPa, Poisson coefficient = 0.30
Anti-buckling tape (Glass filament)	Thickness = 0.81 mm, Young modulus = 350 MPa, Poisson coefficient = 0
Antiwear tape (Polypropylene)	Thickness = 0.3 mm, Young modulus = 350 MPa, Poisson coefficient = 0
Outer sheath (HDPE)	Thickness = 6.55 mm Young modulus = 622 MPa, Poisson coefficient = 0.45

The layers' thicknesses were measured directly from samples extracted from the pipe. The properties of the metallic materials and the polymeric tapes were determined based on data available in the literature. However, the material properties of both the inner and outer sheaths play a crucial role in predicting the pipe's response to bending, especially in the full-slip regime. Therefore, experimental tests were conducted to assess the stress–strain curves for these materials, following the ASTM D638-22 [49] standard. Figure 3 illustrates the experimental setup and the samples before and after the tests, while Figure 4 shows the resulting curves for the inner and outer sheaths.

According to ASTM D638-22 [49], at least five tensile tests are required, and the Young's modulus for each test is the slope of the straight part of the graph close to the origin. Seven tests were performed using inner sheath samples, and six tests were performed using outer sheaths. Table 2 indicates average Young moduli for the inner and outer sheaths of 372 MPa and 622 MPa, respectively. Finally, a Poisson coefficient of 0.45 was adopted for both layers [50].

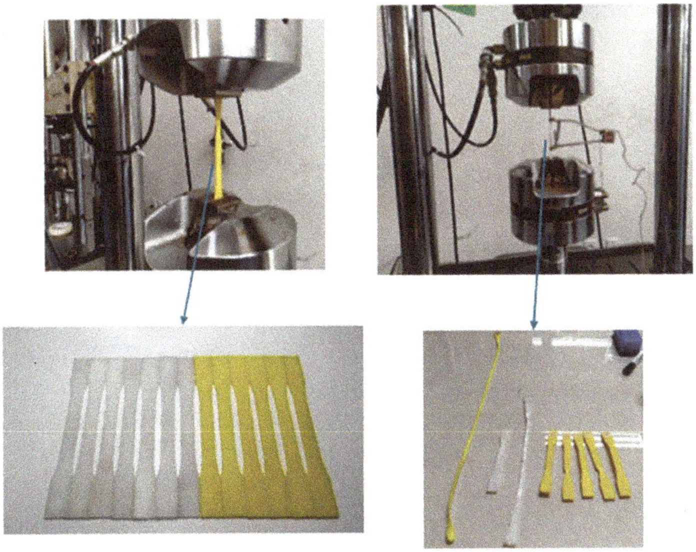

(a) (b)

Figure 3. Experimental tests with the polymeric samples: (**a**) undeformed samples in the experimental tests and (**b**) deformed samples after the experimental tests.

Table 2. Experimental Young moduli of the polymeric sheaths.

Test	Young Modulus (MPa)	
	Inner Sheath (PA)	Outer Sheath (PE)
1	358	601
2	373	650
3	408	619
4	372	577
5	405	644
6	323	640
7	361	-
Average	372	622
Std. deviation	29.1	28.5

2.2. Experimental Apparatus

The experimental tests employed a sample of the 2.5″ flexible pipe described in Section 3.1. The sample had a total length of 4.70 m, including two end-fittings (EF) with a length of 0.35 m each. Hence, the sample's total free span was 4.00 m, which is higher than the minimum values of 10xID (=635 mm) or 3.0 m, as suggested by API RP 17B [51] for full-size pipes.

The experimental apparatus was developed to conduct a four-point bending test, as shown in Figure 5. One end of the specimen was fixed to restrict the axial translation. In contrast, the opposite end was connected to an axial actuator (20 kN capacity) equipped with a load cell, allowing free axial movement. This apparatus ensures that any specimen elongation, whether due to internal pressurization or bending, does not result in additional tensile forces, i.e., the axial load remains constant during the test cycles.

Moreover, both ends of the specimen were free to rotate, a condition achieved through a pivot mechanism integrated with the EF grooves. A bidirectional transverse actuator with a maximum capacity of 100 kN, monitored by a load cell with a measure range up to

25 kN, was mounted to a rigid fork system that distributed the load across two application points via a roller mechanism. This configuration prevented the introduction of unwanted bending moments or secondary loads into the specimen.

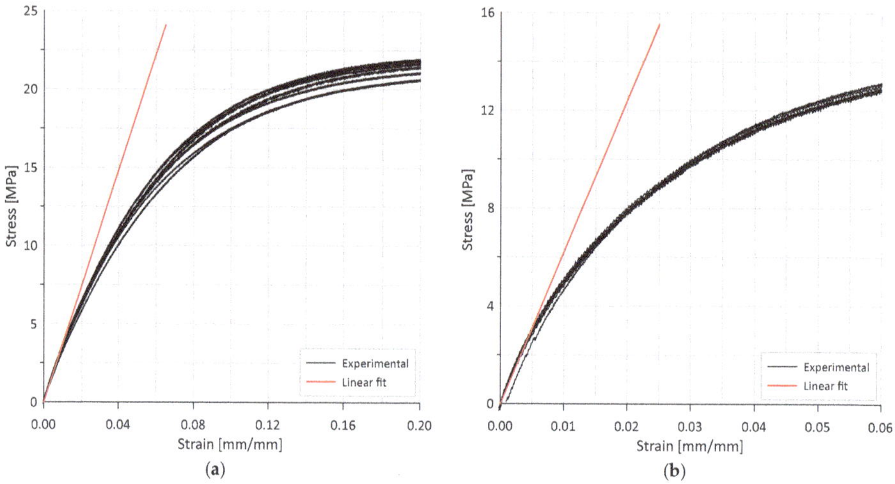

Figure 4. Experimental stress vs. strain curves: (**a**) inner and (**b**) outer sheath.

Displacements were measured at seven distinct points along the sample using linear transducers equipped with potentiometers, as shown in Figure 5. The sensors were fixed to the apparatus and connected to clamps positioned along the sample via steel wires. Five clamps were employed between the contact points of the transverse actuators, with one clamp located in each region adjacent to the connectors. The distance between the contact points of the transverse actuators was 1.78 m, corresponding to approximately four pitches of the external tensile armors.

2.3. Test Execution

The test procedure involved pressurizing the sample, applying an axial load, and executing bending cycles with transverse actuator displacements at a maximum amplitude of 150 mm. The apparatus was designed to record the parameters measured during the cyclic loading. Five cycles were imposed in each test. Another essential aspect of the performed tests was evaluating the impact of the tensile armors' wires degradation on the bending response of the flexible pipe.

The degradation of the wires was induced by flooding the annular space of the pipe with a mixture of synthetic seawater and CO_2. Initially, a heating system was implemented to maintain a constant temperature of 40 °C during the degradation process. An electronically controlled electric trace was helically wrapped around the outer sheath of the pipe with a 50 mm pitch and secured using aluminum tape to enhance thermal conductivity, as shown in Figure 6a. After that, two PT100 temperature sensors were attached to the specimen and connected to a control panel for real-time temperature regulation. Thermal blankets were wrapped around the pipe body to improve thermal insulation, as shown in Figure 6b.

After the heating system was assembled, the annular flooding system, shown in Figure 7, was installed. The test was conducted with the annular space flooded using synthetic seawater and continuous CO_2 circulation at 1 bar. Before initiating the test, the annulus was pre-filled with CO_2-saturated water. To ensure stable and uniform conditions throughout the experiment, CO_2 was continuously introduced through one end-fitting, and its discharge was monitored at the end-fitting on the opposite side. The complete testing procedure followed these sequential steps: initial inertization, reservoir filling and fluid saturation, reservoir fluid saturation, sample filling, and CO_2 bubbling. Upon completion of the test, two post-test procedures were executed, i.e., N_2 bubbling and final inertization.

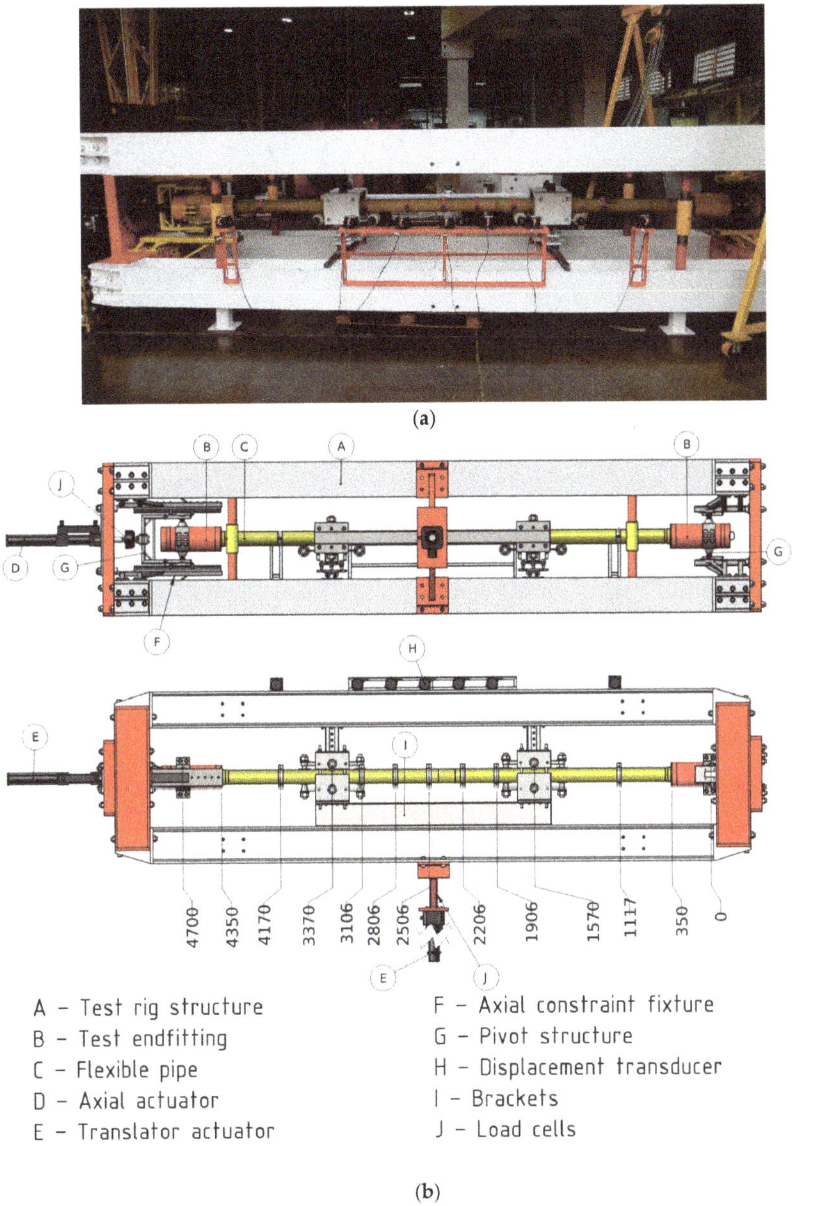

Figure 5. Experimental apparatus: (**a**) 2.5″ flexible pipe mounted in the apparatus and (**b**) Schematic showing the main components.

The initial test T0 imposed only axisymmetric loads on the pipe. Then, the bending test campaign (T1) was conducted using the sample with its original characteristics. Following this, the sample underwent an 18-month degradation period. Subsequently, the second test campaign (T2) was carried out. A summary of the test conditions is provided in Table 3.

(a) (b)

Figure 6. Heating system: (**a**) Electric trace positioning and (**b**) Thermal blankets.

Figure 7. Piping for annulus flooding.

Table 3. Test matrix.

Campaign	Test	Internal Pressure, P_{int} [Bar]	Axial Load, T [kN]	Annulus Degradation
T0	0	200	2	No degradation
T1	1	0	0	No degradation
	2	200	0	
	3	100	5	
	4	100	10	
	5	100	15	
	6	200	5	
	7	200	10	
	8	200	15	
T2	9	0	0	18 months—Flooded annulus with saturated water 1 bar CO_2 at 40 °C
	10	200	0	
	11	100	5	
	12	100	10	
	13	100	15	
	14	200	5	
	15	200	10	
	16	200	15	

3. Theoretical Approach

3.1. Overview

The main challenge in capturing the bending behavior of flexible pipes lies in the nonlinear relationship between bending stiffness and the imposed curvature. As discussed in Section 1, this nonlinearity primarily arises from the frictional forces generated by the axisymmetric loading, which limit the relative sliding between layers. Modeling this behavior requires a detailed representation of the pipe's internal structure, or the development of an equivalent approach.

In the first approach, the computational cost is generally high, requiring a detailed representation of the pipe's internal layers, and is only feasible for the analysis of flexible pipes with short lengths. However, this approach enables a direct representation of both global and local responses, allowing for the accurate capture of the relationship between the pipe's bending moment and curvature. In contrast, the second approach relies on prior knowledge of this relationship under various loading conditions to model the pipe's global behavior. Such information is typically obtained through local analytical or numerical models and incorporated into the formulation of two- or three-dimensional beams, whose structural stiffness is updated incrementally as the applied load varies in the global analyses. Although this method decouples global and local responses, it significantly reduces computational cost by modeling the pipe as an equivalent beam, enabling the simulation of long pipe segments, as required in the experimental tests described. Additionally, these equivalent models can be developed using either the FE method or the FD method. This work chose an FD-based model for its simplicity and efficiency in simulating the experimental tests.

Hence, next, the models employed to analyze the local response of the studied flexible pipe under axisymmetric and bending loads (local models) are described. Following this, a numerical global model is established to analyze the global response of the flexible pipe under the considered experimental loads. Altogether, these models establish the theoretical approach proposed in this work.

3.2. Local Models

3.2.1. Axisymmetric Loads

The initial step in determining the bending response of flexible pipes involves assessing their response to axisymmetric loads, including tension, torsion, and both internal and external pressures. As discussed in Section 1, various models have been developed for this purpose, leading to similar results. Hence, in this study, the model presented by Feret and Bournazel [6] is applied. In this model, a simplified solution to determine the effects of axisymmetric loads may be obtained by considering the following premises:

- Small displacements.
- The same elongation and twist in all layers.
- Negligible contribution of plastic layers to the pipe's mechanical resistance. These layers only transmit pressure.
- No gaps occur between layers.

Considering these assumptions, a linear system of equations may be used to calculate the response for a particular flexible pipe structure. The axial, radial, and torsional balances are, respectively, described as:

$$\sum_{i=1}^{N_a} n_i \sigma_i A_i \cos(\alpha_i) = T + \pi P_{int} r_{int}^2 - \pi P_{ext} r_{ext}^2 = T_{re}, \tag{1}$$

$$\sum_{i=1}^{N_a} \frac{n_i \sigma_i A_i \sin(\alpha_i)\tan(\alpha_i)}{2\pi r_i} = P_{int} r_{int} - P_{ext} r_{ext}, \tag{2}$$

$$\sum_{i=1}^{N_a} n_i \sigma_i A_i \sin(\alpha_i) r_i = T_O. \tag{3}$$

where T and T_{re} are the effective and real axial tensions [52], P_{int} and P_{ext} are the acting internal and external pressures, T_O is the imposed torsional moment, r_{int} and r_{ext} are the radii where the internal and external pressure act, N_a is the number of metallic layers in the pipe, n is the number of wires in the considered layer, A is the cross-section area of the wire (or tendon), α is the laying angle of the wire as depicted in Figure 8, and r is the layer mean radius.

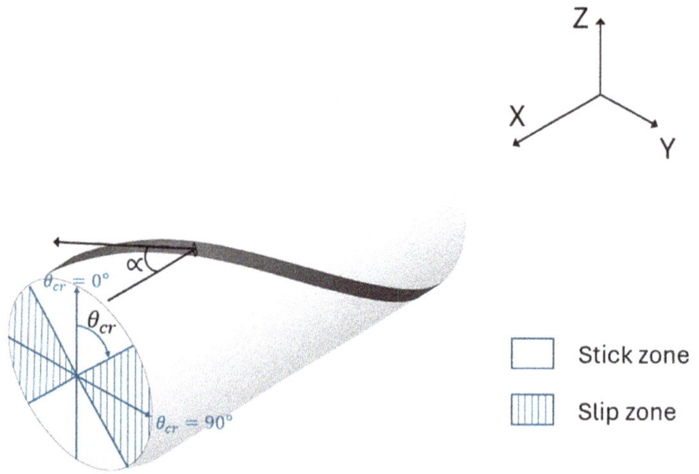

Figure 8. Tensile armor angle α; and slip and stick zones in a flexible pipe cross-section.

In layer i, $i = 1 \cdots N_a$, the normal stress σ_i induced in the wires' cross-sections may be stated as:

$$\frac{\sigma_i}{E_i} = \cos^2(\alpha_i)\frac{\Delta L}{L} + \sin^2(\alpha_i)\frac{\Delta r}{r_i} + r_i \sin(\alpha_i)\cos(\alpha_i)\frac{\Delta\theta}{L} \tag{4}$$

where E is the material Young modulus of the considered metallic layer, L is the length of the pipe, ΔL is the pipe axial elongation, Δr is the radial variation, and $\Delta\theta$ is the pipe twist.

The interlayer contact pressure P_c, is given by:

$$P_{c_{i+1}} = P_{c_i} - \Delta P_{c_i} \tag{5}$$

where $P_{c_1} = P_{int}$, $P_{c_{N_a+1}} = P_{ext}$ and:

$$\Delta P_{c_i} = \frac{n_i \sigma_i A_i \sin(\alpha_i)\tan(\alpha_i)}{2\pi r_i^2} \tag{6}$$

Finally, the apparent axial stiffness EA_{ap} of the pipe may be defined as:

$$EA_{ap} = T_{re} \cdot \frac{L}{\Delta L} = \left(T + \pi P_{int} r_{int}^2 - \pi P_{ext} r_{ext}^2\right) \cdot \frac{L}{\Delta L} \tag{7}$$

In this study, the apparent axial stiffness is employed to verify the accuracy of the axisymmetric model before conducting the bending tests.

3.2.2. Bending Loads

Monotonic Loading

As discussed in Section 2, the response of a flexible pipe to bending is nonlinearly associated with the imposed curvature. In this study, the model proposed by Wang et al. [41] was implemented with a few modifications.

Figure 9 depicts the monotonic bending moment–curvature relation for a tensile armor layer. In the first zone, the wires are attached to their adjacent layers, and no slip is observed (no-slip or stick zone). After reaching the critical curvature κ_{cr} The friction is insufficient to prevent interlayer relative movement, and slip starts in part of the pipe section (partial slip zone). When the imposed curvature reaches κ_f, the slip is fully developed (full-slip zone).

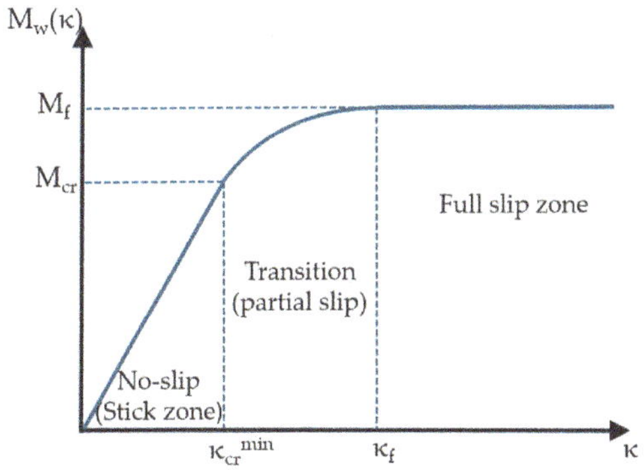

Figure 9. Moment vs. curvature relation in tensile armor.

The initial critical curvature κ_{cr} and the related bending moment M_{cr} for the onset of the tensile armors' slippage are given by [41]:

$$\kappa_{cr} = \frac{\tau_0 + \mu P_{ct}}{F_f E \sin(\alpha)\cos^2(\alpha)}(1+\xi) \tag{8}$$

$$M_{cr} = \frac{\pi r_a^2 \cos^2(\alpha)(\tau_0 + \mu \cdot P_{ct})}{\sin(\alpha)} \tag{9}$$

where τ_0 is the adhesion or initial contact pressure between the layers, μ is the friction coefficient, P_{ct} is the total contact pressure (sum of the acting pressures at the interfaces) between the layers due to the axisymmetric loads, F_f is the shape factor of the tensile armor wire (typically, 0.95), r_a is the mean radius of the tensile armor, and ξ is given by:

$$\xi = \frac{E_s b}{2(1+\nu_s)t_s} \tag{10}$$

where E_s and ν_s are the Young modulus and the Poisson ratio of the polymeric layer, t_s is the thickness of the polymeric layer, and b is the width of the tensile armor wire

The adhesion τ_0 is a modification to the model proposed in this study to account for the effect of an initial interlayer contact pressure in the pipe, which is not attributable to axisymmetric loading. Its influence is substantiated by the experimental results presented in Section 4, where a pronounced hysteretic response is observed even in the absence of axisymmetric loads. This condition has also been documented in previous investigations, wherein an additional contact pressure was incorporated into the calculated P_{ct} for model calibration purposes [36,38,41].

The curvature and the moment at which the tensile armor reaches the full-slip state are [41]:

$$\kappa_f = \frac{\pi \cdot \kappa_{cr}^{min}}{2}, \tag{11}$$

$$M_f = \frac{4\, M_{cr}}{\pi}. \tag{12}$$

When curvature is imposed on the pipe, the polymeric layers deform in shear, which alters the critical curvature at which the tensile armor wires start to slip. To account for this effect, a correction factor ξ was introduced in Equation (8). Furthermore, this study proposes that when multiple polymeric layers are arranged in series beneath or above the tensile armors, an equivalent factor ξ_{eq} should be considered to estimate the tangent stiffness of the combined layers:

$$\frac{1}{\xi_{eq}} = \sum_{i=1}^{N_{pl}} \frac{1}{\xi_i}, \tag{13}$$

where N_{pl} is the number of polymeric layers considered.

Then, for a particular tensile armor j in layer i, the bending moment $\mathrm{M_w}$ as a function of the curvature can be described as:

$$\mathrm{M}_{\mathrm{w}_j}(\kappa) = \begin{cases} \frac{M_{crj}}{\kappa_{crj}} \cdot \kappa, & 0 \le \kappa \le \kappa_{crj} \\ 4F_f r_j^3 t_j \cos^2(\alpha_j) \left\{ \frac{E_i \cos(\alpha_j)\left[\theta_{crj} - \cos(\theta_{crj})\sin(\theta_{crj})\right]\kappa}{2\xi_i} + \frac{(\tau_0 + \mu P_{cti})\cos(\theta_{crj})}{F_f t_j \sin(\alpha)} \right\}, & \kappa_{crj} < \kappa \le \kappa_{f_j} \\ M_{f_j}, & \kappa_{f_j} < \kappa \end{cases} \tag{14}$$

where θ_{cr} is the critical circumferential angle that defines the boundaries in the pipe cross-section between the slip and no-slip zones according to Figure 8, where $\theta = 0°$ represent extrados and $\theta = 180°$ the intrados, and for each curvature κ, θ_{crj} is calculated with:

$$\frac{\frac{\pi}{2} - \theta_{crj}}{\sin\left(\frac{\pi}{2} - \theta_{crj}\right)} = \frac{\kappa}{\kappa_{crj}}. \tag{15}$$

So, for each tensile armor present in the structures, a relation $\mathrm{M_t}(\kappa)$ is to be calculated, and the overall contribution of the tensile armors $\mathrm{M_t}$ in layer i is given by:

$$\mathrm{M}_{\mathrm{t}i}(\kappa) = \sum_{j=1}^{n_i} \mathrm{M}_{\mathrm{w}j}(\kappa), \tag{16}$$

During the bending of the pipe, local torsion and bending effects are observed in the tensile armor wires. These effects cause an increase in the stiffness of the pipe, represented by the expression [41]:

$$EI_{bt} = \frac{1}{2}\sum_{i=1}^{N_a} n_j \left[E_i I_{ni} \left(1 + \sin^2(\alpha_i)\right)^2 \cos(\alpha_i) + E_i I_{bi} \cos^7(\alpha_i) + G_i J_i \sin^2(\alpha_i) \cos^5(\alpha_i) \right], \tag{17}$$

where I_n, I_b, and J are the normal and binormal moments of inertia and the torsional constant of the wire cross-section, and G is the shear modulus of the wire material.

Hence, the moment–curvature relation for a flexible pipe subjected to bending and axisymmetric loading is given by:

$$\mathrm{M}(\kappa) = \sum_{i=1}^{N_a} \mathrm{M}_{\mathrm{t}i}(\kappa) + (EI_{ps} + EI_{bt})\kappa, \tag{18}$$

where the polymeric sheaths' contribution to the stiffness EI_{ps} is:

$$EI_{ps} = \frac{\pi}{64}\sum_{i=1}^{N_{ps}}\left\{E_{p_i}\left[D_{ep_i}^4 - D_{ip_i}^4\right]\right\}, \tag{19}$$

and N_{ps} is the number of polymeric sheaths in the pipe, E_p is the Young modulus of the polymeric sheath, D_{ep} and D_{ip} are the external and the internal diameter of the sheath. Analogously to EI_{bt}, this contribution remains constant regardless of the imposed load.

When the direction of bending is reversed, the layers that were sliding re-adhere, and the stiffness returns to no-slip. This effect causes the hysteresis loop that characterizes the behavior of flexible pipes subject to bending. The modeling of this cyclic response is presented next.

Cyclic Loading

Aiming to model the hysteresis loop described in Section 1, two hypotheses are assumed.

- The bending response does not depend on the load rate: the pipe's response to bending only depends on the acting bending moment, curvature, and contact pressures resulting from the axisymmetric loading at a given time. This hypothesis is reasonable to represent the experimental tests in this study, as the loads were imposed to minimize dynamic effects, and it finds support in previous experimental and numerical studies [11,36,38,41].
- The sliding surface translates without changing its size: this hypothesis ensures that the pipe retains memory of the loading direction, allowing the position of the critical curvature to adjust with the cyclic load. As qualitatively shown in Figure 2, the critical curvature varies with load reversal, exhibiting a response similar to the Bauschinger effect observed in metals.

Hence, by relying on these two hypotheses, the cyclic bending response can be compared to a rate-independent plasticity problem with kinematic hardening. In this case, the cyclic bending response of a flexible pipe can be represented by adapting the return-mapping algorithm proposed by Simo and Hughes [53] as shown in Figure 10.

In the proposed algorithm, EI represents the tangent stiffness at a given curvature. However, the algorithm presented by Simo and Hughes [53] considers a bilinear relation between the quantities (stresses and strains). In contrast, the bending moment–curvature described in the previous section exhibits a nonlinear transition between the no-slip and full-slip regimes. Hence, the tangent stiffness EI employed is given by:

$$\mathrm{EI}(\kappa) = \begin{cases} \left|\frac{\mathrm{dM}(\kappa)}{\mathrm{d}\kappa}\right|, & \text{before first load reversal} \\ \left|\frac{\mathrm{dM_d}(\kappa)}{\mathrm{d}\kappa}\right|, & \text{otherwise} \end{cases}, \tag{20}$$

where M_d is the unloading bending moment vs. curvature curve:

$$\mathrm{M_d}(\kappa) = \begin{cases} EI_{ns}\cdot\kappa, & \kappa \leq 2\kappa_{cr} \\ M(\kappa - \kappa_r) + EI_{ns}\cdot\kappa_r, & \kappa > 2\kappa_{cr} \end{cases}, \tag{21}$$

and κ_r is the curvature at each load reversal, and the no-slip bending stiffness EI_{ns} is stated as:

$$EI_{ns} = \mathrm{EI}(0). \tag{22}$$

1. Given curvature $\kappa_{l,k}$ at a point l ($l = 1 \cdots n_t$) and load step k ($k = 1 \cdots n_{ls}$), where n_t is the number of discrete points in the pipe (see Section 3.3) and n_{ls} is the number of load steps.
2. Initialize variables if load step $k = 1$ and go to step 5:
$$M_{l,1} = 0;\ \kappa^p_{l,1} = 0\ ;\ q_{l,1} = 0; \kappa_{r_{l,1}} = 0.$$
3. Compute the no-slip trial bending moment and test for sliding:
$$M^{trial} = EI_{ns}\left(\kappa_{l,k} - \kappa^p_{l,k-1}\right)$$
$$\xi^{trial} = M^{trial} - q_{k-1}$$
$$f^{trial} = \left|\xi^{trial}\right| - M_{cr}$$
If $f^{trial} \leq 0$, there is no sliding and $(*)_{l,k} \leq (*)^{trial}$, where $(*)$ represents M, κ^p and q. Instead, if $f^{trial} > 0$, slip occurs, and the procedure described in step 4 is necessary.
4. Return mapping:
$$EI_{l,k} == \begin{cases} \dfrac{EI_{ns}\mathrm{EI}(\kappa_{l,k} - \kappa_{r_{l,k}})}{EI_{ns} - \mathrm{EI}(\kappa_{l,k} - \kappa_{r_{l,k}})}, & EI_{ns} \neq \mathrm{EI}(\kappa_{l,k} - \kappa_{r_{l,k}}) \\ 0, & EI_{ns} = \mathrm{EI}(\kappa_{l,k} - \kappa_{r_{l,k}}) \end{cases}$$
$$\Delta\gamma = \frac{f^{trial}}{EI_{ns} + EI_{l,k}},$$
$$M_{l,k} = M^{trial} - \Delta\gamma EI_{ns}\mathrm{sign}\left(\xi^{trial}\right),$$
$$\kappa^p_k = \kappa^p_{k-1} + \Delta\gamma\mathrm{sign}\left(\xi^{trial}\right),$$
$$q_k = q_{k-1} + \Delta\gamma\mathrm{sign}\left(\xi^{trial}\right).$$
5. Cycle or end the algorithm if $k = n_{ls}$, storing $EI_{l,k}$ and $M_{l,k}$.

Figure 10. Adapted return-mapping algorithm for bending moment vs. curvature hysteretic response.

3.3. Global Model

3.3.1. Differential Equation

The flexible pipe sample is assumed to be a beam subjected to axial and transverse loads, as illustrated in Figure 11a. The coordinate system is positioned at the fixed end of the pipe, and the length L corresponds to the distance between the pivot points where the sample is fixed to the test bench. The transverse forces applied by the actuators are denoted by F_A and F_B, while R_1 and R_2 represent the reaction forces at the supports. The axial effective load is represented by T, distances x_A and x_B indicate the locations where the actuators contact the sample, and the transverse displacement imposed by the actuators at points A and B are assumed equal by hypothesis. In this model, the self-weight of the sample and its effects are neglected, and the assumption of small incremental displacements is considered valid.

The bending stiffness of a flexible pipe (either the no-slip or the full slip values) is much lower than its axial stiffness. Moreover, the relation between the free span of the pipe in the experimental apparatus (4.0 m) is more than 10 times higher than the pipe's outer diameter (121.18 mm). Hence, the beam proposed in Figure 11a may be assumed to be slender, and shear effects can be neglected. By relying on this hypothesis, the bending moment M acting in the beam is related to the bending strain energy U as in Equation (23).

$$\mathrm{M}(\kappa) = \frac{\mathrm{dU}(\kappa)}{\mathrm{d}\kappa}. \tag{23}$$

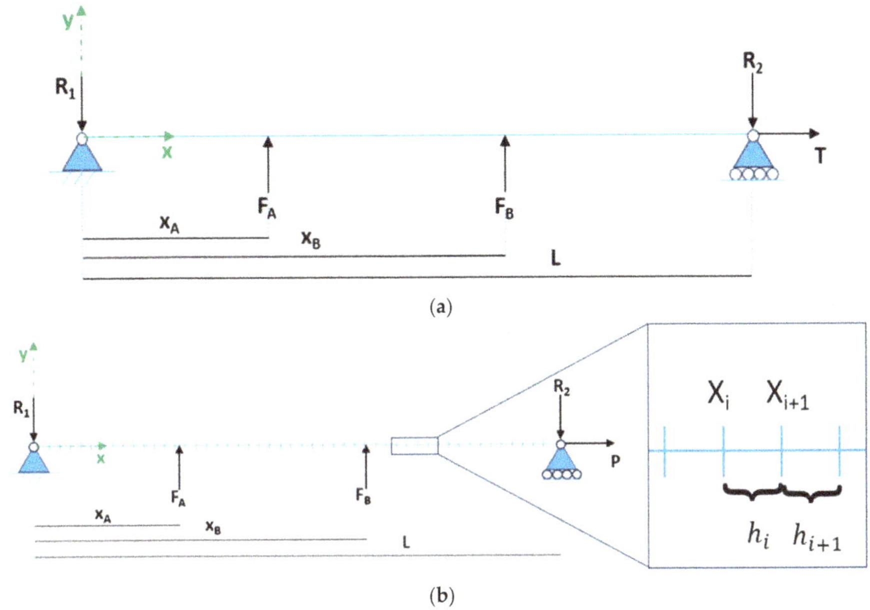

Figure 11. Numerical model representation: (**a**) General view; (**b**) Discretization.

The bending strain energy U is expressed as:

$$U(\kappa) = \frac{1}{2}\text{EI}(\kappa)\kappa^2, \tag{24}$$

By considering Equation (23) in Equation (24), the bending moment is given by

$$M(\kappa) = \text{EI}(\kappa)\kappa + \frac{\kappa^2}{2}\frac{\text{dEI}(\kappa)}{\text{d}\kappa} \tag{25}$$

As discussed in Section 1, the bending stiffness of a flexible pipe progressively reduces from the no-slip value to the full-slip value. Hence, if sufficiently slight curvature variations are imposed on the pipe, the change in the bending stiffness value will also be small. Moreover, relatively low curvatures were achieved in the experimental tests (maximum values were close to 0.1 m^{-1}). Altogether, these aspects indicate a negligible contribution of the quadratic term in Equation (25), which may be rewritten as:

$$M(\kappa) = \text{EI}(\kappa)\kappa \tag{26}$$

The curvature κ at a cross-section distant x from the origin (Figure 11a) depends on the vertical displacement y imposed by the transverse forces F_A and F_B. By representing forces F_A and F_B as $\boldsymbol{F} = \{F_A, F_B\}$ to simplify the notation, and assuming small angle variations ($\frac{\text{d}y(x,\boldsymbol{F})}{\text{d}x} \ll 1$), the curvature variation is expressed as:

$$\kappa(x, \boldsymbol{F}) = \frac{\frac{\text{d}^2 y(x,\boldsymbol{F})}{\text{d}x^2}}{\left\{1 + \left[\frac{\text{d}y(x,\boldsymbol{F})}{\text{d}x}\right]^2\right\}^{3/2}} \cong \frac{\text{d}^2 y(x,\boldsymbol{F})}{\text{d}x^2} \tag{27}$$

Moreover, the total bending moment is the sum of the first-order moments imposed by the transverse forces $\boldsymbol{F}$ and second-order moments related to the beam's transverse displacement and the acting axial load. Therefore, the total bending moment M is:

$$\mathrm{M}(x,\boldsymbol{F}) = \mathrm{M}^{(1)}(x,\boldsymbol{F}) - Ty(x,\boldsymbol{F}), \tag{28}$$

where

$$\mathrm{M}^{(1)}(x,\boldsymbol{F}) = \begin{cases} \left(1-\frac{x_A}{L}\right)\cdot x\cdot F_A + \left(1-\frac{x_B}{L}\right)\cdot x\cdot F_B, & 0 \le x \le x_A \\ \left(1-\frac{x}{L}\right)\cdot x_A\cdot F_A + \left(1-\frac{x_B}{L}\right)\cdot x\cdot F_B, & x_A < x \le x_B \\ \left(1-\frac{x}{L}\right)\cdot x_A\cdot F_A + \left(1-\frac{x}{L}\right)\cdot x_B\cdot F_B, & x_B < x \le L \end{cases}, \tag{29}$$

By rewriting Equation (26) considering Equations (27) and (28), a second-order differential equation is obtained:

$$\frac{\mathrm{d}^2 y(x,\boldsymbol{F})}{\mathrm{d}x^2} - \beta^2(x,\boldsymbol{F})y(x,\boldsymbol{F}) + \frac{\mathrm{M}^{(1)}(x,\boldsymbol{F})}{\mathrm{EI}(x,\boldsymbol{F})} = 0, \tag{30}$$

where the bending stiffness EI depends on the analyzed cross-section in the beam and the imposed transverse load, and β is given by:

$$\beta(x,\boldsymbol{F}) = \sqrt{\frac{T}{\mathrm{EI}(x,\boldsymbol{F})}} \tag{31}$$

For the described test setup, the boundary conditions consist of fixed supports at both ends, such that $y(0) = y(L) = 0$. In addition, displacement compatibility requires equal displacements at both actuator points, $y(x_A) = y(x_B)$.

Equation (30) indicates that the lateral displacement of the beam depends on the bending stiffness at each of its cross-sections, which, in turn, is a function of the imposed curvature. Therefore, the local model proposed in Section 3.2 is employed in this work. However, the nonlinear variation of EI with the imposed loading does not allow a closed-form solution of Equation (30), and a numerical procedure is required. This procedure is described next.

3.3.2. Discretization

In this work, Equation (30), which represents the structural model presented in Figure 11a, is solved relying on the FD method. Initially, the beam is discretized as indicated in Figure 11b, in which the axial position x_l of a point l is

$$x_{l+1} = x_l + lh_l,\ l = 1\cdots n_t - 1, \tag{32}$$

where $x_1 = 0$, n_t is the total number of discretization points in the beam, and h_l is the spacing between two consecutive points x_l and x_{l+1}, i.e.:

$$h_l = \begin{cases} \frac{x_A}{n_A - 1}, & 0 \le x_{l+1} \le x_A \\ \frac{x_B - x_A}{n_B - 1}, & x_A < x_{l+1} \le x_B \\ \frac{L - x_B}{n_C - 1}, & x_B < x_{l+1} \le L \end{cases}, \tag{33}$$

and n_A, n_B, and n_C are the number of discretization points (nodes) in segments A, B, and C, with $n_t = n_A + n_B + n_C - 2$.

Moreover, the effective axial force T is constant throughout the analysis, while the transverse force is incrementally applied, i.e.:

$$F_{Ak} = F_{Ak-1} + \Delta F_{Ak},\ F_{Bk} = F_{Bk-1} + \Delta F_{Bk}, k = 2\cdots n_{ls}, \tag{34}$$

where $F_{A1} = F_{B1} = 0$, n_{ls} is the number of load steps, and ΔF_{Ak} and ΔF_{Bk} are the load increments at points A and B related to load step k.

By relying on Equations (32) and (34), the transverse displacement y at point x_l caused by the transverse force F_k is stated as $y_{l,k}$. However, Equation (30) demands the discrete form of a second-order derivative, which represents the curvature of the beam. As the spacing between consecutive points in the discretized beam may differ, depending on the values of n_A, n_B, and n_C, the curvature of the beam may be approximated following the equation proposed by Singh and Bhadauria [54], i.e.:

$$\frac{\mathrm{d}^2 y(x,F)}{\mathrm{d}x^2} \cong \kappa_{l,k} \cong 2\frac{h_l y_{l-1,k} - (h_l + h_{l-1})y_{l,k} + h_{l-1}y_{l+1,k}}{h_{l-1}h_l(h_l + h_{l-1})}. \tag{35}$$

Now, by considering the discrete notations of x, y, and F, and Equation (35), the discrete form of Equation (30) is given by:

$$h_l y_{l-1,k} - \left[(h_l + h_{l-1}) + \beta_{l,k}^2 \overline{h_l}\right] y_{l,k} + h_{l-1} y_{l+1,k} + \frac{M_{l,k}^{(1)}}{EI_{l,k}} \cdot \overline{h_l} = 0, \tag{36}$$

where $l = 2 \cdots n_t - 1, k = 2 \cdots n_{ls}$, and:

$$\overline{h_l} = \frac{1}{2} \cdot h_{l-1} h_l (h_l + h_{l-1}). \tag{37}$$

Nevertheless, Equation (36) must be solved incrementally, as the bending stiffness of the pipe varies nonlinearly with the imposed loading, as discussed in Section 3.2.2. Thus, the transverse displacement $y_{l,k}$ is given by

$$y_{l,k} = y_{l,k-1} + \Delta y_{l,k},\ l = 1 \cdots n_t,\ k = 2 \cdots n_{ls}. \tag{38}$$

Considering Equations (37) and (38) in Equation (36), the following incremental equation is obtained:

$$h_l \Delta y_{l-1,k} - \left[(h_l + h_{l-1}) + \beta_{l,k}^2 \cdot \overline{h_l}\right] \Delta y_{l,k} + h_{l-1} \Delta y_{l+1,k} + \frac{M_{l,k}^{(1)}}{EI_{l,k}} \overline{h_l} = 0, \tag{39}$$

where $l = 2 \cdots n_t - 1, k = 2 \cdots n_{ls}$.

The transverse displacements Δy and the load increments ΔF_A and ΔF_B are unknown at each load step, totaling n_t variables. Equation (39) gives $n_t - 2$ relations between the transverse displacements. However, the transverse displacements at the extremities of the beam are restrained:

$$\Delta y_{1,k} = \Delta y_{n_t,k} = 0\,, \quad k = 2 \cdots n_{ls}. \tag{40}$$

Finally, as the experimental tests are displacement-controlled, and the fork employed to impose the load can be assumed as rigid, the transverse displacements at points A and B are considered equal to the displacement imposed by the actuators, i.e.:

$$\Delta y_{A,k} = \Delta y_{B,k} = \Delta y_t\,, \quad k = 2 \cdots n_{ls}. \tag{41}$$

where $\Delta y_t = y_t / n_{ls}$ and y_t is the total transverse load imposed by the actuators.

The system of equations formed by Equations (39)–(41) is organized in the matrix form:

$$(\mathbf{K}_k)_{(n_t+2 \text{ x } n_t+2)} \cdot (\mathbf{Y}_k)_{(n_t+2 \text{ x}1)} = (\mathbf{F}_k)_{(n_t+2 \text{ x}1)}\,, \quad k = 2 \cdots n_{ls}, \tag{42}$$

where the non-null terms K, Y, and F related to **K**, **Y** and **F**, respectively, are:

$$\mathrm{K}_{1,1} = \mathrm{K}_{n_A,n_A} = \mathrm{K}_{n_A+n_B-1,n_A+n_B-1} = \mathrm{K}_{n_t,n_t} = 1 \tag{43}$$

$$\begin{gathered}\mathrm{K}_{l,l-1} = h_l,\ \mathrm{K}_{l,l} = -\left[(h_l + h_{l-1}) + \beta_{l,k}^2 \cdot \overline{h_l}\right],\ \mathrm{K}_{l,l+1} = h_{l-1}, \\ l = 2 \cdots n_t - 1,\ l \neq n_A,\ n_A + n_B - 1\end{gathered} \tag{44}$$

$$\mathrm{K}_{l,n_t+1} = \begin{cases} \left(1 - \frac{x_A}{L}\right) \cdot x_l, & 2 \leq l < n_A \\ \left(1 - \frac{x_l}{L}\right) \cdot x_A, & n_A + 1 \leq l < n_A + n_B - 1 \\ \left(1 - \frac{x_l}{L}\right) \cdot x_A, & n_A + n_B \leq l < n_t \\ \left(1 - \frac{x_A}{L}\right) \cdot x_A, & l = n_t + 1 \\ \left(1 - \frac{x_B}{L}\right) \cdot x_A, & l = n_t + 2 \end{cases} \tag{45}$$

$$\mathrm{K}_{l,n_t+2} = \begin{cases} \left(1 - \frac{x_B}{L}\right) \cdot x_l, & 2 \leq l < n_A \\ \left(1 - \frac{x_B}{L}\right) \cdot x_l, & n_A + 1 \leq l < n_A + n_B - 1 \\ \left(1 - \frac{x_l}{L}\right) \cdot x_B, & n_A + n_B \leq l < n_t \\ \left(1 - \frac{x_B}{L}\right) \cdot x_A, & l = n_t + 1 \\ \left(1 - \frac{x_B}{L}\right) \cdot x_B, & l = n_t + 2 \end{cases} \tag{46}$$

$$\mathrm{Y}_l = \begin{cases} \Delta y_{l,k}, & 1 \leq l \leq n_t \\ \Delta F_{Ak}, & l = n_t + 1 \\ \Delta F_{Bk}, & l = n_t + 2 \end{cases} \tag{47}$$

$$\mathrm{F}_l = \Delta y_t, \quad l = n_A \vee l = n_A + n_B - 1 \tag{48}$$

By determining vector **Y**, the transverse displacement increments $\Delta y_{l,k}$ at each point l related to the load increment k and the reaction loads ΔF_{Ak} and ΔF_{Bk}, the bending moment increment ΔM at each point can be calculated with Equation (49) by considering Equation (28).

$$\Delta M_{l,k} = \Delta M_{l,k}^{(1)} - T\Delta y_{l,k}. \tag{49}$$

The total bending moment is then given by:

$$M_{l,k} = M_{l,k-1} + \Delta M_{l,k}. \tag{50}$$

Moreover, the curvature at a point is given by Equation (35), except for the points where the loads are imposed. At these points, a backward approximation for the second derivative of the displacements is considered to avoid numerical instabilities (observed during the case study analyses discussed in Section 4), i.e.:

$$\kappa_{n_A,k} = \frac{y_{n_A,k} - 2y_{n_A-1,k} + y_{n_A-2,k}}{h_{n_A-1}^2} \tag{51}$$

$$\kappa_{n_A+n_B-1,k} = \frac{y_{n_A+n_B-1,k} - 2y_{n_A+n_B-2,k} + y_{n_A+n_B-3,k}}{h_{n_A+n_B-2}^2} \tag{52}$$

3.3.3. Solution

The algorithm used to calculate the sample's response during the bending test is illustrated in Figure 12. Initially, moments, curvatures, and displacements are set to zero.

For points located within the length of the end fittings (EFs), the stiffness EI_{ef} remains constant throughout the test and is equal to $EI_{ef} = 10^3 EI_{ns}$. In the pipe section, an initial no-slip stiffness, Equation (22), is assumed. During the iteration loop, matrix **K** and vector **F** are computed, and the linear system described in Equation (41) is solved. Incremental displacements for the nodes are determined based on their previous positions to establish the current position, as well as variations in bending moments and curvatures.

1. Initialize variables if $k = 1$ and go to step 4:

$$M_1 = 0; \kappa_1^p = 0; q_1 = 0$$

2. Discretize the imposed load using Equation (41).
3. Iterate through load steps $k = 2 \cdots n_{ls}$
   3.1. Calculate the bending stiffness along the beam.
   3.2. Compute matrix **K** using Equations (43) to (46).
   3.3. Compute vector **F** using Equation (48).
   3.4. Solve the system of equations given by Equation (42) to obtain **Y** and the increment of transverse displacement $\Delta y_{l,k}$ at each point l and the load increments ΔF_{A_k} and ΔF_{B_k}, as stated in Equation (47).
   3.5. For $l = 1 \cdots n_t$:
      3.5.1. Calculate the total transverse displacement $y_{k,l}$ with Equation (38).
      3.5.2. Calculate the bending moment increment $\Delta M_{l,k}$ and total bending moment $M_{l,k}$ with Equations (49) and (50), respectively.
      3.5.3. Calculate the resulting curvatures $\kappa_{l,k}$ with Equations. (35), (51) and (52).
      3.5.4. Calculate the bending moments $M_{l,k}^*$ and the tangent bending stiffness $EI_{l,k}^*$ at each point l using the algorithm presented in Figure 10 (bending moments and bending stiffness correspond to the stored values at end of the algorithm).
   3.6. Evaluate the model convergence by comparing the bending moments calculated at step 3.5.2 (M) and those estimated with the algorithm presented in Figure 10 (called, here, M^*):

$$||M|| = \sqrt{\sum_{l=1}^{n_t}\left(\frac{M_{l,k} - M_{l,k}^*}{M_{l,k}}\right)^2}$$

   If $||M|| \leq 10^{-6}$, then the load step converged. Otherwise, go to step 3.2 using $EI_{l,k} = EI_{l,k}^*$.
4. Cycle or end the algorithm when $k = n_{ls}$

Figure 12. Algorithm for response considering the hysteretic behavior of the flexible pipe.

The new stiffness and estimated bending moments are subsequently calculated as a function of the curvature change using the algorithm presented in Figure 10. If the bending moments estimated using the given algorithm match those calculated, the solution has converged, and it proceeds to the next step. If a discrepancy occurs, the loop restarts, replacing the assumed stiffness with the newly calculated value. The described solution procedure was implemented in an electronic spreadsheet in Mathcad (version 15) [55].

4. Results

4.1. Axisymmetric Loading Test (T0)

Before analyzing the bending response of the pipe, the axisymmetric model must be verified, as the evaluation of the pipe bending stiffness relies on the contact pressures induced by the axisymmetric loading.

Accordingly, an experimental test was carried out to determine the apparent axial stiffness of the pipe sample (Section 3.2.1). Initially, an axial force of 2 kN, T_i, was applied to straighten the specimen. Subsequently, the pipe was pressurized up to 200 bar, while the axial displacement of its free end was recorded. The axial force associated with the internal pressure acting on the closed end of the pipe (end-cap effect) was estimated by assuming that the inner sheath of the pipe is sealed at the sample's EF. Under this condition, the real axial force, Equation (53), is expressed as:

$$T_{re} = T_i + \pi P_{int} r_{int_{is}}^2. \quad (53)$$

where $r_{int_{is}}$ is the internal radius of the inner sheath.

Figure 13 presents the relationship between the experimentally measured axial displacement and the axial force induced by internal pressure, along with the prediction of the analytical model (Section 3.2.1). The experimental results exhibit an almost linear trend, with a linear regression yielding a coefficient of determination R^2 = 0.999. The axisymmetric model predicts a linear response that closely matches the experimental data. The slope of the regression line corresponds to an experimental apparent axial stiffness of 90,904 kN, which is only 2.2% higher than the analytical value of 88,931 kN. These results demonstrate that the model provides a reasonable prediction of the pipe's axisymmetric response.

The axisymmetric model was subsequently employed to calculate the contact pressures associated with the axisymmetric loading conditions applied in the bending tests (Table 3). Table 4 reports the total contact pressures acting on the inner and outer tensile armor layers of the pipe (P_{cITA} and P_{cOTA}, respectively), obtained as the sum of the contact pressures at the corresponding layer interfaces.

Table 4. Contact pressures per test condition.

Internal Pressure, P_{int} [bar]	Axial Load, F [kN]	Contact Pressures [1]		Corresponding Tests
		Inner Armor, P_{cITA} [MPa]	Outer Armor, P_{cOTA} [MPa]	
0	0	0	0	1, 9
100	5	3.34	1.22	3, 11
100	10	3.63	1.31	4, 12
100	15	3.91	1.40	5, 13
200	0	5.93	2.05	2, 10
200	5	6.21	2.14	6, 14
200	10	6.50	2.23	7, 15
200	15	6.79	2.32	8, 16

[1] Sum of the contact pressures on the internal and external surfaces of the wires.

4.2. Pure Bending Test (T1-1)

4.2.1. Mesh and Load Increment Sensitivities

This section addresses the definition of mesh parameters and load increments required to reproduce the bending tests with both accuracy and computational efficiency. An initial adhesion τ_0 of 0.9 MPa was assumed, as further justified in the following section, and the transverse displacement imposed by the actuators was equal to 150 mm (full stroke).

Mesh convergence was assessed by progressively increasing the number of nodes until additional refinement produced negligible variations in the model response. Since the simulations were performed under displacement control, the evaluation focused on the forces F_A, F_B, and the total force F_{total}. The displacement increment per step was initially set to 3 mm. A comparison of the results obtained for the different mesh configurations is presented in Table 5.

Starting from the initial mesh, increasing the number of nodes produced only marginal variations in the model predictions. In particular, the total force estimated with the fourth mesh differed by less than 0.1% from that obtained with the sixth mesh. Based on these results, the fourth mesh configuration (60–80–60 nodes) was identified as the most suitable, providing an appropriate balance between computational efficiency and accuracy.

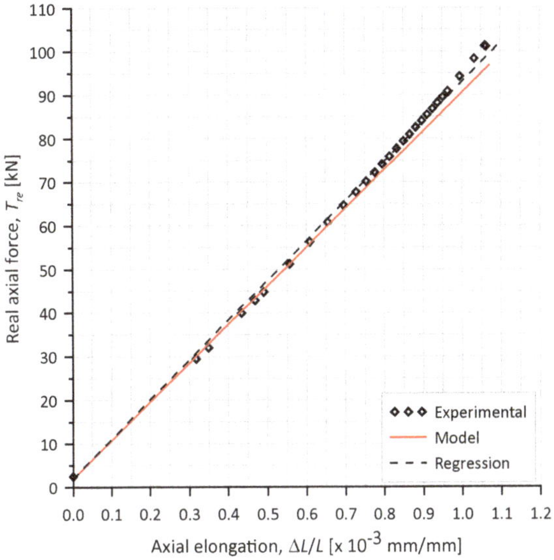

Figure 13. Variation in the pipe's elongation with the resulting real axial tension: internal pressure test.

Table 5. Mesh sensitivity results.

Mesh	n_A	n_B	n_C	n_t	F_A [N]	F_B [N]	F_{total} [N]
1	10	15	10	33	582.1	851.5	1433.6
2	15	20	15	48	582.9	845.8	1428.7
3	30	40	30	98	584.6	838.2	1422.8
4	60	80	60	198	585.4	834.7	1420.1
5	90	120	90	298	585.6	833.6	1419.2
6	120	160	120	398	585.7	833.0	1418.7

A similar procedure was employed to assess the impact of the displacement increment, utilizing the defined mesh. The results obtained are summarized in Table 6. Increments were tested within a range from 6 mm to 0.25 mm. Beyond the 1 mm increment, even when the load per step was reduced by a factor of four (to 0.25 mm), the total force varied by less than 0.5%. Consequently, the 1 mm load increment was chosen as the reference for this study.

Table 6. Load increment sensitivity results.

Δy_t [mm]	F_A [N]	F_B [N]	F_{total} [N]
6.00	584.9	809	1393.9
3.00	585.4	834.7	1420.1
1.50	585.7	847.2	1432.9
1.00	585.8	851.4	1437.2
0.50	586.0	855.5	1441.5
0.25	586.0	857.5	1443.5

4.2.2. Adhesion Investigation

Test T1-1 (Table 3) was conducted without any internal pressure or axial load (pure bend test), but demonstrated notable hysteresis, as seen in Figure 14, where the total transversal force acting on the sample and the actuator displacement during one cycle are depicted. In the absence of an axisymmetric load to create contact pressures between the layers, this behavior can be attributed to an initial adhesion between the tensile armors and the adjacent layers. This initial adhesion restricts the relative movement between layers, resulting in energy dissipation, which is reflected in the area enclosed by the force vs. displacement curve during a cycle, as shown in Figure 14. Since this adhesion cannot be measured directly, it must be calibrated.

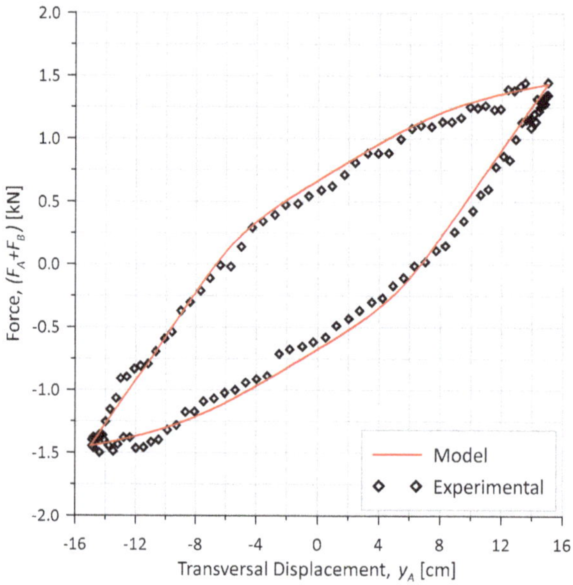

Figure 14. Force vs. displacement relation: pure bending test (T1-1).

A parametric study was conducted to determine which value of adhesion best aligns with the experimental data, specifically focusing on the variation in total force and the area of the hysteresis loop.

The deflection of the sample corresponding to the full stroke of the transversal actuator (150 mm) is presented in Figure 15, illustrating various adhesion conditions along with the full-slip stiffness (analogous to zero adhesion) and the no-slip stiffness (where adhesion is sufficiently high to prevent any sliding during the cycle). The experimentally measured points are also included. As anticipated, due to displacement control, the response of the pipe remains quite similar across the different conditions, although the occurrence of slippage and the associated forces vary from one case to another. The values for positive and negative peak deflections for measurement points between actuators are presented in Table 7.

Table 7. Displacement comparison—Load increment sensitivity results.

Positive Stroke [mm]					
Sensor	**#2**	**#3**	**#4**	**#5**	**#6**
Experimental	164	181	183	179	166
Full slip	167	177	179	175	160
No slip	167	177	179	175	160
τ_0 = 0.5 MPa	170	182	186	181	162
τ_0 = 0.9 MPa	168	179	183	179	162
τ_0 = 1.2 MPa	168	179	182	178	161

Table 7. *Cont.*

Negative Stroke [mm]					
Sensor	**#2**	**#3**	**#4**	**#5**	**#6**
Experimental	158	172	178	174	165
Full slip	160	175	179	176	167
No slip	160	175	179	176	167
τ_0 = 0.5 MPa	162	181	186	182	170
τ_0 = 0.9 MPa	161	179	183	179	168
τ_0 = 1.2 MPa	161	178	182	177	167

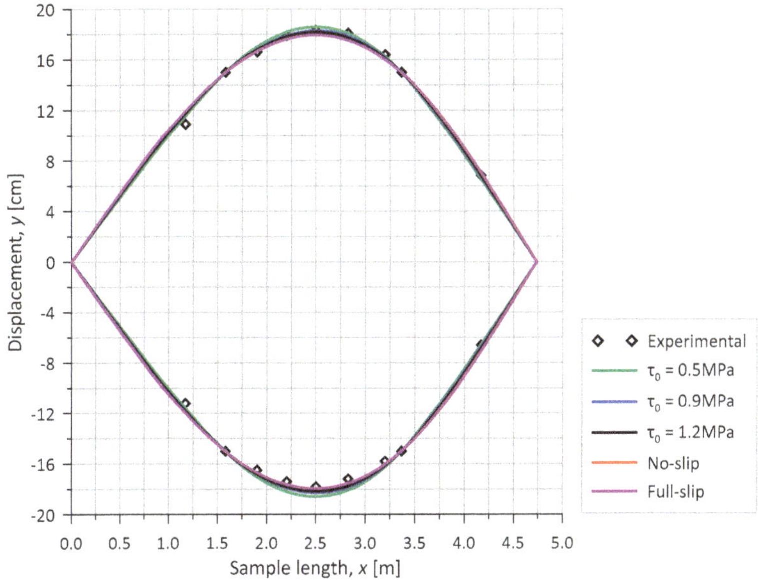

Figure 15. Transverse displacements in the flexible pipe sample: pure bending test (T1-1), full stroke loading.

The curvatures observed under each condition are illustrated in Figure 16. It is noted that the analyses with full-slip and no-slip stiffness yielded identical results, as the tests were conducted under displacement control and, therefore, the final deformed shapes of the pipe were similar. In these cases, three distinct regions can be observed (disregarding those related to the EFs at the extremities), where the curvatures vary linearly.

On the other hand, the initial adhesion leads to a variation in the bending stiffness of the pipe as slip occurs between the layers. In Figure 16, the lengths depicted with a solid line indicate no-slip between layers, while the dashed lines represent segments of the pipe where slippage occurred. In the regions between the extremities (EFs) and the actuators, the curvatures vary nonlinearly due to the bending stiffness variation along the length of these regions. Regarding the region between the actuators, the response obtained with an adhesion of 0.5 MPa is linear, whereas the responses obtained for adhesions of 0.9 MPa and 1.2 MPa are nonlinear. Again, this response is related to the variation in bending stiffness of the pipe. In the first condition (0.5 MPa), full slip occurs between layers, while in the other conditions, stiffness still varies along the length. Finally, it is noted that the maximum curvatures are obtained for adhesions of 0.5 MPa and 0.9 MPa. In contrast, the adhesion of 1.2 MPa indicates a lower value, with the curve tending to approach that obtained with no-slip stiffness.

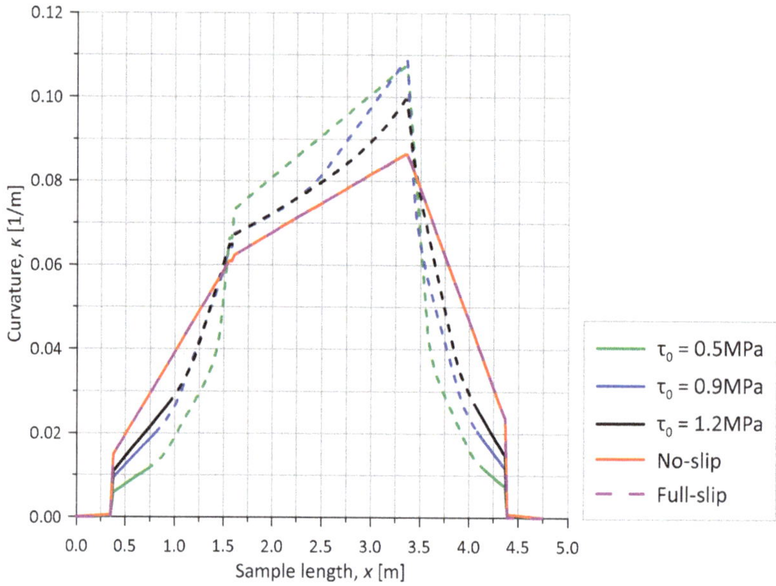

Figure 16. Curvature and slippage occurrence (T1-1): full stroke.

The impact of varying the adhesion between layers, and consequently the relationship between moment and curvature, can be more clearly observed in the force–displacement diagram shown in Figure 17. The graph depicts full-slip stiffness and no-slip stiffness as straight lines. In cases with adhesion, the loops represent cycles where the response initially resembles that of the no-slip condition, transitioning to a full-slip behavior. As adhesion increases, the response loop rotates counterclockwise and becomes narrower, illustrating a shift from the full-slip scenario to the no-slip scenario.

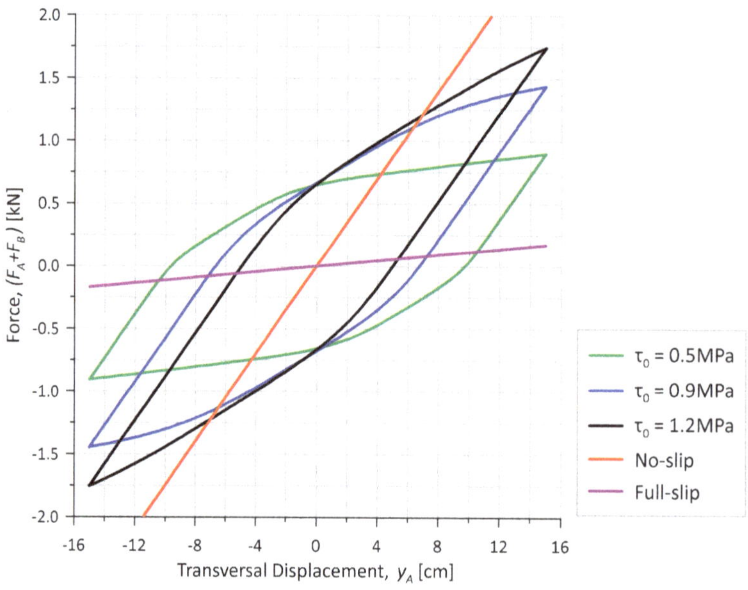

Figure 17. Force vs. displacement relation: pure bending test (T1-1) with different adhesion values.

The bending responses for various adhesion values are presented in Table 8. For each case, the critical curvature κ_{cr} at which slippage begins is listed, along with the ratio $\Delta F_{model}/\Delta F_{exp}$ (named imposed force ratio), which compares the model and experimental forces required to transition the sample from the most negative to the most positive deflections (representing half a cycle). Additionally, A_{model}/A_{exp} (named energy dissipation ratio) provides a comparison of the areas of the loops, while R^2 indicates the coefficient of determination between the model and experimental measurements (Figure 14). The cycle was divided into two segments, upward and downward, and the results from both segments were used to compute R^2.

Table 8. Adhesion sensitivity results.

τ_0 [MPa]	κ_{cr} [1/m]	$\frac{\Delta F_{model}}{\Delta F_{exp}}$	$\frac{A_{model}}{A_{exp}}$	R^2
0.5	0.0107	0.592	1.03	0.848
0.7	0.0150	0.772	1.131	0.963
0.8	0.0171	0.86	1.117	0.988
0.9	0.0193	0.946	1.074	0.993
1.0	0.0214	1.023	1.02	0.983
1.1	0.0236	1.091	0.967	0.961
1.2	0.0257	1.148	0.927	0.933

The critical curvature increases linearly with the considered adhesion, as can be inferred in Equation (8). However, there is a nonlinear variation in the imposed forces and the enclosed areas with this parameter.

Regarding the forces required to perform the cycle, they increase as adhesion increases, as presented in Figure 18, until the adhesion reaches a level at which no slippage occurs during the cycles (approximately 3.5 MPa). After that, the forces are the same regardless of the adhesion value.

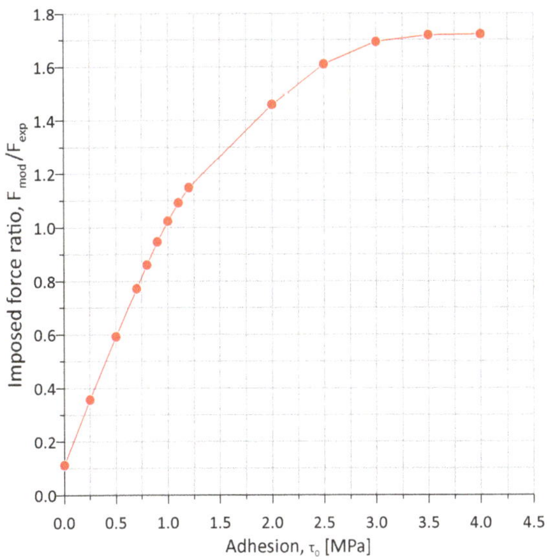

Figure 18. Required force due to adhesion.

The areas represented in the loops of Figure 17 are related to energy dissipation during the load cycles due to the stick–slip effect. Therefore, no dissipation is observed in scenarios without a change in stiffness, resulting in zero area for both no-slip and full-slip conditions. As adhesion, and consequently stiffness, increases from the full-slip scenario, the areas of the loops rise from zero. However, as stiffness continues to increase and the curve approaches the no-slip stiffness condition, the area begins to decrease again, ultimately reaching zero when no slip occurs. Hence, a specific level of adhesion exists that maximizes energy dissipation. By exploring a range of adhesion values, adhesions around 0.7 MPa yield the highest energy dissipation in the test setup, as illustrated in Figure 19, which presents the variation in enclosed area with the considered adhesion. It is essential to note that two different adhesion levels can result in the same amount of dissipated energy, albeit with varying responses. This explains why the 0.5 MPa case exhibits nearly identical energy dissipation to that of the test, despite a noticeably different response. Consequently, the area cannot serve as a standalone parameter for fitting test results.

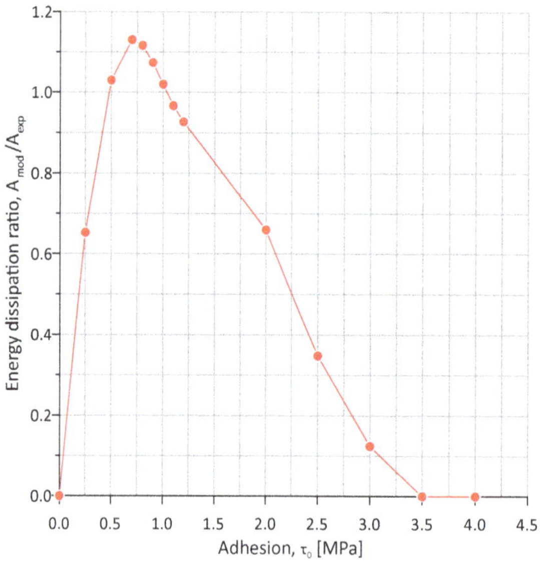

Figure 19. Energy dissipation ratio vs. adhesion.

Additionally, it is essential to acknowledge that other sources of energy dissipation may impact the overall dissipation levels. For instance, in flexible pipes, viscous effects play a role; however, due to the low frequency of load imposition in the test, their impact tends to be minimal. Factors such as test bench equipment, vibrations, and noise may also contribute to energy dissipation. These secondary effects become more pronounced when the energy associated with the stick–slip effect approaches zero.

Hence, the evaluation of the optimal fit was determined by calculating the coefficient of determination (R^2). The best fit was achieved at 0.9 MPa, resulting in an R^2 value of 0.993, which also exhibited similar force and area characteristics as those observed experimentally. Both the experimental and modeled loops of forces vs. actuator displacements are presented in Figure 14.

4.3. Bending Combined with Internal Pressure (T1-2)

The internal pressure and the related real axial tension T_{re}, Equation (53), are crucial for assessing the critical curvature of the pipe, as they induce significant interlayer contact pressures in the pipe. In contrast, the effects of pressure do not influence the global response of the pipe, i.e., only the effective axial load should be considered in the global model [55]. Furthermore, the axial load remains constant in the experimental tests, because one extremity of the pipe was free to elongate.

In this scenario, a sample was tested at an internal pressure of 200 bar and without an axial load. The only difference from the previous case is the increase in stiffness resulting from the higher contact pressures caused by pipe pressurization, as shown in Table 3.

The linear dependence between critical curvature and contact pressure, as expressed in Equation (8), exhibits a slope corresponding to the interlayer friction coefficient. This test, therefore, constitutes an appropriate procedure for calibrating the friction coefficient μ. Numerical results obtained for a range of μ values are reported in Table 9. This table indicates that the increasing friction coefficients require higher transverse forces to bend the pipe, as the no-slip bending stiffness extends to higher curvatures with increasing critical curvature. On the other hand, as discussed in Section 4.2, the maximum energy dissipation occurs for an adhesion of 0.7 MPa. Consequently, an increase in the friction coefficient leads to a decrease in energy dissipation.

Table 9. Friction coefficient sensitivity analysis.

μ	κ_{cr} [1/m]	$\frac{\Delta F_{model}}{\Delta F_{exp}}$	$\frac{A_{model}}{A_{exp}}$	R^2
0	0.019	0.81	0.919	0.982
0.03	0.023	0.890	0.916	0.993
0.04	0.024	0.917	0.910	0.992
0.05	0.026	0.942	0.901	0.993
0.06	0.027	0.968	0.889	0.984
0.07	0.028	0.992	0.875	0.977
0.10	0.032	1.060	0.820	0.945

Moreover, by inspecting Table 8, the friction coefficient that provided the best overall agreement among the three parameters is 0.05. At this value, the energy dissipation and the imposed forces ratios are close to the experimental values, and R^2 tends to 1.0.

Hence, by considering an initial adhesion of 0.9 MPa and a friction coefficient of 0.05, Figure 20 compares the measured force versus transverse displacement relation with the numerically predicted curve, evidencing good agreement between them. Moreover, Figure 21a shows the transverse displacements measured along the pipe sample during the test and the numerically predicted deformed configuration, which also agreed quite well. Finally, Figure 21b illustrates the variation in curvature along the pipe, indicating the presence of no-slip and slip zones. A pronounced nonlinear variation in curvature can be depicted.

In the literature, some studies address the modeling of bending in flexible pipes, but they typically rely on a limited number of experimental tests. For example, Dong et al. [36] applied an analytical bending model proposed by Kebadze and Kraincanic [37] to evaluate the response of a bending test performed by Witz [11] on a 2.5″ flexible pipe under internal pressures of 0 and 30 MPa. A friction coefficient of 0.1 was used, but to better match the experimental results, particularly at low internal pressures, an initial contact pressure of 1.5 MPa was introduced. Both Dai et al. [40] and Wang et al. [41] developed numerical models based on the test reported by Saevik [24], involving a 4″ pipe subjected to internal pressures of 0.7, 10, and 20 MPa. In Dai et al. [40], different friction coefficients were fitted for each pressure level: 0.26, 0.23, and 0.21 for 0.7, 10, and 20 MPa, respectively. Additionally, an initial strain of 0.02 was applied to the tensile armor tendons to simulate the pre-tension between layers and better replicate the hysteresis observed in the experiments. Using the same friction values as Dai et al. [40], Wang et al. [41] investigated the effects of residual strain in the tensile armor, as well as the influence of an initial contact pressure of 2 MPa applied to the outer sheath, aiming to improve the simulation of initial interlayer friction. This study also included analytical model results incorporating the 2 MPa outer sheath pressure. In another study, Ye et al. [38] compared the results of a numerical model with experimental data for a 4″ pipe, determining an interlayer friction coefficient of 0.088 through regression analysis. Although discrepancies were noted under low contact pressure conditions, no specific mechanism was proposed to account for them in the model. Lastly, de Sousa et al. [30] employed a numerical model to study torsion in a 9.13″ flexible pipe. To improve agreement with experimental data, an adhesion value of 5 MPa was applied to all interfaces, alongside a friction coefficient of 0.1.

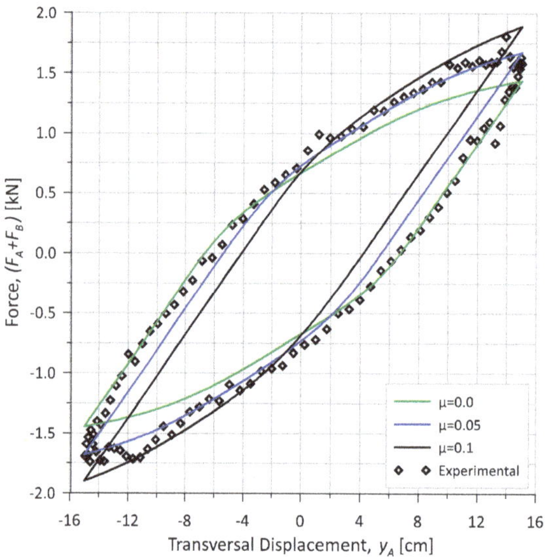

Figure 20. Force vs. displacement in T1-2 test.

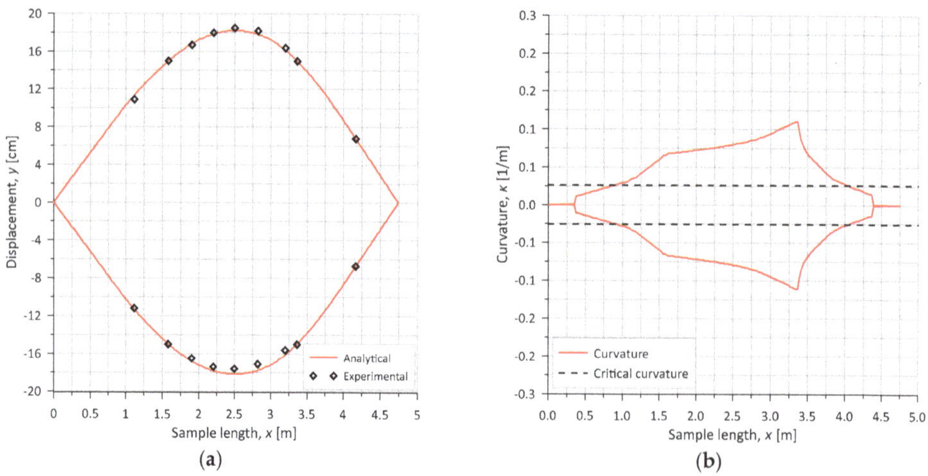

Figure 21. Displacements (**a**) and curvatures (**b**) in T1-2 test.

Consequently, there is significant variation in the values assigned to the interlayer friction coefficient. Due to the limited number of tests, as well as differences in pipe diameter, materials, and manufacturing processes, these values are not directly comparable. Nonetheless, they offer a general indication of the expected range. It is also important to highlight that the adjustment of the friction coefficient depends on how the initial contact pressure is implemented in the model, or whether it is considered at all. For example, a relatively low friction coefficient of 0.05 may be used in conjunction with a high adhesion value of 0.9 MPa. In the proposed model, as described in Equation (8), adhesion is incorporated directly into the stiffness calculation, rather than being treated as an increase in contact pressure (which, in some other models, is multiplied by the friction coefficient).

4.4. Bending Combined with Internal Pressure and Tension (T1-3 to T1-8)

This section presents the results from tests conducted on the sample under six combinations of bending, internal pressure, and tension. Figure 22 shows that the deformed shape of the sample remains relatively unchanged under different test conditions, as the tests were displacement-controlled, with the sample subjected to the same transverse displacement at the force application points. The measured transverse displacements show good correlation with the model predictions.

Regarding the curvatures, the inclusion of an axial load in the tests reveals a distinct change in the curvature profile, as shown in Figure 23. The axial force tends to straighten the sample, resulting in higher curvatures in the sections near the points of force application (actuators), where stiffness transitions occur. In contrast, the curvatures exhibit a more gradual change in scenarios without axial load, as illustrated in Figures 16 and 21b.

Figure 24 compares the relationship between force and transversal displacement obtained in the experimental tests with those predicted by the proposed model, while Table 10 indicates, for each of the conditions simulated, the imposed force ratio, the energy dissipation ratio, and the correlation coefficient R^2. The results obtained show a very good agreement between the experimental tests and the model predictions.

Table 10. Adjusted model parameters vs. experimental results.

Test	$\frac{\Delta F_{model}}{\Delta F_{exp}}$	$\frac{A_{model}}{A_{exp}}$	R^2
T1-1	0.946	1.074	0.993
T1-2	0.942	0.901	0.989
T1-3	0.885	0.997	0.990
T1-4	0.918	1.008	0.996
T1-5	0.950	0.954	0.997
T1-6	0.874	0.975	0.990
T1-7	0.930	0.992	0.995
T1-8	0.938	0.933	0.996
T2-9	1.010	0.965	0.984
T2-10	1.082	1.015	0.983
T2-11	1.031	1.025	0.994
T2-12	1.020	0.990	0.994
T2-13	1.021	0.982	0.995
T2-14	1.057	1.081	0.988
T2-15	1.023	0.948	0.993
T2-16	1.031	0.898	0.992

In contrast to the cases without axial load (as shown in Figures 14 and 20), the forces required to cycle the sample are notably higher due to the significant influence of the axial load, which outweighs the impact of internal pressure. Furthermore, the forces observed in cases with the same axial load remain nearly constant, even with the increase in pressure from 100 bar to 200 bar. This can be attributed to the low friction coefficient used in the model, which is associated with a high adhesion value.

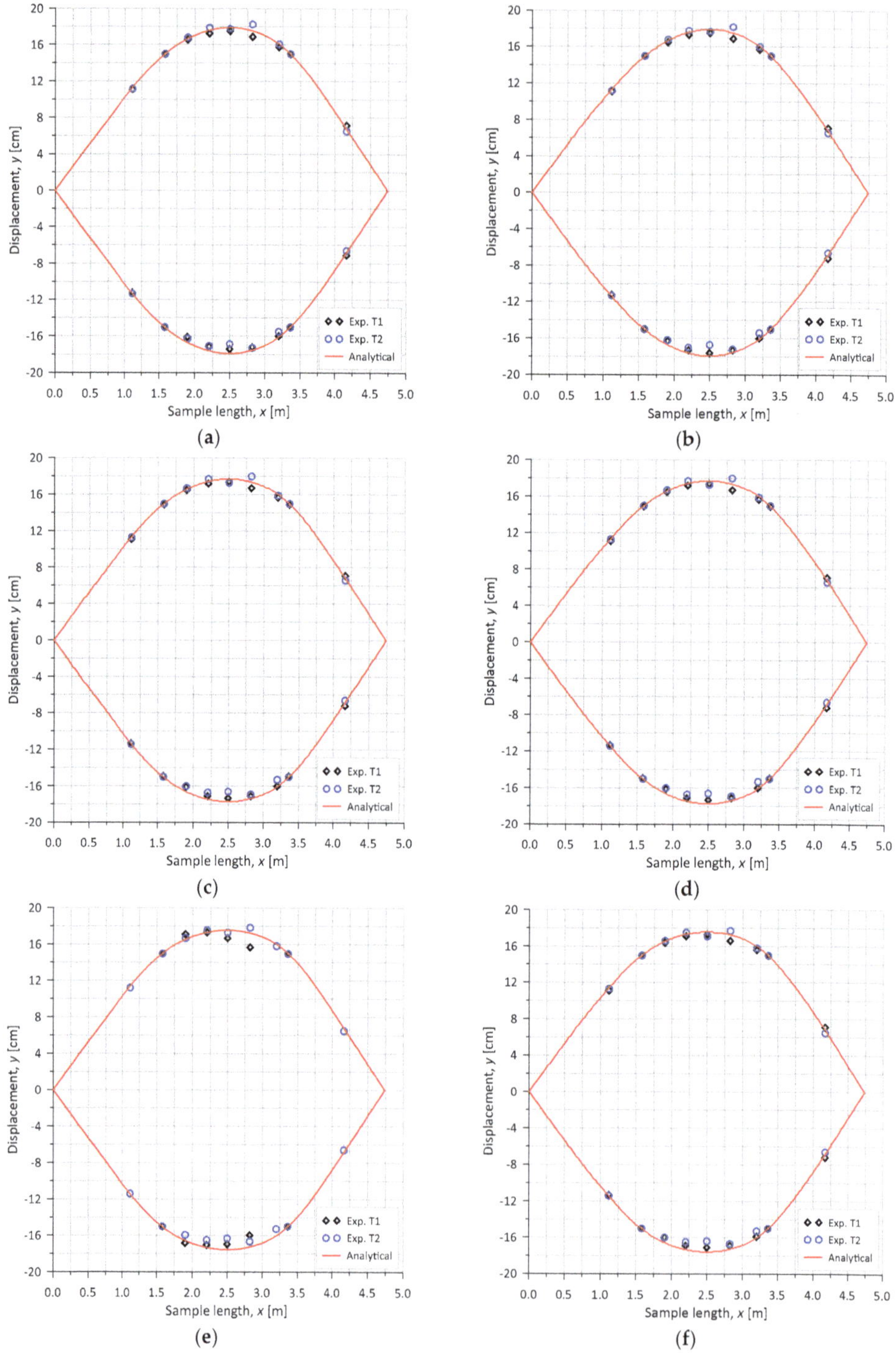

Figure 22. Maximum displacements in tests T1-3 to T1-8 and T2-9 to T2-16 for (internal pressure/tension): (**a**) 100 bar/5 kN; (**b**) 200 bar/5 kN; (**c**) 100 bar/10 kN; (**d**) 200 bar/10 kN; (**e**) 100 bar/15 kN; (**f**) 200 bar/15 kN.

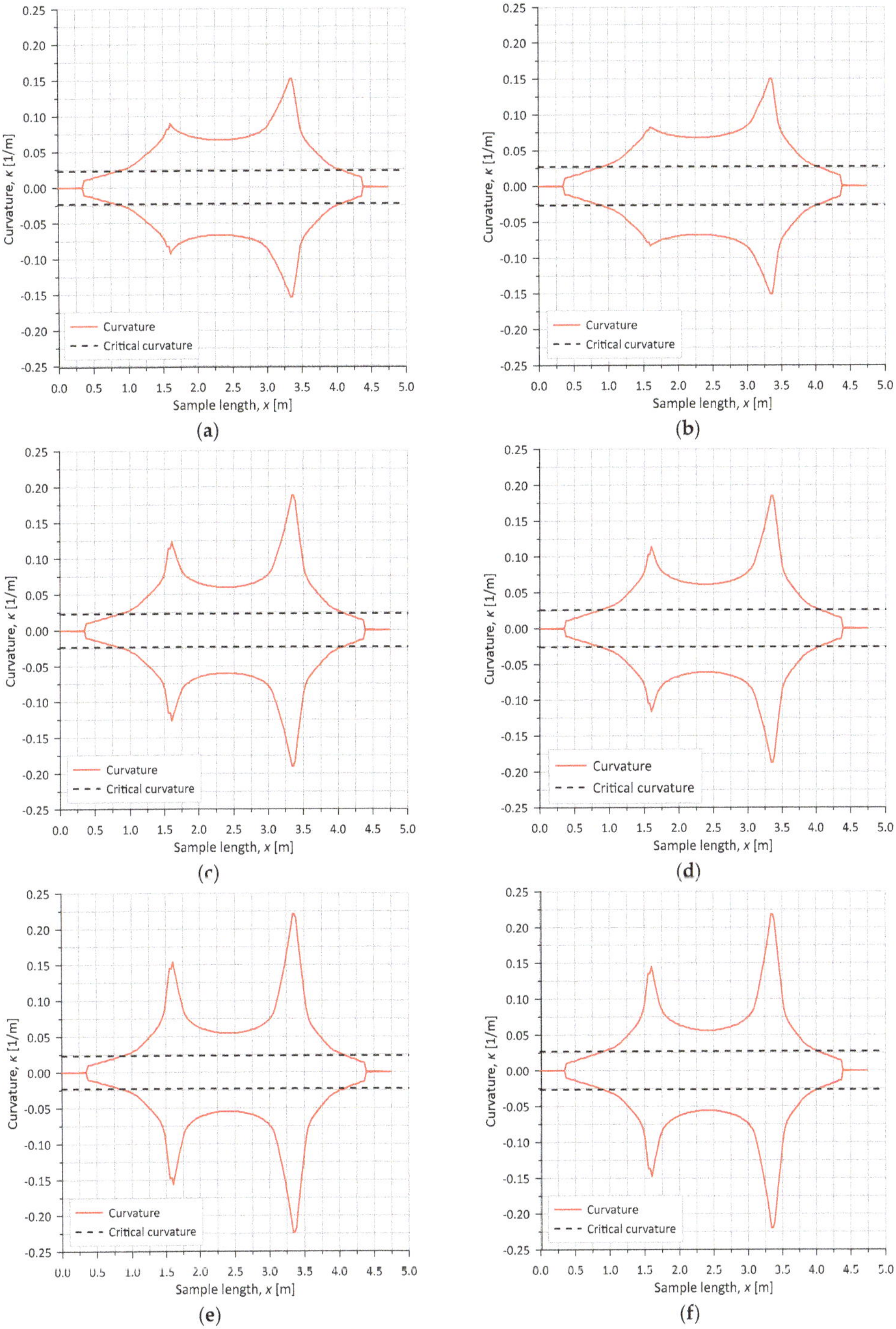

Figure 23. Curvatures in tests T1-3 to T1-8 and T2-9 to T2-16 for (internal pressure/tension): (**a**) 100 bar/5 kN; (**b**) 200 bar/5 kN; (**c**) 100 bar/10 kN; (**d**) 200 bar/10 kN; (**e**) 100 bar/15 kN; (**f**) 200 bar/15 kN.

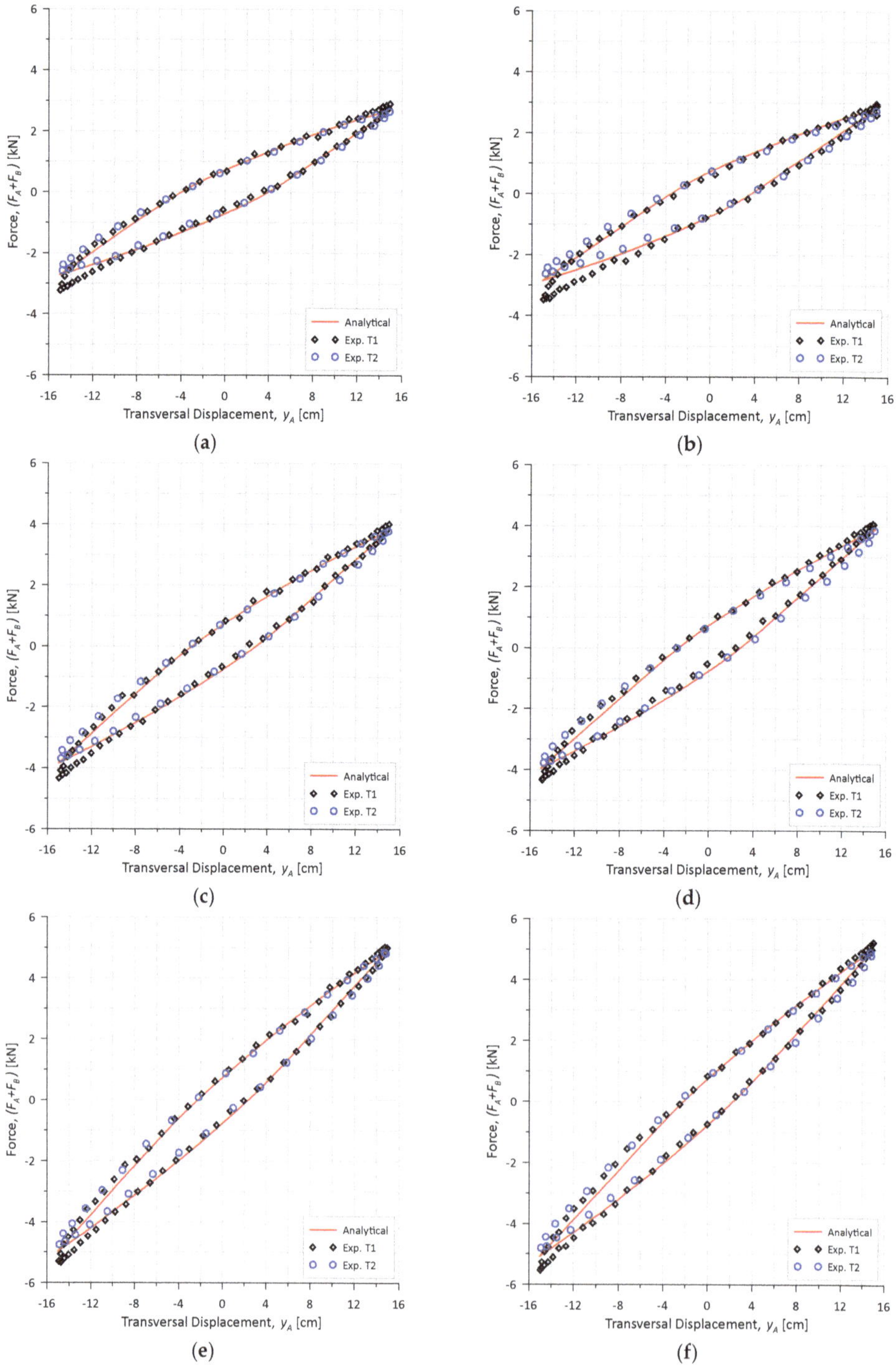

Figure 24. Force vs. transverse displacement in tests T1-3 to T1-8 and T2-9 to T2-16 for (internal pressure/tension): (**a**) 100 bar/5 kN; (**b**) 200 bar/5 kN; (**c**) 100 bar/10 kN; (**d**) 200 bar/10 kN; (**e**) 100 bar/15 kN; (**f**) 200 bar/15 kN.

4.5. Bending Tests After Degradation (T2-9 to T2-16)

After the T1 tests (1 to 8) were conducted, the sample underwent an 18-month degradation period during which the pipe was maintained with a flooded annulus containing saturated water at 1 bar CO_2 and 40 °C. Tests were then repeated (T2—9 to 16), and the response was found to be very similar. The same parameters were used to calibrate the model. Figures 22 and 24 present results before and after degradation.

Despite being subjected to relatively harsh conditions designed to accelerate the degradation of both the tensile armor and polymeric layers, the impact of degradation over time on stiffness appears to be negligible. This indicates that the hysteresis curve employed, e.g., for fatigue analysis (assuming an environment that is equally severe or less severe than the one tested), does not necessitate any updates and remains valid throughout the service life of the pipe. Therefore, the experimental tests suggest that only a single bending stiffness vs. curvature curve can be used, consistent with current design practices, to determine the overall response of the flexible pipe.

However, degradation of the wires results in changes to their material condition, necessitating the consideration of new fatigue curves that indicate a shorter service life for the same overall response. In Coser et al. [56], wire samples were subjected to similar conditions that led to the degradation of the test samples, followed by four-point bending fatigue tests. A comparison of the results with those from non-degraded wires clearly revealed a significant reduction in fatigue resistance. A full-scale test further validated these findings. Additionally, Krishman et al. [57] demonstrated the reduction in fatigue life by inducing controlled pitting corrosion through an electromechanical method, comparing the resulting curves of pitted wires with those from new wires. The curves for the corroded wires exhibited a reduction in fatigue life of approximately 2.6 to 2.7 times.

5. Conclusions

In this study, the global response of a flexible pipe under combined axisymmetric and bending loads was investigated through experimental tests and theoretical modeling.

The experiments were performed on a 2.5″ flexible pipe. The materials of its inner and outer sheaths were first characterized through tensile tests. The pipe was then subjected to combined axisymmetric and cyclic bending loads using a controlled displacement four-point bending setup, under both non-degraded and degraded conditions. The latter involved flooding the annulus and exposing the tensile armors to CO_2-saturated water at 1 bar and 40 °C for 18 months. In all cases, transverse displacements were measured under constant axisymmetric loads, revealing significant hysteretic responses.

Theoretical analyses incorporated existing approaches to predict the monotonic response of flexible pipes under combined axisymmetric and bending loading. The bending formulation was modified to include initial interlayer adhesion, and the models were implemented in a return algorithm to reproduce hysteretic behavior. A finite-difference (FD)-based model was then developed, coupling axisymmetric and cyclic bending responses and integrating the proposed algorithm for cyclic loading. The incorporation of the local into the global model was also discussed in detail.

Theoretical and experimental results showed excellent agreement, accurately reproducing the cyclic response of the pipe. The analyses confirmed that the global response is strongly influenced by interlayer parameters such as adhesion and friction. Increasing these parameters raised the required bending forces until a no-slip condition was reached, beyond which the response stabilized. Energy dissipation varied nonlinearly with friction, peaking at partial slip and vanishing under full-slip conditions, while the adopted parameters (0.9 MPa of initial adhesion and a friction coefficient of 0.05) provided a reasonable approximation. Moreover, tests under degraded conditions showed that, despite wire corrosion, the overall mechanical response remained essentially unchanged, indicating that the global behavior of the riser is not significantly affected under the imposed degradation scenario.

The proposed methodology can be implemented in finite element (FE) models using conventional software, providing an alternative to traditional global analysis tools for flexible pipes and offering the potential for extension to three-dimensional applications. The findings also supply valuable data for model calibration and design guidance, highlighting the importance of accounting for initial adhesion and interlayer friction effects. Nonetheless, further research is needed to improve the quantification of interface parameters, incorporate possible viscous contributions to energy dissipation, and evaluate the influence of interface degradation on the fatigue life of flexible pipes.

Author Contributions: Conceptualization, B.d.P.M.d.C., N.R. and J.R.M.d.S.; methodology, B.d.P.M.d.C., J.R.M.d.S. and N.R.; validation, B.d.P.M.d.C., N.R. and J.R.M.d.S.; formal analysis, J.R.M.d.S.; investigation, E.R. and B.d.P.M.d.C.; data curation, E.R. and B.d.P.M.d.C.; writing—original draft preparation, B.d.P.M.d.C. and E.R.; writing—review and editing, N.R. and J.R.M.d.S.; supervision, N.R. and J.R.M.d.S. All authors have read and agreed to the published version of the manuscript.

Funding: This work was partially supported by the Brazilian funding agencies CNPq (Conselho Nacional de Desenvolvimento Científico e Tecnológico), grant number 308625/2022-6 and CAPES (Coordenação de Aperfeiçoamento de Pessoal de Nível Superior), finance code 001.

Data Availability Statement: Restrictions apply to the availability of the input data employed to generate the numerical models. Such data was obtained from Petrobras (the Brazilian state oil company) and cannot be disclosed due to confidentiality issues. The output data from the results of the analyses are available from the corresponding author upon reasonable request.

Acknowledgments: The authors would like to acknowledge the support of Petrobras (the Brazilian state oil company), which actively boosts research activities and encourages using innovative computational tools in real-life situations.

Conflicts of Interest: Author Bernardo de P. M. da Costa is employed by Petrobras. The remaining authors declare that the research was conducted in the absence of any commercial or financial relation that could be construed as a potential conflict of interest.

References

1. *API SP 17J*; Specification for Unbonded Flexible Pipe, 5th ed. American Petroleum Institute: Washington, DC, USA, 2024.
2. Lobo, C.V.F.; Ribeiro, M.L. SCC-CO_2 Failure Mode: Literature Review and Available Technologies Used to Inspect and Access the Integrity of the Flexible Pipes. In Proceedings of the Offshore Technology Conference (OTC), Houston, TX, USA, 1–4 May 2023.
3. MacLeod, I.; Picksley, J.; Cruickshank, K. *Flexible Pipe Integrity Management Guidance & Good Practice*; Rev. 05; Wood: Aberdeen, Scotland, UK, 2023.
4. Larsen, C.M.; Saevik, S.; Qvist, J. Design analysis. In *Handbook on Design and Operation of Flexible Pipes*; Part B; Fergestad, D., Lotveit, S.A., Eds.; NTNU/4Subsea/SINTEF Ocean: Trondheim, Norway, 2017.
5. Custódio, A.B.; Vaz, M.A. A Nonlinear Formulation for the Axisymmetric Response of Umbilical Cables and Flexible Pipes. *Appl. Ocean Res.* **2002**, *24*, 21–29. [CrossRef]
6. Féret, J.J.; Bournazel, C.L. Calculation of Stresses and Slip in Structural Layers of Unbonded Flexible Pipes. *J. Offshore Mech. Arct. Eng.* **1987**, *109*, 263–269. [CrossRef]
7. Batista, R.C.; Bogarin, J.A.; Ebecken, N.F.F. Local Mechanical Behaviour of Multilayered Flexible Risers. In Proceedings of the 7th Symposium on Offshore Engineering, Rio de Janeiro, Brazil, 21–25 August 1989; pp. 494–510.
8. Love, A.E.H. *A Treatise on the Mathematical Theory of Elasticity*, 4th ed.; Cambridge University Press: London, UK, 2017; ISBN 978-1107618091.
9. Witz, J.A.; Tan, Z. On the Axial-Torsional Structural Behaviour of Flexible Pipes, Umbilicals and Marine Cables. *Mar. Struct.* **1992**, *5*, 205–227. [CrossRef]
10. McIver, D.B. A Method of Modelling the Detailed Component and Overall Structural Behaviour of Flexible Pipe Sections. *Eng. Struct.* **1995**, *17*, 254–266. [CrossRef]
11. Witz, J.A. A Case Study in the Cross-Section Analysis of Flexible Risers. *Mar. Struct.* **1996**, *9*, 885–904. [CrossRef]

12. Ramos, R., Jr.; Martins, C.A.; Pesce, C.P.; Roveri, F.E. A Case Study on the Axial-Torsional Behavior of Flexible Risers. In Proceedings of the 27th International Conference on Offshore Mechanics and Arctic Engineering (OMAE 2008), Estoril, Portugal, 15–20 June 2008.
13. de Sousa, J.R.M.; Magluta, C.; Roitman, N.; Campello, G.C. On the Extensional-Torsional Response of a Flexible Pipe with Damaged Tensile Armor Wires. *Ocean Eng.* **2018**, *161*, 350–383. [CrossRef]
14. Liu, J.; Tan, Z.; Sheldrake, T. An Analytical Model for the Unbonded Flexible Pipe Stress Analysis with Consideration of Nonlinear Material Properties of Metal Layers. In Proceedings of the 29th International Conference on Offshore Mechanics and Arctic Engineering (OMAE 2010), Shanghai, China, 6–11 June 2010.
15. Liu, J.; Ma, J.; Vaz, M.A.; Duan, M. Viscoelastic Damping of Flexible Risers: Theoretical Model. *Ocean Eng.* **2020**, *217*, 107974. [CrossRef]
16. Ramos, R., Jr.; Kawano, A. Local Structural Analysis of Flexible Pipes Subjected to Traction, Torsion and Pressure Loads. *Mar. Struct.* **2015**, *42*, 95–114. [CrossRef]
17. Lu, H.; Vaz, M.A.; Caire, M. Alternative Analytical and Finite Element Models for Unbonded Flexible Pipes under Axisymmetric Loads. *Ocean Eng.* **2021**, *225*, 108766. [CrossRef]
18. de Sousa, J.R.M.; dos Santos, A.C. Symbolic Regression Equations to Predict the Maximum Stress Concentration Factors in Flexible Pipes with Damaged Tensile Armors. In Proceedings of the ASME 43rd International Conference on Offshore Mechanics and Arctic Engineering (OMAE 2024), Singapore, 9–14 June 2024.
19. Li, X.; Jiang, X.; Hopman, H. Predicting the Wet Collapse Pressure for Flexible Risers with Initial Ovalization and Gap: An Analytical Solution. *Mar. Struct.* **2020**, *71*, 102732. [CrossRef]
20. Gay Neto, A.; Martins, C.A.; Pesce, C.P.; Meirelles, C.O.C.; Malta, E.R.; Barbosa Neto, T.F.; Godinho, C.A.F. Prediction of Burst in Flexible Pipes. *J. Offshore Mech. Arct. Eng.* **2013**, *135*, 011401. [CrossRef]
21. Liu, Y.; Wu, Q.; He, J.; Liu, Q.; Li, M.; Wang, G. Coupled Burst and Fracture Failure Characteristics of Unbonded Flexible Riser under Internal Pressure and Axial Tension. *J. Mar. Sci. Eng.* **2025**, *13*, 1866. [CrossRef]
22. Saevik, S.; Thorsen, M.J. Techniques for Predicting Tensile Armour Buckling and Fatigue in Deep Water Flexible Pipes. In Proceedings of the ASME 31st International Conference on Offshore Mechanics and Arctic Engineering (OMAE 2012), Rio de Janeiro, Brazil, 1–6 July 2012.
23. Li, X.; Vaz, M.A.; Custódio, A.B. Analytical Model for Tensile Armors Lateral Deflections and Buckling in Flexible Pipes. *Mar. Struct.* **2019**, *64*, 211–228. [CrossRef]
24. Saevik, S. Theoretical and Experimental Studies of Stresses in Flexible Pipes. *Comput. Struct.* **2011**, *89*, 2273–2291. [CrossRef]
25. Muñoz, H.E.M.; de Sousa, J.R.M.; Magluta, C.; Roitman, N. Improvements on the Numerical Analysis of the Coupled Extensional-Torsional Response of a Flexible Pipe. *J. Offshore Mech. Arct. Eng.* **2016**, *138*, 011701. [CrossRef]
26. Yoo, D.H.; Jang, B.S.; Yim, K.H. Nonlinear Finite Element Analysis of Failure Modes and Ultimate Strength of Flexible Pipes. *Mar. Struct.* **2017**, *54*, 50–72. [CrossRef]
27. Malta, E.R.; Martins, C.A. Finite Element Analysis of Flexible Pipes under Axial Compression: Influence of the Sample Length. *J. Offshore Mech. Arct. Eng.* **2017**, *139*, 011701. [CrossRef]
28. Lukassen, T.V.; Gunnarsson, E.; Krenk, S.; Glejbol, K.; Lyckegaard, A.; Berggreen, C. Tension-Bending Analysis of Flexible Pipe by a Repeated Unit Cell Finite Element. *Mar. Struct.* **2019**, *64*, 401–420. [CrossRef]
29. Provasi, R.; Toni, F.G.; Martins, C.A. Friction Coefficient Influence in a Flexible Pipe: A Macroelement Model. *Ocean Eng.* **2022**, *266*, 112719. [CrossRef]
30. de Sousa, J.R.M.; Magluta, C.; Roitman, N.; Londoño, T.V.; Campello, G.C. A Study on the Response of a Flexible Pipe to Combined Axisymmetric Loads. In Proceedings of the ASME 2013 32nd International Conference on Ocean, Offshore and Arctic Engineering (OMAE 2013), Nantes, France, 9–14 June 2013.
31. de Sousa, J.R.M.; Sagrilo, L.V.S.; Rocha, D.M. Semi-Empirical Equations to Predict the Bisymmetric Dry and Wet Hydrostatic Collapse Strengths of Flexible Pipes. *Appl. Ocean Res.* **2022**, *122*, 1013129. [CrossRef]
32. Cuamatzi-Melendez, R.; Castillo-Hernández, O.; Vázquez-Hernández, A.O.; Albiter, A.; Vaz, M.A. Finite Element Modeling of Burst Failure in Unbonded Flexible Risers. *Eng. Struct.* **2015**, *87*, 58–69. [CrossRef]
33. Gay Neto, A.; Martins, C.A. Flexible Pipes: Influence of the Pressure Armor in the Wet Collapse Resistance. *J. Offshore Mech. Arct. Eng.* **2014**, *136*, 031401. [CrossRef]
34. Axelsson, G.; Skjerve, H. Flexible Riser Carcass Collapse Analyses—Sensitivity on Radial Gaps and Bending. In Proceedings of the ASME 2014 33rd International Conference on Ocean, Offshore and Arctic Engineering (OMAE2014), San Francisco, CA, USA, 8–13 June 2014.
35. Cuamatzi-Melendez, R.; Castillo-Hernández, O.; Vásquez-Hernández, A.O. Finite Element and Theoretical Analyses of Bisymmetric Collapses in Flexible Risers for Deepwaters Developments. *Ocean Eng.* **2017**, *140*, 195–208. [CrossRef]
36. Dong, L.; Huang, Y.; Zhang, Q.; Liu, G. An Analytical Model to Predict the Bending Behavior of Unbonded Flexible Pipes. *J. Ship Res.* **2013**, *57*, 171–177. [CrossRef]

37. Kebadze, E.; Krainkanic, I. Nonlinear Bending Behaviour of Offshore Flexible Pipes. In Proceedings of the Ninth International Offshore and Polar Engineering Conference (ISOPE), Brest, France, 30 May–4 June 1999.
38. Ye, N.; Saevik, S.; Zhou, C. Investigation of Antiwear Tape's Influence on Bending Behavior of One Flexible Riser. In Proceedings of the Twenty-fourth International Ocean and Polar Engineering Conference (ISOPE), Busan, South Korea, 15–20 June 2014.
39. Kim, J.D.; Jang, B.S.; Yun, R.H. Application of Analytical Model in the Prediction of Dynamic Responses and Fatigue Damage of Flexible Risers: Part I—Improvement of Analytical Model Considering Shear Deformation and Varying Tension Effects. *Mar. Struct.* **2021**, *79*, 103044. [CrossRef]
40. Dai, T.; Sævik, S.; Ye, N. Friction Models for Evaluating Dynamic Stresses in Non-Bonded Flexible Risers. *Mar. Struct.* **2017**, *55*, 137–161. [CrossRef]
41. Wang, L.; Ye, N.; Yue, Q. A Novel Helix Contact Model for Predicting Hysteretic Behavior of Unbonded Flexible Pipes. *Ocean Eng.* **2022**, *264*, 112407. [CrossRef]
42. Fang, P.; Li, X.; Jiang, X.; Hopman, H.; Bai, Y. Study of the Effects of Cross Section Shape on the Structural Behaviour of Helical Wires on a Bent Flexible Structure. In Proceedings of the 33rd International Ocean and Polar Engineering Conference (ISOPE), Ottawa, ON, Canada, 19–23 June 2023.
43. Tang, M.; Li, S.; Zhang, H.; Bian, X.; Zhao, X. Monitoring the Slip of Helical Wires in a Flexible Riser under Combined Tension and Bending. *Ocean Eng.* **2022**, *256*, 111512. [CrossRef]
44. Tan, Z.; Quiggin, P.; Sheldrake, T. Time Domain Simulation of the 3D Bending Hysteresis Behaviour of an Unbonded Flexible Riser. *J. Offshore Mech. Arct. Eng.* **2009**, *131*, 031301. [CrossRef]
45. He, Y.; Lu, H.; Vaz, M.A.; Caire, M. Flexible Riser Connection Analysis with I-Tube Interface and Bending Hysteresis Effect. In Proceedings of the ASME 2019 38th International Conference on Ocean, Offshore and Arctic Engineering (OMAE2019), Glasgow, Scotland, UK, 9–14 June 2019.
46. Dai, T.; Zhang, J.; Ren, C.; Xing, Y.; Saevik, S.; Ye, N.; Jin, X.; Wu, J. Application of a Data-Driven Approach for Maximum Fatigue Damage Prediction of an Unbonded Flexible Riser. *Ocean Eng.* **2024**, *256*, 118053. [CrossRef]
47. Gundersen, P.; Doynov, K.; Andersen, T.; Haakonsen, R. Methodology for Determining Remnant Fatigue Life of Flexible Risers Subjected to Slugging and Irregular Waves. In Proceedings of the ASME 2012 31st International Conference on Ocean, Offshore and Arctic Engineering (OMAE 2012), Rio de Janeiro, Brazil, 1–6 June 2012.
48. de Sousa, J.R.M.; de Sousa, F.J.M.; de Siqueira, M.Q.; Sagrilo, L.V.S.; de Lemos, C.A.D. A Theoretical Approach to Predict the Fatigue Life of Flexible Pipes. *J. Appl. Math.* **2012**, *2012*, 983819. [CrossRef]
49. *ASTM D638-22*; Standard Test Method for Tensile Porperties of Plastics. American Society for Testing and Materials: West Conshohocken, PA, USA, 2022.
50. Nitta, K.; Yam, M. Poisson's Ratio and Mechanical Nonlinearity Under Tensile Deformation in Crystalline Polymers. In *Rheology*; de Vicente, J., Ed.; InTech: Vienna, Austria, 2012. [CrossRef]
51. *API RP 17B*; Recommended Practice for Flexible Pipe, 6th ed. American Petroleum Institute: Washington, DC, USA, 2024.
52. Sparks, C.P. *Fundamentals of Marine Riser Mechanics: Basic Principles and Simplified Analyses*, 1st ed.; PennWell Corp: Tulsa, OK, USA, 2007; ISBN 978-1-59370-070-6.
53. Simo, J.C.; Hughes, T.J.R. *Computational Ineslasticity*; Springer: New York, NY, USA, 1998; ISBN 0-387-97520-0.
54. Singh, A.K.; Bhadauria, B.S. Finite Difference Formulae for Unequal Sub-Intervals Using Lagrange's Interpolation Formula. *Int. J. Math. Anal.* **2009**, *3*, 815–827.
55. *Mathcad*, version 15.0; PTC Inc: Woburn, MA, USA, 2015.
56. Coser, T.B.; de Negreiros, L.A.S.; Campelo, G.C.; Radke, E.D.; Rosa, C.D.; de Souza, D.F.; Rodrigues, L.F.V.; Renck, T.S.; Tagliari, M.R.; Borges, M.F. Flexible Pipe Fatigue Life Evaluation Based on a Dynamic Full-Scale Test Considering a Preconditioning Corrosion Procedure. In Proceedings of the ASME 2023 42nd International Conference on Ocean, Offshore and Artic Engineering (OMAE2023), Melbourne, Australia, 11–16 June 2023.
57. Krishman, V.R.; Asher, S.; Kan, W.C.; Popelar, C. Effect of Piting Corrosion on Fatigue LIfe of Flexible Armor Wires. In Proceedings of the Twenty-sixth International Ocean and Polar Engineering Conference (ISOPE), Rhodes, Greece, 26 June–1 July 2016.

Journal of
*Marine Science
and Engineering*

MDPI

Article

Coupled Burst and Fracture Failure Characteristics of Unbonded Flexible Riser Under Internal Pressure and Axial Tension

Yi Liu [1,†], Qitao Wu [1,†], Jiawei He [2], Qingsheng Liu [1,*], Ming Li [1,*] and Gang Wang [1]

1 School of Mechanical and Electric Engineering, Soochow University, Suzhou 215131, China; yliu1214@stu.suda.edu.cn (Y.L.); 2329403116@stu.suda.edu.cn (Q.W.); wanggang81@suda.edu.cn (G.W.)
2 Naval Research Institute (NVRI), Beijing 100161, China; hjw_sjtu@163.com
* Correspondence: qsliu@suda.edu.cn (Q.L.); liming24@suda.edu.cn (M.L.); Tel.: +86-512-65790196 (Q.L. & M.L.)
† These authors contributed equally to this work.

Abstract: Unbonded flexible risers, which can experience large bending deformation, are key equipment in advancing deep-sea exploration for marine resources. However, the riser experiences coupled loading effects from ocean environment. This results in complex response characteristics, leading to potential damage or even destruction. By presenting an analytical–numerical framework, this study uncovers the mechanism underlying the coupled failure of the pressure- and tensile-armor layers, furnishes a new tension–pressure coupled failure boundary for the ultimate-limit-state design of deep-water risers, and supplies the corresponding theoretical verification. Firstly, based on the axisymmetric load assumption, a theoretical model is proposed based on principle of functionality; afterwards, the failure model is defined by considering the material elastoplasticity. Secondly, a full-layered numerical model with detailed geometric properties is established; meanwhile, a simplified 7-layer model without a carcass layer is constructed for comparison. Finally, after verified through experimental data and interactive verification of theoretical and numerical methods, the simplified numerical model is proved to have calculation accuracy and validity. The characteristics are studied by the proposed methods. The comparison results show that the pre-applied internal pressure has limited influence on the axial stiffness of unbonded flexible rise. The initial axial tension would enhance the anti-burst failure ability of unbonded flexible riser, the failure pressure increases by 35% when the tensile force is 500 kN.

Keywords: unbonded flexible riser; axial tension; internal pressure; coupled failure; pressure armor; tensile armor

1. Introduction

Riser systems are pipelines that connect surface facilities to submarine equipment, and has the functionality to with stand high pressure, corrosive media and harsh environmental loads to ensure the safe transportation of marine oil, gas and mineral resources. Among the many different types of production risers, unbonded flexible risers possess several merits, including the capacity to withstand large deformations, resistance to corrosion, simplicity in installation, and recyclable characteristics, and are prioritized or mandatory for the extraction of offshore oil and gas resources. An unbonded flexible riser is typically assembled of a series of functional interlayers (see Figure 1), which can be divided into helical layers and cylindrical layers according to the geometric properties.

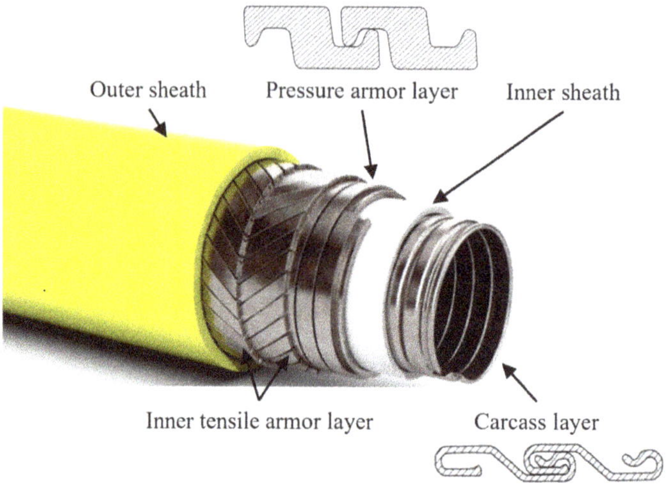

Figure 1. Unbonded flexible riser of rough bore type.

Among the different type of interlayers of unbonded flexible riser, the helical layers are the main load-bearing structures, and the direct cause of the special mechanical properties of the unbonded flexible riser. The tensile-armor layer is the most crucial element of the unbonded flexible riser, which mainly provides axial stiffness, is subjected to axial tension and torque about the axial direction, and ensures the riser's safety due to self-weighting and other environmental loads. The pressure-armor layer constitutes steel self-locking structures with high laying angles and irregular cross-section geometries, typically 'Z' type (see Figure 2a), 'C' type (see Figure 2b), and 'T' type (see Figure 2c,d) et al., which mainly provide radial stiffness and support the internal pressure-armor layer. As with the asymmetric cross-sectional properties of the pressure-armor layer, the burst failure manifests as local stress concentration and structural damage when the internal pressure surpasses the ultimate bearing capacity. However, when exposed to a harsh marine environment, the failure of the riser system occurs from time to time, especially for the unbonded flexible riser, where coupled failure physics might occur, causing major environmental accidents such as offshore oil spill, bringing immeasurable economic and ecological losses [1]. Reliable prediction of their failure behavior is therefore essential for safe, economic, and sustainable deep-water oil and gas production. Recent advances in multifunctional materials offer potential routes to enhance layer durability: for example, hybrid systems based on thermally stable amorphous porous silicon nanoparticles have demonstrated improved thermal and chemical stability [2,3], while nano-fibrous scaffolds have exhibited tailored antimicrobial and repair functionalities [4,5]. Although these material concepts have yet to be transferred to riser technology, they highlight future possibilities for mitigating the coupled failure physics identified above.

Predicting the unbonded flexible risers' behavior commonly involves using numerical and theoretical research techniques. Experimental methods can accurately indicate unbonded flexible riser's mechanical features and help in understanding their cross-sectional mechanical properties; however, experimental studies are costly, leading to a scarcity of experimental data due to the intricate structure of the risers. From the published experimental data related to unbonded flexible risers, Witz [6] experimented on a 2.5-inch, 8-layer unbonded flexible riser under axial tension and torque loads. Yue et al. [7] experimentally confirmed the mechanical characteristics of unbonded flexible risers under axial tension within shallow marine areas. Vargas-Londoño et al. [8] discovered in later experimental studies that the bearing capacities of various spiral steel strips within the tensile-armor layer exhibit inconsistencies. Zhou et al. [9] also found that there is a nonlinear correlation between the load and the strain under axisymmetric load through experimental research and analysis of the anti-friction zone in the unbonded flexible riser.

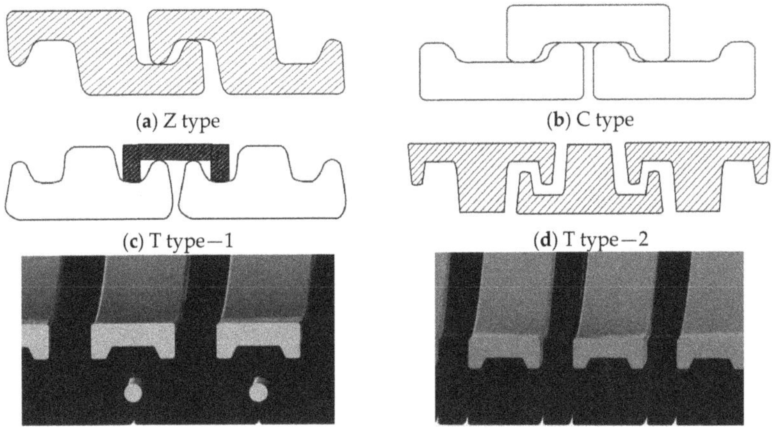
(**a**) Z type (**b**) C type (**c**) T type—1 (**d**) T type—2 (**e**) K type (**f**) X type

Figure 2. A typical sketch of a deep-water unbonded flexible riser. (**a**) Z type; (**b**) C type; (**c**) T type—1; (**d**) T type—2; (**e**) K type; (**f**) X type.

To the date, numerical simulation and theoretical analysis remain the core methods for exploring the mechanical properties of unbonded flexible risers. The theoretical models of unbonded flexible risers mainly consider the stress–strain relationships of each layer structure separately, and also take into account the inter-layer contact interactions [10]. Among them, Kebadze [11] divided all layer structures of unbonded flexible risers into different layers, including cylindrical shell layers and helical layers. He has contributed significantly to research into the theoretical model of such risers under axisymmetric loads. Subsequently, Dong [12] fully considered the spatial geometric relationships of the helical layers and made corresponding corrections to Kebadze's model. In addition, with Kebadze's theoretical model as the foundation, helical layer structures with complex cross-sectional geometries—including the carcass layer and tensile-armor layer—were converted by Bahtui [13] into orthotropic cylindrical shells. However, theoretical approaches have limitations in capturing complex load behavior.

With regard to numerical analysis, Ren et al. [14] established a finite-element method-based model to investigate stress distribution in damaged tensile-armor layers of unbonded flexible risers under elastic tension. Cornacchia et al. [15] conducted numerical simulations to examine the tensile mechanical characteristics of flexible pipes. De Sousa et al. [16] studied the structural deformational of damaged tensile-armor layers under pure axial tension. Under individual loads, the mechanical properties of tensile-armor layers were primarily the focus of these studies. Recent investigations has increasingly directed their attention toward the mechanical response of unbonded flexible risers subjected to combined loading conditions. Liu et al. [17] found that effective tension enhances the capacity of unbonded flexible risers to withstand bending moments when studying pressure-armor layers under coupled loads. Jiang et al. [18] and Wang et al. [19] examined the mechanical behavior of unbonded flexible risers under combined loads. Under axial tension and internal pressure, Sun et al. [20] used theoretical and numerical methods to predict the structural stiffness of risers. Tang et al. [21] investigated stress–strain responses and failure modes within tensile-armor layers of unbonded flexible pipelines subjected to multi-axial loading conditions. However, these studies did not delve into the failure characteristics of the armor layers.

In terms of failure characteristics, Yoo et al. [22] evaluated the ultimate strength and failure modes of unbonded flexible risers under axial tension and compression, but they did not investigate the role of internal pressure in tensile failure under such loading. Berge et al. [23] and Antoine et al. [24] examined the influence of initial manufacturing defects on tensile failure, yet the underlying mechanism remains largely unexplored. Ebrahimi et al. [25] employed finite-element analyses to characterize tensile-armor failure under axisymmetric loading, but their study was conducted without external pressure and thus lacks a corresponding investigation. Similarly, Zhu et al. [26], Kamaya et al. [27] and Neto et al. [28] concentrated solely on predicting burst pressure without addressing the factors that govern failure. Zeng et al. [29] proposed a finite-element-based method for burst-pressure evaluation, yet they did not examine how axial tension influences burst failure. Liu et al. [30] and Wei et al. [31] incorporated bending–internal-pressure coupling, yet they still ignored axial tension, leading to a potentially inaccurate identification of the burst initiation site and propagation path.

Despite progress in research, under axial tension and internal pressure, the mixed failure characteristics of unbonded flexible risers have not been thoroughly explored in the existing literature.

According to a conventional 2.5-inch unbonded flexible riser model that considers the elastoplastic behavior of the steel armor layer, this paper discusses the mixed failure characteristics of the riser's pressure-armor layer and tensile-armor layer under the coupling action of complex loads. The organization of the present paper is outlined below. In Section 2, the theoretical model established in this paper is explained. Leveraging the work–energy reversibility principle, this study derives an analytical framework for both the helical strip component and the cylindrical shell structure subjected to axisymmetric loading conditions, explicitly accounting for material elastoplastic behavior. Section 3 details the development of the numerical simulation using ABAQUS (2022) and the specification of key simulation parameters. Section 4 rigorously assesses the accuracy of both the numerical simulation framework and the analytical formulation, alongside the validity of the simplified 7-layer abstraction. Subsequently, on the basis of the simplified 7-layer abstraction, this section systematically investigates the complex failure mechanisms within the armor system resulting from the concurrent application of axial tensile forces and internal pressurization. Then, on the basis of the simplified model, the mixed failure characteristics of the armor layer under the coupling action of axial tensile load and internal pressure are studied. This study provides a novel tension–pressure coupled failure boundary for the ultimate-limit-state design of deep-water risers, breaking through the single-load limitation of the existing literature and offering a theoretical basis for related engineering practice.

2. Theoretical Model of Mixed Failure in Unbonded Flexible Risers Under Axisymmetric Loads

Unbonded flexible risers are made up of multi-layer complex structures. Each layer has its own functional features, material properties and geometric cross-section characteristics. These layers interact with one another and contribute to the overall characteristics of the unbonded flexible riser [32]. Leveraging the geometric attributes of individual unbonded flexible riser strata, this research classifies layer structures into two broad structural families: cylindrical shell layers and helical layers.

2.1. Limitations of the Model

Given that the inter-layer gap in a new unbonded flexible riser is less than 0.05 mm, the energy dissipated by inter-layer friction is negligible, and the numerical model is only intended to predict the first-yield elastic limit, the theoretical model in this paper is obtained with the following theoretical assumptions:

1. Both ends of each layer are fixedly installed on the terminal device to form an integral whole, and it is postulated that the axial elongation rates of each layer are the same. It ensures that all layers are synchronized in axial deformation, so that the multi-layer structure can be regarded as a whole for mechanical analysis, which greatly simplifies the establishment of control equations.
2. The theoretical model does not consider the gap between layers in the initial calculation stage, so that the classical elastic mechanics theory can be applied to calculate the initial stress state.
3. The materials of each layer are homogeneous and change within the linear elastic range, permitting the application of the generalized Hooke's law.
4. Neglect the initial defects formed in the production process. The aim is to analyze the performance of an idealized, flawless structure so as to establish a performance upper bound benchmark.
5. Neglect the impact of the boundary effect and consider that the thickness of each layer changes consistently along the pipe's length and in the radial direction. It is beneficial to analyze the mechanical behavior of any cross-section of the pipeline independently without considering the influence of end constraints.
6. Ignore the energy caused by friction from the relative movement between the layers. It can simplify the derivation process of the energy method.
7. It is assumed that the pressure-armor layer can still support a certain level of internal pressure when it reaches yield stress, making the model closer to the actual engineering judgment.

2.2. Equilibrium Equations of Cylindrical Shell Layers Under Axisymmetric Loads

A cylindrical shell consists of the inner and outer sheath layers made of polymer materials. This structure serves primarily to segregate fluids inside and outside the riser, minimize inter-layer friction, and ward off wear on steel-layer components. For a more realistic description of the geometric shape and stress–strain characteristics, this paper takes into account the thickness change in the cylindrical shell layer and solves it based on the thick-walled cylinder theory. As shown in Figure 3, there is the cylindrical shell layer before and after deformation. It is assumed that the axisymmetric loads acting on the cylindrical shell layer include torque T, axial force F, and external and internal pressures P_o and P_i. The Cartesian coordinate system xyz represents the spatial position before deformation (the solid-line part), while the x′y′z′ coordinate system represents the spatial position after deformation (the dashed-line part). The corresponding variation amounts are expressed as the axial variation amount ΔL, the radial variation amount ΔR, the torsion angle $\Delta\phi$ around the central axis z, and the thickness variation amount Δt.

For homogeneous materials, $\gamma = \Delta\phi/L$, the calculation formula for the strain energy of the cylindrical shell layer, can be expressed as:

$$\begin{aligned} U &= \frac{1}{2}\int_{v}(\sigma_1\varepsilon_1+\sigma_2\varepsilon_2+\sigma_3\varepsilon_3+\tau_{12}\gamma_{12})dv \\ &= \frac{\pi}{2}[(\lambda+2G)(\varepsilon_1^2+\varepsilon_2^2+\varepsilon_3^2)+2\lambda(\varepsilon_1\varepsilon_2+\varepsilon_1\varepsilon_3+\varepsilon_2\varepsilon_3)+GR_m^2\gamma^2](R_o^2-R_i^2)L \end{aligned} \tag{1}$$

where the corresponding geometric strains include axial strain ε_1, radial strain ε_2, thickness strain ε_3 and torsional angle strain γ_{12}, which are defined as follows, as well as the stress expressions in each direction:

$$\begin{aligned} \varepsilon_1 &= \tfrac{\Delta L}{L} \\ \varepsilon_2 &= \tfrac{\Delta R}{R_m} \\ \varepsilon_3 &= \tfrac{\Delta t}{t} \\ \gamma_{12} &= R_m \tfrac{\Delta \phi}{L} \\ \sigma_1 &= \lambda e + 2G\varepsilon_1 \\ \sigma_2 &= \lambda e + 2G\varepsilon_2 \\ \sigma_3 &= \lambda e + 2G\varepsilon_3 \\ \tau_{12} &= G\gamma_{12} \end{aligned} \tag{2}$$

The calculation formula for the work performed by external forces is as follows, where ΔV_i and ΔV_o are the volume change amounts inside and outside the cylindrical shell, respectively:

$$W = F\Delta L + T\Delta\phi + P_i\Delta V_i - P_o\Delta V_o \tag{3}$$

Let the internal and external volumes before deformation be V_i and V_o, respectively, and let the volumes after being deformed by axisymmetric loads be V_i' and V_o', respectively. According to the spatial geometric relation, we can obtain [11]:

$$\begin{cases} V_i = \pi R_i^2 L \\ V_o = \pi R_o^2 L \\ V_i' = \pi(R_i + \Delta R - \frac{\Delta t}{2})^2(L + \Delta L) \\ V_o' = \pi(R_o + \Delta R + \frac{\Delta t}{2})^2(L + \Delta L) \end{cases} \tag{4}$$

According to Equation (3), ΔV_i and ΔV_o can be expressed as:

$$\begin{aligned} \Delta V_i &= \pi\left(R_i + \Delta R - \tfrac{\Delta t}{2}\right)^2(L + \Delta L) - \pi R_i^2 L \\ &\approx \pi(2R_i\Delta R - R_i\Delta t)L + \pi R_i^2\Delta L \\ &= \pi R_i L(2R_m\varepsilon_2 - t\varepsilon_3 + R_i\varepsilon_1) \end{aligned} \tag{5}$$

$$\begin{aligned} \Delta V_o &= \pi\left(R_o + \Delta R + \tfrac{\Delta t}{2}\right)^2(L + \Delta L) - \pi R_o^2 L \\ &\approx \pi(2R_o\Delta R + R_o\Delta t)L + \pi R_o^2\Delta L \\ &= \pi R_o L(2R_m\varepsilon_2 + t\varepsilon_3 + R_o\varepsilon_1) \end{aligned} \tag{6}$$

Substituting Equations (4) and (5) into Equation (2), the expression for the external force work can be obtained:

$$W = (F\varepsilon_1 + T\gamma)L + \pi P_i R_i L(2R_m\varepsilon_2 - t\varepsilon_3 + R_i\varepsilon_1) - \pi P_o R_o L(2R_m\varepsilon_2 + t\varepsilon_3 + R_o\varepsilon_1) \tag{7}$$

According to the principle of reversibility of work and energy [27] $\partial\Pi = \partial W - \partial U = 0$, the following can be obtained:

$$\frac{\partial U}{\partial\varepsilon_1} = \frac{\partial W}{\partial\varepsilon_1}\frac{\partial U}{\partial\gamma} = \frac{\partial W}{\partial\gamma}\frac{\partial U}{\partial\varepsilon_2} = \frac{\partial W}{\partial\varepsilon_2}\frac{\partial U}{\partial\varepsilon_3} = \frac{\partial W}{\partial\varepsilon_3} \tag{8}$$

The equilibrium equations of the cylindrical shell layer under axisymmetric loads are obtained from Equations (1), (6), and (7):

$$\begin{bmatrix} k_{11} & k_{12} & k_{13} & k_{14} \\ k_{21} & k_{22} & k_{23} & k_{24} \\ k_{31} & k_{32} & k_{33} & k_{34} \\ k_{41} & k_{42} & k_{43} & k_{44} \end{bmatrix} \begin{bmatrix} \Delta L/L \\ \Delta\phi/L \\ \Delta R/R_m \\ \Delta t/t \end{bmatrix} = \begin{bmatrix} F + \pi P_i R_i^2 - \pi P_o R_o^2 \\ T \\ 2\pi R_m (P_i R_i - P_o R_o) \\ -\pi t (P_i R_i + P_o R_o) \end{bmatrix} \quad (9)$$

where

$$\begin{array}{ll} k_{11} = \frac{\nu EA}{(1+\nu)(1-2\nu)} + \frac{EA}{1+\nu} & k_{12} = k_{21} = 0 \\ k_{13} = k_{31} = \frac{\nu EA}{(1+\nu)(1-2\nu)} & k_{14} = k_{41} = \frac{\nu EA}{(1+\mu)(1-2\nu)} \\ k_{22} = \frac{E}{2(1+\nu)} \cdot \frac{\pi}{2}(R_o^4 - R_i^4) & k_{23} = k_{32} = 0 \\ k_{24} = k_{42} = 0 & k_{33} = \frac{\nu EA}{(1+\nu)(1-2\nu)} + \frac{EA}{1+\nu} \\ k_{34} = k_{43} = \frac{\nu EA}{(1+\nu)(1-2\mu)} & k_{44} = \frac{\nu EA}{(1+\nu)(1-2\nu)} + \frac{EA}{1+\nu} \end{array}$$

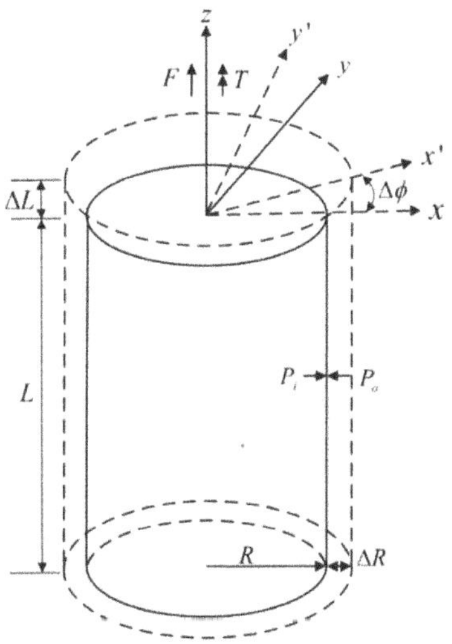

Figure 3. Diagram of the cylindrical layer under axisymmetric loads.

2.3. *Equilibrium Equation of Helical Layer Under Axisymmetric Load*

The helical layer serves as the primary structural configuration within unbonded flexible risers. This structural form includes the anti-friction layer made of polymer materials, as well as the carcass layer, the pressure-armor layer, and the tensile-armor layer made of metal materials. The force relationship and spatial geometric deformation of the spiral belt under the action of external loads are shown in Figures 4 and 5.

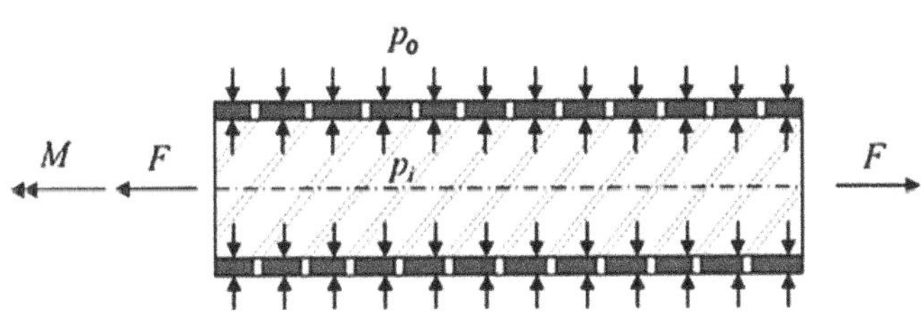

Figure 4. Loading condition of helical layer.

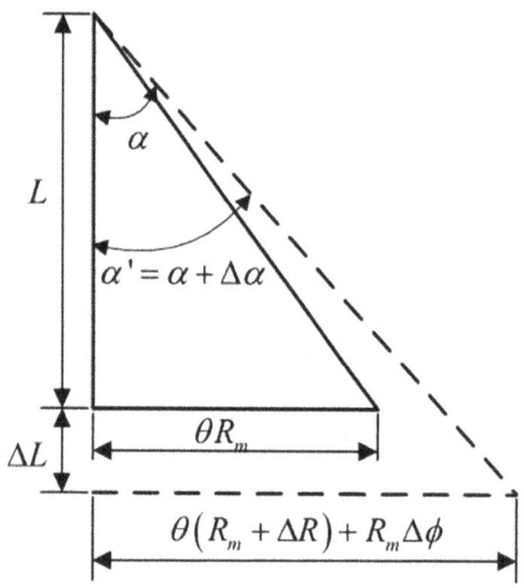

Figure 5. Profiles of helical layer.

According to Equation $\varepsilon = (S' - S)/S$, the strain of the spiral belt along its axial direction can be expressed as [11,32]:

$$\varepsilon_a = \cos^2\alpha \cdot \frac{\Delta L}{L} + R_m \sin\alpha\cos\alpha \cdot \frac{\Delta\phi}{L} + \sin^2\alpha \cdot \frac{\Delta R}{R_m} \tag{10}$$

When subjected to bending loads or large axisymmetric loads, the radial deformation of the spiral belt is sometimes non-negligible, so the cylindrical shell layer model is used for simulation [11,12]. For structures with regular rectangular cross-sections such as tensile-armor layers, the strain change in the thickness direction also needs to be considered, as shown in Figure 6 [11].

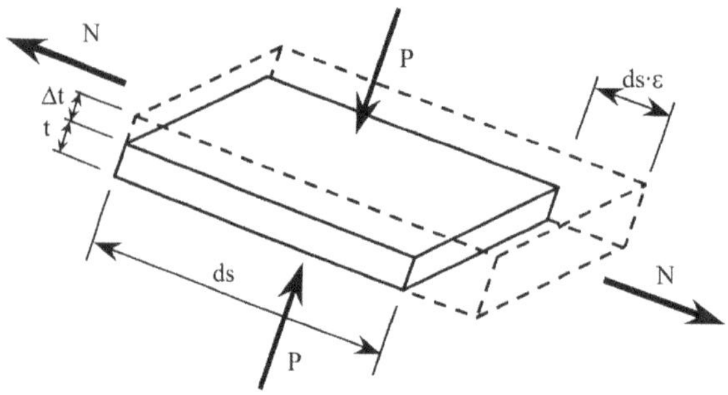

Figure 6. Helical element with axial and radial strains.

The strain of the spiral belt in the thickness direction can be expressed as [6]:

$$\varepsilon_r = \frac{\Delta t}{t} \tag{11}$$

According to the calculation formula of strain energy for a single helical belt when considering the thickness change [33], the internal energy calculation formula for a helical layer with n helical belt can be obtained:

$$\begin{aligned}
U &= nU_s \\
&= \frac{1}{2}\frac{An}{\cos\alpha}\int_L (\sigma_a\varepsilon_a + \sigma_r\varepsilon_r)dz \\
&= \frac{1}{2}\frac{An}{\cos\alpha}\frac{E}{1-\nu^2}\begin{bmatrix} \cos^4\alpha\cdot\varepsilon_1^2 + R_m^2\sin^2\alpha\cos^2\alpha\cdot\gamma^2 + \sin^4\alpha\cdot\varepsilon_2^2 + \\ 2R_m\sin\alpha\cos^3\alpha\cdot\varepsilon_1\gamma + 2\sin^2\alpha\cos^2\alpha\cdot\varepsilon_1\varepsilon_2 + \\ 2R_m\sin^3\alpha\cos\alpha\cdot\gamma\varepsilon_2 + 2\nu\cos^2\alpha\cdot\varepsilon_1\varepsilon_3 + \\ 2\nu R_m\sin\alpha\cos\alpha\cdot\gamma\varepsilon_3 + 2\nu\sin^2\alpha\cdot\varepsilon_2\varepsilon_3 + \varepsilon_3^2 \end{bmatrix} L
\end{aligned} \quad (12)$$

The volume changes inside and outside the tensile-armor layer can be equivalent to the volume changes in the cylindrical shell layer, and the work performed by external forces can be expressed by Equation (7). According to the principle of reversibility of work and energy, the equilibrium equation of the helical layer considering thickness changes under axisymmetric loads can be obtained [11]:

$$\begin{bmatrix} k_{11} & k_{12} & k_{13} & k_{14} \\ k_{21} & k_{22} & k_{23} & k_{24} \\ k_{31} & k_{32} & k_{33} & k_{34} \\ k_{41} & k_{42} & k_{43} & k_{44} \end{bmatrix}\begin{bmatrix} \Delta L/L \\ \Delta\phi/L \\ \Delta R/R_m \\ \Delta t/t \end{bmatrix} = \begin{bmatrix} F + \pi P_i R_i^2 - \pi P_o R_o^2 \\ T \\ 2\pi R_m(P_i R_i - P_o R_o) \\ -\pi t(P_i R_i + P_o R_o) \end{bmatrix} \quad (13)$$

where

$$\begin{array}{ll}
k_{11} = \frac{nAE}{1-\nu^2}\cos^3\alpha & k_{12} = k_{21} = \frac{nEAR_m}{1-\nu^2}\sin\alpha\cos^2\alpha \\
k_{13} = k_{31} = \frac{nEA}{1-\nu^2}\sin^2\alpha\cos\alpha & k_{14} = k_{41} = \frac{nEA\nu}{1-\nu^2}\cos\alpha \\
k_{22} = \frac{nEAR_m^2}{1-\nu^2}\sin^2\alpha\cos\alpha & k_{23} = k_{32} = \frac{nEAR_m}{1-\nu^2}\sin^3\alpha \\
k_{24} = k_{42} = \frac{nEAR_m\nu}{1-\nu^2}\sin\alpha & k_{33} = \frac{nEA}{1-\nu^2}\frac{\sin^4\alpha}{\cos\alpha} \\
k_{34} = k_{43} = \frac{nEA\nu\sin^2\alpha}{1-\nu^2}\cos\alpha & k_{44} = \frac{nEA}{(1-\nu^2)\cos\alpha}
\end{array}$$

2.4. Overall Equilibrium Equations Under Axisymmetric Loads

Since the two ends of the riser are fixedly installed on the terminal devices, forming an integral whole, it is assumed that each layer has the same amount of change. Consequently, the equilibrium equations of the cylindrical shell layer and the helical layer, considering the thickness change under the action of axisymmetric loads, can be obtained:

$$\begin{cases} k_{31}^n\frac{\Delta L}{L} + k_{32}^n\frac{\Delta\phi}{L} + k_{33}^n\frac{\Delta R^n}{R_m^n} + k_{34}^n\frac{\Delta t^n}{t^n} = 2\pi R_m^n(P_i^n R_i^n - P_o^n R_o^n) \\ k_{41}^n\frac{\Delta L}{L} + k_{42}^n\frac{\Delta\phi}{L} + k_{43}^n\frac{\Delta R^n}{R_m^n} + k_{44}^n\frac{\Delta t^n}{t^n} = -\pi t_p^n(P_i^n R_i^n + P_o^n R_o^n) \end{cases} \quad (14)$$

When the n-th layer and the (n + 1)-th layer are in contact under axisymmetric load, the spatial geometric relationship of adjacent layers can be obtained according to the geometric continuity condition [11,12]:

$$R_m^n + \Delta R^n + \frac{t^n + \Delta t^n}{2} = R_m^{n+1} + \Delta R^{n+1} - \frac{t^{n+1} + \Delta t^{n+1}}{2} \quad (15)$$

According to the cylindrical shell layer, helical layer, overall equilibrium equations and the relationship between adjacent layers, the inter-layer contact pressure and strain of unbonded flexible risers can be solved. When the obtained inter-layer contact pressure is negative, the corresponding relationship between adjacent layers needs to be modified until all inter-layer contact pressures are not negative before the solution can be obtained.

2.5. Failure Model

2.5.1. Tensile Failure Model

Both the positive stress on the section caused by the bending of the steel wire and the sectional shear stress caused by the torsion of the steel wire are far less than the normal stress σ_N caused by the tension of the steel wire. Therefore, it can be approximately considered that the stress on the cross-section of the armored steel wire during the pipeline tension is only caused by the tension. When subjected to tensile load, the strain is:

$$\varepsilon_N = \sqrt{1 + \frac{\Delta h}{\pi R} \sin\alpha \cos\alpha - 1} \tag{16}$$

The normal stress induced by tension is:

$$\sigma_N = E(\sqrt{1+\delta} - 1) \tag{17}$$

The ultimate tension of unbonded flexible risers can be obtained:

$$F = \sum_{i=1}^{m} n_i \sigma_i A_i \frac{1}{\sqrt{1 + \frac{\tan^2 \alpha_i}{\left(1 + \frac{\Delta h_{\max}}{2\pi R_i} \tan\alpha_i\right)^2}}} \tag{18}$$

2.5.2. Internal Pressure Failure Model

Considering the spiral-band structure characteristics of unbonded flexible risers, when the tensile and internal pressure conditions are comprehensively considered, the calculation formulas for the stress and radial pressure of unbonded flexible risers are as follows, where σ_T represents the section stress of the pressure-armor layer:

$$p = \frac{n\sigma_T A \sin^2\alpha}{2\pi r_i r_m \cos\alpha} \tag{19}$$

The ultimate internal pressure of pressure-armor layer can bear can be obtained:

$$p = \frac{t}{r_m} F_f \sigma_u \tag{20}$$

3. Numerical Simulation

3.1. Finite Element Model of Unbonded Flexible Riser

The 2.5-inch 8-layer unbonded flexible riser model proposed by Witz [2] is adopted as the research subject in this paper. A full-scale finite-element model is established in Abaqus (2022) software of 6.13 version using a layered modeling approach, where both the carcass layer and the pressure-armor layer are incorporated with detailed geometric configurations. Specifically, Figure 7 depicts the comprehensive cross-section geometry of the pressure-armor layer, the total length is approximately 14.5 mm with the other corresponding geometric properties presented in Table 1. In addition, the cross-sectional shape of the spiral strips in the tensile-armor layer is a regular rectangle, and the model in this paper is designed with two layers of tensile armor, where the inner and outer tensile-armor strata exhibit counter-directional lay angles. Meanwhile, technical specifications detailing constitutive and dimensional attributes for all individual layers within the unbonded flexible riser appear in Table 2.

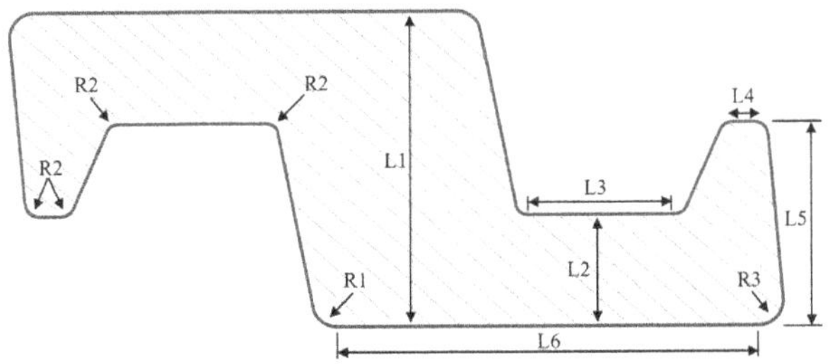

Figure 7. Cross-section properties of pressure-armor layer.

Table 1. Geometric parameters of pressure-armor layer.

Parameter	Value	Parameter	Value	Parameter	Value
L_1	6.2 mm	L_4	0.5 mm	R_1	1.0 mm
L_2	2.2 mm	L_5	4.0 mm	R_2	0.5 mm
L_3	3.0 mm	L_6	8.5 mm	R_3	1.0 mm

Table 2. Geometric and material parameters of an 8-layer unbonded flexible riser [6].

Layer Number	Layer Type	Section Size (mm^2)	Inner Radius (mm)	Outer Radius (mm)	Number of Tendons	Laying Angle (°)	Material
1	Carcass layer	28 × 0.7	31.60	35.10	1	−87.5	AISI 304
2	Pressure sheath	-	35.10	40.00	-	-	Nylon 12
3	Pressure armor layer	9.25 × 0.6	40.05	46.25	2	−85.5	FI-15
4	Anti-friction tape	-	46.25	47.75	-	-	Nylon 11
5	Inner tensile armor	6 × 3	47.75	50.75	40	−35.0	FI-41
6	Anti-friction tape	-	50.75	52.25	-	-	Nylon 11
7	Outer tensile armor	6 × 3	52.25	55.25	44	35.0	FI-41
8	Fabric tape	-	55.25	55.75	-	-	-

The carcass layer, allowing oil and gas resources as well as seawater to flow freely in and out of this layer, is a self-locking, non-watertight structure formed by winding S-shaped spiral steel strips. It mainly bears the external liquid pressure [12]. In order to prevent the riser from crushing failure under pressure [34], oil and gas will have a direct impact on the inner sheath layer instead of the carcass layer. Therefore, the model can be simplified into a 7-layer model with the carcass layer removed for numerical simulation (as shown in Figure 8). The computational validity of the simplified model will be analyzed in the following sections. Meanwhile, the finite-element representation of the unbonded flexible riser must incorporate adequate length to effectively minimize boundary disturbance effects. Since the tensile-armor layer is the component that bears the primary mechanical properties in the riser, the length of the numerical model in this study is set to approximately twice the pitch length of the outer tensile-armor layer of the unbonded flexible riser, specifically 1.0 m.

To precisely document the stress distribution throughout the entire riser model and the specific deformation of each layer, while eliminating the impact of the hourglass issue on computational outcomes, an 8-node linear incompatible element (C3D8I) is employed for meshing the models of each layer. This element is more capable of managing plasticity, stress hardening, significant deformation, and substantial strain. The numerical model in this paper contains the pressure-armor layer structure considering the actual cross-sectional geometric shape, which involves a large number of geometric nonlinearity and contact nonlinearity problems. Therefore, in order to make the numerical calculation results of the finite-element software converge effectively, the explicit algorithm (ABAQUS) is adopted for solution in this paper.

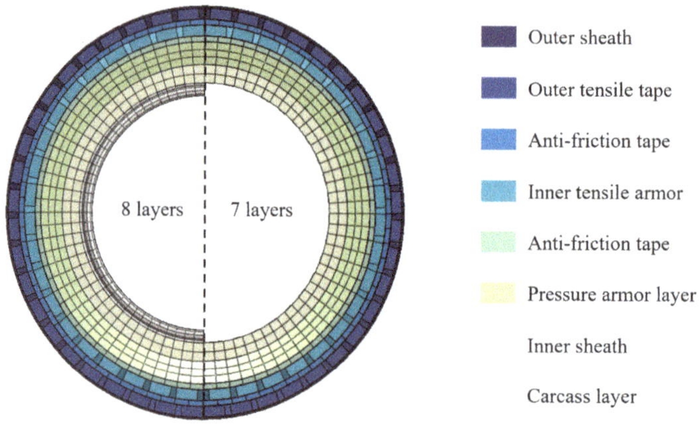

Figure 8. Cross-section of an unbonded flexible riser FE model.

Two geometric reference points—RP-1 and RP-2—are positioned at both ends of the model's central axis (see Figure 9) to aid in applying external loads to the riser's numerical model and controlling the model's boundary conditions with greater ease. Kinematic constraints are imposed across all degrees of freedom for individual layer components at prescribed end-section locations. Additionally, comprehensive constraints on the end faces are imposed on the two reference points: RP-1 is restricted in the torsional direction around the Z-axis, whereas RP-2 is fully constrained across all degrees of freedom (see Figure 10). For load application, the model's axial tensile load is directly exerted on RP-1, and through applying a uniformly distributed radial pressure to the inner surface of the inner sheath, the internal pressure load is realized.

Figure 9. Reference points of finite-element model of unbonded flexible riser.

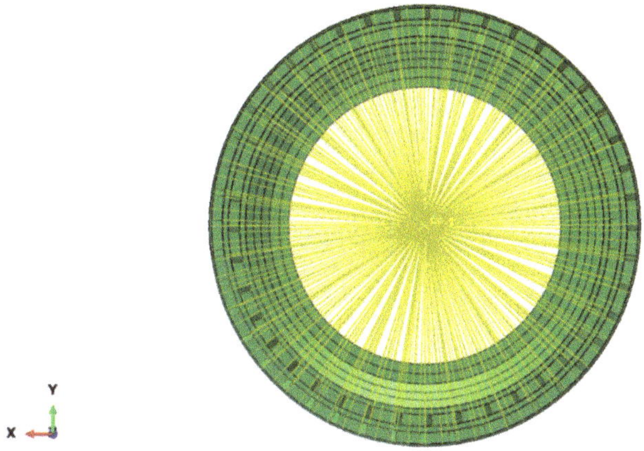

Figure 10. Kinematic constraints applied to the riser ends at RP-2.

Given the presence of extensive nonlinear contacts both between layers and within layer structures in the numerical model, and in consideration of the model's convergence, the explicit solver in ABAQUS is employed for computational analysis. Meanwhile, since inertial effects may affect the calculation results, it is necessary to regulate the kinetic energy during the computation process. In this study, the influence of inertial effects is mitigated through quasi-static loading. Specifically, the loading duration is extended, and a smooth step amplitude curve is adopted for loading, with the loading time set to 0.3 s.

As contact problems lack continuity-contact constraints and can only be introduced when surfaces are in contact with each other—the general contact algorithm [35] is utilized herein. This algorithm features a relatively simple definition and can automatically search for mutually contacting surfaces during computation. The normal contact property is set to hard contact, and the general function method is employed to enhance constraints, meaning separation occurs when the contact pressure between interfaces is no greater than zero. The tangential property is simulated using Coulomb friction.

3.2. Failure Criterion

3.2.1. Determination of Failure of Tensile-Armor Layer

When subjected to excessive axial tensile force, the tensile-armor layer experiences failure. Its yield strength is set as 650 MPa, with the ultimate strength specified as 1100 MPa. Using the simplified 7-layer riser model, various internal pressures are imposed, alongside a sufficiently high axial tensile load, to ensure the tensile-armor layer undergoes tensile failure. Failure of the tensile-armor component under tension is characterized by its axial load-elongation curve, stress contours, and strain contours. During the elastic phase, the stress distribution within the tensile-armor layer is relatively uniform; as the stress enters the plastic region, it gradually becomes concentrated. It is considered that once the stress in the tensile-armor layer reaches the plastic region, the layer stays in a compressed state until the element attains the ultimate stress and undergoes significant deformation.

3.2.2. Determination of Failure of Pressure–Tensile-Armor Layer

Internal pressure-induced bursting refers to the failure of the pressure-armor layer caused by excessive internal pressure. The pressure-armor layer is assigned a yield strength of 300 MPa and an ultimate strength of 600 MPa [36]. Employing the simplified 7-layer riser model, under varying axial tensile loads, a sufficiently high internal pressure is imposed to induce bursting failure in the pressure-armor layer. This bursting failure is characterized by the pressure-armor layer's internal pressure-axial tensile deformation curve, stress contours, and strain contours.

In the linear stress phase of the pressure armor layer, the distribution of stress variations is relatively uniform. During the elastic phase, the stress distribution within the pressure-armor layer remains relatively even; as stress transitions into the plastic phase, it gradually accumulates in localized regions. Assumptions hold that once stress in the pressure-armor layer enters the plastic phase, the layer retains a compressed state—though to a far smaller degree—until the element attains ultimate stress and experiences significant deformation.

4. Discussion

The working conditions of unbonded flexible risers in the deep-sea environment are extremely complex. Their structures need to bear not only a single load but also the combined effects of multiple combined loads and internal and external pressures simultaneously. Thus, it is crucial to systematically study the mechanical responses of risers under combined loads. In the following paragraphs, the response characteristics and mechanical performance of risers under the action of internal pressure and axial tensile loads will be introduced in detail.

4.1. Model Verification

Building on Witz's experiments, this study calculated the numerical and theoretical results of the unbonded flexible riser under tensile loading, with the outcomes presented in Figure 11. Moreover, Figure 11 also presents the average axial tensile stiffness that other scholars and organizations have predicted. The variations between the theoretical and numerical approaches are nearly linear, exhibiting a high degree of consistency. Meanwhile, with the average tensile stiffness derived from experimental methods as the benchmark, the numerical approach is more capable of predicting the axial tensile stiffness of the unbonded flexible pipe. Specifically, the relative deviations of the results obtained via the numerical and theoretical approaches (against the experimental data) are 8.07% and 14.81%, respectively.

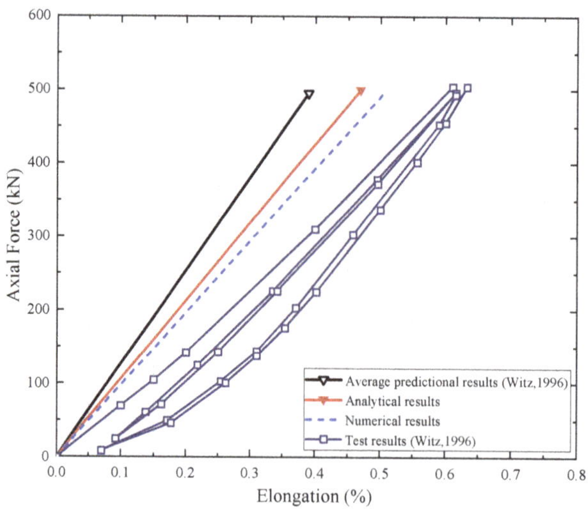

Figure 11. Axial force-elongation curves of different techniques [6].

Subsequent studies by various scholars have reported tensile stiffness values of unbonded flexible risers obtained using different methods. A comparison between these values and the finite-element model stiffness of 99.12 MN proposed in this study is presented in Table 3. The results show relatively small deviations, further demonstrating the validity of the model proposed in this study.

Table 3. Axial tensile stiffness values obtained from different organizations and scholars.

Axial Tensile Stiffness	Relative Deviation	References
86.8 MN	12.4%	Numerical (Lei et al., 2023) [37]
108.7 MN	9.6%	Numerical (Tang et al., 2019) [21]
105.1 MN	6%	Experiment (de Sousa et al., 2012) [38]
127 MN	28.1%	Analytical (Ramos, 2008) [39]

Building upon the verification of the validity of numerical and theoretical models when subjected to tensile loads, this paper compares the theoretical and numerical results of the axial elongation rate of the pressure-armor layer of the riser under the action of internal pressure (as shown in Figure 12). It is observed that the two match well in the linear stage. Consequently, under the action of internal pressure, the behavior of unbonded flexible risers can be effectively predicted through the numerical and theoretical methods put forward in this paper.

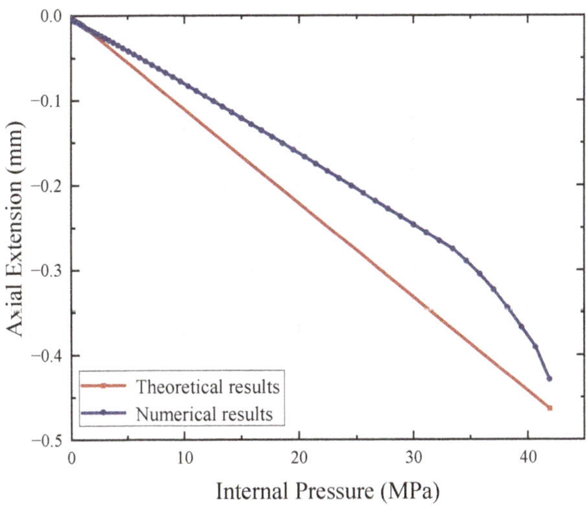

Figure 12. Relationship between internal pressure and axial shift.

Furthermore, this study contrasts the internal pressure–axial elongation ratios of the full-layer model and the simplified numerical model. Analysis reveals that, prior to the yielding of the pressure-armor layer, the computational outcomes of the two models exhibit a near-linear trend (as shown in Figure 13) and show high consistency. As internal pressure continues to rise, the pressure-armor layer undergoes plastic deformation, with deviations between the two models emerging once the internal pressure reaches 59 MPa. The primary cause lies in the self-locking properties of the pressure-armor layer: its deformation induces corresponding deformation in the carcass layer, yet the theoretical model fails to account for the pressure-armor layer's self-locking structure, thereby giving rise to discrepancies between the models. Overall, under internal pressure, the carcass layer's deformation exerts minimal influence on the overall performance of the unbonded flexible riser, which confirms the validity of the 7-layer simplified numerical model.

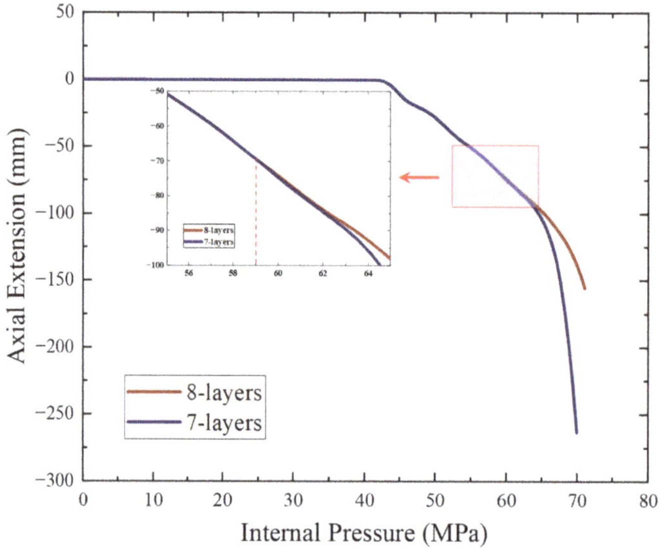

Figure 13. Comparison of internal pressure–axial extension elongation curves by full-layer model and simplified numerical model.

Based on the above research, the 7-layer simplified model has significant computational efficiency advantages in the engineering linear design stage and the research focusing on the coupling failure of the sheath, and the accuracy meets the requirements. The 8-layer full model is a necessary choice for nonlinear limit analysis, fine failure mechanism research, and the simulation of lining behavior.

4.2. Failure Characteristic

4.2.1. Tensile Failure Characteristics

The ability of unbonded flexible risers to resist deformation under loads can be evaluated by their equivalent tensile, compressive, torsional, and bending stiffnesses. Axial tensile stiffness is a crucial mechanical property of unbonded flexible risers. When subjected to tensile loads, if the axial elongation of the flexible pipe is ε, the expression for the equivalent tensile stiffness K is as follows:

$$K = F/\varepsilon \tag{21}$$

Figure 14 shows the cross-sectional stress distribution in the riser when the internal tensile-armor layer in Case 1 reaches the ultimate stress. Critically, the results reveal a highly non-uniform stress profile. As expected, the vast majority of the tensile load is carried by the inner and outer tensile-armor layers, which validates their primary function within the flexible pipe structure. More importantly, the identification of the stress concentration at the end of the tensile-armor wire is a key finding. This observation not only aligns with the established structural design principles of flexible risers but, crucially, confirms the predictive capability and scientific rationale of our simplified FE model. It demonstrates that the model successfully captures the critical failure mechanism, thereby justifying its use for this type of analysis and providing confidence in the subsequent results derived from it.

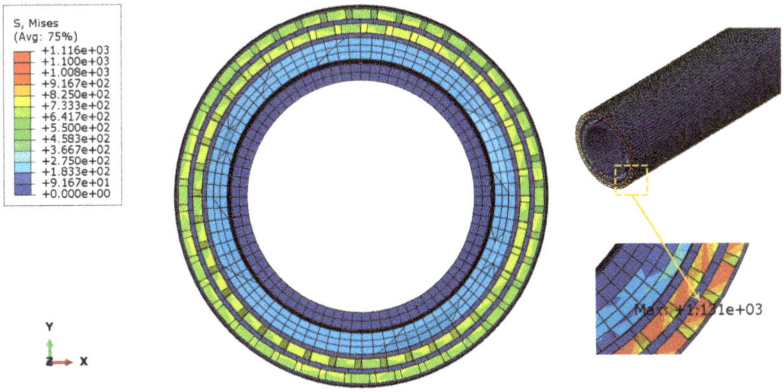

Figure 14. Map of von Mises stresses on the cross-section of simplified FE model.

Figure 15a,b, respectively, present the Mises stress states of the internal and external tensile armors in the unbonded flexible riser in Case 1. Evidently, the stress in the internal tensile armor is greater than that in the external one, making the internal tensile armor more prone to failure. This may result in the initial instability of the inner layer [10]. This insight into the failure sequence is critical for prioritizing integrity monitoring and designing layered protection systems. Furthermore, when subjected to axial tension, the stress of the helical layer cannot change strictly and evenly along the riser's axial direction. The stress distribution at the end of the helical layer is non-uniform, and especially more obvious in the external tensile armor, as shown in Figure 15b. In contrast, the stress distribution in the middle part of the external tensile armor is relatively uniform, with an average stress of approximately 550 MPa, while the average stress of some end elements is only about 360 MPa. This suggests that a portion of the load is transferred to adjacent layers or components at the terminus, creating stress concentrations elsewhere. Therefore, these results underscore the paramount importance of optimizing end-fitting design to mitigate stress concentrations.

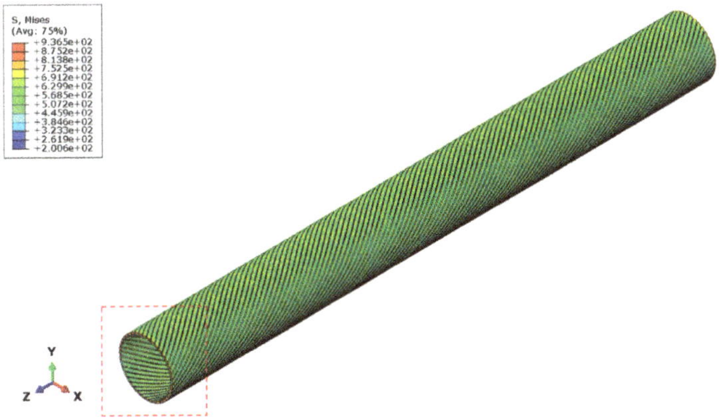

(a)

Figure 15. *Cont.*

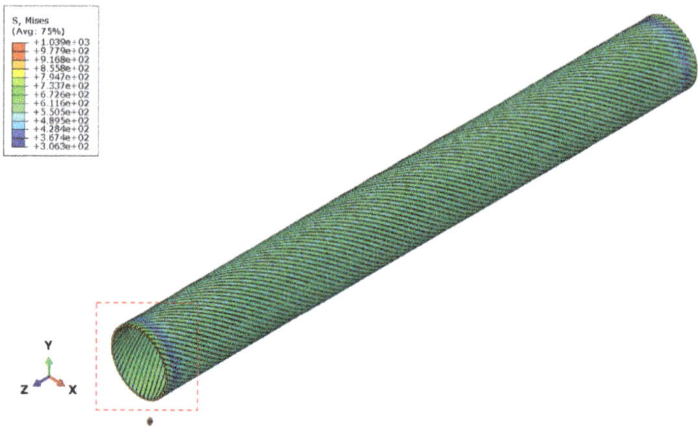

(**b**)

Figure 15. Distribution of von Mises stress in the tensile-armor layer under tensile load. (**a**) The distribution of von Mises stress in the internal tensile armor; (**b**) the distribution of von Mises stress in the external tensile armor.

4.2.2. Characteristics of Burst Failure

The primary component in unbonded flexible risers responsible for withstanding internal pressure is pressure-armor layer. With variations in other loads, the effect of internal pressure on the riser alters accordingly. Additionally, unlocking constitutes the most prevalent failure pattern of the pressure-armor layer.

Figure 16 reveals substantially greater stress magnitudes along the pressure-armor stratum's internal contour compared to its external surface. Moreover, the stress of the Z-shaped cross-section gradually decreases from bottom to top, which conforms to the general law of stress change along the cross-section thickness direction when the internal pressure acts. This stress distribution is critical, as it indicates that failure initiation is most likely to occur at the inner surface, guiding future inspection and failure analysis focus.

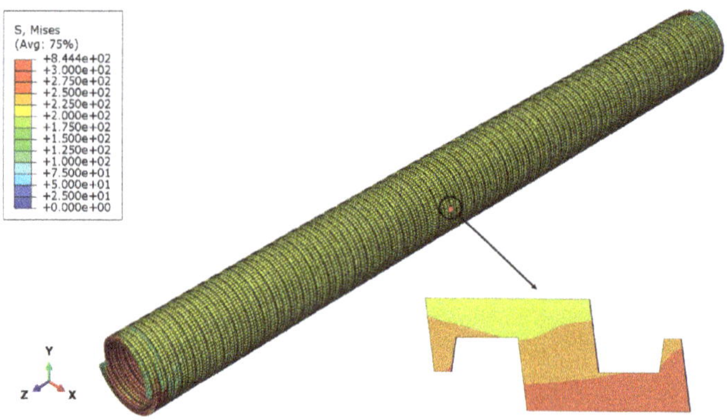

Figure 16. Contours of the pressure-armor layer.

When the internal pressure reaches 42 MPa, the entire cross-section of the pressure-armor layer reaches the yield stress (as shown in Figure 17). The pressure-armor layer, with its self-locking structure, maintains a uniform axial deformation of the pipeline after basically reaching the yield stress, unlike the tensile-armor layer, which prevents stress concentration or large local deformations. This behavior is fundamentally different from a conventional ductile pipe and highlights the ingenious design of the Z-shaped profile: even in a plastic state, the interlocking mechanism prevents a catastrophic rupture by redistributing the load and maintaining structural integrity.

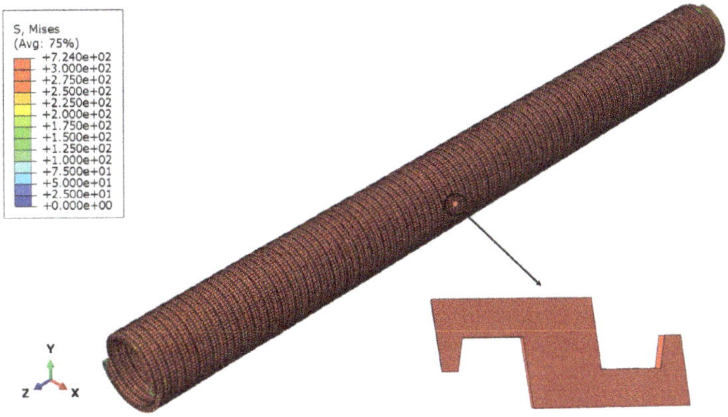

Figure 17. Contours of the pressure-armor layer at yield stress.

Figure 18 illustrates the evolution of von Mises stress in the inner and outer tensile-armor layers as a function of increasing internal pressure. When the pressure-armor layer yields (i.e., when the internal pressure reaches 42 MPa), the von Mises stresses of the tensile-armor layer increases sharply. This abrupt stress escalation signifies a fundamental shift in the load path: after the failure of the primary pressure-containing component (the pressure-armor layer), the tensile-armor layers are forced to absorb the additional hoop stress through their helical geometry. This verification further establishes that the tensile-armor layer assumes primary responsibility for resisting internal pressurization following failure of the pressure-armor layer. Furthermore, under sustained internal pressure loading, the internal tensile-armor layer continues to carry significantly greater loads than its external counterpart. This sudden load shift underscores the importance of considering this failure cascade in safety assessments. The system's integrity post pressure-armor failure relies entirely on the residual capacity of the tensile armor, a scenario that must be accounted for in defining ultimate limit states and assessing overall system safety factors.

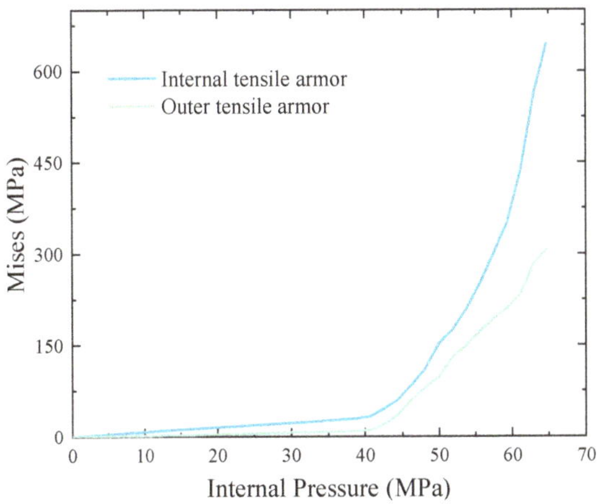

Figure 18. von Mises–internal pressure relation diagram of tensile-armor layer.

4.3. Parameter Analysis

4.3.1. Tensile Mechanical Response of Risers Under Combined Loads

To study the influence of internal pressure on axial tensile failure, excessive axial tensile force is applied under different constant internal pressures to cause the tensile-armor layer to stretch and fail. The failure of the tensile-armor layer can be identified from the axial-load versus axial-elongation curve by locating the abrupt inflection point where the axial load changes significantly during loading. The working conditions are set as follows: the internal pressure ranges from 0 to 40 MPa, with an interval of 10 MPa; an axial tensile displacement load is applied.

Figure 19 presents the tensile characteristics of the riser under different initial internal pressure conditions. The results show that the application of initial internal pressure will make the riser experience axial shortening first. Then, when an axial tensile force is applied, risers under different internal pressures start to elongate linearly from different starting points. When the internal pressure is 40 MPa, the initial axial elongation rate is −0.034% until it returns to its original length. Notably, the pre-applied internal pressure has no significant impact on the slope of the axial elongation rate of the riser. This means that the initial internal pressure has almost no influence on the stiffness of the flexible pipe, and its stiffness remains 99.4 MN. The ability of the riser to resist deformation under axial loads has not changed. After reaching the ultimate stress, the riser fails. This observation—that pre-applied internal pressure has limited influence on the axial stiffness—is consistent with the findings of Sun et al. [20] and Tang et al. [21], who also reported that the axial tensile stiffness of unbonded flexible risers remains relatively stable under varying internal pressures.

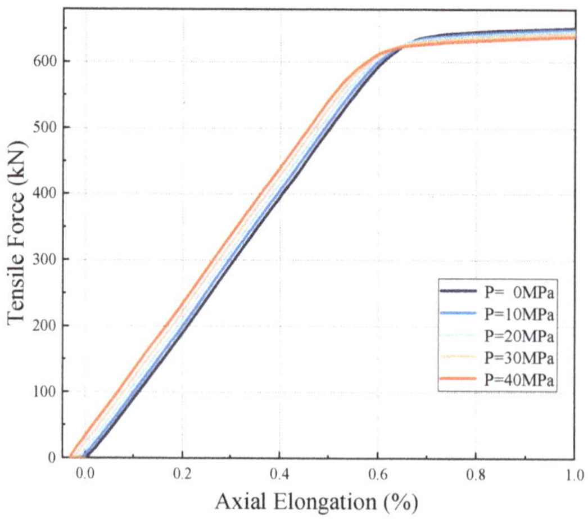

Figure 19. Tensile force versus axial relative elongation of the flexible pipe in different circumstances.

Figure 20 presents the axial tensile forces which lead to the yielding of the inner tensile-armor layer under diverse initial internal pressure conditions. The analysis results show that with the increase in the initial internal pressure, the axial tensile force that the inner tensile-armor layer in the riser can endure exhibits a gradually decreasing trend, dropping from 593 kN when the internal pressure is 0 MPa to 574 kN when the internal pressure is 40 MPa, which decreases by 3.2%. Although the application of the initial internal pressure decreases the deformation amount of the riser to a certain degree, it also renders the riser more likely to experience axial failure simultaneously.

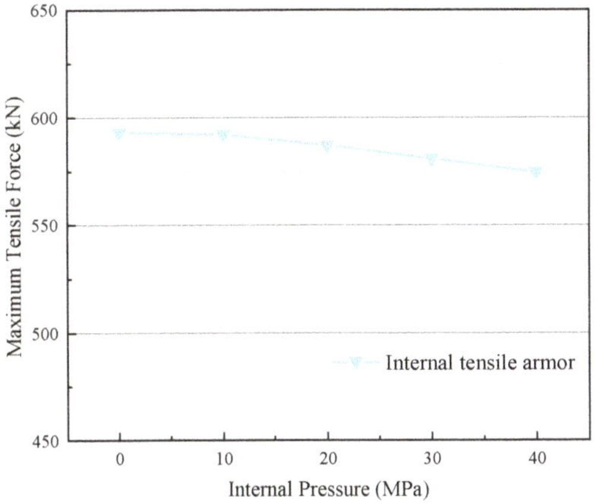

Figure 20. The maximum tensile force that the internal tensile armor can withstand under different internal pressure.

4.3.2. Pressure Mechanical Response of Risers Under Combined Loads

To study the influence of axial tensile load on internal burst failure, specific calculation working conditions were set as follows: the internal pressure was P = 100 MPa and the axial tensile load was T = 0–500 kN, with an interval of 100 kN. Meanwhile, the stress concentration caused by the end-boundary effect was neglected, and the stress distribution is consistent and relatively uniform during the internal pressure loading process.

Figure 21 indicates that, under the initial axial tensile force, the riser experiences axial elongation. Once the internal pressure begins to act, the riser then shows a trend of linear shortening. Evidently, the initially applied axial tensile force has no substantial impact on the slope of the axial shortening rate of the riser. Moreover, under different tensile forces, the change in the axial shortening rate of the unbonded flexible riser as the internal pressure rises remains the same.

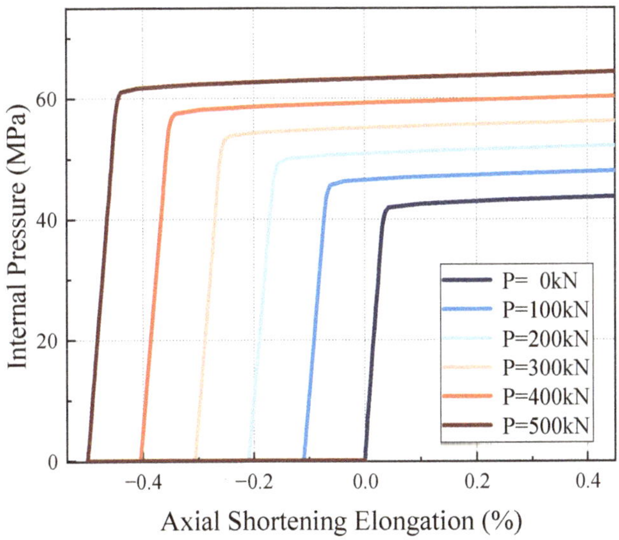

Figure 21. Internal pressure versus axial shortening elongation of the flexible pipe in different circumstances.

Figure 22 extracts the stress at the intermediate position when the pressure-armor layer yields. The analysis results demonstrate that the maximum internal pressure that the pressure-armor layer can bear when it yields is significantly affected by the initial axial tensile force. Compared with the situation of no initial tensile force, the failure pressure increases by 35% when the tensile force is 500 kN. As the axial tensile force keeps increasing, the internal pressure value that the pressure-armor layer can endure rises accordingly. This, in turn, increases its failure threshold and exhibits a higher anti-failure ability. This significant enhancement reveals a previously underappreciated synergistic effect between tension and pressure in the failure regime. While recent studies have advanced our understanding of stiffness under combined loads [20,21], they did not extrapolate these findings to predict the ultimate failure boundary. Our work demonstrates that the initial axial tensile force pre-tensions and tightens the helical layers, improving their radial stiffness and resistance to the ovalization that precedes burst failure.

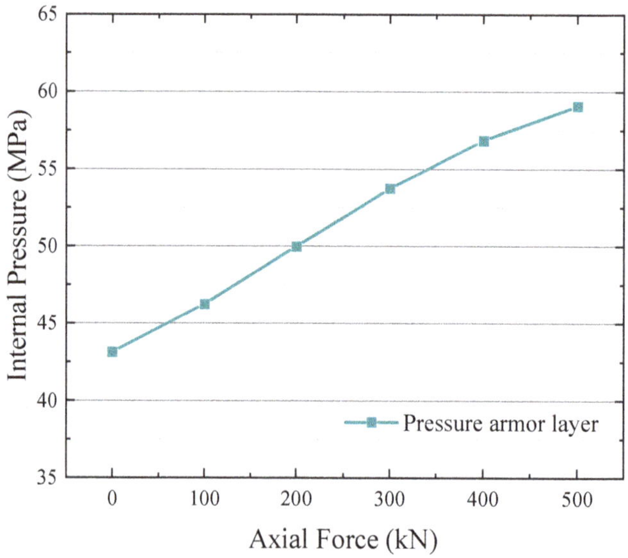

Figure 22. Internal pressure versus axial force at yield.

Current specifications generally adopt an "internal-pressure-dominant" curve for burst-limit verification, treating tension only as an additional membrane force that reduces the allowable value. The discovery that axial tension can enhance burst capacity by up to 35% challenges this paradigm. This suggests that for risers operating in high-tension environments (e.g., deep-water applications with large hang-off weights), the current design margins for burst failure may be overly conservative. Our proposed tension–pressure coupled failure boundary (as illustrated in Figure 22) provides a more accurate and potentially more efficient design envelope. By adopting this new boundary, engineers could optimize the cross-section design, potentially reducing material usage and overall weight without compromising safety, leading to significant cost savings in material and installation.

Figure 23 presents the stress contours of the pressure-armor layer at the time of riser failure under different initial axial tensile forces (in the x-y plane). It is observable that with the increase in internal pressure, the two ends of the pressure-armor layer are the first to reach the ultimate stress (600 MPa). As the initial axial tensile load rises, the two ends gradually reach the ultimate stress, and the area range reaching the ultimate stress also expands accordingly, ultimately leading to burst failure. Notably, when the initial tensile force is at 200 kN, the left end reaches the ultimate stress first. When it is at 400 kN, the right end first reaches the ultimate stress. Even at 500 kN, while the stress at the left end has not increased significantly, the right end has already burst and failed. This phenomenon implies that as the initial axial tensile force increases, the area that first reaches the ultimate stress shifts between the two ends of the armor layer, which is of great significance for the research on the riser failure mode. The migrating failure point can be rationalized by considering the boundary constraints imposed by the end fittings and the evolution of inter-layer contact pressure. Increasing axial tension redistributes the radial contact forces between the armor and the underlying carcass/pressure sheath, thereby shifting the position where local bending plus membrane stress is maximized. Simultaneously, the end fittings introduce highly localized radial restraint; the superposition of this fixed boundary effect with the moving "hoop-versus-bending" balance causes the hotspot to drift from one end to the other.

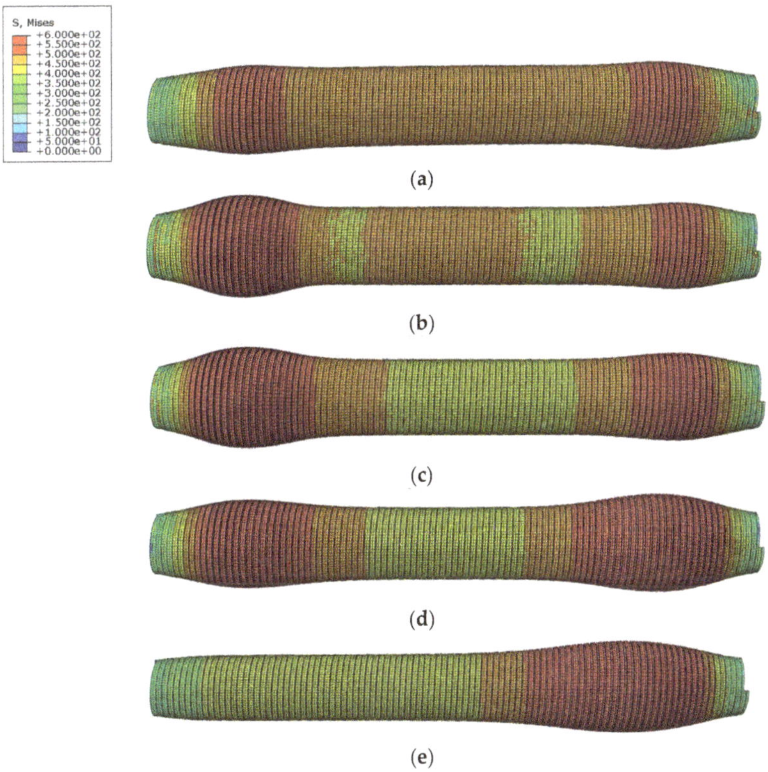

Figure 23. Distribution of von Mises stress in the pressure-armor layer. (**a**) T = 100 kN; (**b**) T = 200 kN; (**c**) T = 300 kN; (**d**) T = 400 kN; (**e**) T = 500 kN.

The observed migration of the failure initiation point is not merely an academic curiosity but has profound consequences for integrity management. In practical terms, this means that the location most susceptible to failure is not fixed but shifts with the load history. This insight critically informs risk-based inspection (RBI) strategies. Traditional inspection plans might focus on a presumed critical zone, but our findings indicate that inspections must account for the prevailing load conditions to correctly identify the current hotspot. For instance, if a riser is experiencing high tensile loads, inspection resources should be prioritized towards the end where the failure initiates under tension (e.g., the right end in this case). This allows for a more intelligent, condition-based allocation of inspection and maintenance resources, ultimately reducing downtime and life-cycle costs while improving safety by focusing on the area of highest actual risk.

Figure 24 illustrates the variation in the von Mises at the same point of the pressure-armor layer with the analysis step under different initial tensile forces. Axial tensile forces are applied in the first 0.3 s, and internal pressure is applied after 0.3 s. Figure 25 shows the stress contours of the pressure-armor layer at four different special moments when the initial applied tensile force is 500 kN. It can be seen intuitively that the stress changes in the riser at different times and under different loads. In the initial stage of applying the axial tensile load, the stress increases nonlinearly until it reaches point b, as shown by (a) and (b) in Figure 24. After the internal pressure is applied (after 0.3 s), the increment of deformation due to the axial tensile force is gradually offset, and the stress also gradually decreases. Until a certain critical value (i.e., point (c) in Figure 24) is reached, that is, when the riser returns to its original length, the stress begins to show a linear increasing trend (d). The application of internal pressure (after 0.3 s) significantly affects the variation trend of the stress of the pressure-armor layer. Moreover, a larger initial axial tensile force causes the unbonded flexible riser to contract inward. At the same time, there is a larger initial axial elongation, which prevents the riser from expanding radially under the action of internal pressure, and also provides a larger adjustment space for the stress reduction. This phenomenon reasonably explains the variation presented in Figure 21; that is, the magnitude of the initial axial tensile force has an important influence on the failure characteristics of the pressure-armor layer. The larger the initial tensile force is, the larger the initial stress of the pressure layer under internal pressure is, and the internal pressure value when finally reaching the yield stress also increases accordingly.

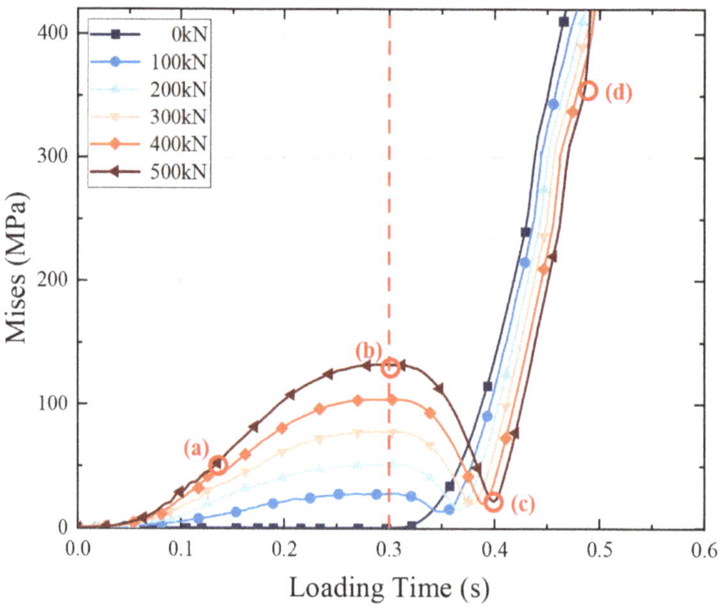

Figure 24. Resistance to pressure at the same point in the armor.

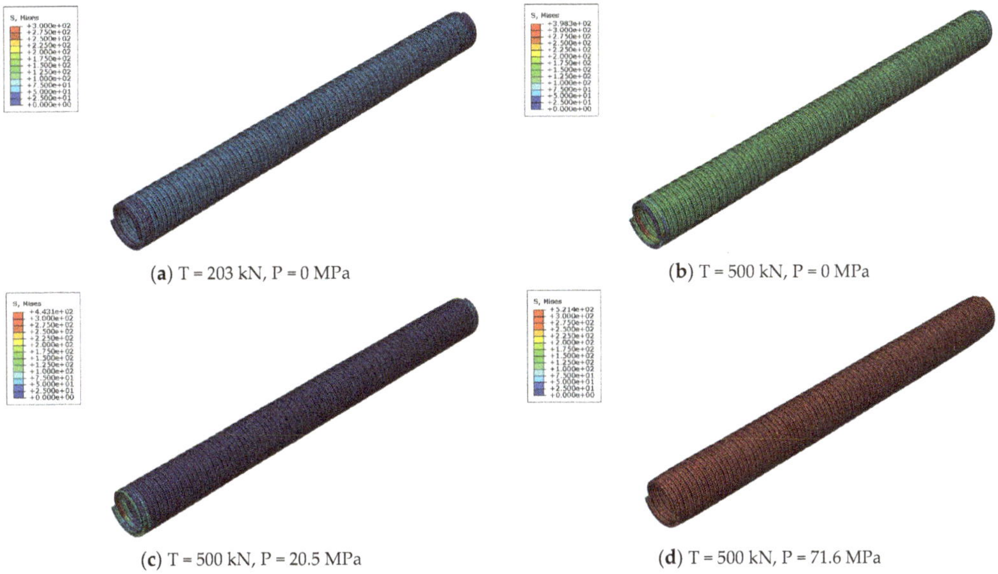

Figure 25. Distribution of von Mises stress in the pressure-armor layer. ((**a**) T = 203 kN, P = 0 MPa; (**b**) T = 500 kN, P = 0 MPa; (**c**) T = 500 kN, P = 20.5 MPa; (**d**) T = 500 kN, P = 71.6 MPa).

5. Conclusions

In this paper, the mechanical properties of a 2.5-inch unbonded flexible riser under internal pressure and axial tensile load are studied numerically and verified analytically. This study fills a critical research gap by systematically investigating the coupled failure physics that previous works have overlooked. Constructed using ABAQUS, a 7-layer simplified model that excludes the carcass layer has been constructed. Then, the mixed failure characteristics of the unbonded flexible riser under initial internal pressure and axial tensile load as well as initial axial tensile load and internal pressure are studied. Beyond the fundamental mechanical insights, this study delivers tangible engineering value. The quantification of the tension-pressure coupling effect provides a scientific basis for moving beyond conservative design rules. Furthermore, the understanding of failure mechanism migration enables the development of more sophisticated and cost-effective integrity management protocols. By translating these findings into practice, the industry can achieve not only enhanced safety but also reduced capital and operational expenditures for deep-water developments. Principal research findings are consolidated below:

1. During tensile failure of unbonded flexible risers, both inner and outer tensile-armor layers assume primary load-bearing responsibility. Stress magnitudes within the internal tensile-armor layer exceed those sustained by the external counterpart. Moreover, when the riser is subjected to axial tension, the stress does not change strictly and evenly along its axial direction. This is especially obvious at the end.
2. Pre-applied internal pressure has no significant impact on the stiffness of unbonded flexible riser. Moreover, as the initial internal pressure increases from 0 MPa to 40 MPa, the axial tensile load capacity of the riser gradually diminishes. Although the deformation amount of the riser is reduced to a certain extent, it is more likely to experience axial failure. This is because the axial tensile load and the internal and external pressures will cause the contact pressure effect between the layers of the non-bonded flexible riser, so increasing the external pressure has a negative effect on the tensile strength of the composite riser.

3. Under combined loading, the pressure-armor layer carries most of the load and its Z-section stress grows from the inner edge to the outer edge. This layer first undergoes a stress drop, then re-loads until burst. After burst, the inner tensile-armor layer immediately takes over as the primary pressure-bearing component, its peak stress exceeding that of the outer layer and serving as a secondary barrier.
4. The initial axial tensile force magnitude significantly influences the failure characteristics of the pressure-armor layer. A higher initial tensile force endows it with a greater anti-burst ability; the failure pressure increases by 35% when the tensile force is 500 kN. This increase far exceeds the material's own strain-hardening effect, primarily because the axial pre-compression suppresses radial ovalisation of the armor coil, thereby delaying the local buckling-yield coupling failure. Moreover, a larger initial axial tensile force results in a higher initial stress.

Author Contributions: Conceptualization, Q.L. and Y.L.; methodology, Q.L.; software, Y.L. and Q.W.; validation, J.H.; formal analysis, G.W.; investigation, M.L.; resources, Q.L. and M.L.; data curation, Y.L.; writing—original draft preparation, Q.L. and Y.L.; writing—review and editing, M.L.; visualization, Q.W.; supervision, Q.L.; project administration, G.W.; funding acquisition, Q.L. and M.L. All authors have read and agreed to the published version of the manuscript.

Funding: This paper was funded by the National Natural Science Foundation of China (Grant No. 12202236), the Natural Science Foundation of Jiangsu Province (Grant No. BK20220500), the China Postdoctoral Science Foundation (Grant No. 2025M772955), the Jiangsu Funding Program for Excellent Postdoctoral Talent (Grant No. 2022ZB560), and the Research Project of State Key Laboratory of Mechanical System and Vibration (Grant No. MSV202501).

Data Availability Statement: The original contributions presented in this study are included in the article. Further inquiries can be directed to the corresponding authors.

Conflicts of Interest: The authors declare no conflicts of interest.

References

1. Liu, Q.S.; Xue, H.X.; Tang, W.Y.; Yuan, Y. Theoretical and numerical methods to predict the behavior of unbonded flexible riser with composite armor layers subjected to axial tension. *Ocean Eng.* **2020**, *199*, 107038. [CrossRef]
2. Tchalala, M.R.; El-Demellawi, J.K.; Kalakonda, P.; Chaieb, S. High thermally stable hybrid materials based on amorphous porous silicon nanoparticles and imidazolium-based ionic liquids: Structural and chemical analysis. *Mater. Today Proc.* **2021**, *39*, 1132–1140. [CrossRef]
3. Kalakonda, P.; Thudumu, S.; Mynepally, S.L.; Mandal, P.; Banne, S.; Kalakonda, P.B.; Podili, B.B. Engineering micro/nano-fibrous scaffolds with silver coating for tailored wound repair applications. *J. Nanoparticle Res.* **2023**, *25*, 254. [CrossRef]
4. Dabbeta, N.K.; Kante, M.; Konda, R.; Devarala, A.; Kalakonda, P.B.; Khanam, S.J.; Savu, R.N.; Banovath, M.; Siddey, L.S.R.; Bommaji, V.N.; et al. ECO-inspired Cu-Zn bimetallic nanoparticles for Enhanced Water Purification Technologies. *Mater. Chem. Phys. Sustain. Energy* **2025**, *2*, 100013. [CrossRef]
5. Hasan, I.; Bassi, A.; Kalakonda, P. CuO surface functionalized β-cyclodextrin grafted PMMA hybrid nanocomposite for advance oxidative mineralization of methyl orange. *Surf. Interfaces* **2025**, *56*, 105456. [CrossRef]
6. Witz, J.A. A case study in the cross-section analysis of flexible risers. *Mar. Struct.* **1996**, *9*, 885–904. [CrossRef]
7. Yue, Q.J.; Lu, Q.Z.; Yan, J.; Zheng, J.; Palmer, A. Tension behavior prediction of flexible pipelines in shallow water. *Ocean Eng.* **2013**, *58*, 201–207. [CrossRef]
8. Vargas-Londoño, T.; de Sousa, J.R.M.; Magluta, C.; Rotiman, N. A theoretical and experimental analysis of the bending behavior of unbonded flexible pipes. In Proceedings of the ASME 2014 33rd International Conference on Ocean, Offshore and Arctic Engineering, San Francisco, CA, USA, 8–13 June 2014; OMAE2014-24247.
9. Zhou, C.; Ye, N.; Sævik, S. Effect of anti-wear tape on behavior of flexible risers. In Proceedings of the ASME 2014 33rd International Conference on Ocean, Offshore and Arctic Engineering, San Francisco, CA, USA, 8–13 June 2014; OMAE2014-24261.
10. Shao, W.Z.; Lin, Y.; Yan, J.H. Modeling the reeling process of offshore pipelines: A theoretical framework based on updated Voce-Chaboche model. *Ocean Eng.* **2025**, *332*, 121383. [CrossRef]
11. Kebadze, E. Theoretical Modelling of Unbonded Flexible Pipe Cross-Sections. Doctoral Thesis, South Bank University, London, UK, 2000.

12. Dong, L.L. Theoretical Calculation of Cross-Sectional Properties and Fatigue Analysis of Unbonded Flexible Risers in BSR Region. Doctoral Thesis, Dalian University of Technology, Dalian, China, 2013.
13. Bahtui, A. Development of a Constitutive Model to Simulate Unbonded Flexible Riser Pipe Elements. Doctoral Thesis, Brunel University, London, UK, 2008.
14. Ren, S.F.; Tang, W.Y.; Kang, Z.; Geng, H. Numerical study on the axial-torsional response of an unbonded flexible riser with damaged tensile armor wires. *Appl. Ocean Res.* **2020**, *97*, 102045. [CrossRef]
15. Cornacchia, F.; Liu, T.; Bai, Y.; Fantuzzi, N. Tensile strength of the unbonded flexible pipes. *Compos. Struct.* **2019**, *218*, 142–151. [CrossRef]
16. De Sousa, J.R.M.; Campello, G.C.; Kwietniewski, C.E.F.; Ellwanger, G.B.; Strohaecker, T.R. Structural response of a flexible pipe with damaged tensile armor wires under pure tension. *Mar. Struct.* **2014**, *39*, 1–38. [CrossRef]
17. Liu, Q.S.; Xue, H.X.; Tang, W.Y. Behavior of unbonded flexible riser with composite armor layers under coupling loads. *Ocean Eng.* **2021**, *229*, 109907. [CrossRef]
18. Jiang, H.; Yang, H.Z. Mechanical analysis of deep water unbonded flexible riser under combined loads. *China Ocean Eng.* **2014**, *32*, 93–99.
19. Wang, B.D.; Zhang, H.; Liu, X.B.; Xu, L.X.; Fu, Y. Analysis of the Mechanical Behaviors of Fiber Reinforced Unbonded Flexible Pipe Under Combined Load. In Proceedings of the 13th International Pipeline Conference, Calgary, AB, Canada, 28–30 September 2020; IPC2020-9346.
20. Sun, H.B.; Wang, Y.Y.; Jia, L.S.; Lin, Z.Q.; Yu, H. Theoretical and numerical methods for predicting the structural stiffness of unbonded flexible riser for deep-sea mining under axial tension and internal pressure. *Ocean Eng.* **2024**, *310*, 118672. [CrossRef]
21. Tang, L.P.; He, W.; Zhu, X.H.; Zhou, Y.L. Mechanical analysis of un-bonded flexible pipe tensile armor under combined loads. *Int. J. Press. Vessel. Pip.* **2019**, *171*, 217–223. [CrossRef]
22. Yoo, D.H.; Jang, B.S.; Yim, K.H. Nonlinear finite element analysis of failure modes and ultimate strength of flexible pipes. *Mar. Struct.* **2017**, *54*, 50–72. [CrossRef]
23. Berge, S.; Wang, T.; Langhelle, N. Surface characterization and fatigue strength of corroded armor wire. In Proceedings of the 33rd International Conference on Offshore Mechanics and Arctic Engineering, San Francisco, CA, USA, 8–13 June 2014; OMAE2014-24140.
24. Antoine, F.H. Prevention and monitoring of fatigue-corrosion of flexible risers' steel reinforcements. In Proceedings of the 21st International Conference on Offshore Mechanics and Arctic Engineering, San Diego, CA, USA, 10–15 June 2007; OMAE2007-29186.
25. Ebrahimi, A.; Kenny, S.; Hussein, A. Finite Element Investigation on the Tensile Armor Wire Response of Flexible Pipe for Axisymmetric Loading Conditions Using an Implicit Solver. *J. Offshore Mech. Arct. Eng.* **2018**, *140*, 041402. [CrossRef]
26. Zhu, X.-K.; Leis, B.N. Theoretical and numerical predictions of burst pressure of pipelines. *J. Press. Vessel Technol.* **2007**, *129*, 644–652. [CrossRef]
27. Kamaya, M.; Suzuki, T.; Meshii, T. Failure pressure of straight pipe with wall thinning under internal pressure. *Int. J. Press. Vessel. Pip.* **2008**, *85*, 628–634. [CrossRef]
28. Neto, A.; Martins, C.; Pesce, C.; Meirelles, C.; Malta, E.; Neto, T.; Godinho, C. Prediction of burst in flexible pipes. *J. Offshore Mech. Arct. Eng.* **2013**, *135*, 011401. [CrossRef]
29. Zeng, D.Z.; Wang, X.; Ming, K.J.; Yu, C.X.; Zhang, Y.H.; Yu, Z.M.; Yan, J.; Li, F.G. Analysis and optimization design of internal pressure resistance of flexible composite pipe. *Int. J. Press. Vessel. Pip.* **2024**, *210*, 105271. [CrossRef]
30. Liu, X.Y.; Qu, Z.Y.; Liu, Y.; He, J.W.; Si, G.J.; Wang, S.C.; Liu, Q.S. Numerical Simulation of Burst Failure in 2.5-Inch Unbonded Flexible Riser Pressure Armor Layers. *Metals* **2024**, *14*, 762. [CrossRef]
31. Wei, R.; Li, X.; Vaz, M.A.; Custódio, A.B. Finite Element Modelling of Pressure Armor Unlocking Failure in Unbonded Flexible Risers. *J. Mar. Sci. Eng.* **2025**, *13*, 411. [CrossRef]
32. Ren, S.F. Mechanical Properties and Typical Failure Characteristics of Unbonded Flexible Riser Cross-Sections. Doctoral Thesis, Shanghai Jiao Tong University, Shanghai, China, 2015.
33. Zhang, X.J.; Chang, Z.Y. Advance of research of unbonded flexible risers. In Proceedings of the International Conference on Applied Mechanics and Materials, Sanya, China, 15–16 November 2014. ICAMM1015–1019.
34. Grytå, G. Fatigue of Flexible Riser in Bend Stiffener Area. Master's Thesis, Norwegian University of Science and Technology, Trondheim, Norway, 2011.
35. Zhang, M.M.; Chen, X.Q.; Fu, S.X.; Guo, Y.; Ma, L. Theoretical and numerical analysis of bending behavior of unbonded flexible risers. *Mar. Struct.* **2015**, *44*, 311–325. [CrossRef]
36. Cuamatzi-Melendez, R.; Castillo-Hernández, O.; Vázquez-Hernández, A.; Albiter5, A.; Váz, M. Finite element modeling of burst failure in unbonded flexible risers. *Eng. Struct.* **2015**, *87*, 58–69. [CrossRef]
37. Lei, Q.L.; Chen, C.Y.; Zhu, X.H. Tensile failure behavior of unbonded flexible riser with damaged tensile armor wires. *Ocean Eng.* **2023**, *267*, 113112. [CrossRef]

38. De Sousa, J.R.M.; Viero, P.F.; Magluta, C.; Roitman, N. An Experimental and Numerical Study on the Axial Compression Response of Flexible Pipes. *J. Offshore Mech. Arct. Eng.* **2012**, *134*, 031703. [CrossRef]
39. Ramos, R., Jr.; Martins, C.d.A.; Pesce, C.P. A case study on the axial-torsional behavior of flexible risers. In Proceedings of the ASME 27th International Conference on Offshore Mechanics and Arctic Engineering, Estoril, Portugal, 15–20 June 2008; OMAE2008-57514.

Journal of
Marine Science
and Engineering

MDPI

Article

The Effect of In-Pipe Fluid States and Types on Axial Stiffness Characteristics of Fiber-Reinforced Flexible Pipes

Jingyue You, Yinglong Zhao * and Ben Zhang

Naval University of Engineering, Wuhan 430000, China
* Correspondence: zhaoyl_zss@163.com

Abstract: As critical components in marine engineering fluid transmission systems, fiber-reinforced flexible (FRF) pipes have static mechanical properties that depend on internal fluid pressure. Current analytical approaches predominantly employ uniformly distributed load (UDL) assumptions to simulate unidirectional fluid pressure effects on pipe surfaces. However, existing methodologies neglect fluid–pipe structure coupling effects. This study investigates the rubber-based FRF pipe by establishing a numerical model incorporating fluid–structure interaction effects and material nonlinearity, aiming to explore how different fluid states (closed or constant pressure) and fluid types (incompressible or compressible) influence the mechanical behavior of the FRF pipe under axial loading. Experimental validation of the numerical model demonstrates that UDL assumptions remain valid for gas-filled pipes (both in the closed and constant pressure states) and the liquid-filled pipe in the constant pressure state. The incompressibility of the filled liquid significantly enhances pipe axial stiffness, invalidating the UDL approximation method in liquid-filled closed states. Furthermore, the asymptotic saturation model proposed effectively quantifies the liquid-induced enhancement in axial stiffness. The developed numerical model and derived conclusions provide valuable insights into structural design optimization, experimental protocol development, and practical engineering applications for FRF pipes.

Keywords: fiber-reinforced flexible pipe; hyperelasticity; fluid–structure interaction; axial stiffness

1. Introduction

Fiber-reinforced flexible (FRF) pipes exhibit exceptional design versatility, where key parameters such as matrix material formulations, fiber reinforcement compositions, and ply stacking sequences can be optimized to meet performance requirements in diverse application scenarios [1,2]. This characteristic grants them broad application prospects in marine engineering [3–6]. Compared with conventional metal pipes, FRF pipes leverage their low stiffness and high compliance to achieve larger displacement adaptability and complex load-bearing capacity [7,8]. Moreover, they can serve as elastic damping elements to effectively mitigate vibration and impact induced pipeline damage [9,10]. Consequently, the mechanical performance of FRF pipes not only governs pipe system efficiency but also critically determines the safety and reliability of marine structures. Given that internal fluid pressure constitutes a dominant operational load during fluid transportation, investigation of its influence on the mechanical behavior of FRF pipes, such as axial stiffness and deformation, has significant engineering implications.

The mechanical performance and failure behavior of FRF pipes under static internal pressure loading have remained a focal research area. In theoretical investigations, Kruijer et al. [11]

applied laminate theory to reinforced thermoplastic pipes, developing a nonlinear extended plane stress model that accounted for fiber reorientation and diameter variation during pressurization, thereby revealing the mechanical response of long pipes under hydrostatic pressure. Xia et al. [12] derived exact elastic solutions for stresses and deformations in multilayered filament-wound pipes under internal pressure using three-dimensional anisotropic elasticity theory. Subsequent work by Xing et al. [13] introduced axisymmetric thick-walled cylinder theory to analyze the mechanical behavior of multi-angle hybrid fiber-reinforced thick-walled cylinders under combined loads (internal/external pressure and axial force). Additionally, Gao et al. [14] investigated deformation responses of closed-end FRF pipes with helical stiffeners under internal pressure by employing Donnell approximations and Euler–Bernoulli beam theory. While these studies represent significant methodological advancements over simplified fiber-only models (e.g., Rogers et al. [15] and Evans et al. [16]), they relied on idealized material constitutive assumptions. Nevertheless, experimental validation by Gu et al. [17] demonstrated that approximating the polymeric matrix as a linear elastic material leads to substantial discrepancies between theoretical predictions and experimental results.

Fang et al. [5] addressed material nonlinearity in their investigation of glass fiber-reinforced flexible pipes under tension and internal pressure by iteratively updating material properties at each load increment. Building on this method, Wang et al. [18] introduced stress–strain curve fitting techniques to parameterize experimental material data, enabling material property updates. However, such methods are limited to scenarios where material nonlinearity is characterized under unidirectional loading conditions. Given the pronounced nonlinear behavior exhibited by polymeric matrices in FRF pipes under multiaxial stress states, hyperelastic constitutive models rooted in strain energy density functions provide an accurate representation of their complex mechanical responses [19]. The capability of numerical simulations to address material nonlinearity has established them as a pivotal tool for investigating FRF pipe mechanics under internal pressure. Representative studies include Tonnato et al. [20], who predicted burst pressures of floating composite pipes; and Pan et al. [4], who quantified the influence of fiber and rubber properties on pressure-deformation responses and failure pressures in helically ribbed composite marine pipes. These investigations collectively demonstrated the accuracy and reliability of numerical methodologies through experimental validation.

The above and other theoretical and numerical studies generally employ constant uniformly distributed load (UDL) to characterize the unidirectional mechanical effect between internal fluids and FRF pipes. To simulate end effects induced by internal pressure, conventional methods typically apply constant UDL or axial tension to pipe end plates. This idealized assumption overlooks fluid pressure redistribution resulting from pipe deformation and varying structural responses caused by different fluid states and types.

The fluid cavity method has emerged as an innovative solution to overcome these limitations and has been successfully implemented across multiple disciplines. Tong et al. [21] modeled blood as an incompressible fluid within cardiac and aortic cavities to investigate trauma mechanisms of aortic rupture during occupant–vehicle interior collisions; Ye et al. [22] simulated gas behavior in marine inflatable membrane structures, elucidating bending and failure characteristics under operational conditions; and Xiao et al. [23] analyzed stiffness–gas mass relationships in dual-wedge safety airbags for medical devices through gas containment simulations. Complementary studies [24,25] further extended this methodology to airbag systems and pneumatic tensegrity domes, respectively. These implementations collectively confirm the method's versatility in simulating fluid–structure coupling phenomena, positioning it as an available tool for modeling various fluid states and types within FRF pipes.

This paper investigates rubber-based FRF pipes, focusing on their static mechanical response characteristics under axial loads. It specifically examines how differences in fluid states (closed state vs. constant pressure state) and fluid types (compressible type vs.

incompressible type) influence the mechanical behavior of the pipes. Section 2 details the numerical modeling methodology for FRF pipes, including the fitting of the hyperelastic constitutive model for rubber materials, the construction of fluid–structure coupling models using the fluid cavity method, and the specification of meshing strategies and boundary conditions. Section 3 outlines the experimental design for testing the axial stiffness of FRF pipes. Section 4 presents a comparison and validation of numerical results against experimental results, analyzing the differential effects of fluid states and types on axial stiffness while exploring the applicability of UDL simulations and the sensitivity of fluid parameters. Finally, Section 5 summarizes the main conclusions, provides recommendations for engineering applications, and suggests directions for future research. It is important to note that the term "closed state" refers to the condition where the fluid is fully sealed within the pipe after pressurization, with no mass exchange occurring between the pipe and its external environment. Conversely, the "constant pressure state" denotes the condition where the fluid inside the pipe maintains a constant pressure during axial loading after pressurization. Additionally, "UDL simulations" refer to the simplification of the fluid force acting on the pipe wall as a uniformly distributed load perpendicular to the inner surface of the pipe.

This study is the first to apply the fluid cavity method to FRF pipe research, constructing a numerical model that incorporates material nonlinearity and fluid–structure coupling effects, thereby overcoming the limitations of the traditional UDL assumption. Through numerical simulations and experimental validation, it is demonstrated that the strong coupling effect induced by the incompressibility of liquid under the closed state renders the UDL assumption invalid. Furthermore, the study quantitatively separates the contributions of structural stiffness and liquid stiffness, elucidating the nonlinear enhancement mechanism of axial stiffness by the liquid bulk modulus. This provides a new reference for static fluid–structure coupling research in FRF pipes.

2. Numerical Modeling of the Fiber-Reinforced Flexible Pipe

2.1. Material Characterization and Hyperelastic Constitutive Modeling

Rubber materials exhibit multiaxial stress coupling characteristics under complex deformation modes. To accurately characterize their mechanical behavior, this study conducts uniaxial tension, equibiaxial tension, planar tension, and volumetric compression tests. And stress–strain data of the material are obtained to determine the material parameters required for input into the numerical model. The experimental setup and rubber specimens are illustrated in Figure 1. All specimens are sectioned from 2 mm thick rubber sheets that share identical curing histories with the matrix material of the FRF pipe.

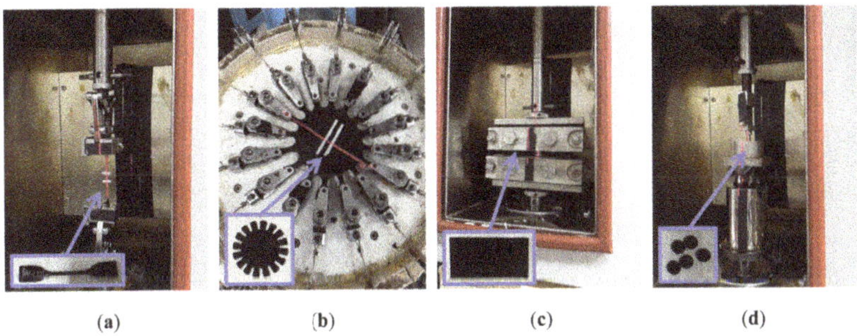

(a) (b) (c) (d)

Figure 1. Rubber material tests and specimens: (**a**) Uniaxial tension test conducted according to ASTM D412 Type [26]. (**b**) Equibiaxial tension test (specimen dimension: Ø60 mm × 2 mm). (**c**) Planar tension test (specimen dimension: 150 mm × 75 mm × 2 mm). (**d**) Volumetric compression test (specimen dimension: Ø8 mm × 2 mm).

All experiments are conducted under controlled ambient temperature (23 ± 2 °C). Uniaxial tension, planar tension, and volumetric compression tests are performed using a WDW-10 testing machine, which is manufactured by E-RUBBER Technology (Beijing, China), with a maximum capacity of 10 kN. Biaxial tension testing employs a specialized system consisting of fixtures, a clamp drive system, and a multi-axis sensor array. After specimen installation, initial clamp adjustments ensure zero-stress states by eliminating residual tension or bending moments. Quasi-static loading is applied at strain rates of 0.01 mm/s (0.02 mm/s for volumetric compression). Deformation measurements are obtained using laser extensometry tracking displacement between two central white fiducial markers, as shown in Figure 1. To mitigate Mullins effects [27,28], the testing strain rates are kept similar to those experienced by FRF pipes produced by the Naval University of Engineering (Wuhan, China).

To characterize the hyperelastic behavior of rubber materials, a parameterized constitutive relationship is established using experimentally obtained stress–strain data. The strain energy function U is formulated based on three strain invariants I_1, I_2, and I_3, which collectively provide a complete deformation characterization [29]:

$$U = f(I_1, I_2, I_3), \tag{1}$$

Due to the incompressibility of rubber, the third strain invariant is constrained as $I_3 \equiv 1$. The remaining invariants are kinematically related to the principal stretch ratios (λ_1, λ_2, λ_3) via the following constitutive definitions:

$$\begin{cases} I_1 = \lambda_1^2 + \lambda_2^2 + \lambda_3^2 \\ I_2 = \lambda_1^2\lambda_2^2 + \lambda_2^2\lambda_3^2 + \lambda_3^2\lambda_1^2 \end{cases}, \tag{2}$$

The Mooney–Rivlin and Neo-Hookean constitutive models are widely employed in the literature to describe strain energy functions [30]. Their mathematical representations are defined as follows:

$$\begin{array}{l} U = C_{10}(I_1 - 3) + C_{01}(I_2 - 3) + \frac{1}{D_1}(J_{el} - 1)^2 \\ U = C_{10}(I_1 - 3) + \frac{1}{D_1}(J_{el} - 1)^2 \end{array}, \tag{3}$$

The Mooney–Rivlin model incorporates the second strain invariant I_2, serving as an extended formulation of the Neo-Hookean model. The material constants C_{10} and C_{01} govern shear behavior, while D_1 denotes the compressibility coefficient and J_{el} represents the elastic volume ratio. The material testing results and model fitting for the rubber are presented in Figure 2, where the fitting process is conducted using Abaqus software. The experimental data indicate that the material does not exhibit a stiffening effect with increasing strain (as evidenced by the absence of upward curvature in the stress–strain curve). Consequently, there is no need to employ models incorporating higher-order strain invariants for fitting (e.g., the Ogden model or Yeoh model). Although the Mooney–Rivlin model introduces the second strain invariant I_2 to enhance adaptability under multi-axial stress conditions, the negative value obtained in the fitting result ($C_{01} = -50.3$) leads to instability in the material's constitutive behavior. This instability is specifically manifested by the anomalous points shown in Figure 2b, where stress decreases with increasing strain. In contrast, all parameters of the Neo-Hookean model ($D_1 = 8.62 \times 10^{-3}$, $C_{10} = 1.98$) are positive. This can be attributed to the relatively minor conformational changes in the rubber molecular chains under small-strain conditions, where entropic elasticity deformation likely dominates as the primary deformation mechanism. Given its ability to provide a simple yet accurate approximation [29], the Neo-Hookean model is selected for the final fitting results.

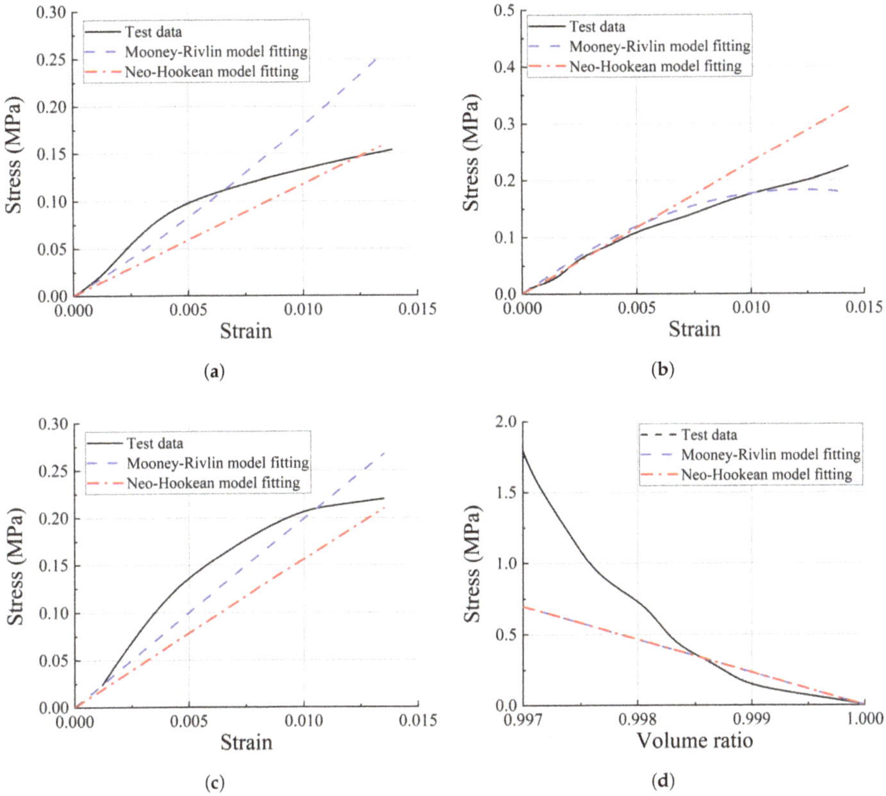

Figure 2. Rubber material testing data and hyperelastic constitutive model fitting results: (**a**) Uniaxial tension. (**b**) Equibiaxial tension. (**c**) Planar tension. (**d**) Volumetric compression.

2.2. Development of the Numerical Model

The FRF pipe is manufactured through a vulcanization process integrating a rubber matrix with fiber reinforcement layers. Featuring a triple-flange configuration at both ends, the pipe comprises four reinforcement layers with complementary winding angles embedded within the rubber matrix (innermost layer winding angle is 53.1° relative to the axial direction). Detailed reinforcement fiber parameters are summarized in Table 1.

Table 1. Reinforced fiber material parameters.

Parameter	Value
Density (t/mm^3)	1.44×10^{-9}
Young's modulus (MPa)	3.28×10^4
Poisson's ratio	0.36

To facilitate installation and ensure post-installation sealing integrity, the flanges are designed with structural features including bolt holes, bosses, and chamfers. For computational efficiency, these detailed features are omitted from the geometry model. Furthermore, geometric simplifications are applied at the intermediate flange region, where the rubber matrix and reinforcement fibers are mechanically constrained, as illustrated in Figure 3.

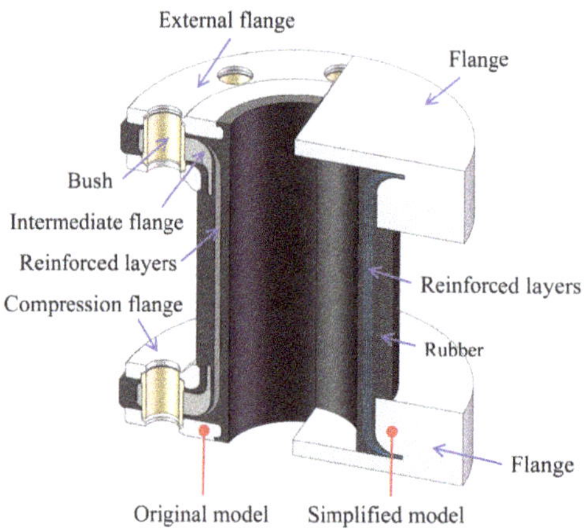

Figure 3. Geometry model of the fiber-reinforced flexible (FRF) pipe.

2.2.1. Fluid–Structure Coupling Model Based on the Fluid Cavity Method

When modeling liquid-filled FRF pipes, the liquid bulk modulus K_{liquid} is introduced to account for fluid incompressibility. Assuming that K_{liquid} remains independent of liquid density, the internal pressure p_{liquid} can be expressed as:

$$p_{liquid} = -K\left(\rho_{liquid}^{-1}(p,\theta) - \rho_{liquid0}^{-1}(\theta)\right) , \tag{4}$$

where ρ_{liquid} and $\rho_{liquid0}$ denote the final and initial liquid densities, respectively, p represents the final pressure, and θ indicates temperature. Assuming ideal gas behavior for the internal gas, the pipe pressure p_{gas} can be expressed using the ideal gas equation of state as:

$$p_{gas} = \rho_{gas}R\left(\theta - \theta^{Z}\right) , \tag{5}$$

where ρ_{gas} denotes gas density, θ^{Z} represents absolute temperature, and R is the gas constant defined as the ratio of the universal gas constant to the molecular weight of the ideal gas. It should be explicitly noted that temperature effects are excluded in this investigation, and in constant pressure states, the internal fluid pressure remains invariant. Additionally, the fluid cavity model is implemented by selecting all inner surfaces of the FRF pipe and defining a cavity reference node (see Figure 4). This reference node has a single degree of freedom governing pressure variation within the cavity. To simulate diverse fluid behaviors, nitrogen and water are selected as representative compressible and incompressible media, with their material parameters summarized in Table 2.

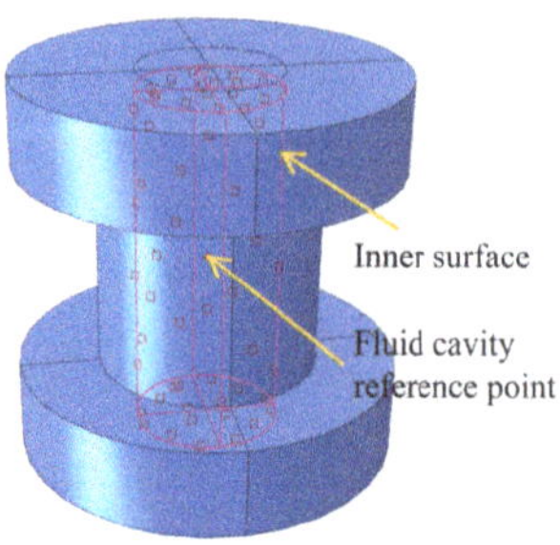

Figure 4. The fluid cavity model in the FRF pipe.

Table 2. Parameters of the fluid cavity model.

	Parameter	Value
Water	Density (t/mm^3) Bulk modulus (MPa)	1×10^{-9} 2100
Nitrogen	The molecular weight of the ideal gas (t/mol)	2.8×10^{-5}

2.2.2. Interaction, Meshing, and Boundary Conditions

The vulcanization process induces dense chemical crosslinking networks at the rubber–fiber interface, thereby enabling efficient stress transfer between matrix and reinforcement. Consequently, the rubber–fiber interfacial interactions are simulated using an embedded region constraint (as illustrated in Figure 5). Within this modeling approach, the nodal displacements of the reinforcement fiber elements are computed by interpolating displacements from adjacent to the rubber matrix elements.

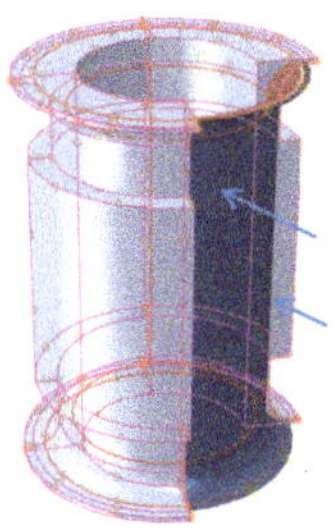

Figure 5. Interfacial bonding mechanism between rubber matrix and reinforcement fibers.

The developed FRF pipe numerical model is illustrated in Figure 6, where the rubber matrix and fiber reinforcement layers are discretized using C3D8R and M3D4R elements, respectively, and the total mesh comprises 131,616 elements. The flange components, manufactured from cast copper alloy and carbon structural steel, exhibit mechanical strength properties that significantly exceed those of the rubber matrix and reinforcement fibers. As a result, these flanges are modeled as rigid bodies and excluded from the computational domain to improve computational efficiency.

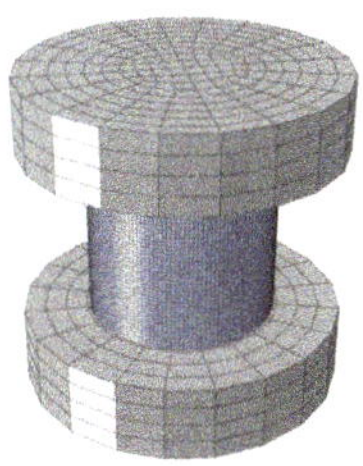

Figure 6. Finite element discretization of the FRF pipe.

Since the material of the flange is cast copper alloy and carbon structural steel, whose strength far exceeds that of the rubber matrix and the reinforcing fiber material, the flange is simplified as a rigid body in the model and does not participate in the calculation. In establishing the numerical model, the rubber matrix and the fiber-reinforced layer are modeled using C3D8R and M3D4R elements, respectively. To ensure the grid independence of the numerical model and determine the computational cost, the axial stiffness of the FRF pipe is calculated for four different grid sizes when the pipe is filled with water and in the closed state (Table 3). The results indicate that the error monotonically converges as the grid is refined. Specifically, the error for the grid with a unit size of 2 mm is only 1.23%. Regarding computational time, the grid with a unit size of 2 mm requires 284 s (AMD Ryzen 9 7940H, 16GB memory), which is only 11.52% of the time required for the grid with a unit size of 1.5 mm. Therefore, considering both accuracy and efficiency, a numerical model established with a grid size of 2 mm is adopted for subsequent analysis (as shown in Figure 6).

Table 3. Grid convergence analysis (water-filled closed state, initial pressure of 1 MPa).

Grid Size (mm)	Total Number of Elements	Axial Stiffness (N/mm)	Relative Error (vs. Grid Size 1.5 mm)	Calculation Time (s)
3	49,233	1389.65	−6.31%	91
2.5	83,081	1437.74	−3.07%	171
2	131,616	1465.04	−1.23%	284
1.5	306,649	1483.29	-	2466

In the numerical model, the motion of both end flanges is kinematically constrained to their respective reference points. To simulate the pressure charging and axial loading processes, two analysis steps are implemented. As shown in Figure 7a, during the first analysis step, both flanges are fixed while fluid pressure is applied by controlling the degree of freedom at the fluid cavity reference node, thereby simulating the pressurization process. In the second analysis step, the bottom flange remains fixed, and axial loading is applied via displacement control at the top flange (see Figure 7b). The fluid pressure control strategy involves two operational conditions:

1. Closed state simulation: The fluid cavity pressure control is deactivated after the first analysis step, effectively trapping initially pressurized fluid within the pipe;
2. Constant pressure state simulation: The fluid pressure applied in the first analysis step is maintained throughout the second analysis step, ensuring invariant internal pressure during axial loading.

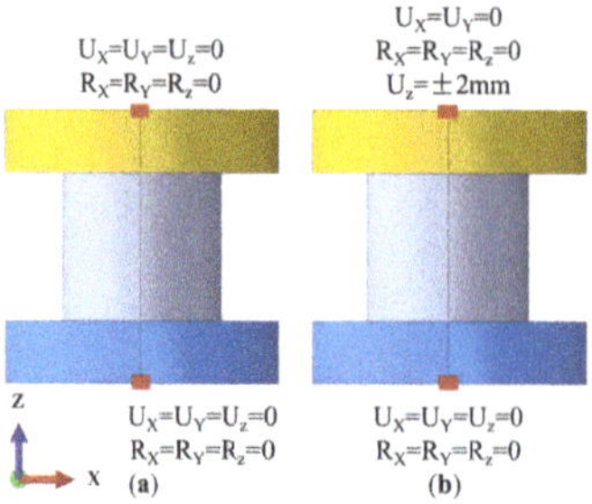

Figure 7. Multi-step boundary conditions: (**a**) Pressurization process. (**b**) Axial displacement process.

3. Axial Stiffness Testing of Fiber-Reinforced Flexible Pipes

The predictive accuracy of the numerical model is validated through axial stiffness testing of the FRF pipe. To ensure the reliability of experimental results, two FRF pipe specimens are prepared for axial stiffness testing (as shown in Figure 8). The specimens have nominal dimensions of 166 mm in length, 65 mm in inner diameter, and 100 mm in outer diameter. Measurement results indicate maximum deviations between actual and nominal dimensions of 0.2 mm (length) and 5.73 mm (outer diameter), corresponding to relative errors of 0.12% and 5.73%, respectively. These minor deviations suggest that the manufacturing process demonstrates satisfactory dimensional consistency.

Figure 8. FRF pipe specimens for axial stiffness testing.

The axial stiffness testing of FRF pipes is conducted using an MTS testing machine, with the experimental setup shown in Figure 9. The FRF pipe is connected to the testing apparatus through end plates, where the lower end plate features a through-port for interfacing with fluid transfer lines that connect the pipe to a pressurization pump and accumulator. The accumulator serves to maintain constant internal pressure. A pressure gauge integrated into the fluid transfer line allows for real-time pressure monitoring. The test configuration further includes a pressure-regulating valve for fluid pressure control. To achieve quasi-static loading conditions, a displacement rate of 0.1 mm/s is applied during axial loading within a ± 2 mm range.

The FRF pipe has a maximum working pressure of 1.6 MPa. To prevent negative pressure development during axial loading, which could impair pressure gauge and valve performance, a target pressure of 1.5 MPa is selected. Prior to axial loading, the FRF pipe is pressurized to the target pressure, after which the valves are closed to simulate mechanical behavior in the closed state. With valves kept open, the accumulator ensures constant fluid pressure, enabling mechanical response characterization in the constant pressure state. Notably, nitrogen and water are employed as working fluids. However, due to the limitations of the accumulator equipment, constant-pressure testing is exclusively conducted with water as the internal medium.

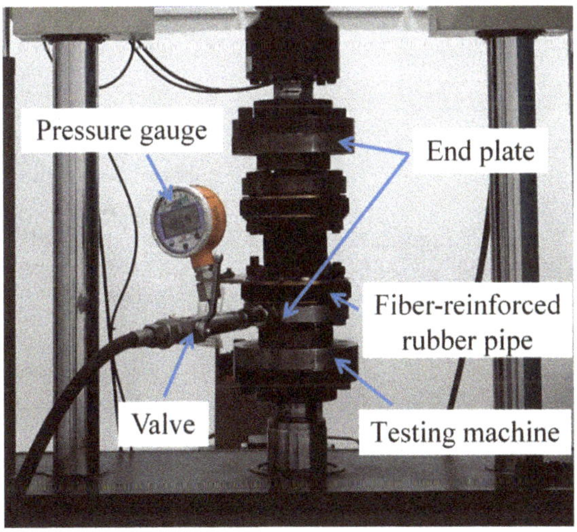

Figure 9. Experimental setup for axial stiffness testing of FRF pipes.

4. Results and Discussion

4.1. Numerical Results and Experimental Validation of FRF Pipe Axial Stiffness

The axial stiffness values averaged from specimen 1 and specimen 2 are compared with numerical results in Table 4, revealing a maximum deviation of 1.32% between numerical and experimental results. The primary error source originates from the manual tensioning process during reinforcement layer fabrication. Workers typically split and stretch the sulfurized cord fabric ends at the intermediate flange (see Figure 10), inadvertently introducing unintended prestress into the structure. Furthermore, discrepancies exist between the fitted hyperelastic constitutive model and material test data (Figure 2). Given the small magnitude of maximum error, the numerical model demonstrates effective predictive capability for FRF pipe axial stiffness under various fluid states and types.

Table 4. Error between experimental mean axial stiffness and numerical results.

State	Fluid	Axial Stiffness (N/mm)		Error
		Numerical	Experimental	
Closed	Water	1567.06	1583.09	−1.01%
	Nitrogen	1248.7	1245.12	0.29%
Constant pressure	Water	1288.23	1271.47	1.32%

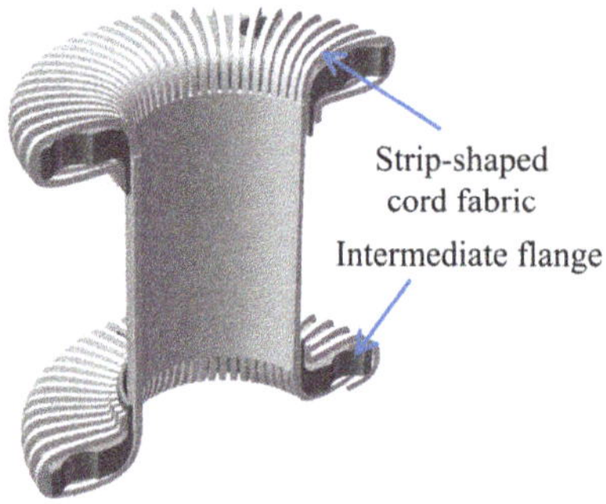

Figure 10. Installation process of sulfurized fiber-reinforced cord fabric.

4.2. Effects of Fluid States and Types on Axial Stiffness

Using the established numerical model, the axial stiffness of the FRF pipe is calculated across an initial internal pressure range of 0 MPa to 1.6 MPa (the maximum working pressure of the test specimens), with additional simulations incorporating UDL application on the inner pipe surface. As shown in Figure 11, when the pipe contains nitrogen (in both closed state and constant pressure state), its axial stiffness aligns with the UDL-based simulation results. This indicates weak fluid–structure coupling between nitrogen and the FRF pipe, as nitrogen primarily exerts uniformly distributed normal pressure on the inner pipe surface. Therefore, regardless of the nitrogen containment state, its pressure effects can be simplified as UDL.

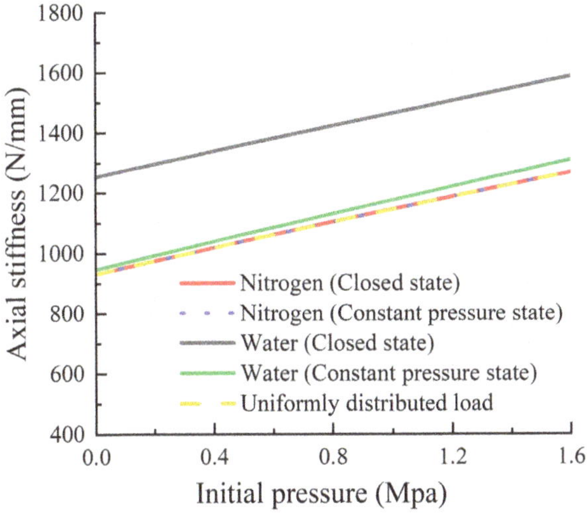

Figure 11. Numerical results of axial stiffness in varied fluid states and types.

Furthermore, Figure 11 demonstrates that the axial stiffness of the nitrogen-filled FRF pipe increases progressively with rising internal pressure. To elucidate this pressure-dependent stiffening mechanism, stress distributions in both the rubber matrix and four-layer cord fabric reinforcement are comparatively analyzed at internal pressures of 0.5 MPa and 1 MPa. For consistency in comparison, all subplots in Figure 12 adopt identical stress contour legends.

As shown in Figure 12, significant stress differentials are observed between the inner and outer pipe surfaces, as well as along the central cross-section toward the axial ends. These spatially heterogeneous stress distributions collectively confirm the establishment of a three-dimensional stress state within the FRF pipe structure. Under identical tensile or compressive loading conditions with the same internal pressure, stress magnitudes progressively decrease with increasing pipe radius, indicating stress redistribution facilitated by the rubber matrix's cushioning effect. Flange constraints effectively suppress deformation in regions I and III. Under tensile loading, region II exhibits concave inward deformation. The higher internal pressure in Figure 12c reduces stress magnitudes in this region compared to Figure 12a, demonstrating enhanced resistance to radial deformation due to increased pressure. Conversely, compressive loading induces outward convexity in region II, where internal pressure elevates stress levels in Figure 12d relative to Figure 12b, highlighting pressure-dependent stress amplification. Notably, irregular stress distributions emerge in regions IV, attributed to angular deviations between cord fabric orientation and pipe axis.

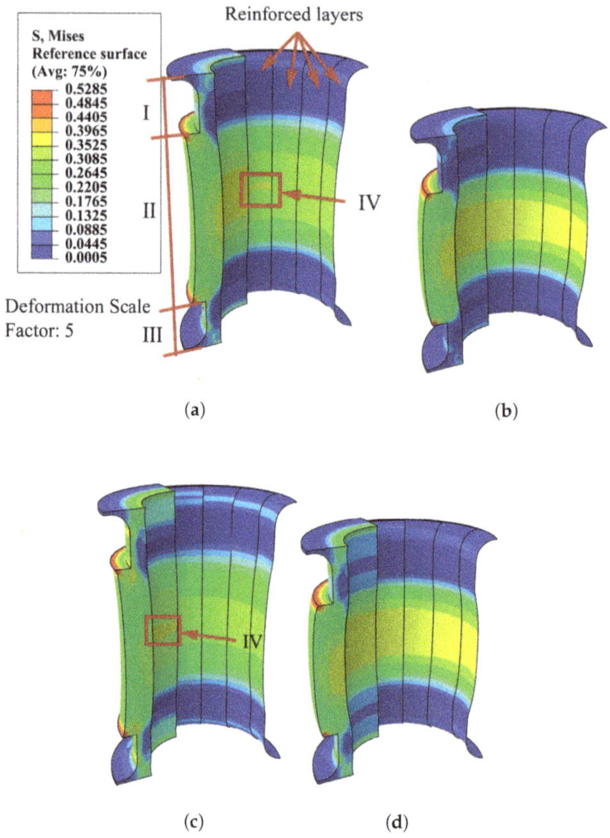

Figure 12. Deformation and stress contours of rubber matrix and fiber-reinforced layers in the nitrogen-filled FRF pipe under axial loading: (**a**) 0.5 MPa internal pressure, tensile load, (**b**) 0.5 MPa internal pressure, compressive load, (**c**) 1 MPa internal pressure, tensile load, (**d**) 1 MPa internal pressure, compressive load.

Figure 11 reveals that in the constant pressure water-filled state, the axial stiffness of FRF pipes marginally exceeds the UDL simulation result. Beyond stiffness variations, Figure 13 comparatively illustrates deformation characteristics of the inner pipe diameter under both loading conditions. During the pressurization and compression loading, the stiffness of the water in the pipe (the pressure required to maintain its state) is higher than under pure mechanical loading, leading to reduced deformation. In the axial tensile loading, the UDL can partially counteract the contraction, resulting in minimal deformation. Tension causes the pipe to contract, which increases the water pressure. Simultaneously, the reduction in pressure weakens the resistance to contraction, thereby causing more pronounced deformation. According to Figure 11, the maximum discrepancy between water pressurization and UDL-based axial stiffness reaches 3.32%. Furthermore, minor deformation discrepancies between both conditions (as shown in Figure 13) validate the mechanical equivalence of simplifying constant pressure water effects as UDL.

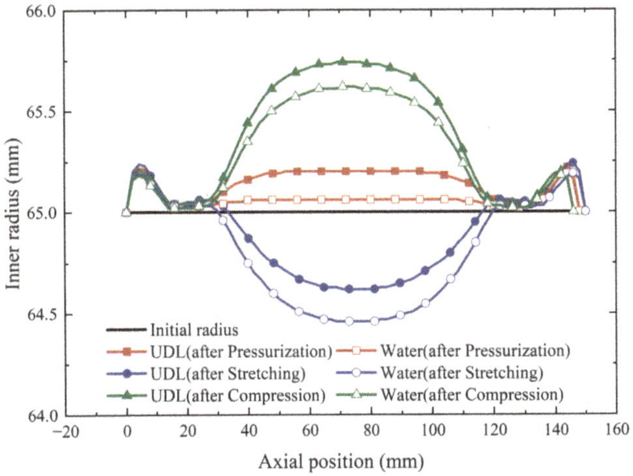

Figure 13. Change in inner diameter of the FRF pipe: water-filled constant pressure state vs. UDL conditions at 1.6 MPa.

In contrast, when the FRF pipe is in a closed state and contains water, its axial stiffness is significantly higher than under other conditions. At zero internal pressure, the axial stiffness is 34.52% greater than that observed under nitrogen-filled conditions. To investigate the underlying mechanisms driving this fluid-type-dependent stiffness variation in closed systems, the variation patterns of fluid volume and pressure during axial loading are analyzed. As shown in Figure 14, the horizontal axis represents computational time, with the intervals 0–1, 1–1.5, and 1.5–2 corresponding to the pressurization, axial tension, and axial compression processes, respectively. Additionally, the legend annotation "initial pressure" refers to the internal pressure stabilized at the end of the 0–1 pressurization step.

Figure 14 illustrates distinct fluid-dependent behaviors under axial loading. Significant cavity volume changes occur under axial loads when the pipe contains nitrogen, while pressure variations remain relatively negligible due to the compressibility of nitrogen. In contrast, when the pipe contains water, smaller volume changes are observed. Specifically, under tensile loading, volume reduction induces a substantial pressure increase (e.g., from an initial pressure of 0.8 MPa to a final pressure of 1.85 MPa, exceeding the specimen's maximum working pressure). Under compressive loading, volume expansion causes a pressure reduction (e.g., from initial pressure of 0.4 MPa to final pressure of −0.08 MPa, resulting in negative pressure). These findings highlight risks of excessive or negative pressure-induced failure in water-filled systems.

It can be assumed that the incompressibility of water inherently constrains pipe deformation, while the resulting deformation conversely induces pressure variations within the water medium. This synergistic interaction between the fluid's incompressibility and structural deformation significantly enhances the axial stiffness of the FRF pipe. Consequently, theoretical analyses and numerical simulations must explicitly account for water's coupling effects rather than the approximating water-filled closed state as UDL. Furthermore, when modeling end forces induced by internal pressure via equivalent axial forces, special attention must be paid to pressure-dependent force fluctuations during axial loading.

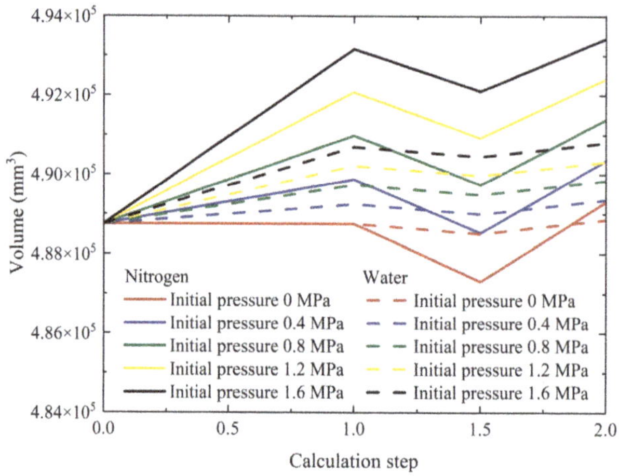

(a)

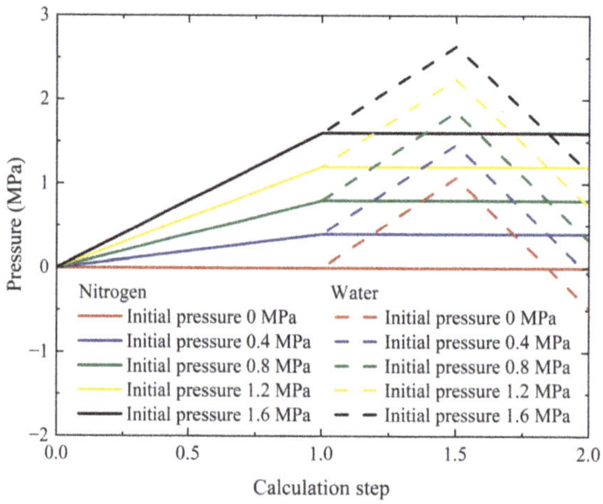

(b)

Figure 14. Pressure and volume variations of the nitrogen-filled and water-filled FRF pipe in the closed state: (**a**) Volume variation. (**b**) Pressure variation.

4.3. Effects of Fluid Parameters on the Axial Stiffness of the FRF Pipe in the Closed State

In the closed state, the axial stiffness of the FRF pipe is intrinsically linked to the physical properties of the filled fluid. As summarized in Table 2, the behavior of gases is characterized by their molecular weights, while the incompressibility of liquids is quantified through their bulk moduli. Based on molecular weight data of common gases from Reference [31], axial stiffness is calculated for gaseous media with molecular weights ranging from 1.5×10^{-5} t/mol to 4.5×10^{-5} t/mol, encompassing common gases such as oxygen, carbon dioxide, and carbon monoxide. For liquids, a bulk modulus range of 500–2500 MPa is selected, covering typical values for crude oil, liquefied natural gas, and methanol. The computational results are graphically presented in Figure 15.

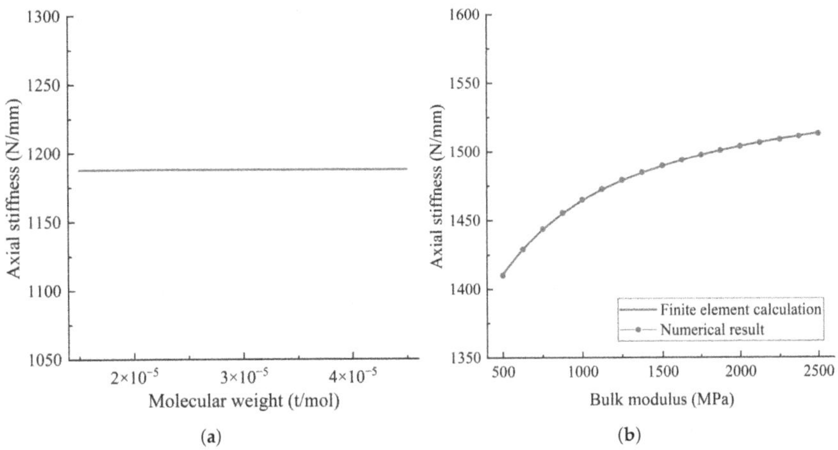

Figure 15. Influence of fluid parameters on axial stiffness of the FRF pipe in the closed state by numerical method (1.5 MPa internal pressure): (**a**) Variation in ideal gas molecular weight. (**b**) Variation in liquid bulk modulus.

As shown in Figure 15, the axial stiffness of the FRF pipe demonstrates no correlation with the molecular weight of ideal gases, indicating that the gas type has a negligible effect on the pipe structure. This observation further validates the conclusion that fluid–structure interactions are weak in gas-filled systems. Conversely, in liquid-filled systems, the axial stiffness increases with bulk modulus, albeit with diminishing enhancement rates. This is because liquids with a higher bulk modulus generate substantial counteracting forces against pipe deformation. However, as bulk modulus continues to increase, the liquid's contribution to stiffness asymptotically saturates, resulting in reduced sensitivity of axial stiffness to further modulus increments.

For liquid-filled pipes, Figure 15 demonstrates that the axial stiffness k_{total} of FRF pipes in the closed state is determined by both structural stiffness and fluid incompressibility. For gas-filled pipes, the incompressibility contribution is negligible, leading to k_{gas} being governed solely by structural properties. Considering the asymptotic saturation of liquid-induced stiffness as the bulk modulus increases, an asymptotic saturation model is proposed:

$$k_{total} = k_{\text{gas}} + \frac{\alpha K_{liquid}}{K_{liquid} + \beta}, \quad (6)$$

where α denotes the saturation stiffness (the limiting value of liquid-induced stiffness), and β represents the bulk modulus at half-saturation. Parameter identification is performed through nonlinear least-squares minimization of the sum of squared errors (SSE):

$$SSE = \sum_{i=1}^{17} \left[k_{total,i} - \left(k_{gas} - \frac{\alpha K_{liquid,i}}{\beta + K_{liquid,i}} \right) \right]^2, \quad (7)$$

Fitting yields $\alpha = 367.59 \pm 0.4$ N/mm and $\beta = 327.3 \pm 1.8$ MPa, as shown in Figure 15b. The model achieves a coefficient of determination $R^2 = 0.99$ with a root mean square error of 0.437 N/mm, demonstrating excellent predictive capability. Using the structural stiffness from Table 4 (nitrogen-filled closed state) and the bulk modulus of water from Table 2, the calculated axial stiffness of the FRF pipe filled with water in the closed state is 1563.14 N/mm, which differs by only −1.26% from the experimental result in Table 4.

5. Conclusions

This study establishes a numerical model that incorporates fluid–structure interactions between static internal fluids and the FRF pipe, with the aim of analyzing mechanical responses under axial loading across diverse fluid conditions. The maximum discrepancy between numerically predicted and experimentally measured axial stiffness is 1.3%, thereby validating the model's reliability. Key findings are summarized as follows:

1. Weak fluid–structure coupling allows gas-induced loads to be simplified as UDL in both the closed and constant pressure states. This simplification remains valid for the water-filled constant pressure state as well;
2. In the closed state, when water serves as the internal fluid of the pipe, the strong coupling effect between water's incompressibility and the pipe structure precludes the simplification of the force exerted by water on the pipe as UDL. Additionally, during axial loading, significant fluctuations in water pressure may result in the internal pipe pressure exceeding or falling below the allowable pressure range specified for the pipe design. Furthermore, particular attention should be given to the fact that end effects induced by water pressure no longer manifest as a constant force;
3. In the closed state, there is a positive correlation between the bulk modulus of the liquid and the axial stiffness of the FRF pipe. By establishing an asymptotic saturation model, the quantitative relationship between the structural stiffness of the pipe and the stiffness contribution from the liquid is elucidated. This formula decouples structural stiffness from liquid effects, ensuring that it degrades to the gas-condition result at low bulk modulus values and asymptotically approaches the upper limit of the combined structural–fluid stiffness at high bulk modulus values.

In summary, this work elucidates the critical influence of fluid states and types on the axial stiffness of the FRF pipe, which can guide the design and optimization of FRF pipes in various engineering applications. Future research will further transcend the limitations of static fluid coupling analysis by focusing on the following directions to enhance engineering applicability: First, under actual complex operating conditions, explore the regulatory mechanisms of fluid viscosity and flow velocity on the axial stiffness of pipes. And under the matched asymptotic expansion, the research on the hydroelastic response can be extended by drawing on the work of Mohapatra et al. [32]. Second, incorporate a multiphysics coupling to comprehensively analyze the influence of factors such as temperature effects on the axial stiffness of FRF pipes, providing a theoretical foundation for their application in complex scenarios, including deep-sea energy transportation and ship power systems.

Author Contributions: Conceptualization, J.Y. and Y.Z.; methodology, J.Y.; software, J.Y.; validation, B.Z.; investigation, J.Y.; resources, Y.Z.; data curation, J.Y.; writing—original draft preparation, J.Y.; writing—review and editing, B.Z.; visualization, J.Y.; supervision, Y.Z.; project administration, Y.Z.; funding acquisition, Y.Z. All authors have read and agreed to the published version of the manuscript.

Funding: This study was funded by the Fund for Key Basic Research Projects of Basic Strengthening Program with a grant number of 2020-JCJQ-ZD-222.

Data Availability Statement: The data sets in this study are available from the corresponding authors for appropriate reasons.

Conflicts of Interest: The authors declare no conflicts of interest.

Abbreviations

The following abbreviations are used in this manuscript:

FRF	Fiber-reinforced flexible
UDL	Uniformly distributed load
MPa	Megapascal
C3D8R	Three-dimensional 8-node linear brick element with reduced integration
M3D4R	Three-dimensional 4-node membrane element with reduced integration
WDW	Microcomputer-controlled Electronic Universal Testing Machine
MTS	Material Testing System
ASTM	American Society for Testing and Materials
SSE	Sum of squared errors

References

1. Alabtah, F.G.; Mahdi, E.; Eliyan, F.F. The use of fiber reinforced polymeric composites in pipelines: A review. *Compos. Struct.* **2021**, *276*, 114595. [CrossRef]
2. Diniță, A.; Ripeanu, R.G.; Ilincă, C.N.; Cursaru, D.; Matei, D.; Naim, R.I.; Tănase, M.; Portoacă, A.I. Advancements in Fiber-Reinforced Polymer Composites: A Comprehensive Analysis. *Polymers* **2023**, *16*, 2. [CrossRef] [PubMed]
3. Wei, D.; An, C.; Wu, C.; Duan, M.; Estefen, S.F. Torsional structural behavior of composite rubber hose for offshore applications. *Appl. Ocean. Res.* **2022**, *128*, 103333. [CrossRef]
4. Gao, Q.; Zhang, P.; Duan, M.; Yang, X.; Shi, W.; An, C.; Li, Z. Investigation on structural behavior of ring-stiffened composite offshore rubber hose under internal pressure. *Appl. Ocean. Res.* **2018**, *79*, 7–19. [CrossRef]
5. Fang, P.; Xu, Y.; Gao, Y.; Ali, L.; Bai, Y. Mechanical responses of a fiberglass flexible pipe subject to tension & internal pressure. *Thin-Walled Struct.* **2022**, *181*, 110107.
6. Liu, Q.; Xue, H.; Tang, W.; Yuan, Y. Theoretical and numerical methods to predict the behaviour of unbonded flexible riser with composite armour layers subjected to axial tension. *Ocean. Eng.* **2020**, *199*, 107038. [CrossRef]
7. De Oliveira, J.; Goto, Y.; Okamoto, T. Theoretical and methodological approaches to flexible pipe design and application. In Proceedings of the Annual Offshore Technology Conference, Houston, TX, USA, 6–9 May 1985; Volume 17, pp. 517–526.
8. Wang, Y.; Lou, M.; Liang, W.; Zhang, C. Numerical and experimental investigation on tensile fatigue performance of reinforced thermoplastic pipes. *Ocean. Eng.* **2023**, *287*, 115814. [CrossRef]
9. Jaszak, P.; Skrzypacz, J.; Adamek, K. The design method of rubber-metallic expansion joint. *Open Eng.* **2018**, *8*, 532–538. [CrossRef]
10. Xu, G.m.; Shuai, C.g. Axial and lateral stiffness of spherical self-balancing fiber reinforced rubber pipes under internal pressure. *Sci. Eng. Compos. Mater.* **2021**, *28*, 96–106. [CrossRef]
11. Kruijer, M.; Warnet, L.; Akkerman, R. Analysis of the mechanical properties of a reinforced thermoplastic pipe (RTP). *Compos. Part A Appl. Sci. Manuf.* **2005**, *36*, 291–300. [CrossRef]
12. Xia, M.; Takayanagi, H.; Kemmochi, K. Analysis of multi-layered filament-wound composite pipes under internal pressure. *Compos. Struct.* **2001**, *53*, 483–491. [CrossRef]
13. Xing, J.; Geng, P.; Yang, T. Stress and deformation of multiple winding angle hybrid filament-wound thick cylinder under axial loading and internal and external pressure. *Compos. Struct.* **2015**, *131*, 868–877. [CrossRef]
14. Gao, P.; Gao, Q.; An, C.; Zeng, J. Analytical modeling for offshore composite rubber hose with spiral stiffeners under internal pressure. *J. Reinf. Plast. Compos.* **2020**, *40*, 352–364. [CrossRef]
15. Rogers, T. Problems for helically wound cylinders. In *Continuum Theory of the Mechanics of Fibre-Reinforced Composites*; Springer: Berlin/Heidelberg, Germany, 1984; pp. 147–178.
16. Evans, J.; Gibson, A. Composite angle ply laminates and netting analysis. *Proc. R. Soc. Lond. Ser. A Math. Phys. Eng. Sci.* **2002**, *458*, 3079–3088. [CrossRef]
17. Gu, F.; Huang, C.k.; Zhou, J.; Li, L.p. Mechanical response of steel wire wound reinforced rubber flexible pipe under internal pressure. *J. Shanghai Jiaotong Univ. Sci.* **2009**, *14*, 747–756. [CrossRef]
18. Wang, Y.; Lou, M.; Yang, L.; Wu, L. Study on the tensile properties of reinforced thermoplastic pipes under different internal pressures and temperatures. *Int. J. Press. Vessel. Pip.* **2022**, *200*, 104820. [CrossRef]
19. Khaniki, H.B.; Ghayesh, M.H.; Chin, R.; Amabili, M. A review on the nonlinear dynamics of hyperelastic structures. *Nonlinear Dyn.* **2022**, *110*, 963–994. [CrossRef]
20. Tonatto, M.L.; Tita, V.; Araujo, R.T.; Forte, M.M.; Amico, S.C. Parametric analysis of an offloading hose under internal pressure via computational modeling. *Mar. Struct.* **2017**, *51*, 174–187. [CrossRef]
21. Tong, F.; Lan, F.; Chen, J.; Li, D.; Li, X. Numerical study on the injury mechanism of blunt aortic rupture of the occupant in frontal and side-impact. *Int. J. Crashworthiness* **2022**, *28*, 270–279. [CrossRef]

22. Ye, Y.; Gan, J.; Liu, H.; Guan, Q.; Zheng, Z.; Ran, X.; Gao, Z. Experimental and numerical studies on bending and failure behaviour of inflated composite fabric membranes for marine applications. *J. Mar. Sci. Eng.* **2023**, *11*, 800. [CrossRef]
23. Xiao, Y.; Liu, T.; Meng, C.; Jiao, Z.; Meng, F.; Guo, S. Numerical simulation modeling and kinematic analysis onto double wedge-shaped airbag of nursing appliance. *Sci. Rep.* **2023**, *13*, 14261. [CrossRef] [PubMed]
24. Sosa, E.M.; Wong, J.C.S.; Adumitroaie, A.; Barbero, E.J.; Thompson, G.J. Finite element simulation of deployment of large-scale confined inflatable structures. *Thin-Walled Struct.* **2016**, *104*, 152–167. [CrossRef]
25. Cao, Z.; Wan, Z.; Yan, J.; Fan, F. Static behaviour and simplified design method of a Tensairity truss with a spindle-shaped airbeam. *J. Constr. Steel Res.* **2018**, *145*, 244–253. [CrossRef]
26. American Society for Testing and Materials. *Standard Test Methods for Vulcanized Rubber and Thermoplastic Elastomers–Tension*, d412-16 ed.; ASTM International: West Conshohocken, PA, USA, 2016.
27. Mullins, L. Softening of rubber by deformation. *Rubber Chem. Technol.* **1969**, *42*, 339–362. [CrossRef]
28. Shahzad, M.; Kamran, A.; Siddiqui, M.Z.; Farhan, M. Mechanical Characterization and FE Modelling of a Hyperelastic Material. *Mater. Res.* **2015**, *18*, 918–924. [CrossRef]
29. Ali, A.; Hosseini, M.; Sahari, B.B. A review of constitutive models for rubber-like materials. *Am. J. Eng. Appl. Sci.* **2010**, *3*, 232–239. [CrossRef]
30. Zhang, B.; Zhao, Y.; You, J.; Zhang, Z. Experimental and numerical analysis of rubber isolator dynamic stiffness under hydrostatic pressure. *Ocean. Eng.* **2024**, *314*, 119650. [CrossRef]
31. Dassault Systèmes. About Surface-Based Fluid Cavities. Online Documentation. Available online: https://help.3ds.com/2022x/simplified_chinese/dsdoc/sima3dxanlrefmap/simaanl-c-surfacebasedcavityover.htm?contextscope=onpremise (accessed on 7 April 2025).
32. Mohapatra, S.C.; Guedes Soares, C. Impact of the compressive force on wave diffraction by a circular elastic floater for offshore aquaculture system. In *Innovations in the Analysis and Design of Marine Structures*; CRC Press: Boca Raton, FL, USA, 2025; pp. 13–19.

Journal of
Marine Science and Engineering

MDPI

Article

The Prediction of the Valve Opening Required for Slugging Control in Offshore Pipeline Risers Based on Empirical Closures and Valve Characteristics

Jiqiang Fu [1], Quanhong Wu [1,2,3], Jie Sun [4], Hanxuan Wang [1] and Suifeng Zou [1,*]

1 State Key Laboratory of Multiphase Flow in Power Engineering, Xi'an Jiaotong University, Xi'an 710049, China; fujiqiang@stu.xjtu.edu.cn (J.F.); 20103578@ceic.com (Q.W.); whx666@stu.xjtu.edu.cn (H.W.)
2 Longyuan (Beijing) New Energy Engineering Technology Co., Ltd., Beijing 100034, China
3 National Energy Wind Power Operation Technology R&D Center, Xi'an 710309, China
4 School of Oil & Natural Gas Engineering, Southwest Petroleum University, Chengdu 610500, China; sunj101@swpu.edu.com
* Correspondence: zou_suifeng@xjtu.edu.cn

Abstract: Topside choking is a common way to eliminate severe slugging flow in pipeline riser systems in offshore oil and gas fields. However, a lack of fundamentals in two-phase flow gives rise to difficulty in the model selection of valves and the effective control of the valves. In this study, the prediction of the valve opening required for slugging control based on measurable parameters is investigated experimentally and theoretically. It is found that the resistance coefficient factor of the valve is almost the same for pipeline risers and simple vertical pipes when severe slugging is eliminated. Therefore, fluid parameters can be approximated by the conditions of a simple vertical pipe. The target of control is to achieve dual-frequency fluctuation, and it is quantitatively converted to the pressure drop of the valve. Based on these two empirical enclosures, the valve opening can be worked out by using the gas fraction model and the theoretical model of valve flow resistance. The non-slip model is found to be better than the drift-flux model in the final prediction of the optimal valve opening. An explicit model for the calculation of the optimal resistance factor and the corresponding valve opening is established, making it more convenient to select the valve in the design stage of offshore oil and gas exploitation. The average absolute error of the proposed model is +0.01%, which is smaller than the simulation performed by OLGA 7.0 software (+4.91% before tuning and +0.08% after tuning). A field case with a flexible S-shape riser proves the good applicability of the model (with a deviation smaller than ±2%). The applications of the prediction model in the model selection of the valve and uncertain factors in the operation are also discussed.

Keywords: two-phase flow; slugging control; topside choking; valve opening; prediction model

1. Introduction

In offshore oil and gas exploitation, construction difficulties and high cost are encountered in the application of single-phase pipelines with underwater gas–liquid separators. Therefore, gas–liquid mixture pipelines are much more common in the transportation of oil and gas from seabed wells to onshore separation equipment [1], and they are the lifeline of offshore energy transportation [2]. In the mixed flow process of gas–liquid, two-phase flow, different flow patterns appear due to the different flow velocity and compressibility of each phase. Different flow patterns sometimes have different effects on pipeline risers and related equipment [3]. A hazardous two-phase flow regime and its control are among the key issues of flow assurance for gas–liquid mixture transporting pipelines in offshore oil and gas fields. The typical appearance of the hazardous two-phase flow regime is severe slugging. This flow regime occurs at small to moderate flow rates as a result of pipeline riser geometry and seabed terrain [4]. Severe slugging is a cyclic flow condition, which is commonly divided into four stages: liquid slug growth, liquid slug production, gas penetration and blowout, and liquid fallback and formation of liquid blockage [5–8], as shown in Figure 1. This flow regime has occurred in oil fields in the North Sea [9,10], Bohai Sea [11,12], and South China Sea [13]. The hazard of this flow regime lies in many aspects of operation processes, which include kinetic force on fittings and vessels, pressure cycling, control instability, and inadequate phase separation [14,15]. As the gas flow at the riser top is intermittent, gas turbine generators on the offshore platform will be shut down frequently, and the switch to diesel generators will result in greater cost and more pollutants [16]. On the other hand, the blowout stage is the fast outflow of gas and liquid at the riser top, which may not only result in separator overflow [17] but also cause pressure-fluctuation-induced vibration of the platform structure [18]. Therefore, this hazardous flow regime must be eliminated or controlled so as to ensure the operational safety of downstream facilities on the platform.

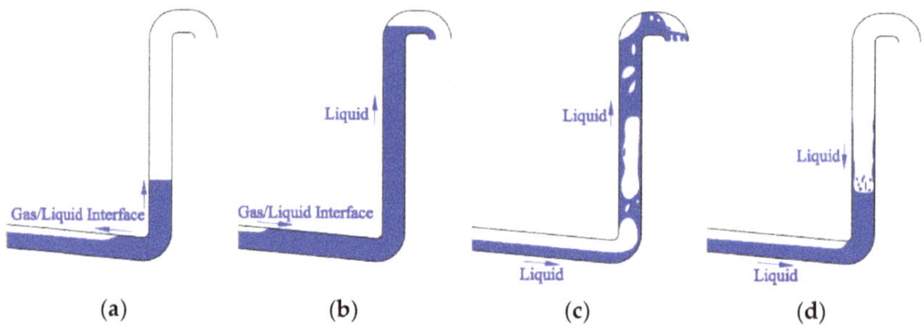

Figure 1. A cycle of typical severe slugging. (**a**) Liquid slug growth; (**b**) liquid slug production; (**c**) gas penetration and blowout; (**d**) liquid fallback.

It is widely agreed that severe slugging cannot be completely avoided through the optimization of pipeline design [19]. So, measures should be taken to control (mitigate) and eliminate the flow regime, which we call slugging control. The most widely applied control method is topside choking [20]. A regulating valve is installed at the connection pipe between the riser top and the separator. By adjusting its opening, the peak value of the transient flow rate during the blowout stage is reduced. Meanwhile, the flow resistance across the valve fluctuates with the alternate passing of gas and liquid slugs. The fluctuation, in turn, enhances the entrance of gas (and suppresses liquid blockage) at the riser bottom as a result of the fluctuating, pressure-driven force.

According to laboratory studies [21,22], dynamic control can achieve better results, but it is less applicable at present, not only due to the difficulties in parameter setting and tuning of the dynamic control model, but also because the dynamic control model works only within a limited interval around the optimal valve opening, while there is no practical prediction method. In conventional studies of slugging control [23], the flow resistance factor is an adjustable parameter in the flow model that is not mapped to the opening. For the conventional proportional–integral (PI) control model, the performance depends on the relation between the process variable and the manipulated variable. If the relation is non-linear, integral control would result in sharp tuning of valve opening and significant overshoot of pressure (which may exceed the design pressure). If integral control is abandoned, the tuning is rather slow, and it would take tens of slugging cycles for elimination, as in the case presented in [13]. If the optimal opening can be estimated, the tuning can be more effective without causing safety problems. Then, the questions become clear—how to select the model of regulating valve, and how to find the optimal opening of the valve. This study focuses on the second issue, which also helps to find the working range of the dynamic model in the first issue.

Manual traversing is an awkward method that can only be applied in a laboratory setting to determine the reference optimal opening. This is because the response of the flow to control is slow in an offshore pipeline with tens of km or more than 100 km. An operator is impatient when spending tens of days seeking the optimal value, and the manual attempt to find the optimal opening was unsuccessful in an offshore oil field in China (hereinafter called Oil Field W) [13]. Based on certain rules, manual traversing can be replaced by an automatic scheme [24,25]. For Oil Field W, it took four days for the absolute change of 7% in valve opening [13]. However, the setting of the manipulated variable and the parameters of the automatic control model require inexpressible experience. For an automatic method without knowledge of pipe flow, the search for the optimal opening value cannot be sped up tremendously. Theoretically, if the flow rate (or daily production) could be metered, and the fluid properties and the factory parameters of the regulating valve were known, then the optimal opening could be determined. Unfortunately, there is no published model within the knowledge of the authors. The closest models are conventional flow regime transition criteria [8,26,27], which use the resistance factor as a known parameter to predict whether the flow regime is severe slugging. In other words, the calculation of the resistance factor is implicit and requires iteration or trials. In these models, the condition for the occurrence of the blowout stage was analyzed and regarded as the criterion. However, the volume of fluids in the riser makes the outflow condition more complex than the assumption of local flow patterns in the models, and the Initiation of a blowout is not always linked to the necessity of activating valve control. As Pedersen et al. [28] pointed out, an unstable flow regime is not always an unstable system. Andreolli et al. [29] attempted to use the discharge coefficient of a choke device, or C_d, to derive a stability criterion. C_d was set based on experience. The relation between the valve opening and flow regime transition is deeply implicit, and the result was qualitative.

There are two major problems to be solved before the prediction model can be established. First, what condition is the optimal condition (i.e., the target of control), and second, what are the values of some macroscopic flow parameters (pressure or pressure drop) at the optimal condition? In published works, whether severe slugging was eliminated or controlled was judged by observing the pressure fluctuation amplitude. If the amplitude was smaller than a given threshold, then severe slugging was regarded as eliminated. However, the threshold was not related to flow appearance or mechanism.

The obstacle In finding the relation between the parameters and the optimal condition lies in the two-phase flow model. Jansen et al. [26], Malekzadeh et al. [8], and Fadairo et al. [27] could predict a choke coefficient for eliminating severe slugging; however, the coefficient is NOT non-dimensional, which can not be mapped to the factory parameters of the regulating valve. Thus, the optimal opening remains unknown. There are two types of mathematical descriptions for gas–liquid two-phase flow in a pipeline riser: one simple model consists of four separated sub-models (corresponding to four stages of severe slugging); another model is based on governing equations of two-fluid flow (like OLGA software, which will be discussed later, and the codes developed by China University of Petroleum; see [30]). However, neither of them can directly solve for the optimal valve opening, and the settings require experience [29]. It is less effective with more effort, as an average-level designer or operator can only tune the diameter and C_d in the default orifice model in OLGA software, while the model cannot describe all valve types, which will be discussed later. A designer or operator has to try a few input opening values before the optimal value can be found, which is estimated to take several days. Moreover, the accuracy of prediction depends on the simulated flow regime or the accuracy of the two-phase flow model itself. If the simulated regime is not severe slugging while the actual condition is, then the opening can never be worked out. Therefore, it is necessary to establish an explicit model to predict the optimal opening with the information in the specification file of the valve and with fewer requirements for experience.

This study combines experiments and theoretical modeling. Choke experiments with different valve openings at different superficial velocities are performed (Section 2). The target of control is determined by the continuity of outflow (Section 3.1). Based on the comparison between the pipeline riser and a simple vertical pipe, the simplification of the two-phase flow model is discussed, and the magnitude of macroscopic flow parameters at the optimal condition is determined (Section 3.2). Then, a prediction model based on factory parameters of the regulating valve is established (Section 3.3). The predicted results are compared with experimental results and those predicted using OLGA software simulation (Section 4.1). Finally, the model is validated by an industrial case, and how to select the model of the regulating valve is briefly discussed (Section 4.2).

2. Experiment

2.1. Setup of Experimental Loop

The sketch of the experimental loop is shown in Figure 2. Although the loop itself is an oil–water–gas three-phase system, only water and air sources are displayed in Figure 2 since oil was not used in the experiments. The test section consists of a horizontal pipeline, a downward inclined pipe with an inclination angle of 5° to horizontal, and a vertical riser. The lengths of the three sections are 114 m, 20.4 m, and 16.3 m, respectively, and the inner diameter of the whole test section is 50 mm. Most of the test section is made of stainless steel, except for a transparent section of 2.75 m in the middle of the riser. The fluids, namely, air and tap water, are metered separately before mixing and entering the pipeline. A gas–liquid separator is installed at the highest position of the experimental facility. The riser top and the entrance of the separator are connected by an electric ball valve, with a nominal diameter of 50 mm. The deviation of the feedback opening from the set opening is ±0.5%. A cylinder with an inner diameter of 0.5 m is installed right below the separator to measure the accumulated liquid outflow, and the outflow rate can therefore be derived. An injection hole is located at the riser elbow, which is 40 cm above the lowest position of the elbow. The pressure and differential pressure transducers that are involved in this study are also marked in Figure 2. For detailed information on transducers and flow meters, see Table 1. More details of the experimental facility are available in [31], which presents the dimensions of the horizontal section.

Table 1. Installation and accuracy of transducers and flow meters.

Transmitter	Manufacturer	Transmitter Model	Range	Accuracy
Orifice Flowmeter (L)	YOKOGAWA	EJA-115	0–1.35 $m^3 \cdot h^{-1}$	±1% F.S.
Orifice Flowmeter (M)	YOKOGAWA	EJA-115	0–0.22 $m^3 \cdot h^{-1}$	±1% F.S.
Orifice Flowmeter (S)	YOKOGAWA	EJA-115	0–0.027 $m^3 \cdot h^{-1}$	±1% F.S.
Electromagnetic Flowmeter (L)	YOKOGAWA	AE204MG	0–42.5 $m^3 \cdot h^{-1}$	±0.5% F.S.
Electromagnetic Flowmeter (S)	YOKOGAWA	AE115MG	0–6.36 $m^3 \cdot h^{-1}$	±0.5% F.S.
Mass Flowmeter (L)	Siemens	Mass 6000	0–80 $kg \cdot h^{-1}$	±0.1% F.S.
Mass Flowmeter (S)	Emerson	CMF 010	0–24 $kg \cdot h^{-1}$	±0.1% F.S.

Table 1. *Cont.*

Transmitter	Manufacturer	Transmitter Model	Range	Accuracy
Pressure (P5/p_b)	Micro	MPM480	0–500 kPa	±0.25% F.S.
Differential Pressure (DP5/Δp_R)	Rosemount	3051CD3	−20–200 kPa	±0.15% F.S.
Differential Pressure (DP11/Δp_v)	Rosemount	3051CD3	0–240 kPa	±0.15% F.S.
Differential Pressure (DP12)	Rosemount	3051CD3	0–240 kPa	±0.15% F.S.

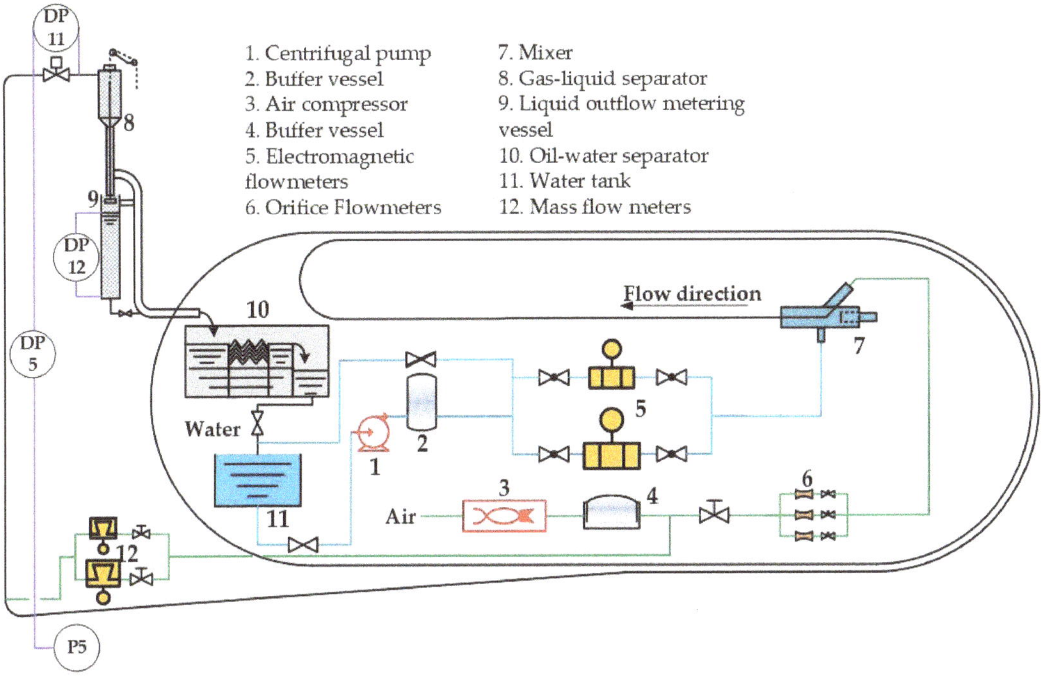

Figure 2. Sketch of experimental loop.

2.2. Experimental Condition

The experiments were performed at room temperature. The pressure of the water supply was maintained at 1.2 Mpa (absolute) by regulating the inlet valve of the metering section and the valve on the return line. The pressure of the air supply was 0.9 Mpa (absolute).

The experimental work consists of choking experiments for pipeline risers and for a simple vertical pipe. Under the former condition, air was injected at the mixer, and the flow regime in the pipeline risers was severe slugging (Figure 3); under the latter condition, air was injected at the riser bottom, i.e., only the riser was in a state of two-phase flow. The opening of the regulating valve ranged from 15% to 100%. Nine cases of choking experiments for pipeline risers and the corresponding cases for the simple vertical pipe were conducted initially (see [31]), and another eleven cases for pipeline risers were performed to expand the dataset for validation. For all experimental cases, the valve opening was adjusted in a decreasing manner. The superficial velocities of the experiments were as follows:

Liquid superficial velocity, u_{SL} (m/s): 0.10, 0.25, 0.45, and 0.60;

Gas superficial velocity, u_{SG0} (m/s): 0.10, 0.25, 0.45, 0.60, and 1.00.

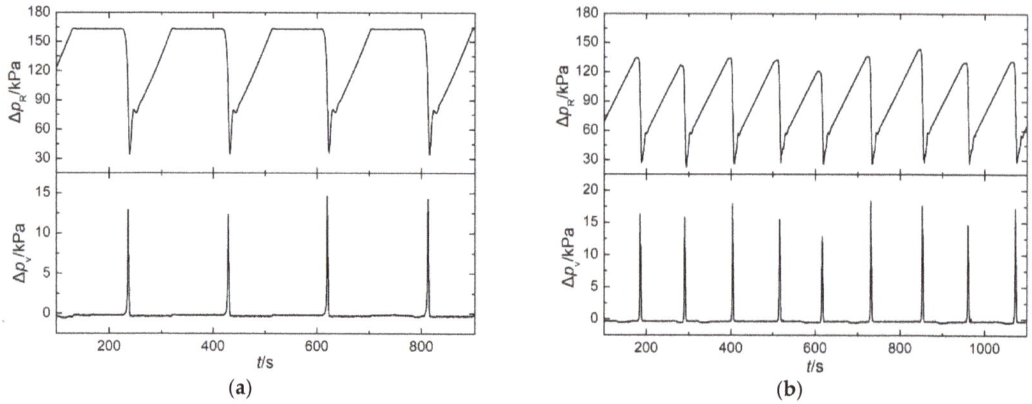

Figure 3. Pressure and differential pressure trend of severe slugging. (**a**) u_{SL} = 0.25 m/s, u_{SG0} = 0.15 m/s, SS1; (**b**) u_{SL} = 0.10 m/s, u_{SG0} = 0.60 m/s, SS2.

2.3. Flow Coefficient

The flow coefficient and the type of inherent flow characteristic of the regulating valve were unavailable because the nameplate corroded after more than ten years. Instead, a calibration test using single-phase water was performed, as shown in Figure 4. For the opening between 10% and 35%, the flow coefficient, K_V, showed a log-linear relation with the valve opening, Z. For Z > 35%, K_V varied significantly with Z. A log-linear regression was performed for Z = 10–35%, and the regression line was applied for prediction. For Z > 35%, the K_V values at u_{SL} = 0.80 m/s were applied.

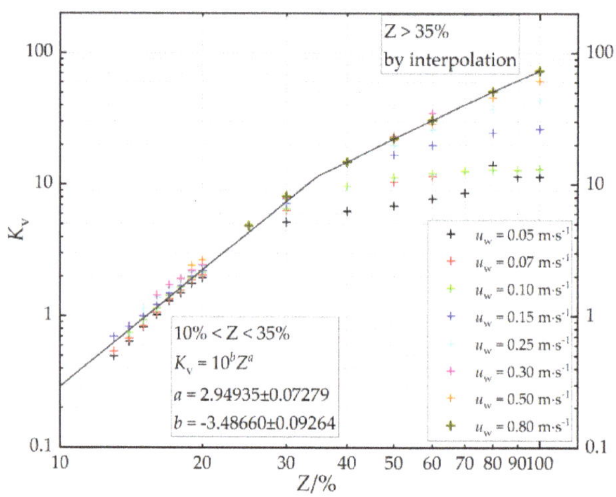

Figure 4. Calibration of flow coefficient for the regulating valve on the experimental loop.

3. Derivation of Optimal Valve Opening

3.1. Target of Control

Figure 5 shows the trend of pressure and differential pressure with the gradual turning down of the regulating valve. Four stages can be divided. When the opening was still large, cyclic flow cut-off was still observed, as indicated by the cyclic zero value of the pressure drop across the valve, Δp_V. With the decrease in opening, the duration of the flow cut-off reduced and reached zero. With a further decrease in opening, dual-frequency fluctuation appeared. Dual frequency is a condition where the long cycles of pressure fluctuation with large fluctuation still exist but are superimposed by short cycles with small fluctuation. Under this condition, the blowout stage is mitigated and turns into the fast outflow of a series of slugs, and the outflow covers the whole cycle of slugging. Finally, the low-frequency component disappeared, and the flow became stable with small pressure fluctuations. Dual-frequency fluctuation and stable flow were observed as slug flow with variable speed.

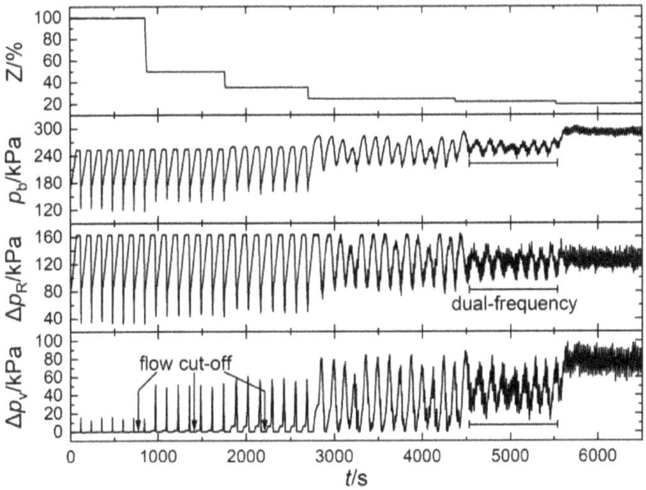

Figure 5. Pressure and differential pressure trend of severe slugging with decrease in valve opening (u_{SL} = 0.25 m/s, u_{SG0} = 0.25 m/s).

The time-averaged pressure at the riser bottom increased monotonically with the decrease in opening. However, the maximum pressure did not show a monotonous relationship with opening. It first increased, then decreased, and finally rose again with the decrease in opening. The flow regime at the time the maximum pressure rose again was dual-frequency fluctuation. Since the minimum value of Δp_V was above zero, the maximum value could be regarded as no greater than twice its average value, as validated by 16 experimental cases listed in Table 2 (DP12 was unavailable for the remaining 4 cases). The outflow rate is smoothed by a 10-second window, as transient values are of no significance. The condition with the ratio of peak to average smaller than 2 was regarded as safe. Therefore, the target of control was set as dual-frequency fluctuation.

Table 2. Peak value of liquid outflow rate when dual-frequency fluctuation appears.

No.	u_{SL} (m/s)	u_{SG0} (m/s)	Peak Value of u_{SL} at Outlet (m/s)	Ratio (Peak Over Average)
1	0.10	0.10	0.176	1.76
2	0.10	0.25	0.188	1.88
3	0.10	0.60	0.154	1.54
4	0.10	1.00	0.171	1.71
5	0.25	0.10	0.355	1.42
6	0.25	0.25	0.398	1.59
7	0.25	0.60	0.387	1.55
8	0.25	1.00	0.447	1.79

Table 2. *Cont.*

No.	u_{SL} (m/s)	u_{SG0} (m/s)	Peak Value of u_{SL} at Outlet (m/s)	Ratio (Peak Over Average)
9	0.45	0.10	0.564	1.25
10	0.45	0.25	0.591	1.31
11	0.45	0.60	0.713	1.58
12	0.45	1.00	0.601	1.34
13	0.60	0.10	0.775	1.29
14	0.60	0.25	0.750	1.25
15	0.60	0.60	0.772	1.29
16	0.60	1.00	0.719	1.20

3.2. Macroscopic Flow Parameters at Optimal Condition

According to Figure 5, the average pressure under the condition of dual-frequency fluctuation was considered roughly the same as the maximum pressure under the condition of Z = 100%, or $p_s + \rho_L g H$. This approximation is not precise, as indicated by Figure 6; however, it is still acceptable, which will be discussed in Section 4.1. Then, an empirical closure could be established:

$$p_b = p_s + \rho_L g H \tag{1}$$

In other words,

$$\Delta p_g + \Delta p_v = \rho_L g H(1 - \alpha_R) + \rho_G g H \alpha_R + \Delta p_v = \rho_L g H \tag{2}$$

Rewrite Equation (2), and we derive

$$\Delta p_v = (\rho_L - \rho_G) g H \alpha_R \tag{3}$$

where

$$\alpha_R = \frac{\alpha_b + \alpha_t}{2} \tag{4}$$

The relation between pressure drop and flow velocity is quadratic, which is expressed by

$$\Delta p_v = \frac{f}{2}\rho_{mix,t} u_{mix,t}^2 = \frac{f}{2}[\rho_L(1 - \alpha_t) + \rho_{G,t}\alpha_t](u_{SL} + u_{SG,t})^2 \tag{5}$$

The next question is how to calculate the resistant factor, f. The average of f for each case was calculated based on the time averaged superficial velocity upstream of the valve and the time-average Δp_v recorded in the experiment. It is interesting to find that the resistant factor for pipeline risers became approximately the same as that for a simple vertical pipe at the same opening after dual-frequency fluctuation appeared, or f ratio ≈ 1, as shown in Figure 7 and Table 3. In Figure 7, the resistant factor, f, is larger for pipeline risers than for a simple vertical pipe before severe slugging is eliminated. The reasons can be explained as follows: before severe slugging is eliminated, severe blowout will cause large peak values of Δp_v. As flow resistance is approximately proportional to the square of flow velocity, the time-averaged Δp_v is significantly larger for pipeline risers than for a simple vertical pipe. While $\rho_{mix,t}$ and $u_{mix,t}$ change little for both pipeline risers and a simple vertical pipe, f is also larger for pipeline risers. With a smaller opening, the blowout becomes less severe, and the time average Δp_v becomes closer to the condition of a simple vertical pipe. Therefore, the deviation of Δp_v or f from those of a simple vertical pipe can act as an indicator of whether severe slugging is eliminated, and it is possible to calculate f based on the time-averaged flow parameters in a simple vertical pipe.

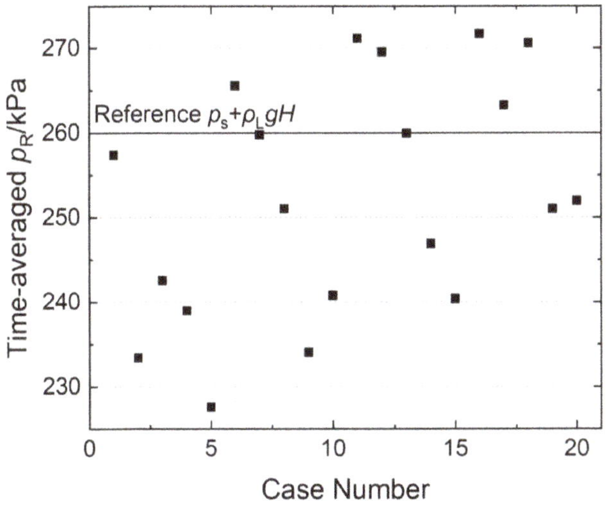

Figure 6. Average pressure at the riser bottom when dual-frequency fluctuation appears. Cases 1–5: u_{SL} = 0.10 m/s; Cases 6–10: u_{SL} = 0.25 m/s; Cases 11–15: u_{SL} = 0.45 m/s; Cases 16–20: u_{SL} = 0.60 m/s. Cases 1, 6, 11, 16: u_{SG} = 0.10 m/s; Cases 2, 7, 12, 17: u_{SG} = 0.25 m/s; Cases 3, 8, 13, 18: u_{SG} = 0.45 m/s; Cases 4, 9, 14, 19: u_{SG} = 0.60 m/s; Cases 5, 10, 15, 20: u_{SG} = 1.00 m/s.

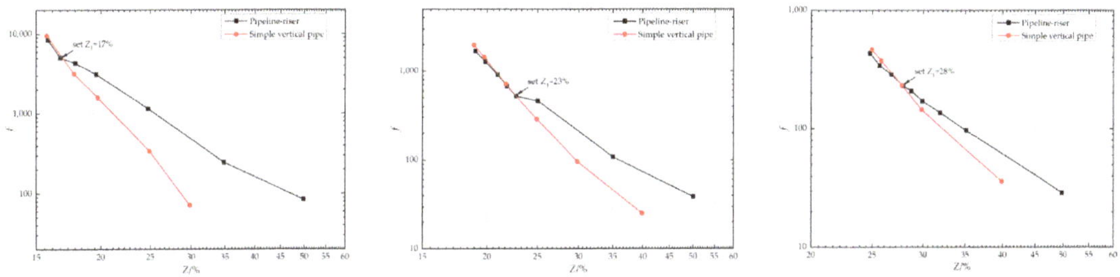

Figure 7. Relation between resistant factor and valve opening. (**a**) u_{SL} = 0.10 m/s, u_{SG0} = 0.10 m/s; (**b**) u_{SL} = 0.25 m/s, u_{SG0} = 0.60 m/s; (**c**) u_{SL} = 0.45 m/s, u_{SG0} = 0.25 m/s.

Table 3. Valve opening values when certain phenomena appear.

No.	u_{SL} (m/s)	u_{SG0} (m/s)	Set Z_1 * %	Set Z_2 * %	f Ratio * at min (Z_1, Z_2)
1	0.10	0.10	17	17	0.90
2	0.10	0.25	18	18	1.03
3	0.10	0.60	18	17	0.90
4	0.25	0.10	22	22	0.98
5	0.25	0.25	22	23	0.92
6	0.25	0.60	23	23	0.99
7	0.45	0.10	28	27	1.06
8	0.45	0.25	28	26	0.95
9	0.45	0.60	26	27	0.86
10	0.25	0.60	23	23	0.99

* Z_1 is the opening value below which the time-averaged resistance factor of the valve is approximately the same for pipeline risers and simple vertical pipes. Z_2 is the opening value below which the flow condition in the riser is dual-frequency fluctuation or stable flow. Deviation within ±2% is acceptable, as explained in Section 4.1. The "f ratio" refers to the time-averaged f for pipeline risers over that for simple vertical pipes.

3.3. Optimal Valve Opening

For a simple vertical pipe, the gas fraction, α, is calculated by non-slip model (Equation (6)), although more accurate drift-flux models are available. The deviation between drift-flux models will be discussed in Section 4.1.

$$\alpha = \frac{u_{SG}}{u_{SG} + u_{SL}} \tag{6}$$

Since u_{SG} is related to local pressure, α_t and Δp_v have to be solved in an iterative manner. After the flow parameters are solved, f can be worked out by Equations (6) and (7). With f calculated, the corresponding opening value, Z, can be determined. The relation between f and Z is connected by the flow coefficient, K_v. The definition of K_v is the flow rate of liquid (usually water, in m^3/h) at the resistance of 1×10^5 Pa, i.e.,:

$$\frac{f}{2}\rho_L\left(\frac{K_v}{3600A}\right)^2 = 10^5 \tag{7}$$

Rewrite Equation (7), and we derive

$$K_v = 900\pi d_v^2\sqrt{\frac{200000}{f\rho_L}} \tag{8}$$

Then, Z can be obtained by the relation between K_v and Z.

4. Validation

4.1. Laboratory Results and Discussion

The experimental results, i.e., the optimal valve opening listed in Table 2, were used to validate the model established in Section 3. The flowchart is shown in Figure 8. The final result of the model was predicted by the K_v-Z relation of the valve shown in Figure 3. The predicted optimal opening versus the reference value is plotted in Figure 9. The absolute deviation ranges from −1.73% to +4.91%, with the average of +0.60%. The cases where obvious positive deviation occurred were those with smaller gas fraction, where more pressure was required (the cases above the reference line in Figure 6). Evidently, the reason is that Δp_v was underpredicted. The deviation of most cases was within an acceptable range of ±2% due to the actual property of the actuator of common automatic valves, whose deviation of feedback opening from the set opening can be as large as ±2%. For negative deviation, our recent related study showed that no obvious "overchoking" (increase in pressure from the value at reference opening) was found; for the specific experimental apparatus, the increase was smaller than 10 kPa for most cases [32]. Although a few cases show larger deviation, the predicted opening could still act as the average opening for dynamic choking once the reliability of the frequent action of the valve could be overcome, since the opening interval within which dynamic choking works could be 10–20% [21,22].

The optimal valve opening was also investigated using OLGA 7.0 software, an industry-standard tool for transient simulation of multiphase petroleum production developed by SPT Group. The geometry and nodes for simulation are presented in Figure 10. It should be noted that the default valve model in OLGA, "Orifice", is unable to describe the specific valve. In the default orifice model, only a fixed orifice diameter and a fixed C_d value are set, and the flow characteristics are also fixed. As a result, different velocities require different C_ds to make the optimal valve opening consistent with experimental results, which is impossible for a single valve. For Cases 1–8 in Table 2, the tuned C_d ranges from 0.12 to 0.35 to obtain the experimentally determined optimal opening, while the default C_d is 0.84 in OLGA 7.0 software, and the recommended C_d by Al-Safran and

Kelkar [33] was 0.75. Hence, the flow characteristics of the valve (Figure 4) have to be input manually. Since the software uses C_V, the flow coefficient in British units, the K_V-Z relation is replaced by the C_V-Z relation, where $C_V = 1.156\ K_V$. The C_V-Z relation was input as a "Table" on "Library", and the "Model" of the valve was set as "Hydrodynamic". As displayed in Figure 9, the optimal valve opening for all cases was overpredicted. The deviation for most cases was above +2%, which is unacceptable. A detailed check on each case finds that the predicted flow regime for the case (u_{SL} = 0.60 m/s, u_{SG0} = 0.10 m/s) at Z = 100% is stable, with continuous gas outflow at the riser top; the predicted flow regime for the case (u_{SL} = 0.60 m/s, u_{SG0} = 0.25 m/s) at Z = 100% is almost stable. So, for the specific facility, the reason why OLGA software overpredicts the optimal opening lies in the deviation of flow regime prediction.

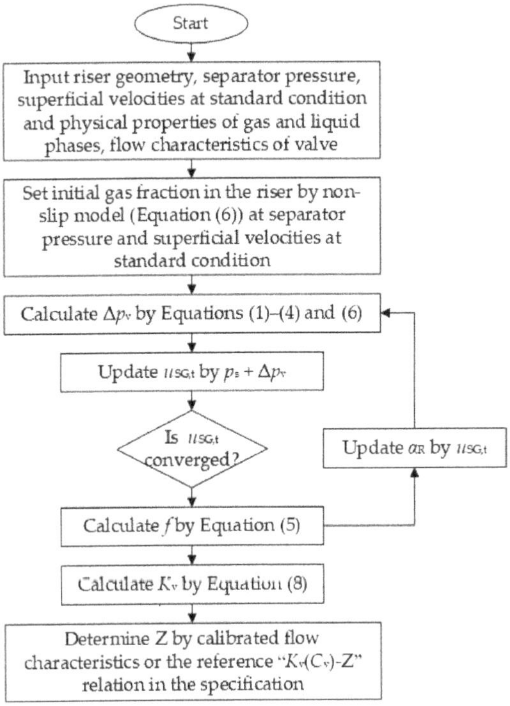

Figure 8. Flowchart of the proposed model.

It is possible to tune the "table" of the C_V-Z relation to obtain better predicted results; see Figure 11. After the "table" is tuned, the average deviation is reduced from +4.91% to +0.08%, and the absolute error of each case is within ±2%. However, in the real application, there are NO test data for tuning the parameter before severe slugging occurs. So, it must follow the factory data or calibration data before its application.

For the above results, the gas fraction was calculated using the non-slip model. In fact, the gas fraction should be smaller because of the drift flow of gas. A widely used drift-flux model [34] was also used to calculate the gas fraction. However, the opening is seriously overpredicted, as shown in Figure 12. Two aspects may lead to these results. One is that the time-averaged gas fraction is larger for pipeline risers than for simple vertical pipes due to severe flow intermittency, as observed in the previous study [31]; so, Δp_V is underpredicted. In addition, smaller gas fractions result in a bigger average mixture density and, hence, a smaller resistant factor and a positive deviation of predicted opening. The other is that the drift flow of gas recedes with the decrease in the valve opening. Direct observation at the transparent section in the riser found that the slug flow at the choking condition was not at a steady velocity. Owing to the pressure fluctuation at the valve induced by the alternate passing of gas and liquid slugs, the velocity rose when gas passed the valve and dropped when liquid passed the valve. In other words, the slug flow "stepped" upwards at the choking condition. With liquid and gas velocities getting closer, the gas fraction should be larger. One more factor in the experimental data processing may also lead to better accuracy of the non-slip model. In this study, the experimental two-phase flow resistant factor was calculated by the mixture velocity, and the average fluid density used to calculate the resistant factor was calculated using the non-slip model. We have no other choice because the actual gas and liquid flow velocities were unmeasurable. We believe this is also an important reason why the non-slip model performs better.

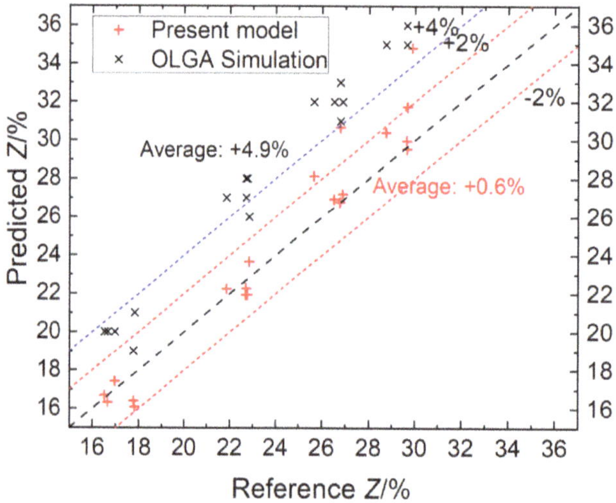

Figure 9. Predicted optimal opening versus the reference value obtained by manual traversing. The dashed red line represents a deviation of 2%, while the dashed blue line represents a deviation of 4%.

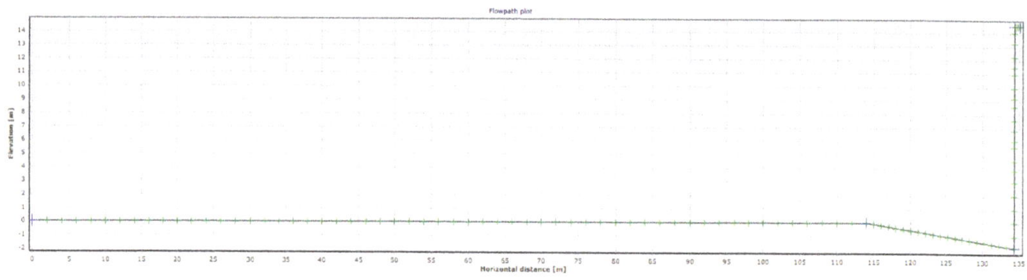

Figure 10. Geometry and nodes of flowpath for OLGA simulation.

However, even if the theoretically largest gas fraction is substituted into Equation (3), Δp_V is still underpredicted at smaller gas–liquid ratios (the cases above the reference line in Figure 6). The underprediction leads to a large deviation of Z because of the small absolute value of Δp_V. For correction, we suggest that α_R in Equation (3) should be substituted by α_t, and we derive

$$\Delta p_V = (\rho_L - \rho_G) g H \alpha_t \quad (9)$$

When Equation (9) is combined with Equations (5), (6), and (8), the positive deviation at a small gas–liquid ratio (GLR) decreases significantly, while the negative deviation at a large GLR shows only a slight increase, as demonstrated in Figure 12. The absolute deviation ranges from −1.93% to +2.87%, with a mean value of +0.01%. Therefore, correction Equation (9) should be applied.

Finally, it should be noted that the proposed model is independent of flow regime determination. The model should only be applied only when necessary and should therefore be used in conjunction with appropriate flow regime transition criteria. These criteria should be selected based on actual control requirements [27].

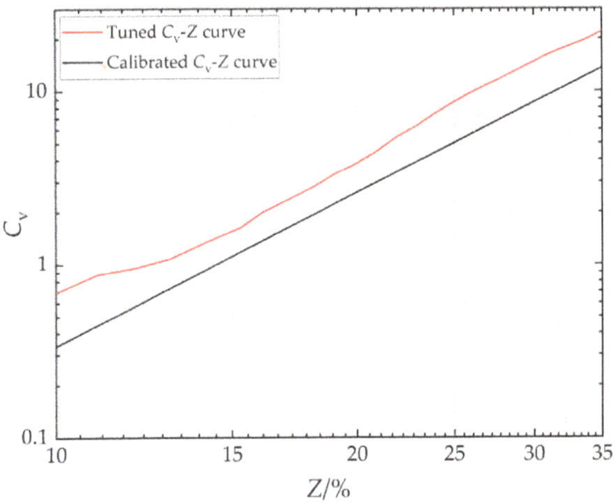

Figure 11. C_V-Z relation before and after tuning for OLGA simulation.

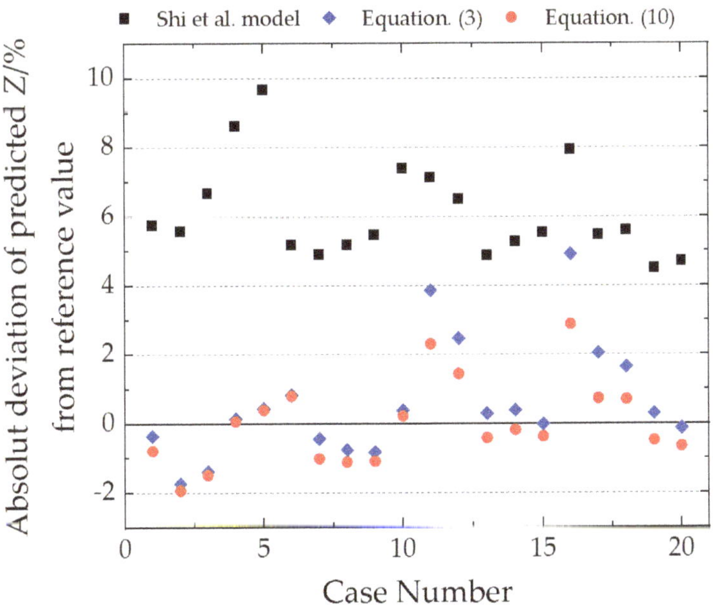

Figure 12. Deviation of predicted optimal opening. Case numbers are the same as those in Figure 6.

4.2. Field Case

Oil Field W consists of one FPSO and five wellhead platforms [13]. The pipelines connecting platform W8-3A to FPSO via platform W14-3A experienced severe slugging issues [35], with the flowpath configuration illustrated in Figure 13. Initially, operators tried manual tuning of the regulating valve with the "equal percentage" characteristic, and set the opening as 27%. Following implementation of an automatic model [24], the value opening was gradually increased over four days and eventually stabilized at 34% [13], which was determined to be the optimal setting at Z = 34%. The production data from this period were used to validate the current model. The known conditions, including pipeline geometry and measurable parameters, are summarized in Table 4. Based on experimental work by Ye et al. [36], the flow pattern of the oil–water mixture was considered to be like an oil-in-water emulsion; consequently, the surface tension was assumed to be equivalent to that of water. The average temperature of the wellhead and ambient was taken as the fluid temperature. The rangeability of the valve, R, which is the ratio of $K_{v,max}$ (or $C_{v,max}$) over $K_{v,min}$ (or $C_{v,min}$), was not recorded on the nameplate; and it was assumed to be 50, which is a common value. Then, the ideal K_v-Z relation of this specific valve was obtained according to the definition of equal percentage characteristic:

$$Z = \frac{100}{\lg R}\lg\frac{K_v}{K_{v,min}} = 58.859191\lg\frac{1.156K_v}{C_{v,min}} \tag{10}$$

The diameter of the regulating valve is different from that of riser. Therefore, the target resistant factor was calculated based on the superficial velocities at the valve location, with the results presented in Table 5. Due to the flow characteristic of "equal percentage", a large deviation of Δp_v does not translate to a significant deviation in Z. The predicted optimal opening of 33.06% shows satisfactory agreement with the reference value, falling within an absolute deviation of ±2%, which demonstrates both the accuracy of the model and its applicability to a flexible S-Shape riser. The case is of a large GLR, and Equation (3) gives a similar predicted opening of 33.31%. However, if Shi et al.'s model [34] is applied, Z is severely overpredicted (51.07%).

Table 4. Known conditions in Oil Field W used for validation.

Parameter	Value	Unit
H	138.9	m
d_R	250.9	mm
$p_{s,avg}$	560	kPa(a)
$T_{wellhead}$	329.25	K
$T_{ambient}$	303.15	K
Q_L	1985	m^3/d
Q_{G0}	102,705	m^3/d
ρ_L	850.7	kg/m^3
ρ_{G0}	1.179	kg/m^3
$\Delta p_{v,avg}$	566	kPa
σ	0.0691	N·m
d_v	250	mm
$C_{v,max}$	1000	$gal{\cdot}min^{-1}{\cdot}psi^{-1}$
valve flow characteristic	equal percentage	-

Under actual operating conditions, the flow characteristics may deviate from the factory specifications, and the calibrated rangeability often differs from ideal conditions. For instance, Chinese-manufactured values with equal percentage characteristics typically exhibit rangeability values between 30 and 50. As defined in the IEC standard 60534-

2-4:2009 [37], the calibrated rangeability value is defined as the K_v at Z = 100% over K_v at Z = 5%. However, the same standard strictly limits deviations from ideal flow characteristics (log-linear relation for equal percentage characteristics) within the operating range of Z = 10–80%. So, it is still reasonable to adopt the ideal flow characteristics. Table 6 shows the results concerning the above factors.

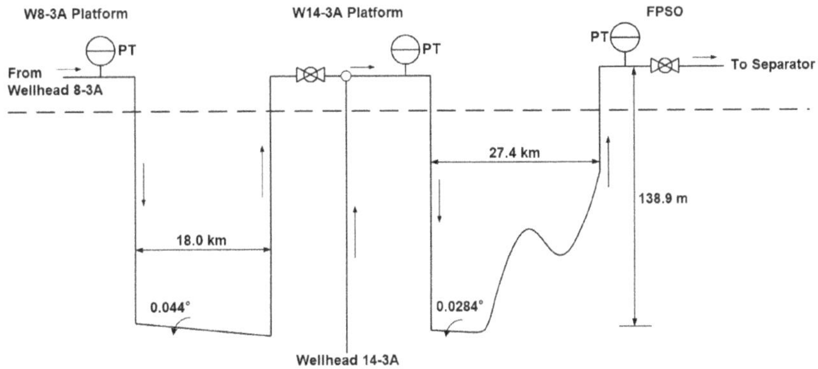

Figure 13. Sketch of flowpath in Oil Field W, where severe slugging occurred.

Table 5. Calculated results for the case of Oil Field W.

Parameter	Value	Unit
α_t (α_v)	0.7965	
α_b	0.7673	
α_R	0.7819	
Δp_v	887.87	kPa
$u_{SL,v}$	0.465	m/s
$u_{SG,v}$	1.819	m/s
$\rho_{mix,v}$	185.52	kg/m^3
f	1810	
K_v	63.69	m^3·h^{-1}·(10^5 Pa)$^{-1}$
Z	33.06% (Equation (9))	

Table 6. Calculated results of Equation (9) for the case of Oil Field W concerning the deviation of valve flow characteristics.

Rangeability	Definition of Rangeability	Predicted Z/%
50	$K_v(Z = 100\%)/K_v(Z = 0\%)$	33.06%
40	$K_v(Z = 100\%)/K_v(Z = 0\%)$	29.01%
30	$K_v(Z = 100\%)/K_v(Z = 0\%)$	23.00%
50	$K_v(Z = 100\%)/K_v(Z = 5\%)$	36.40%
40	$K_v(Z = 100\%)/K_v(Z = 5\%)$	32.55%
30	$K_v(Z = 100\%)/K_v(Z = 5\%)$	26.85%

4.3. Discussion on Application

The proposed model provides valuable support for regulating valve selection. The selection process should begin by calculating the minimum $K_{v,min}$ or $C_{v,min}$ at the lowest estimated production. The corresponding opening value should be maintained at 10% or greater; otherwise, the operating condition may exceed the calibration range. Z = 10% is the second smallest calibrated opening suggested by IEC Standard 60534-2-4:2009 [37]. The smallest is Z = 5%, but the deviation of ±2% of the feedback opening should also be considered. Although the calculated $K_{v,min}$ or $C_{v,min}$ is independent of d_v, the actual

$K_{v,min}$ or $C_{v,min}$ is dependent on d_v due to the structure and manufacturing technique of the valve, or more exactly, its closure member. Generally, $K_{v,min}$ or $C_{v,min}$ is larger for bigger valves. Then, we choose an initial d_v of the rangeability, R, and the maximum $K_{v,max}$ or $C_{v,max}$ can be determined. If this value is smaller than K_v or C_v corresponding to the largest production rate within the occurrence of severe slugging, a valve with larger R must be selected; if not possible, another alternate valve with larger $K_{v,max}$ has to be laid in a parallel branch pipe. According to the discussion above, d_v, R, and $K_{v,max}$ are constrained by each other. The selection of the valve may also be subjected to other restrictions like the operation of the gas–liquid separator or slug catcher and the dead band, as well as the deviation of the feedback value from the set opening of the valve. Hence, the optimization model for valve selection is worthy of study in future work.

Based on the experiments and field data, we believe that the model works within the pressure range (separator pressure) from atmospheric to 1 MPa. We also consider that the pressure range can be extended. For bigger p_s, the slugging itself is less severe, as indicated by the reduced velocity range of occurrence on the flow regime map [38]. With a less severe blowout, the difference of average Δp_v between pipeline risers and a simple vertical pipe is smaller, resulting in a bigger optimal opening, which is consistent with the experimental observation [38]. The temperature range of common fields is several dozen degrees centigrade, which does not differ much from experimental conditions and should be within the application range. As for the gas–liquid ratio, its application range is related to the actual necessity of control, as mentioned in Section 4.1.

In the field configuration, solid particles in the products may attach to the valve core, reducing the actual cross-sectional area of the flow channel in the valve, and the actual optimal opening should be larger as a result of the smaller actual K_v. However, the deviation will not shake the fundamentals of the model, which is that the two-phase flow resistance factor is the same for pipeline risers and simple vertical pipes. As for the valve type, it is not a direct parameter. Each type should provide the characteristics of the K_v-Z relation. Once K_v-Z is determined, the influence of valve type only lies in the aspect of the precision of travel control, which is out of the scope of this study.

Another uncertainty of application lies in the complex pipe geometry from the riser top to the separator. On the deck of an FPSO, the pipe from the single point mooring to the separator includes a series of upward, horizontal, and downward sections and may exhibit a concave geometry. According to the OLGA simulation, the optimal opening varies with the valve position on the pipe. If the valve is installed on the concave part of the pipe, the actual optimal opening should be smaller than the predicted result. Future work can focus on the modification of the model under these conditions.

5. Conclusions

In this study, a prediction model of valve opening required for slugging control in offshore pipeline risers is proposed. The model is based on macroscopic phenomena of slugging elimination and the flow characteristics of the regulating valve. Two empirical closures are added based on experimental observation. The proposed model is validated by laboratory and field cases. The major findings and contributions of the study are as follows:

1. The target of slugging control can be set as the occurrence of dual-frequency fluctuation. At the time, the sum of the time-averaged total pressure drop along the riser and that across the valve is approximately the same as the gravity pressure drop of single-phase water in the riser without choking. So, the conservation of total pressure drop can be assumed.

2. The resistance factor of the valve is approximately the same as slug flow in a simple vertical pipe at the time severe slugging is eliminated, indicating the feasibility of calculating the resistance factor by a two-phase flow model in a simple vertical pipe.
3. Because of severe flow intermittency in pipeline risers and weaker gas drift flow, and in order to be consistent with the processing of measured data, the non-slip assumption is recommended in the calculation of the gas fraction in the riser.
4. The average absolute error of the proposed model is +0.01%, which is smaller than the simulation performed by OLGA software (+4.91% before tuning and +0.08% after tuning). The validation of the field case also proves the good applicability of the model (with deviation smaller than ±2%).
5. Future work can focus on the application of the model to the valve selection at the design stage and the modification of the model for different valve installation positions on a real platform at the operation stage.

Author Contributions: Conceptualization, S.Z.; methodology, J.F.; software, J.S.; validation, J.S.; formal analysis, J.F., Q.W. and H.W.; data curation, Q.W. and S.Z.; writing—original draft preparation, J.F.; writing—review and editing, S.Z.; supervision, S.Z.; project administration, S.Z.; funding acquisition, S.Z. All authors have read and agreed to the published version of the manuscript.

Funding: This work is financially supported by the China National Key Research and Development Project (grant number 2022YFC2806202), the Fundamental Research Funds for the Central Universities (grant number xtr042024002), and the Central Government Guided Local Science and Technology Development Fund Project (grant number 2024ZYD0123).

Data Availability Statement: The datasets generated and analyzed during the current study are available from the corresponding author upon reasonable request.

Acknowledgments: Data support from CNOOC China Limited—Zhanjiang Branch is gratefully acknowledged.

Conflicts of Interest: Author Q.W. was employed by Longyuan (Beijing) New Energy Engineering Technology Co., Ltd. The remaining authors declare that the research was conducted in the absence of any commercial or financial relationships that could be construed as a potential conflict of interest.

Abbreviations

The following abbreviations are used in this manuscript:

A	Area, m^2.
C_v	Flow coefficient, $gal \cdot min^{-1} \cdot psi^{-1}$ (usually omitted).
d	Diameter, m.
f	Resistance factor, dimensionless.
g	Gravitational acceleration, 9.80665 $m \cdot s^{-2}$.
H	Height of riser, m.
K_v	Flow coefficient, $m^3 \cdot h^{-1} \cdot (10^5 Pa)^{-1}$ (usually omitted).
p	Pressure, Pa.
Δp	Differential pressure, Pa.
Q	Mass flow rate, m^3/d.
T	Temperature, K.
u	Velocity, m/s.
Z	Valve opening, %.
Greek letters	
ρ	Density, kg/m^3.
α	Gas fraction, dimensionless.
σ	Surface tension, N·m.

Subscripts	
avg	Average.
b	Riser bottom.
G	Gas.
L	Liquid.
mix	Gas–liquid mixture.
R	Riser.
S	Superficial.
s	Separator.
t	Riser top.
0	Standard condition: at 101,325 Pa and 273.15 K for laboratorial experiments and at 100,000 Pa and 288.75 K for the field case.

References

1. Li, X.; Yue, Y.; Du, H.; Yue, L.; Ding, R.; Liu, C.; Li, Y. Investigation on leakage detection and localization in gas-liquid stratified flow pipelines based on acoustic method. *J. Pipeline Sci. Eng.* **2022**, *3*, 100127. [CrossRef]
2. Sun, W.; Zhang, J.; Mukhtar, Y.; Zuo, L.; Dong, S. Safety Evaluation Method for Submarine Pipelines Based on a Radial Basis Neural Network. *Sustainability* **2023**, *15*, 12724. [CrossRef]
3. Qiao, W.; Guo, H.; Huang, E.; Chen, H.; Lian, C. Two-Phase Flow Pattern Identification by Embedding Double Attention Mechanisms into a Convolutional Neural Network. *J. Mar. Sci. Eng.* **2023**, *11*, 793. [CrossRef]
4. Yu, H.; Xu, Q.; Cao, Y.; Huang, B.; Guo, L. Effects of riser height and pipeline length on the transition boundary and frequency of severe slugging in offshore pipelines. *Ocean. Eng.* **2024**, *298*, 117253. [CrossRef]
5. Banjara, V.; Pereyra, E.; Avila, C.; Murugavel, D.; Sarica, C. Improved severe slugging modeling and mapping. *Geoenergy Sci. Eng.* **2025**, *244*, 213446. [CrossRef]
6. Schmidt, Z. Experimental Study of Two-Phase Slug Flow in a Pipeline-Riser Pipe System. Ph.D. Dissertation, University of Tulsa, Tulsa, OK, USA, 1977.
7. Tengesdal, J.Ø. Investigation of Self-Lifting Concept for Severe Slugging Elimination in Deep-Water Pipeline/Riser Systems. Ph.D. Thesis, University of Tulsa, Tulsa, OK, USA, 2002.
8. Malekzadeh, R.; Henkes, R.A.W.M.; Mudde. Severe slugging in along pipeline–riser system: Experiments and predictions. *Int. J. Multiph. Flow* **2012**, *46*, 9–21. [CrossRef]
9. Henriot, V.; Courbot, A.; Heintzé, E.; Moyeux, L. Simulation of Process to Control Severe Slugging: Application to the Dunbar Pipeline. In Proceedings of the 1999 SPE Annual Technical Conference and Exhibition, Houston, TX, USA, 3–6 October 1999. [CrossRef]
10. Havre, K.; Stornes, K.O.; Stray, H. Taming slug flow in pipelines. *ABB Rev.* **2000**, *46*, 55–63.
11. Li, Z.; Luo, J.; Lu, J. Application and Analysis about Affection and Treatment of Slug Flow in BZ25-1 Oil Field. *Shipbuild. China* **2012**, *53*, 310–315. (In Chinese)
12. Li, C.; Wu, W.; Zhang, L. Numerical simulation on control measures of severe slug flow in Boxi Oilfield. *Oil Gas Storage Transp.* **2018**, *37*, 95–100. (In Chinese) [CrossRef]
13. Zhang, X.; Li, N.; Chen, B. Analysis and treatment on the severe slug flow in pipeline S-shaped riser of Wenchang Oilfield. *Oil Gas Storage Transp.* **2019**, *38*, 472–480. (In Chinese) [CrossRef]
14. Bai, Y.; Bai, Q. *Subsea Pipelines and Risers*, 1st ed.; Elsevier Ltd.: Oxford, UK, 2005; pp. 277–316. [CrossRef]
15. Pedersen, S.; Bram, M.V. The Impact of Riser-Induced Slugs on the Downstream Deoiling Efficiency. *J. Mar. Sci. Eng.* **2021**, *9*, 391. [CrossRef]
16. Li, X.; Liu, Y.; Abbassi, R.; Khan, F.; Zhang, R. A Copula-Bayesian approach for risk assessment of decommissioning operation of aging subsea pipelines. *Process Saf. Environ.* **2022**, *167*, 412–422. [CrossRef]
17. Liu, Y.; Zhu, X.; Song, J.; Wu, H.; Zhang, S.; Zhang, S. Research on bypass pigging in offshore riser system to mitigate severe slugging. *Ocean. Eng.* **2022**, *246*, 110606. [CrossRef]
18. Zheng, G.; Xu, P.; Li, L.; Fan, X. Investigations of the Formation Mechanism and Pressure Pulsation Characteristics of Pipeline Gas-Liquid Slug Flows. *J. Mar. Sci. Eng.* **2024**, *12*, 590. [CrossRef]
19. Banjara, V.; Pereyra, E.; Avila, C.; Sarica, C. Critical review & comparison of severe slugging mitigation techniques. *Geoenergy Sci. Eng.* **2023**, *228*, 212082. [CrossRef]
20. Nnabuife, S.G.; Tandoh, H.; Whidbore, J.F. Slug flow control using topside measurements: A review. *Chem. Eng. J. Adv.* **2022**, *9*, 100204. [CrossRef]
21. Ogazi, A.I. Multiphase Severe Slugging Flow Control. Ph.D. Thesis, Cranfield University, Cranfield, UK, 2011.

22. Park, K.-H.; Kim, T.-W.; Kim, Y.-J.; Lee, N.; Seo, Y. Experimental investigation of model-based IMC control of severe slugging. *J. Petrol. Sci. Eng.* **2021**, *204*, 108732. [CrossRef]
23. Jahanshahi, E.; Skogestad, S. Simplified Dynamical Models for Control of Severe Slugging in Multiphase Risers. *IFAC Proc. Vol.* **2011**, *44*, 1634–1639. [CrossRef]
24. Guo, L.; Li, N.; Ye, J.; Chen, S. Choking Device and Method for Real-Time Eliminating Severe Slugging in a Pipeline-Riser System. Patent CN102410391B, 10 July 2013.
25. Pedersen, S.; Durdevic, P.; Yang, Z. Learning control for riser-slug elimination and production-rate optimization for an offshore oil and gas production process. *IFAC Proc.* **2014**, *47*, 8522–8527. [CrossRef]
26. Jansen, F.E.; Shoham, O.; Taitel, Y. The elimination of severe slugging: Experiment and modeling. *Int. J. Multiph. Flow.* **1996**, *22*, 1055–1072. [CrossRef]
27. Fadairo, A.; Obi, C.; Li, K.; Rasouli, V.; Aboaba, O.; Gbadamosi, O. An improved model for severe slugging stability criteria in offshore pipeline-riser systems. *Petrol. Res.* **2022**, *7*, 318–328. [CrossRef]
28. Pedersen, S.; Durdevic, P.; Yang, Z. Challenges in slug modeling and control for offshore oil and gas productions: A review study. *Int. J. Multiph Flow* **2017**, *88*, 270–284. [CrossRef]
29. Andreolli, I.; Azevedo, G.R.; Baliño, J.L. Application of stability solver for severe slugging considering a choking device and a tubing connected to a Wet Christmas Tree. *J. Petrol. Sci. Eng.* **2020**, *192*, 107316. [CrossRef]
30. Gong, J.; Kang, Q.; Wu, H.; Li, X.; Shi, B.; Song, S. Application and prospects of multi-phase pipeline simulation technology in empowering the intelligent oil and gas fields. *J. Pipeline Sci. Eng.* **2023**, *3*, 100127. [CrossRef]
31. Zou, S.; Yao, T.; Guo, L.; Li, W.; Wu, Q.; Zhou, H.; Xie, C.; Liu, W.; Kuang, S. Non-Uniformity of Gas/Liquid Flow in a Riser and Impacts of Pipe Configuration and Operation on Slugging Characteristics. Exp. *Therm. Fluid Sci.* **2018**, *96*, 329–346. [CrossRef]
32. Wang, H.; Zou, S.; Liu, T.; Xu, L.; Du, Y.; Yan, Y.; Guo, L. Experimental investigation of online solving for critical valve opening for eliminating severe slugging in pipeline-riser systems. *Int. J. Heat Fluid Flow* **2024**, *110*, 109584. [CrossRef]
33. Al-Safran, E.M.; Kelkar, M. Predictions of Two-Phase Critical-Flow Boundary and Mass-Flow Rate Across Chokes. *SPE Prod. Oper.* **2009**, *24*, 249–256. [CrossRef]
34. Shi, H.; Holmes, J.A.; Durlofsky, L.J.; Aziz, K.; Diaz, L.R.; Alkaya, B.; Oddie, G. Drift-Flux Modeling of Two-Phase Flow in Wellbores. *SPE J.* **2005**, *10*, 24–33. [CrossRef]
35. Ye, J.; Guo, L. Multiphase flow pattern recognition in pipeline–riser system by statistical feature clustering of pressure fluctuations. *Chem. Eng. Sci.* **2013**, *102*, 486–501. [CrossRef]
36. Ye, J.; Guo, L.; Xie, C.; Xu, Q. Experiment Study on Phase Inversion of Oil-water Two Phase Flow Using Double-parallel-ring Conductance Probe. *J. Eng. Thermophys.* **2010**, *31*, 983–986. (In Chinese)
37. *IEC 60534-2-4:2009*; Industrial-Process Control Valves. Part 2-4: Flow Capacity-Inherent Flow Characteristics and Rangeability. International Electrotechnical Commission: Geneva, Switzerland, 2009.
38. Zhao, X. Study on Transition Mechanism and Control Method of Severe Slugging in High-pressure Pipeline-riser System. Ph.D. Dissertation, Xi'an Jiaotong University, Xi'an, China, 2024. (In Chinese).

Journal of
Marine Science and Engineering

Article

Experimental and Numerical Investigation on Mechanical Properties of Large-Diameter Metallurgically Clad Pipes

Feng Wang [1,2], Yanan Gao [1], Zhiguo Hu [1], Shuo Yang [1], Zhenying Cui [1], Rui Fu [3] and Lin Yuan [3,*]

1 PipeChina Engineering Technology Innovation Co., Ltd., Tianjin 300450, China
2 School of Materials Science and Engineering, Tianjin University, Tianjin 300072, China
3 School of Civil Engineering, Tianjin University, Tianjin 300072, China
* Correspondence: lin_yuan@tju.edu.cn

Abstract: Bimetallic-clad pipes demonstrate exceptional advantages in transporting corrosive oil and gas through the combination of the load-carrying capacity of the base material and the anti-corrosive function of the thin layer of corrosion-resistant alloy. This study investigates the mechanical properties of 24-inch X65 + Alloy625 metallurgically clad pipes through experimental tests and finite element analysis. Uniaxial tensile testing with digital image correlation reveals uniform deformation between the base and clad layers until interfacial failure initiates at an average strain threshold of 34.17%. Microstructural characterization shows continuous metallurgical bonding, with the X65 layer exhibiting polygonal ferrite and bainitic phases, contrasting with the austenitic equiaxed grain structure of Alloy625. In terms of numerical modeling, finite element analyses that consider both initial geometric imperfections and manufacturing-induced residual stresses are performed to evaluate the bending response of the clad pipe. The effect of initial ovality and residual stresses on its bending capacity is also studied.

Keywords: metallurgical clad pipe; interface bonding; bending capacity; microstructural characterization; offshore pipelines

1. Introduction

With sustained global economic development and growing energy demands, the exploitation of strategic resources such as oil and natural gas has progressively shifted from terrestrial and shallow-water environments to more challenging deep-water and high-corrosion conditions [1–3]. Bimetallic clad pipes have emerged as effective solutions in offshore oil and gas extraction, combining the structural integrity of carbon steel base pipes with the corrosion resistance of stainless steel or nickel-based alloy liners through advanced metallurgical bonding techniques [4–7]. These composite pipes offer superior service life compared to conventional carbon steel pipelines while maintaining cost-effectiveness relative to solid corrosion-resistant alloy pipes [8].

Two primary categories of bimetallic pipes exist based on interfacial bonding mechanisms: mechanical and metallurgical clad pipes. Metallurgically bonded pipes demonstrate enhanced interfacial strength and load-bearing capacity compared to their mechanically bonded counterparts, making them particularly suitable for marine engineering applications [9,10]. However, their complex mechanical behavior and failure mechanisms under operational conditions pose significant challenges to reliability assessment. Although previous studies have made progress in understanding composite pipe behavior, critical knowledge gaps remain regarding the mechanical properties of large-diameter metallurgically clad pipes.

Extensive research has been conducted on mechanically lined pipes through both experimental and numerical approaches. Full-scale tests under axial compression, bending, and reeling loads have been carried out by Hilberink et al. [11–13], while Tkaczyk et al. [14,15] investigated the evolution of liner buckling during cyclic bending with and without internal pressure. Yuan et al. [16] performed small-scale four-point bending tests to analyze the effect of local indentation on structural response. Numerical studies by Vasilikis and Karamanos [17], and Yuan and Kyriakides [18–20], explored liner wrinkling and local buckling under bending and axial compression, revealing the influence of geometric imperfections. Further investigations [21–23] addressed the effects of internal pressure and tension on structural stability during reeling and combined loading, showing good agreement with experimental observations.

Recent investigations by Yoshitake [24] employed centrifugal casting to produce metallurgically bonded pipes with uniform wall thickness. Slany et al. [25] conducted comprehensive analyses of Inconel 625-clad pipes subjected to 0.7D bending deformation. Yang et al. [26] developed a three-dimensional finite element (FE) model incorporating girth welds to evaluate strain capacity under combined internal pressure and bending moments, identifying the heat-affected zone as a critical failure location. Subsequent research by the same group [27] investigated cyclic loading effects on welded metallurgical clad pipes, revealing 120% higher strain concentrations in weld regions during installation. Yuan et al. [28] employed machine learning algorithms to predict collapse pressures of 12-inch mechanically lined pipes, examining the influence of diameter-to-thickness ratios, initial ovality, and material hardening. Overall, prior studies primarily focused on relatively small-diameter pipes (<12 inches), leaving the mechanical behavior of large-diameter clad pipes under deep-sea conditions still poorly understood.

This study focuses on 24-inch X65 + Alloy625 metallurgically clad pipes through experimental characterization and numerical simulation. Unlike most previous studies that concentrate on small-diameter pipes (<12 inches) or simplified boundary conditions, this work investigates large-diameter composite pipes under realistic offshore loading scenarios. The investigation encompasses: (1) geometric characterization via laser scanning and residual stress measurement through the ring-cutting method; (2) uniaxial tensile testing of the base material (X65 steel), clad material (Alloy625), and composite specimens using digital image correlation (DIC) for full-field strain measurement; (3) microstructural analysis through metallography and electron backscatter diffraction (EBSD); and (4) development of a detailed three-dimensional FE model in ABAQUS, incorporating initial ovality and residual stresses to simulate the bending behavior. This comprehensive approach provides new insights into the failure mechanisms and design optimization of large-diameter metallurgical clad pipes in deep-sea environments.

2. Materials and Experimental Methods

2.1. Sample Preparation

A 1.5 m long 24-inch metallurgically clad pipe was sectioned into six segments (340 mm × 1 and 230 mm × 5), as illustrated in Figure 1a,b. Geometric parameters, including length, outer diameter, and wall thickness, were measured at multiple angular positions for each segment, with averaged values presented in Table 1. Specifically, the outer diameter and wall thickness were measured along the circumferential direction at 0°, ±8°, ±30°, ±60°, ±90°, ±120°, ±150°, ±172°, and 180°, taking the weld seam as the 0° reference. The FE model incorporated measured ovality values calculated according to Det Norske Veritas (DNV) specifications, yielding a mean ovality of 0.407%, as detailed in Table 2.

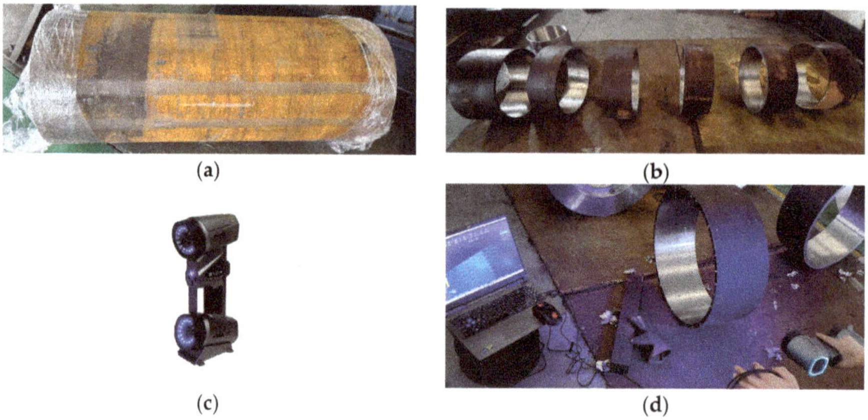

Figure 1. Pipe segments and 3D scanning: (**a**) A 1.5 m long 24-inch metallurgically clad pipe; (**b**) Six segments of the clad pipe; (**c**) AlphaScan metrology-grade handheld scanner; (**d**) Scanning process.

Table 1. Geometric dimensions of pipe segments.

Segment	No. 1	No. 2	No. 3	No. 4	No. 5	No. 6
L/mm	346.50	229.20	225.80	226.40	227.20	233.60
D/mm	610.31	610.50	609.81	610.02	610.32	609.78
t/mm	20.44	20.47	20.19	20.14	20.37	20.01

Table 2. Ovality measurement of pipe segments.

Segment	No. 1	No. 2	No. 3	No. 4	No. 5	No. 6
D_{max}/mm	610.70	612.36	611.88	612.28	612.76	611.98
D_{min}/mm	609.80	610.00	609.20	609.32	609.64	609.12
Δ	0.1475%	0.3869%	0.4393%	0.4852%	0.5115%	0.4689%
Average	0.4066%					

Three-dimensional laser scanning was then performed using an AlphaScan metrology-grade handheld scanner (Insvision Technology Co., Ltd., Hangzhou, China) (Figure 1c) following a standardized protocol: (1) scanner calibration using reference artifacts; (2) validation of measurement accuracy through 300 mm standard specimen testing; and (3) pipeline scanning with pre-applied fiducial markers. The scanning process is depicted in Figure 1d.

2.2. Residual Stress Measurement

Hoop residual stresses were determined through the ring-cutting method following the analytical formulation:

$$|\sigma_R| = \frac{Et}{4\pi R}\alpha \tag{1}$$

where α represents the opening angle after longitudinal cutting, E is the elastic modulus, t is the wall thickness, and R is the pipe radius. The cutting procedure and angular measurement are shown in Figure 2a,b.

Figure 2. Longitudinal cutting of pipe segments: (**a**) Longitudinal cutting procedure; (**b**) Schematic diagram of the opening angle measurement.

2.3. Tensile Testing

Uniaxial tensile tests were then conducted according to GB/T 228.1-2010 standards [29] for base material (X65), clad material (Alloy625), and composite specimens using a Shimadzu AGX-600 kN universal testing machine (SHIMADZU Corporation, Kyoto, Japan) (Figure 3a).

Figure 3. Tensile testing equipment: (**a**) Shimadzu AGX-600 kN universal testing machine; (**b**) XTDIC-5M digital image correlation system.

Four X65 steel specimens (86 mm parallel length × 20 mm width × 16 mm thickness) were tested at a 2.0 mm/min loading rate, with strain measurement performed using the XTDIC-5M digital image correlation system (XTOP 3D Technology (Shenzhen) Co., Ltd., Shenzhen, China) (Figure 3b) at a 1000 ms acquisition interval, requiring a speckle pattern application on specimen surfaces. Similarly, four Alloy 625 specimens (50 mm × 12.5 mm × 2.7 mm) were tested at a 1.0 mm/min displacement rate using a Y25/10-N extensometer (LISHI (Shanghai) Instrument Co., Ltd., Shanghai, China) with a 25 mm gauge length. To assess strain compatibility between layers, 12 composite specimens (86 mm × 20 mm × 20 mm) were tested at a 1.5 mm/min loading rate, with full-field strain distributions captured by the DIC system, while virtual extensometers recorded engineering strains.

2.4. Microstructural Analysis

Metallographic testing and electron backscatter diffraction (EBSD) scanning analysis were also performed to investigate the microstructures of the base layer, clad layer, and interfacial bonding region. Three specimens with dimensions of 8 × 8 × 4 mm were extracted from the pipe using a wire-cutting method. These specimens were subsequently gradually ground and polished with metallographic sandpaper. One specimen was etched with a corrosive agent after mechanical polishing to prepare a metallographic specimen for microscopic observation, while the other two underwent argon ion polishing for EBSD scanning analysis following mechanical polishing.

3. Experimental Results

3.1. Three-Dimensional Scanning

The six pipe segments were subjected to three-dimensional laser scanning, with Segment 1 selected as the representative case for detailed presentation. Figure 4a displays its three-dimensional scanning results, while Figure 4b shows the two-dimensional cross-sectional profiles with 20× magnification of surface defects. The red contours indicate amplified inner wall imperfections, and the purple contours represent magnified outer wall irregularities. As shown, radial deviations from nominal circular profiles reveal distinct contact zones of the indentation tool on the inner surface during the JCO forming steps. The scanning accuracy was verified through calibration tests on a 300 mm standard specimen, achieving a measurement precision of 0.047%. In addition, the scanned profile was also checked against manual measurements at different angles of the pipe.

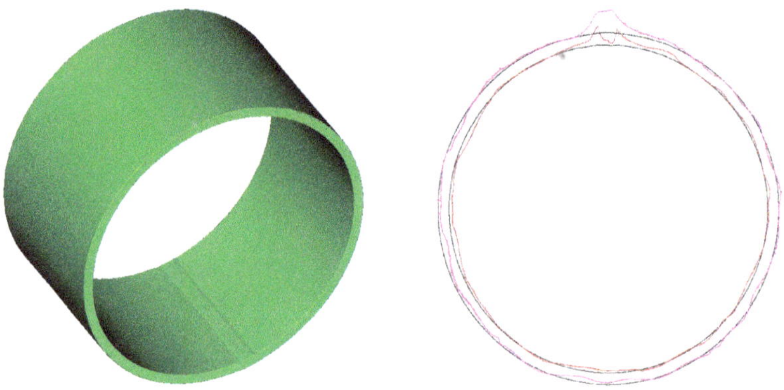

(a) (b)

Figure 4. Scanning Results: (**a**) Three-dimensional laser scanning result of Segment 1; (**b**) Cross-sectional profiles of Segment 1 with defects magnified by a factor of 20.

3.2. Residual Stress Measurement Results

Hoop residual stresses were calculated using the ring-cutting method according to Equation (1), with a nominal Young's modulus of 200 GPa used in the calculation. The measured opening angles and corresponding residual stress values for all six pipe segments are summarized in Table 3. The results indicate consistent residual stress distribution along the pipe circumference, with an average magnitude of 54.58 MPa and a maximum variation of ±15.24 MPa between different pipe sections relative to the mean value.

Table 3. Measurement of ring-cutting opening and calculated residual stress.

Segment	R/mm	t/mm	α/rad	$\lvert\sigma_R\rvert$/MPa
No. 1	294.94	20.44	0.04370	48.20
No. 2	295.02	20.47	0.03562	39.34
No. 3	294.81	20.19	0.05182	56.48
No. 4	294.94	20.14	0.06411	69.67
No. 5	294.98	20.37	0.05130	56.38
No. 6	294.89	20.01	0.05318	57.43

3.3. Tensile Testing Results

3.3.1. Base Material (X65 Steel)

The X65 specimens exhibited typical ductile fracture characteristics during tensile testing. Figure 5a shows the deformed specimens with significant necking. The engineering stress-strain curves obtained through DIC measurement (Figure 5b) reveal a yield strength of 552.08 MPa and an ultimate tensile strength of 618.03 MPa. Detailed mechanical properties are provided in Table 4.

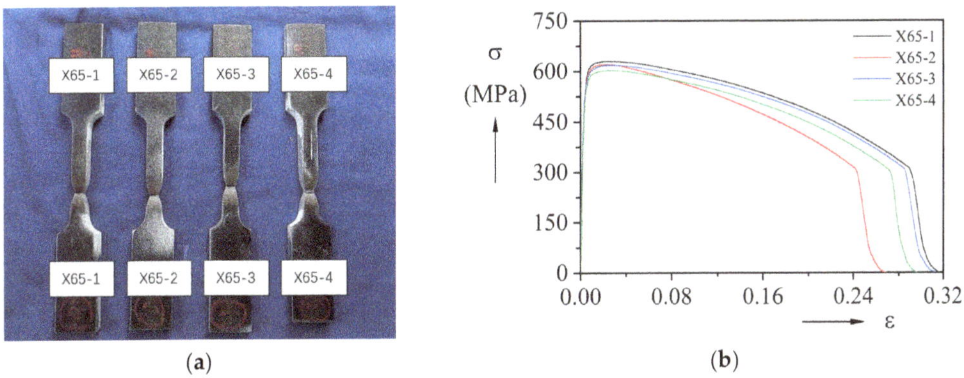

Figure 5. Tensile testing results of the base material: (**a**) Specimen of base material after tensile test; (**b**) Stress-strain curves.

Table 4. Main material parameters of base material.

Specimen	E/GPa	f_y/MPa	f_u/MPa	A_{gt}/%
X65-1	208	570.10	630.04	2.48
X65-2	207	556.53	620.83	2.22
X65-3	208	550.26	619.01	2.83
X65-4	210	531.43	602.24	2.63

3.3.2. Clad Material (Alloy625)

Alloy625 specimens demonstrated superior ductility with significantly higher fracture strains than the X65 counterparts. The deformed specimens are shown in Figure 6a, and the stress-strain curve in Figure 6b displays continuous strain hardening without a distinct yield plateau, characteristic of austenitic alloys. The yield strength reached 679.09 MPa with an ultimate strength of 965.46 MPa and an average strain of 27.90% at peak load, as detailed in Table 5.

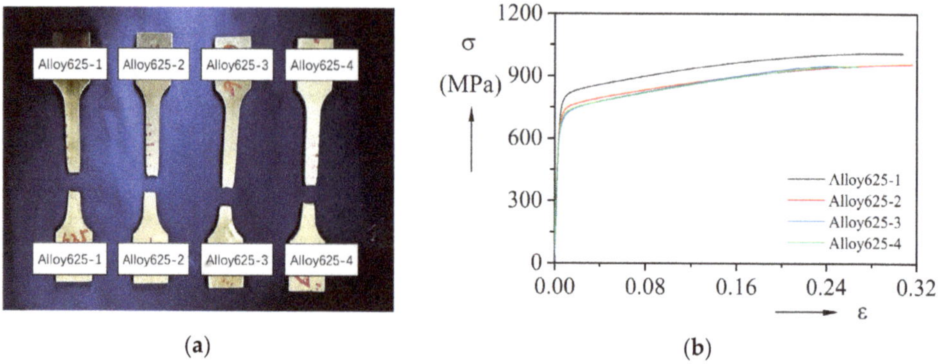

Figure 6. Tensile testing results of the clad material: (**a**) Specimen of clad material after tensile test; (**b**) Stress-strain curves.

Table 5. Main material parameters of clad material.

Specimen	*E*/GPa	f_y/MPa	f_u/MPa	A_{gt}/%
Alloy625-1	205	736.17	1010.67	29.54%
Alloy625-2	207	678.13	956.66	31.98%
Alloy625-3	211	641.66	948.03	24.34%
Alloy625-4	203	660.41	946.49	25.85%

3.3.3. Composite Material (X65 + Alloy625)

As to the composite material, the side view of the fractured specimens is presented in Figure 7a (specimen Clad-01 to Clad-12 from left to right). The corresponding nominal stress-strain curves are in Figure 7b. Here, the nominal stress was calculated based on the total area of the cross-section. Interestingly, the composite specimens demonstrated sequential failure characteristics during tensile testing. Initially, the tensile loading induced uniform deformation of the base and clad material. Subsequently, necking originated at peak load, which was followed by audible interface delamination sound. This phenomenon is consistent with the delamination behavior observed in previous studies on bimetallic materials during tensile testing [6]. After continuous loading, a second fracture sound marked clad layer failure after several seconds, while additional loading eventually led to progressive base layer necking until final rupture with complete load drop. This study employed DIC to measure engineering strain during experiments. Unlike previous studies, this approach allows for the acquisition of strain data over the full tensile process, yielding a more complete dataset.

For illustration, the stress-strain curve and the corresponding DIC measurement of Clad-01 are specifically displayed in Figures 8 and 9. As shown, the curve exhibits a progressive decline after reaching peak stress. At point a (33.73% strain), interfacial delamination occurs with a sharp stress drop, followed by gradual stress recovery. Subsequently, the clad material Alloy625 fractures at point b (34.23% strain), causing a secondary abrupt decline in stress. Thereafter, only the base layer sustains loading until point c (40.19% strain), where a third stress drop coincides with base material fracture. Key mechanical parameters are summarized in Table 6.

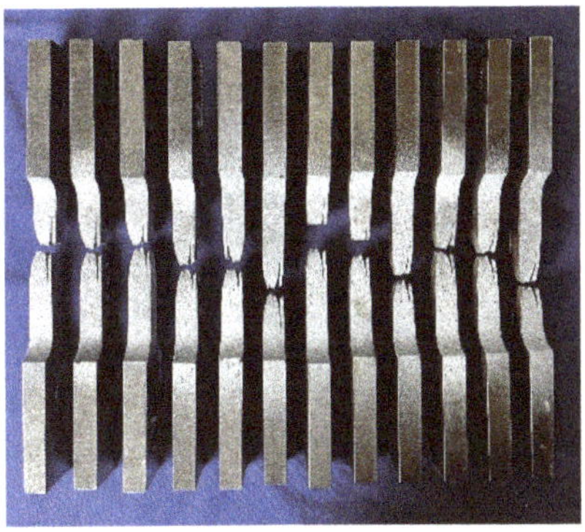

(a)

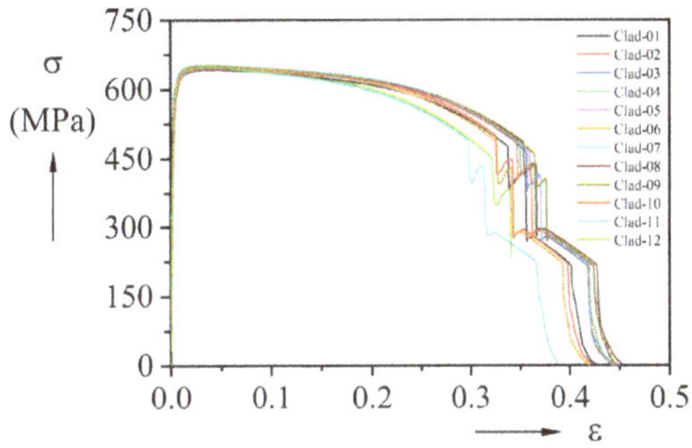

(b)

Figure 7. Tensile testing results of the composite material: (**a**) Specimen of composite material after tensile test; (**b**) Stress-strain curves.

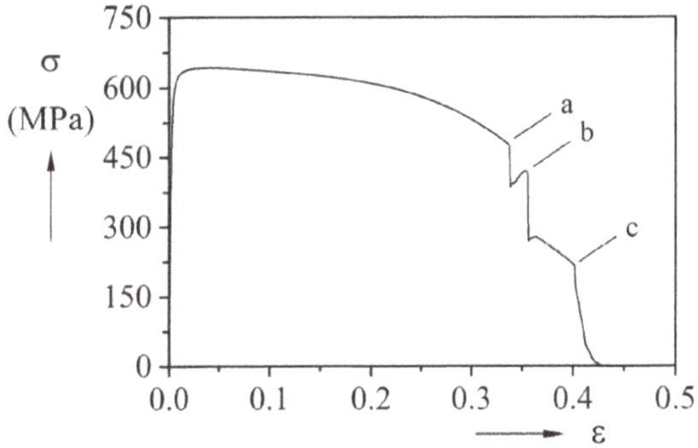

Figure 8. Stress-strain curve of specimen Clad-01.

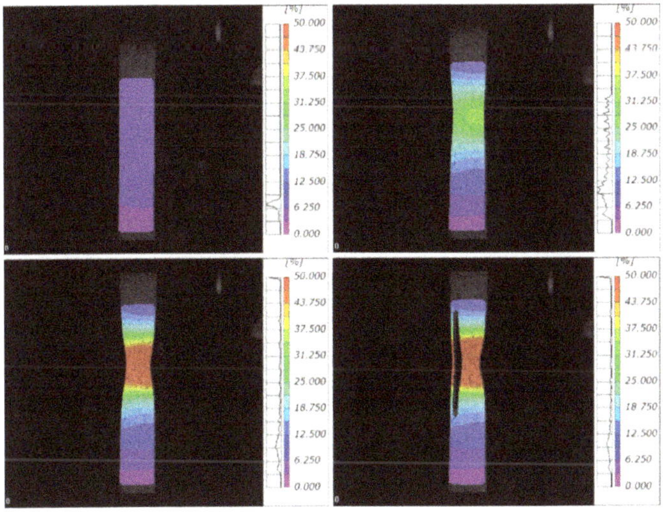

Figure 9. Measured strain distribution map from DIC.

Table 6. Main material parameters of composite material.

Specimen	E/GPa	f_y/MPa	f_u/MPa	A_{gt}/%	ε_d/%	ε_f/%
Clad-01	203	586.73	642.09	3.81	33.72	34.23
Clad-02	204	582.96	650.23	3.43	32.92	34.73
Clad-03	201	580.26	642.60	4.42	35.71	37.36
Clad-04	200	586.23	649.18	3.74	34.93	35.55
Clad-05	196	565.93	643.63	3.99	35.92	36.56
Clad-06	200	586.21	650.39	3.30	32.59	33.17
Clad-07	191	595.34	645.31	3.26	30.00	31.33
Clad-08	190	604.59	642.43	4.01	35.30	35.64
Clad-09	200	557.93	651.25	4.18	36.49	37.89
Clad-10	191	572.66	645.31	3.96	34.74	36.37
Clad-11	203	602.53	653.41	3.42	35.67	36.24
Clad-12	204	594.29	647.51	3.24	32.10	32.91

The strain contour maps obtained through DIC processing are presented in Figure 9. These maps demonstrate a gradual strain localization after necking from both specimen ends toward the central region. The strain initially exhibits uniform distribution across transverse sections, and these identical strain patterns between the base and clad layers confirm their synchronous deformation during tensile loading. Then, delamination takes place, and an extended length of the clad layer separates from the base material, but the strain level of both layers is still almost the same.

3.4. Microstructural Analysis Results

Optical microscopy (OM) observations of X65 steel are presented in Figure 10a (four representative locations). The interface between the two materials appears smooth with minor undulations, but without detectable defects. The X65 region near the interface primarily consists of polygonal ferrite and ferritic bainite structures, with dispersed black M/A (martensite-austenite) constituents within the ferritic bainite. While these M/A phases may induce localized stress concentrations, their limited volume fraction suggests negligible impact on overall toughness.

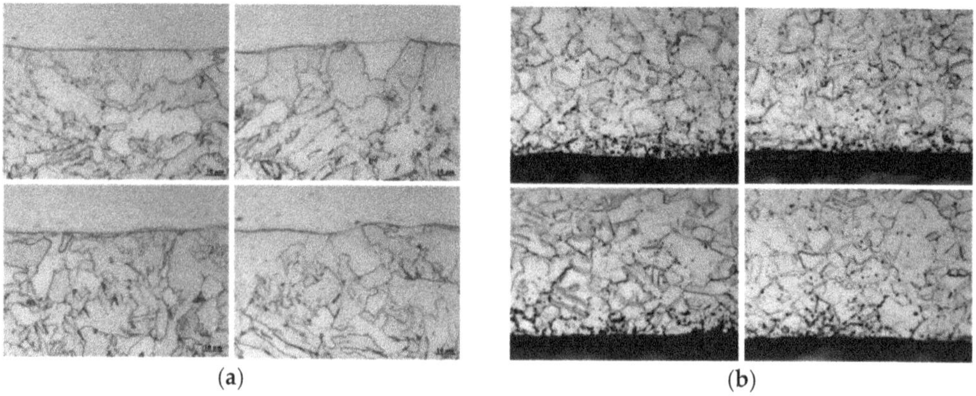

Figure 10. OM observations results: (**a**) OM observations of X65 Steel; (**b**) OM observations of Alloy625.

OM examination of Alloy625 (Figure 10b, four representative locations) reveals a homogeneous single-phase structure composed of fine equiaxed grains at the interface region. The distinct thermal conductivity between ferrite and austenite inevitably induces grain refinement and solid solution strengthening in specific regions of the clad layer, resulting in nanoscale grains and solid solution structures. Dark precipitates are observed within grains and along boundaries, exerting grain boundary-pinning effects that further refine the microstructure. Notably, irregularly distributed dark aggregates concentrate along the bimaterial interface, particularly at edge regions.

Electron backscatter diffraction (EBSD) analysis results are shown in Figure 11. The grain orientation map (Figure 11a) demonstrates clustered fine grains with similar crystallographic orientations in X65 pipeline steel, contrasting with coarse equiaxed grains containing annealing twins in Alloy625. This twin formation likely originates from structural reorganization during recrystallization, where low stacking fault energy promotes twin generation through stacking fault interactions. Near the fusion line, X65 grains exhibit agglomerated distributions resulting from incomplete austenite transformation at elevated temperatures, while regions distant from the interface show typical as-rolled microstructures due to insufficient heating. The phase distribution map (Figure 11b) confirms the dominant ferritic matrix in X65 with minor austenite phases at grain boundaries, whereas Alloy625 maintains primarily austenitic structure containing trace ferrite phases along grain boundaries.

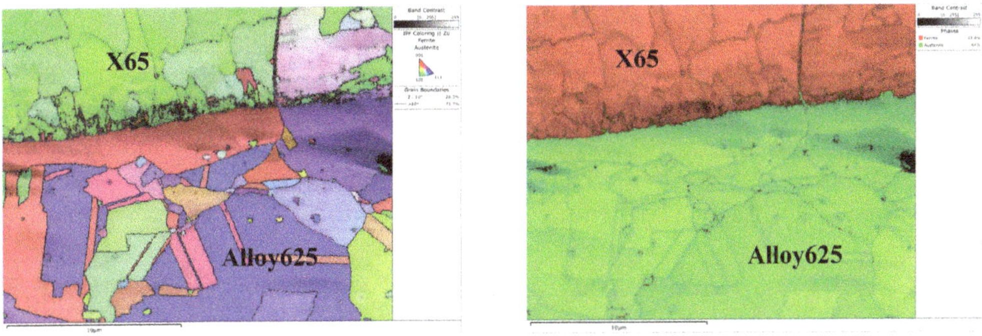

(a) (b)

Figure 11. EBSD analysis results: (**a**) Grain orientation map; (**b**) Phase distribution map

4. Finite Element Analysis

4.1. Model Setup

Bending load represents the most frequent loading condition for subsea pipelines in practical engineering applications. A three-dimensional FE model was developed using ABAQUS 2019 (Dassault Systèmes, Vélizy-Villacoublay, France) commercial software to simulate pipe bending behavior under quasi-static conditions, employing the Static General module.

The FE model replicates a 24-inch metallurgical clad pipe with 609.6 mm outer diameter, 17 mm base layer thickness, 3 mm clad layer thickness, and 1828.8 mm length (equals 3 diameters). Considering dual symmetry characteristics, a quarter model was established with symmetric boundary conditions applied on z = 0 and y = 0 planes (Figure 12). The model employs C3D8 solid elements for the base pipe and S4 shell elements for the clad layer, with four elements through the base wall thickness and a single-layer shell for the clad. This element configuration is consistent with the modeling strategy adopted by Yuan et al. [18] in their study on clad pipes, ensuring accurate representation of mechanical behavior while maintaining computational efficiency. Circumferential and axial directions contain 108 and 99 elements, respectively, incorporating graded axial meshing with refined elements at mid-span regions to balance computational efficiency and accuracy.

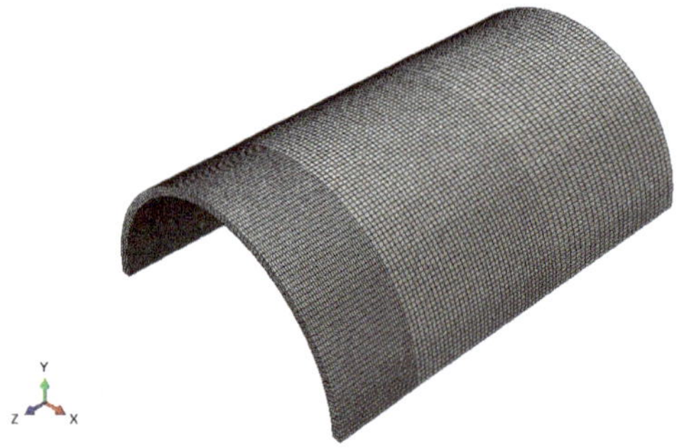

Figure 12. Finite element model.

Experimentally measured material properties were assigned to corresponding inner and outer tubes. In addition, considering the interface delamination takes place at a strain level of 34.17%, the bonding between the two material layers was simulated through tie constraints. Bending moments were applied via reference point rotation coupled with pipe end surfaces. The cross-section at the end plane is free to ovalize while ensuring the plane-remains-plane assumption.

It is worth pointing out that the FE model explicitly incorporates initial ovality of 0.407% and residual stresses of 54.58 MPa from manufacturing processes, utilizing experimentally measured values for both parameters. The residual stress field was introduced via the INITIAL CONDITIONS, TYPE = STRESS option in ABAQUS.

4.2. Bending Simulation

The Mises stress contours of the metallurgically clad pipe before and after bending are presented in Figure 13a,b. The stress distribution along the axial direction remains uniform during bending deformation, while cross-sectional ovalization becomes evident. The moment-curvature and ovality-curvature curves extracted from the pipe's symmetric plane are shown in Figure 14a,b. Specifically, the moment of the base and clad layer, as well as the value of the composite pipe, are plotted, respectively. As expected, the base material layer carries most of the load. To facilitate comparison, the results are presented in dimensionless form. The normalizing variables are based on the parameters of the base pipe as follows:

$$M_o = \sigma_o {D_o}^2 t_o,\ \kappa_1 = \frac{t_o}{{D_o}^2},\ D_o = D - t_o$$

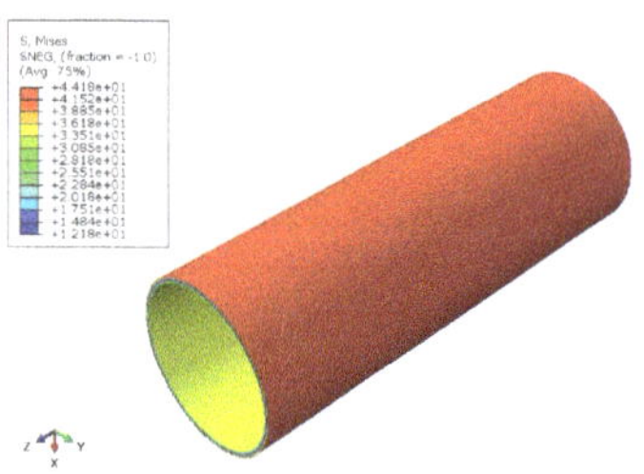

(a)

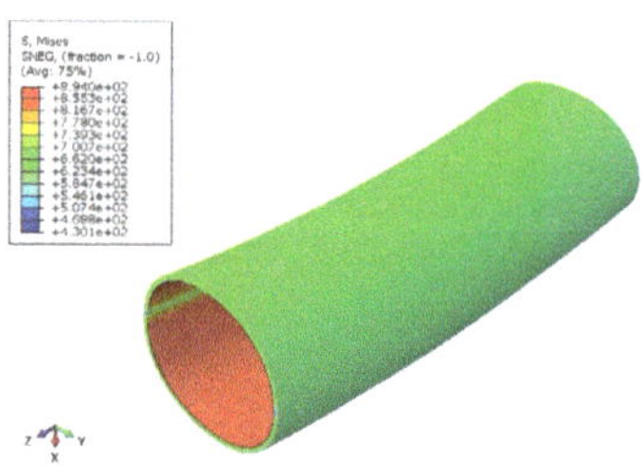

(b)

Figure 13. Mises stress contour of the base case: (**a**) After the introduction of residual stresses; (**b**) After bending.

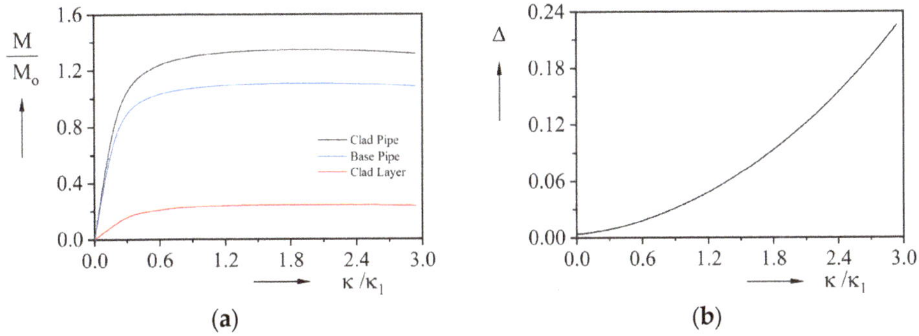

Figure 14. Finite element results of the base case: (**a**) Moment-curvature curves (**b**) Ovality-curvature curve.

As demonstrated in Figure 14a, the bending moment increases linearly with curvature during the elastic stage, followed by a reduced growth rate when plastic yielding occurs. Then, the normalized bending moment reaches its peak value of 1.35 at a normalized curvature of 1.92, followed by a gradual decline. Subsequently, the normalized bending moment gradually decreases with increasing curvature. The normalized curvature at the maximum bending moment is considerably greater than that of mechanically lined pipes [18], indicating enhanced bending resistance. At the same time, the normalized maximum bending moment is also significantly larger. Figure 14b reveals progressive ovality accumulation with curvature, where the curve's increasing slope indicates an accelerated ovalization rate at higher curvatures.

Comparative analyses of pipes with different initial ovalities are shown in Figures 15 and 16 to further investigate the mechanical behavior under bending loads. Ovalities of 0%, 0.41%, 0.81%, and 1.22% are assigned to the FE model. As observed, Figure 15a demonstrates that higher initial ovality reduces the peak bending moment, indicating weakened load-bearing capacity. However, the reduction is approximately 0.5%, which is relatively small and can be considered negligible. In addition, Figure 15b confirms that initial ovality tends to amplify cross-sectional deformation under identical curvature, which further leads to a reduced moment of inertia and consequent capacity reduction—consistent with findings in Figure 15a.

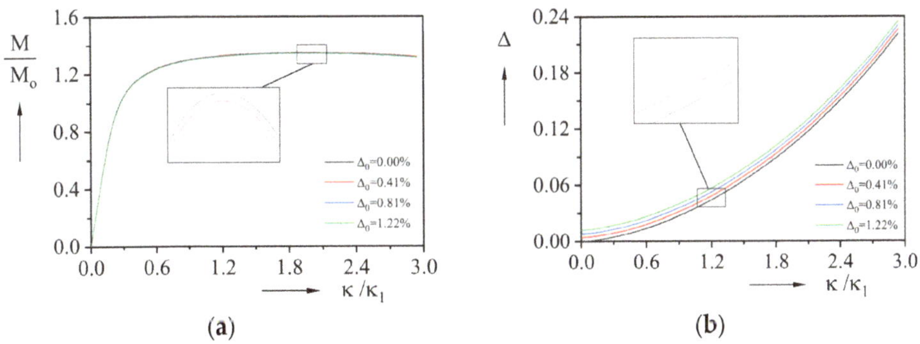

Figure 15. Finite element results of pipes with different initial ovalities: (**a**) Moment-curvature curves; (**b**) Ovality-curvature curves.

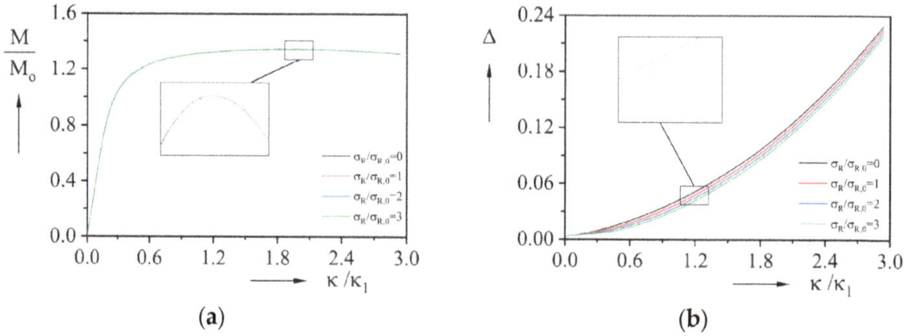

Figure 16. Finite element results of pipes with different residual stresses: (**a**) Moment-curvature curves; (**b**) Ovality-curvature curves.

The residual stress effects were studied by assigning $\sigma_R/\sigma_{R,0}$ = {0, 1.0, 2.0, 3.0} to the FE model. Figure 16a shows negligible variations in peak moments across different residual stress levels, suggesting a limited influence on ultimate bending capacity. Notably, higher residual stresses suppress ovalization growth rates during bending (Figure 16b), despite identical initial ovality conditions. This interesting phenomenon aligns with Kyriakides' observations of bending-type residual stress-suppressing effects in single-wall pipes [30]. The introduction of residual stress initiates reverse ovalization development during bending-induced plasticization, with strain localization concentrating preferentially in the cross-section's 0° and 180° orientations.

5. Conclusions

This paper investigates the mechanical properties of X65 + Alloy625 metallurgically clad pipes through experimental and numerical methods. Uniaxial tensile tests demonstrate that the base material (X65 steel) and clad material (Alloy625) exhibit excellent coordinated deformation capability. Interface delamination occurs when the average strain reaches 34.17%, followed by sequential fracture of the clad and base layers. Microstructural analysis reveals well-bonded interfaces with X65, which are primarily composed of polygonal ferrite and ferritic bainite, while Alloy625 consists of austenitic equiaxed grains. Finite element simulations indicate that larger initial ovality reduces ultimate bending moment, whereas residual stresses show limited influence on both load-bearing capacity and ovality progression. In this study, the actual measured initial ovality of the pipe was 0.4066%, and the residual stress was 54.58 MPa. The simulation yielded a maximum normalized bending moment of 1.35, corresponding to a normalized curvature of 1.92. These findings can provide some references for the safe design and engineering application of metallurgically clad pipes in deep-sea environments. Future investigations will address the effects of cyclic loading on interface degradation mechanisms. Moreover, subsequent modeling efforts would also integrate weld joint details, enabling a comprehensive assessment of strain localization and failure risks in girth-welded clad pipe systems.

Author Contributions: Conceptualization, F.W. and L.Y.; Methodology, F.W. and S.Y.; Software, S.Y.; Validation, Y.G. and Z.C.; Formal analysis, Y.G. and Z.C.; Investigation, Y.G., Z.H., S.Y., Z.C. and R.F.; Resources, Z.H. and R.F.; Data curation, Y.G. and Z.H.; Writing—original draft, F.W., S.Y. and R.F.; Writing—review and editing, F.W., Y.G., Z.H., S.Y., Z.C., R.F. and L.Y.; Visualization, Y.G.; Supervision, L.Y.; Project administration, R.F. and L.Y.; Funding acquisition, L.Y. All authors have read and agreed to the published version of the manuscript.

Funding: This work was supported by the National Natural Science Foundation of China (Grant number 52271288).

Data Availability Statement: The original contributions presented in this study are included in the article. Further inquiries can be directed to the corresponding author.

Conflicts of Interest: Authors Feng Wang, Yanan Gao, Zhiguo Hu, Shuo Yang and Zhenying Cui were employed by the company PipeChina Engineering Technology Innovation Co., Ltd. The remaining authors declare that the research was conducted in the absence of any commercial or financial relationships that could be construed as a potential conflict of interest.

Nomenclature

Symbol	Definition
α	Opening angle after longitudinal cutting
Δ	Ovality
Δ_0	Initial ovality
ε	Strain
ε_d	Delamination strain
ε_f	Clad fracture strain
σ	Stress
σ_R	Residual stress
σ_o	Yield strength of the base pipe
$\sigma_{R,0}$	Measured average residual stress
A_{gt}	Elongation at maximum force
D	Diameter
E	Elastic modulus
f_y	Yield strength
f_u	Ultimate tensile strength
L	Length
M	Moment
R	Pipe radius
t	Wall thickness
t_o	Base pipe thickness

References

1. Chee, J.; Walker, A.; White, D. Controlling lateral buckling of subsea pipeline with sinusoidal shape pre-deformation. *Ocean Eng.* **2018**, *151*, 170–190. [CrossRef]
2. Gao, Z.; Han, B.; Li, L.; Ma, G.; Niu, S. Numerical simulation of residual stress in post internal-welding process of bimetal composite pipe and optimization of welding sequence. *Int. J. Press. Vessel. Pip.* **2022**, *199*, 104730. [CrossRef]
3. Hou, D.; Cheng, X.; Wei, L.; Li, Y.; Li, K.; Zhang, Z.; Zhang, H.; Bian, J.; Li, Y.; Si, W. Oil and natural gas geochemistry in offshore China: Advances and challenges. *Nat. Gas Ind. B* **2024**, *11*, 631–644. [CrossRef]
4. Gao, J.; Li, C.-L.; Zhao, S. Optimization of Welding Process for 825 Bimetallic Composite Pipes Based on Corrosion Testing. *J. Phys. Conf. Ser.* **2024**, *2834*, 012177.
5. Wang, F.-C.; Li, W.; Han, L.-H. Interaction behavior between outer pipe and liner within offshore lined pipeline under axial compression. *Ocean Eng.* **2019**, *175*, 103–112. [CrossRef]
6. Liu, X.; Bai, R.; Uy, B.; Ban, H. Material properties and stress-strain curves for titanium-clad bimetallic steels. *J. Constr. Steel Res.* **2019**, *162*, 105756. [CrossRef]
7. Eiselstein, H.; Tillack, D. The invention and definition of alloy 625. *Superalloys* **1991**, *718*, 1–14.
8. Dong, J.; Wang, D.; Guo, Y. Preparation and Performance Evaluation of Erosion Resistant Lining of Bimetallic Composite Pipe. In Proceedings of the International Pipeline Conference, Calgary, AB, Canada, 26–30 September 2022; p. V002T003A021.
9. Sun, Y.; Yu, S.; Wang, B.; Liu, L.; Liu, E.; Feng, T. Interface Microstructure and Mechanical Properties of Nickel/Steel Bimetal Composite Pipe Fabricated by Explosive Welding. *Metals* **2024**, *14*, 1253. [CrossRef]
10. Withers, P.J.; Bhadeshia, H. Residual stress. Part 1–measurement techniques. *Mater. Sci. Technol.* **2001**, *17*, 355–365. [CrossRef]

11. Hilberink, A.; Gresnigt, A.; Sluys, L. Mechanical behaviour of lined pipe during bending: Numerical and experimental results compared. In Proceedings of the International Conference on Offshore Mechanics and Arctic Engineering, Rotterdam, The Netherlands, 19–24 June 2011; pp. 401–412.
12. Hilberink, A.; Gresnigt, A.; Sluys, L. Liner wrinkling of lined pipe under compression: A numerical and experimental investigation. In Proceedings of the International Conference on Offshore Mechanics and Arctic Engineering, Shanghai, China, 6–11 June 2010; pp. 311–322.
13. Hilberink, A.; Gresnigt, A.M.; Sluys, L.J. *Mechanical Behaviour of Lined Pipe*; Civil Engineering & Geoences: Delft, The Netherlands, 2011; pp. 401–412.
14. Tkaczyk, T.; Pepin, A. Methods of Reel-Laying a Mechanically Lined Pipe. US8864417B2, 21 October 2014.
15. Tkaczyk, T.; Pepin, A.L.; Denniel, S. Integrity of mechanically lined pipes subjected to multi-cycle plastic bending. In Proceedings of the International Conference on Offshore Mechanics and Arctic Engineering, Rotterdam, The Netherlands, 19–24 June 2011; pp. 255–265.
16. Yuan, L.; Zhou, J.; Liu, H.; Chen, N.-Z. Buckling of dented bi-material pipe under bending Part I: Experiments and simulation. *Int. J. Solids Struct.* **2022**, *257*, 111705. [CrossRef]
17. Vasilikis, D.; Karamanos, S.A. Mechanical behavior and wrinkling of lined pipes. *Int. J. Solids Struct.* **2012**, *49*, 3432–3446. [CrossRef]
18. Yuan, L.; Kyriakides, S. Liner wrinkling and collapse of bi-material pipe under bending. *Int. J. Solids Struct.* **2014**, *51*, 599–611. [CrossRef]
19. Yuan, L.; Kyriakides, S. Liner wrinkling and collapse of bi-material pipe under axial compression. *Int. J. Solids Struct.* **2015**, *60*, 48–59. [CrossRef]
20. Yuan, L.; Kyriakides, S. Plastic bifurcation buckling of lined pipe under bending. *Eur. J. Mech.-A/Solids* **2014**, *47*, 288–297. [CrossRef]
21. Yuan, L.; Liu, Z.; Chen, N. On the buckling of mechanically lined pipes under combined tension and bending. *Ocean Eng.* **2022**, *262*, 111991. [CrossRef]
22. Yuan, L.; Kyriakides, S. Liner buckling during reeling of lined pipe. *Int. J. Solids Struct.* **2020**, *185*, 1–13. [CrossRef]
23. Focke, E.S.; Gresnigt, A.M.; Hilberink, A. Local buckling of tight fit liner pipe. *J. Press. Vessel Technol.* **2011**, *133*, 011207. [CrossRef]
24. Yoshitake, A.; Torigoe, T. Centrifugally cast bimetallic pipe for offshore corrosion resistant pipelines. In Proceedings of the ISOPE International Ocean and Polar Engineering Conference, Osaka, Japan, 10–15 April 1994; p. ISOPE-I-94-115.
25. Slany, M.; Sedlak, J.; Zouhar, J.; Zemcik, O.; Kouril, K.; Polzer, A.; Pokorny, Z.; Joska, Z.; Dobrocky, D.; Studeny, Z. Analysis of bimetal pipe bends with a bend of 0.7 D with a cladding layer of Inconel 625. *Int. J. Adv. Manuf. Technol.* **2021**, *117*, 3859–3871. [CrossRef]
26. Yang, R.; He, Y.; Zhang, Y.; Duan, M. Strain Capability Analysis of Metallurgically Clad Pipe With Girth Welds Under Combined Loads. In Proceedings of the International Conference on Offshore Mechanics and Arctic Engineering, Singapore, 9–14 June 2024; p. V003T004A053.
27. Yang, R.; He, Y.; Zhang, Y.; Duan, M.; Wang, S.; Wei, S. Strain capacity of metallurgically clad pipes with girth welds during deep-sea reel-lay installation. *Ocean Eng.* **2025**, *325*, 120743. [CrossRef]
28. Yuan, L.; Wang, C.; Luo, Q.; Chen, N. Collapse pressure prediction of mechanically lined pipes using FEM and machine learning techniques. *Ocean Eng.* **2023**, *268*, 113418. [CrossRef]
29. *GB/T 228.1-2010*; Metallic Materials—Tensile Testing—Part 1: Method of Test at Room Temperature. Standards Press of China: Beijing, China, 2010.
30. Kyriakides, S.; Corona, E. *Mechanics of Offshore Pipelines: Volume i: Buckling and Collapse*; Gulf Professional Publishing: Oxford, UK, 2023.

Journal of
Marine Science and Engineering

MDPI

Article

Fatigue Life Prediction of Submarine Pipelines with Varying Span Length and Position

Daoyu Jiang [1], Xiaowei Huang [1], Deping Zhao [1], Haijing Yang [2] and Guoqiang Tang [2,*]

[1] Haikou Sub-Bureau of Guangzhou Bureau, EHV Transmission Company of China Southern Power Grid Co., Ltd., Haikou 570216, China; jdyisjdy@163.com (D.J.); huangxw126@126.com (X.H.); zhaodeping1989@163.com (D.Z.)
[2] State Key Laboratory of Coastal and Offshore Engineering, Dalian University of Technology, Dalian 116024, China; yanghaijing@mail.dlut.edu.cn
* Correspondence: guoqiang.t@dlut.edu.cn

Abstract: Free spans of submarine pipelines are prone to be subjected to vortex-induced vibration (VIV) under the action of currents, leading to fatigue damage of submarine pipelines. In the traditional method, the fatigue damage is predicted assuming that the length of free span is a constant. However, the free-span length may vary in time and space due to local scour and sand wave migration in engineering practice. This study proposed probabilistic methods to predict the fatigue life of the free spans by considering the effect of variant span length and span position. Truncated Gaussian, Raileigh and Uniform distributions of span length due to local scour, and a sinusoidal pattern with a constant migration rate is assumed for the sand wave due to the lack of field scan data. The fatigue life of a 120 m long span under a constant current-induced flow with the velocity of 0.7 m/s has been assessed. Results show that comparing with the fatigue life of a fixed span, the present method leads to an increase in the fatigue life by about ten times.

Keywords: submarine pipeline; free span; fatigue life; probability density function

1. Introduction

Submarine pipelines are widely used to transport oil and gas in ocean engineering. Free spans might form due to the local scour, sand wave migration, etc. When the currents flow through the free-spanning pipelines, periodic displacement responses will occur if VIV happens. VIV might lead to the fatigue failure of submarine pipelines, causing severe economic losses and environmental disasters.

Many studies have focused on the fatigue failure of free-spanning pipelines. Kapurial et al. [1] studied the fatigue failure problem caused by the cross-flow VIV of free-spanning pipelines supported on semi-infinite elastic soil seabed at the ends and gave the formula for calculating the fatigue damage of pipelines under the combined action of waves and currents. Fyrileiv and Mørk [2] studied the dynamic responses of the free-spanning pipelines based on the Euler–Bernoulli beam theory and predicted the fatigue life of the pipelines under different boundary conditions. Gamino [3] used the Euler–Lagrange coupling method to simulate the interaction between pipelines and seabed. There are many seabed—pipeline interaction models used to predict the VIV, and accurate prediction of VIV is necessary to reasonably predict the fatigue life of submarine pipelines. The realistic model is the nonlinear model due to the varying seabed contact. Ulveseter et al. [4] used the time domain model to calculate the cross-flow VIV and used discrete dampers and springs to model the seabed–pipeline contact. The fatigue analysis of the pipelines is affected not only by the seabed–pipeline interaction, there are many other factors that can also affect the fatigue analysis. Yttervik et al. [5] analyzed the influence of current velocity and direction on the fatigue failure of submarine pipelines. Esplin and Stappenbelt [6] evaluated the maximum allowable span length and fatigue damage characteristics of the submarine pipeline through the S-N curve. Gazis [7] presented a Monte Carlo simulation methodology for a reliability assessment of free-spanning pipeline systems subjected to random waves. Dong et al. [8] calculated the fatigue life under different span lengths using a finite element numerical model.

Numerous factors influence pipeline fatigue, and the seabed environment is intricate, jointly contributing to significant uncertainties in submarine pipeline fatigue. The fatigue of submarine pipelines has great uncertainty. Therefore, some studies have analyzed the fatigue reliability of pipelines from the perspective of probability. Palmer [9] discussed the definition of failure probabilities in pipeline reliability analysis. Shabani et al. [10] used the theory of Probability of Failure (POF) to estimate the reliability of free-spanning submarine pipelines. They used the Monte Carlo method to calculate the POF of pipelines with different ratios of span length to pipeline diameter. Shabani et al. [11] used Finite-Element Reliability based on the Artificial Neural Network method to evaluate the fatigue probability of free-spanning pipelines. Monte Carlo Sampling was used to optimize the Artificial Neural Network to obtain the POF, which was used to evaluate the fatigue reliability level. They also investigated the effect of different span lengths and operating pressure on POF. Li et al. [12] considered the dynamic characteristics and uncertainties of free-spanning submarine pipeline fatigue failure. The Dynamic Bayesian Network was used to model dynamic fatigue failure scenarios. And then, analyzing the dynamic probability of pipeline fatigue failure, the time-dependent occurrence probability of fatigue failure causes is obtained. The above probabilistic analysis used the Monte Carlo method, which requires considerable work.

The above research on pipeline fatigue was based on fixed free span. However, engineering investigations confirm that free-spanning pipelines are dynamic because of local scour or sand waves. Some pipelines may be re-buried by the transported sediment or may continue to be eroded by the current to form a longer span (Pinna et al. [13]). With the movement of sand waves, the original free span is buried, and a new free span is simultaneously formed at both ends of the sand wave (Pu et al. [14]). This study considers the varying span length and position due to the local scour and sand wave migration, respectively. The fatigue life of the pipeline is predicted from the perspective of probability. The proposed method can quickly solve the fatigue life's PDF under the various PDF of span length. More field or experimental data related to the damage determined from pipeline monitoring (c.f., Worden et al. [15], West et al. [16], Ceravolo et al. [17]) are helpful to improve the accuracy and robustness of the proposed model.

The rest of this study is organized as follows: The numerical model used in this study is briefly introduced in Section 2. Then, the numerical model is validated by considering the VIV response of a free span. Section 3 investigates the PDF of the fatigue life of the pipeline under varying span lengths due to local scour, and then the fatigue life of the free-spanning pipeline in the situation where the suspension position changes due to sand wave migration is predicted; Finally, conclusions are drawn in Section 4.

2. Numerical Model

In the present study, a time domain numerical model, proposed by Ulveseter et al. [4], is employed to estimate the cross-flow VIV dynamic response of the free span due to local scour. The sketch of the formation of the free span is shown in Figure 1. The sketch of the formation of free span due to local scour. The coordinate origin locates at the leftmost side of the pipeline. The x-axis is along the axis of the pipeline and the y-axis is perpendicular to the seabed. Two parts of the pipeline contacting with the seabed are defined as the span shoulder, with their lengths (L_s) both 130 m. The starting point of the free span is 130 m away from the coordinate origin, and the total length of the span (L) is 120 m.

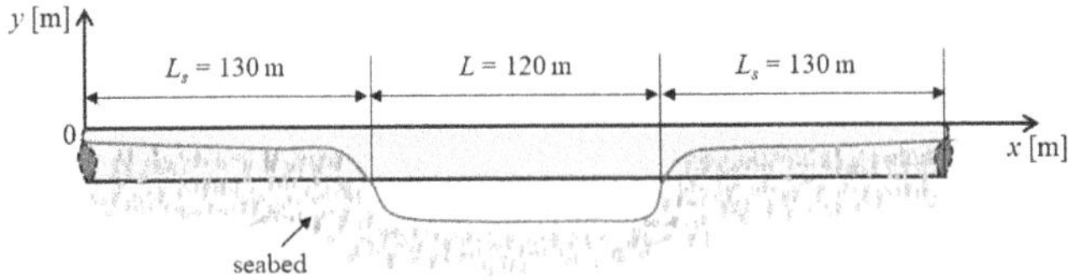

Figure 1. The sketch of the formation of free span due to local scour.

In the present study, the pipeline is idealized as an Euler–Bernoulli beam. The governing equation for the oscillation of the free span subjects to VIV is:

$$EI\frac{\partial^4 u_y}{\partial x^4} - N\frac{\partial^2 u_y}{\partial x^2} + C\frac{\partial u_y}{\partial t} + m_p\frac{\partial^2 u_y}{\partial t^2} = f(x,t) \tag{1}$$

where EI is the bending stiffness, N is the pretension in the pipeline, C is the total damping, including the soil damping and structural damping, u_y is the displacement of the pipeline in the y direction, m_p is the specific mass per unit length, and $f(x, t)$ is the hydrodynamic force on the pipeline.

In the present study, the governing Equation (1) is solved by the finite element method. The finite element method is used to discretize Equation (1) using the Hermit interpolation function. The dynamic response of the free span can be obtained by solving the following dynamic equilibrium equation:

$$M\ddot{s}(t) + (C_s + C_b)\dot{s}(t) + (K_s + K_b)s(t) = F_y \tag{2}$$

where M is the consistent mass matrix including added mass, C_s and C_b are the structural damping matrix and the soil damping matrix, K_s and K_b are the structural stiffness matrix and the soil stiffness matrix, $s(t)$ is the response displacement of the pipeline, and F_y is the hydrodynamic force from Thorsen's model (Thorsen et al. [18]). A Rayleigh-damping, $C_s = \alpha_1 M + \alpha_2 K_s$, is employed to model the structural damping with $\alpha_1 = 1.18$ and $\alpha_2 = 1.42 \times 10^{-4}$ as proposed by Thorsen et al. [18].

The pipeline is idealized as an assembly of a set of beam elements. Each element involves four degrees of freedom, including one cross-flow translation and one rotation at each end. Employing the Hermit shape function, the element mass matrix m_e and the element structural stiffness matrix k_s are illustrated by Equations (3) and (4), respectively.

$$m_e = \frac{m}{420}\begin{bmatrix} 156 & 22l & 54 & -13l \\ 22l & 4l^2 & 13l & -3l^2 \\ 54 & 13l & 156 & -22l \\ -13l & -3l^2 & -22l & 4l^2 \end{bmatrix} \tag{3}$$

$$k_s = \frac{EI}{l^3}\begin{bmatrix} 12 & 6l & 12 & 6l \\ 6l & 4l^2 & -6l & 2l^2 \\ -12 & -6l & 12 & -6l \\ 6l & 2l^2 & -6l & 4l^2 \end{bmatrix} + \frac{N}{30l}\begin{bmatrix} 36 & 3l & -36 & 3l \\ 3l & 4l^2 & -3l & -l^2 \\ -36 & -3l & 36 & -3l \\ -3l & -l^2 & 3l & 4l^2 \end{bmatrix} \tag{4}$$

where m is the sum of specific mass and added mass per unit length, and l is the element length.

Complex contact between the pipeline and the surrounding seabed can occur when VIV takes place, and significant changes in the contact region are possible including separation and recontact, hence a nonlinear seabed–pipeline contact model is introduced as shown in Figure 2. The seabed–pipeline interaction is piecewise nonlinear, which is modeled as a series of discrete soil dampers and soil springs. Nonlinear seabed–pipeline interaction analysis requires the update of the stiffness matrix and damping matrix for each time step during time integration. The vertical soil stiffness k_b = 40,000 N/m, at each time step, using the information associated with the seabed position and the pipeline displacement to check whether the pipeline is in contact with the seabed. If there is contact, the value of stiffness and damping is added. If there is no contact, we check the next node along the pipeline without adding stiffness and damping terms (Ulveseter et al. [4]).

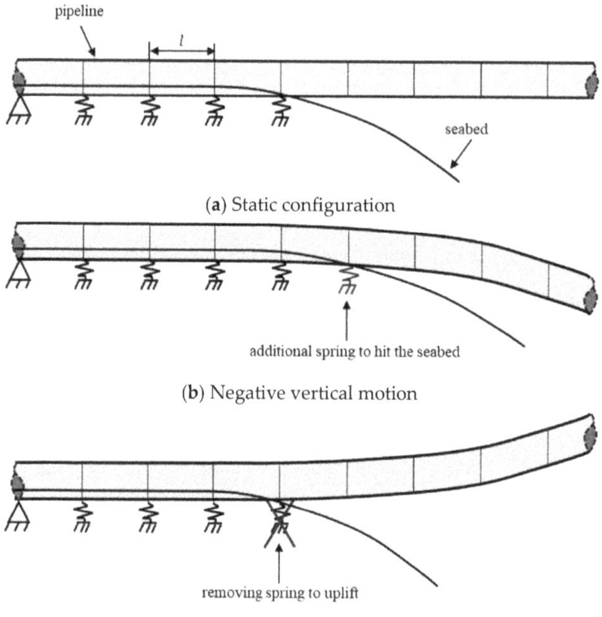

(**a**) Static configuration

(**b**) Negative vertical motion

(**c**) Positive vertical motion

Figure 2. Seabed–pipe interaction model.

The hydrodynamic force perpendicular to the incoming flow direction consists of the added mass term, the excitation force term, and the hydrodynamic damping term (Thorsen et al. [18]):

$$F_y = -\rho\frac{\pi D^2}{4}\ddot{s} + \frac{1}{2}\rho D U^2 C_v \cos(\varphi_{exc,y}) - \frac{1}{2}\rho D C_{d,y}|\dot{s}|\dot{s} \tag{5}$$

where C_v is an empirical coefficient for amplitude-dependent force (Thorsen et al. [18], Figure 2), which is a function of the displacement amplitude ratio s_0/D determined from the empirical database used in the frequency-domain model for predicting the VIV response (Passano et al. [19]), $\varphi_{exc,y} = 2\pi f_s t + \theta$ is the instantaneous phase of F_y with $f_s = S_t U/D$ being the vortex shedding frequency, where $S_t \approx 0.2$ is the Strouhal number (Summer [20]), θ is the initial phase at $t = 0$, and $C_{d,y} = 0.31 + 0.89\, s_0/D$. It worth noting that the wall-proximity effect has not been considered by Thorsen's hydrodynamic model. For spans with the gap ratio $e/D < 1$, adopting this model will overestimate the fatigue damage, because the vortex shedding will be noticeably suppressed when the pipeline gets closer to the wall.

The model parameters of the mentioned numerical model are listed in Table 1.

Table 1. Model parameters.

Parameter	Symbol	Value
Outside diameter/(m)	D	0.55
Wall thickness/(m)	d	0.02
Pipeline mass for unit length/(kg/m)	m_p	315
Young's modulus/(Pa)	E	2.08×10^{11}
Bending stiffness of pipeline/(N/m)	EI	2.9×10^8
Span length/(m)	L	120
Shoulder length/(m)	L_S	130
Current velocity/(m/s)	U	0.7
Water density/(kg/m^3)	ρ	1025
Vertical soil stiffness/(N/m)	k_b	40,000

Using Newmark-β time integration, the response cross-flow displacement along the axis of the free span (x-axis) is calculated. Then, the stress is obtained according to the following relationship:

$$\sigma = \frac{Ds_{xx}}{2}E \tag{6}$$

where s_{xx} is the second derivative of the response.

Figures 3 and 4 show the simulated response amplitude ratio s_0/D and stress amplitude σ_a comparisons among the present numerical model, the nonlinear model of seabed–pipeline interaction developed by Ulveseter et al. [4], and the linear model in VIVANA, respectively. The present results agree well with those predicted by Ulveseter et al. [4]. However, the results derived from the VIVANA give a much larger response and stress because the linear model in VIVANV cannot consider the nonlinear contact between the pipeline and the seabed.

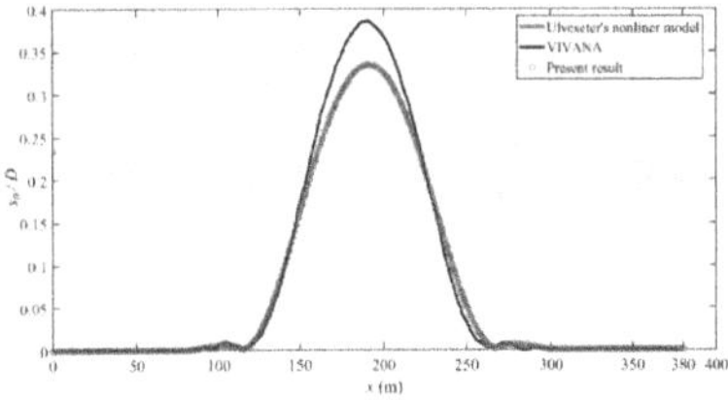

Figure 3. Comparison of response amplitude ratio distribution.

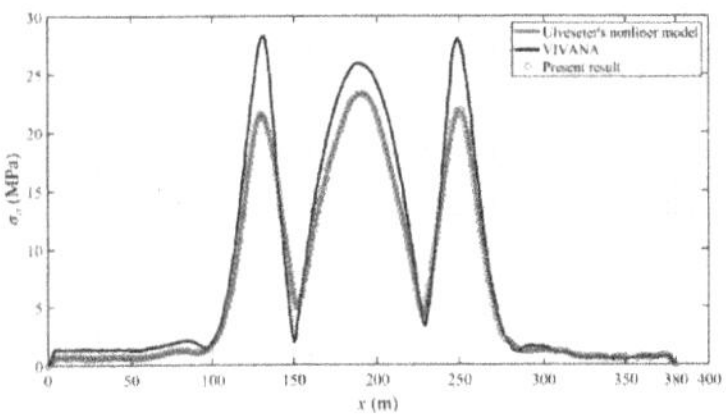

Figure 4. Comparison of the stress amplitude distribution.

The rainflow counting method is used for the stress statistics to obtain the number of stress cycles n_i under the i-th order stress. From the S-N curve, under the stress range S_i, the predicted number of failure cycles N_i can be obtained. Noting that the S-N curve is suitable for the zero-mean stress case, which cannot be directly used in this study. For handling the non-zero mean stress, an additional equation $\sigma_{ar} = f\ (\sigma_a, \sigma_m)$ is needed to modify the stress (Dowling et al. [21]):

$$\sigma_{ar} = \sqrt{\sigma_a \sigma_{\max}} \tag{7}$$

where σ_m is the mean stress, and $\sigma_{\max} = \sigma_a + \sigma_m$ is the maximum stress.

The fatigue damage of a submarine pipeline is calculated by Miner's linear damage accumulation rule (Palmgren [22]; Miner [23]).

$$D_f = \sum \frac{n_i}{N_i} \tag{8}$$

where $1 < D_f < 0$ is the cumulative fatigue damage.

When the variable of the cumulative fatigue damage D_f is obtained, the fatigue damage over one year can be calculated as:

$$D_a = D_f \times t_s / (\Delta t \times t_1) \tag{9}$$

where $t_s = 3600 \times 24 \times 365$ is the number of seconds in a year, Δt is the time interval, t_1 is the time history measured by the rain flow method. The fatigue life of the pipeline can be calculated as:

$$T = \frac{1}{D_a} \tag{10}$$

The result of the fatigue damage of the pipeline over one year is shown in Figure 5. It can be seen from Figure 5 that from the starting point of the span, the pipeline begins to have fatigue damage, and the largest fatigue damage is found at 190 m, near the mid-span. Its fatigue damage is 0.00563 1/year, and the predicted fatigue life of the pipeline is about 177 years.

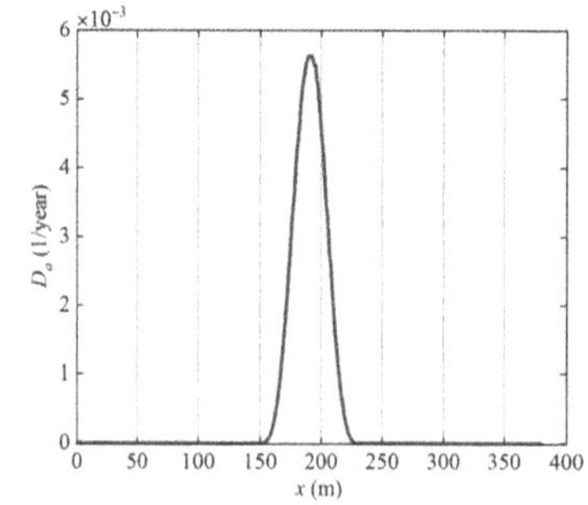

Figure 5. Fatigue damage in one year.

3. Numerical Results

3.1. Fatigue Life Analysis of the Pipeline Under Varying Span Length Due to Local Scour

When the pipeline is installed on the seabed, the current may erode the seabed to form larger spans. With the increase in the span length, the pipeline will gradually sag. Eventually, the center of the spanning section of the pipeline will contact with the seabed again, and the soil will accrete around the touchdown section, forming a new span. The burial and suspension caused by the scouring are dynamic, so the span length varies with time.

For investigating the fatigue failure of the pipeline under varying free spans, the model introduced in Section 2 is chosen. The parameters are shown in Table 1. The span length caused by the local scour varies from 0 to $L_{\max}$. The maximum span length $L_{\max}$

is defined as the pipeline deflecting one diameter (Leckie et al. [24]), which is estimated according to:

$$L_{\max} = \left(\frac{384DEI}{3\omega_{op}}\right)^{\frac{1}{4}} = 120\ \text{m} \tag{11}$$

where ω_{op} is the submerged weight per unit pipeline length.

The varying span length L can be described by a PDF $P(L)$. The uncertainty of L leads to the uncertainty of fatigue life T. Following Lin [25] and Cao et al. [26], if L and T have a one-to-one transformation relationship, given the PDF $P(L)$ of L, one computes the PDF $P(T)$ as:

$$P(T) = P(L)\left|\frac{dL}{dT}\right| \tag{12}$$

There is a limited amount of information regarding observations of the span length in the field. Therefore, some assumed distributions of the span length are studied to demonstrate the procedure of the proposed method. It is noteworthy that in a specific engineering design, the PDF of span length is better to be determined based on field scans.

Because L must be nonnegative in physics, a truncated Gaussian PDF between non-negative values 0 m and 120 m is first investigated:

$$P_1(L) = \frac{1}{\sigma\sqrt{2\pi}}e^{-\frac{(L-\mu)^2}{2\sigma^2}} \tag{13}$$

where μ = 60 m is the mathematic expectation and σ = 25 m is the standard deviation in the truncated Gaussian PDF.

According to Equation (12), the key to obtaining $P(T)$ lies in deriving the mathematical relationship between the span length L and the fatigue life T. For this purpose, an array of L values ranging from 0 m to 120 m with an equal interval of 3 m can be generated, and the corresponding fatigue life T for each L value is computed using the method proposed in Section 2. According to the DNV-RP-F105 design code (DNV [27]), when the span length is smaller than 30 D, the fatigue damage induced by VIV is negligible. Hence, the fatigue life is infinite, as shown in Table 2 when the span length is less than 16.5 m.

Table 2. The data of span length and fatigue life.

Span length (m)	120	117	......	63	60	......	15	12	......	0
Fatigue life (year)	177	315	......	1893	2494	......	∞	∞	......	∞

According to the results associated with (T, L) in Table 2, an empirical formula with the form written in Equation (14) is proposed to describe the correlation between L and T, where the empirical parameters $a_1 = -3.083$, $a_2 = -3.724$, $a_3 = 396.4$ and $a_4 = -0.2252$ are determined based on the least square method.

$$\begin{aligned} &L = a_1 e^{a_2 \ln(T)} + a_3 e^{a_4 \ln(T)} \quad 16.5\ \text{m} \le L \le 120\ \text{m} \\ &T = \infty \qquad 0\ \text{m} \le L \le 16.5\ \text{m} \end{aligned} \tag{14}$$

Taking the partial derivative of L in Equation (14) with respect to T, one obtains:

$$\left|\frac{dL}{dT}\right| = \left|11.48e^{-3.724\ln(T)}\frac{1}{T} - 89.27e^{-0.2252\ln(T)}\frac{1}{T}\right| \tag{15}$$

Substituting Equation (15) into Equation (12) yields the analytical expression of $P(T)$:

$$P(T) = P(L)\left|11.48e^{-3.724\ln(T)}\frac{1}{T} - 89.27e^{-0.2252\ln(T)}\frac{1}{T}\right| \quad (16)$$

Figure 6 is the plot of Equation (16) with $P_1(L)$.

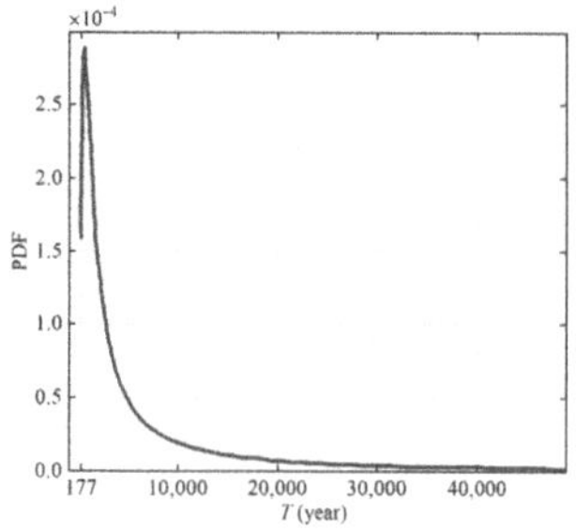

Figure 6. PDF of $P(T)$ with $P_1(L)$.

The span length satisfying Rayleigh distribution (Equation (17)) and Uniform distribution (Equation (18)) are also studied. Following the same procedure, the analytical expression of $P(T)$ is obtained. Figures 7 and 8 plot Equation (12) when $P(L)$ is modeled as Equations (17) and (18), respectively.

$$P_2(L) = \frac{L}{53^2}e^{-\frac{L^2}{2\times 53^2}} \quad (17)$$

$$P_3(L) = \frac{1}{120} \quad (18)$$

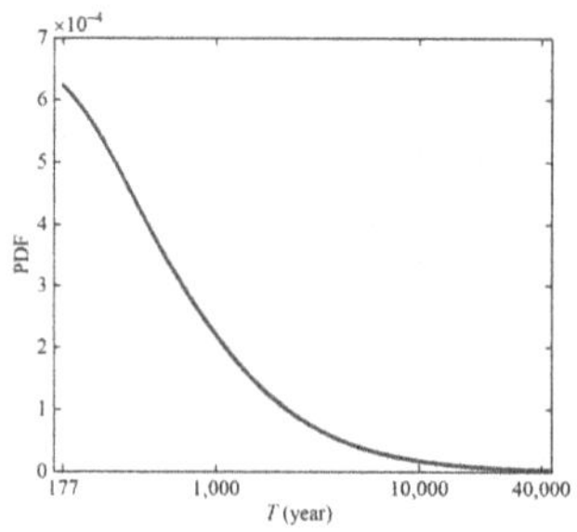

Figure 7. PDF of $P(T)$ with $P_2(L)$.

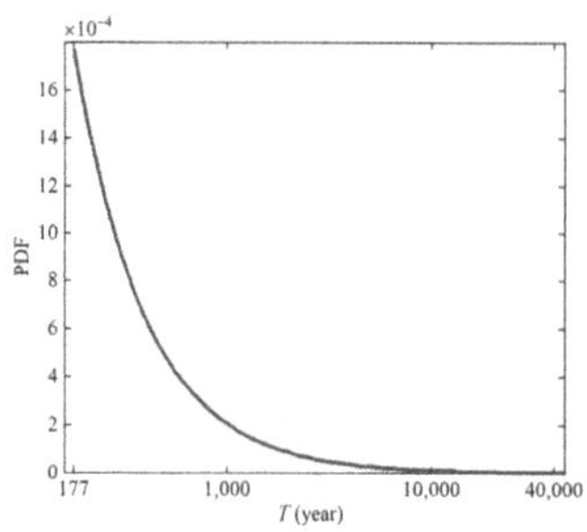

Figure 8. PDF of $P(T)$ with $P_3(L)$.

The traditional calculation method assumes that the span length of the model remains unchanged, and the maximum length is usually used as the basis of VIV and fatigue analysis (He et al. [28]; Esplin [6]). According to Section 2, the fatigue life of the pipeline with a maximum span length of 120 m is 177 years. In this section, the average fatigue life is obtained by $E(T)$ when the fatigue life PDF is known.

$$E(T) = \int_{0}^{T_{\max}} TP(T)dT \tag{19}$$

where $T_{\max}$ is the fatigue life of the pipeline when the span length is 18 m.

From Table 3, the fatigue life calculated by the proposed method is 1766 years or more, about ten times even more than that of the traditional method. The traditional fatigue life calculation method based on fixed span length could be modified to consider the effect of the variant span length. However, it should be noted that the distribution of span length in the field conditions might not strictly follow the three PDFs examined in this paper.

Table 3. The average fatigue life calculated by the different PDFs.

Probability Density Curve	Truncated Gaussian	Rayleigh	Uniform
E(*T*)/(year)	1837	1766	1876

The Coefficient of Variation (COV) is a helpful statistic for comparing the degree of variation between two data series, even if their means are significantly different. It is the ratio of standard deviation to the expected value. The comparison of the COVs between the span length (COV_L) and fatigue life (COV_T) is shown in Table 4. From Table 4, the COV of fatigue life is about five times the span length. The slight uncertainty of the span length leads to considerable uncertainty of the fatigue life. Thus, it is necessary to consider the effect of the varying span length on fatigue life.

Table 4. The average fatigue life calculated by the probability density curve.

COV / Probability Density Curve	COV_L	COV_T
Truncated Gaussian distribution	0.4723	2.9459
Rayleigh distribution	0.4649	2.7337
Uniform distribution	0.5774	2.8033

3.2. Fatigue Life Analysis of the Pipeline Under Changing Suspension Position Due to Sand Wave Migration

The seabed is not entirely flat. There are various seabed forms in many different areas (Knaapen [29]). Sand wave is a kind of seabed form, and its migration may lead to the formation and elimination of free spans. As the sand wave migrates, the suspension position will change correspondingly (Pu et al. [14]). This section aims to propose a method to compute fatigue damage and fatigue life of pipelines caused by sand wave migration. A sinusoidal sand wave migration is assumed.

Sand waves are widely distributed along the pipeline route areas. Engineering investigations show that their crest-to-crest distances vary from 200 m to 800 m, and the observed migration rates range from zero to tens of meters per year (Hulscher and Van Den Brink [30]; Bijker et al. [31]). In this section, the selected wavelength is 312 m, and the sand wave migration rate is 6 m per year. The parameters of the model are shown in Table 1. The model of this section is shown in Figure 9. The origin of the coordinates is located at the starting position of the span, and the span length is 120 m. The x-axis represents the direction of sand wave migration, and the y-axis represents the direction perpendicular to the seabed.

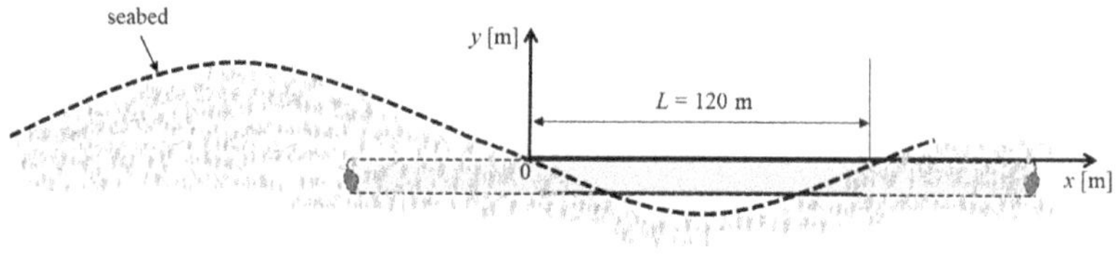

Figure 9. Pipeline model.

With the migration of sand waves, the buried and exposed parts of the pipeline change; that is, the suspension position changes and the span length of the studied pipeline remains unchanged. The migration process is shown in Figure 10, where the black dotted line indicates the initial position of the sand wave, and the red curve represents the moving sand wave. The red dot represents the suspension position. It can be seen from Figure 10 that the exposure and burial of the pipeline change periodically. The burial and exposure states of the pipeline due to sand wave migration are closely related to the fatigue damage of the pipeline.

At the initial moment, the mid-span position is at 60 m (Figure 10a). The chosen pipeline section is completely exposed, and the entire pipeline is considered when calculating the fatigue damage. One year later, the sand wave moves a certain distance, and the mid-span position moves a corresponding distance at 66 m (Figure 10b). During this period, part of the model is buried, and a new part of the pipeline will be exposed. Since we only focus on the chosen pipeline section, the new part is not considered when calculating the fatigue damage. In the 22nd year, the model is completely buried. At this moment, the buried model does not have fatigue damage (Figure 10c). As the sand wave continues to move, the model is exposed again, and only the fatigue damage of the exposed part of the model needs to be calculated (Figure 10d). After 53 years, all the chosen section is exposed again (Figure 10e). From Figure 10, the process is a complete cycle of 53 years. During this migration process, the pipeline is not vibrating all the time, so fatigue damage does not always occur.

For computing the fatigue life of the chosen pipeline section model during the migration process, the fatigue damage at each position of the pipeline is calculated once a year, and the fatigue analysis is carried out according to Section 2. In this way, the fatigue damages of the pipeline section at various points in different periods are calculated. The cumulative fatigue damage in 53 years of each point and its fatigue life are shown in Table 5. The fatigue life is about 1437 years. According to the traditional calculation method, the span position of the model remains unchanged. The pipeline is always in the suspended span state, and the pipeline vibrates continuously. According to the method proposed in Section 2, the fatigue life of the pipeline under a fixed span position is 177 years. Compared with the approach proposed in this study, the fatigue damage calculated by the traditional method is much larger. Accordingly, the fatigue life calculated by the proposed method is about eight times that of the traditional method. It is further verified that the traditional fatigue life analysis method is conservative.

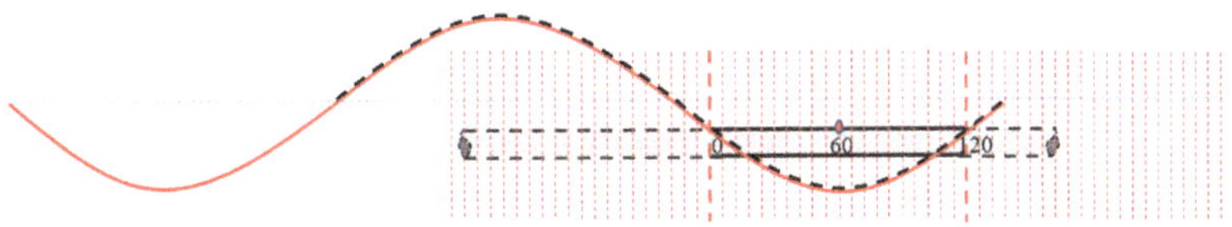

(**a**) The first year

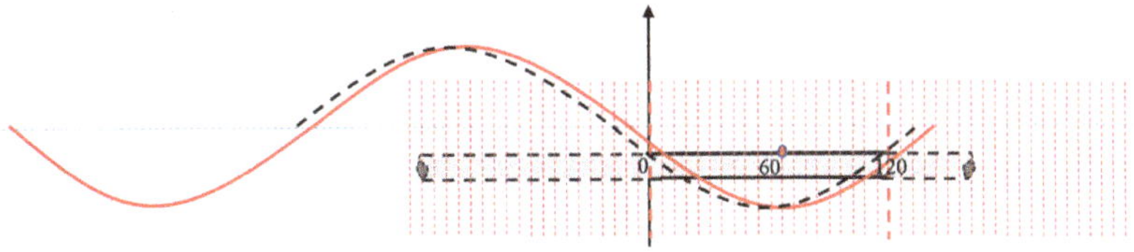

(**b**) The second year

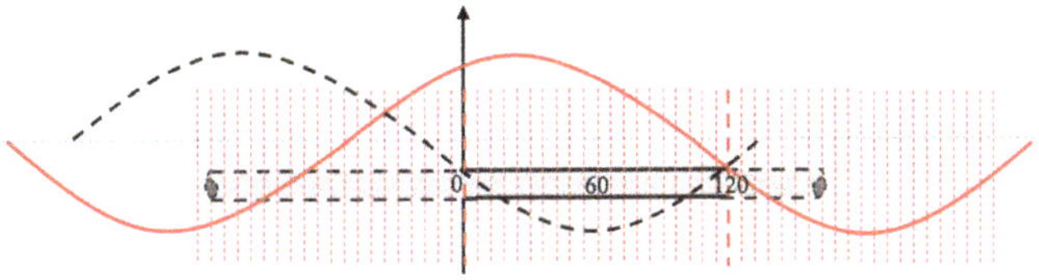

(**c**) The 22nd year

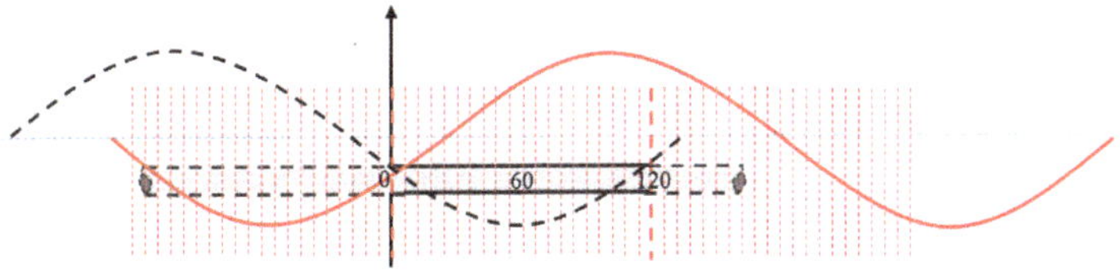

(**d**) The 34th year

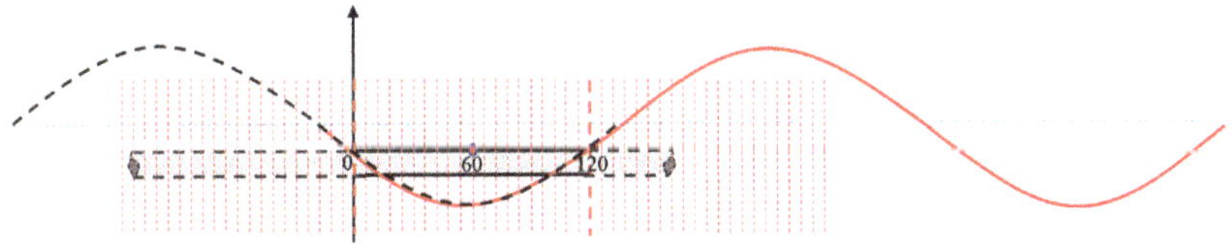

(**e**) The 53rd year

Figure 10. The change process of pipeline suspension position.

Table 5. The calculation results of fatigue damage at each point under the varying suspension position.

	Position (m)					
In Year	**0**	**24**	**48**	**72**	**96**	**120**
1	1.24×10^{-10}	1.58×10^{-5}	3.70×10^{-3}	4.13×10^{-3}	3.20×10^{-5}	7.28×10^{-11}
2		1.33×10^{-7}	2.14×10^{-3}	5.29×10^{-3}	3.10×10^{-4}	6.81×10^{-11}
3		1.13×10^{-10}	8.95×10^{-4}	5.63×10^{-3}	1.13×10^{-3}	1.33×10^{-10}
4		6.57×10^{-11}	2.20×10^{-4}	5.16×10^{-3}	2.58×10^{-3}	1.33×10^{-7}
5		1.24×10^{-10}	1.58×10^{-5}	3.70×10^{-3}	4.13×10^{-3}	3.20×10^{-5}
6			1.33×10^{-7}	2.14×10^{-3}	5.29×10^{-3}	3.10×10^{-4}
7			1.13×10^{-10}	8.95×10^{-4}	5.63×10^{-3}	1.13×10^{-3}
8			6.57×10^{-11}	2.20×10^{-4}	5.16×10^{-3}	2.58×10^{-3}
9			1.24×10^{-10}	1.58×10^{-5}	3.70×10^{-3}	4.13×10^{-3}
10				1.33×10^{-7}	2.14×10^{-3}	5.29×10^{-3}
11				1.13×10^{-10}	8.95×10^{-4}	5.63×10^{-3}

Table 5. *Cont.*

In Year	Position (m)					
	0	24	48	72	96	120
12				6.57×10^{-11}	2.20×10^{-4}	5.16×10^{-3}
13				1.24×10^{-10}	1.58×10^{-5}	3.70×10^{-3}
14					1.33×10^{-7}	2.14×10^{-3}
15					1.13×10^{-10}	8.95×10^{-4}
16					6.57×10^{-11}	2.20×10^{-4}
17					1.24×10^{-10}	1.58×10^{-5}
18						1.33×10^{-7}
19						1.13×10^{-10}
20						6.57×10^{-11}
21						1.24×10^{-10}
……						
33	7.28×10^{-11}					
34	6.81×10^{-11}					
35	1.33×10^{-10}					
36	1.33×10^{-7}					
37	3.20×10^{-5}	7.28×10^{-11}				
38	3.10×10^{-4}	6.81×10^{-11}				
39	1.13×10^{-3}	1.33×10^{-10}				
40	2.58×10^{-3}	1.33×10^{-7}				
41	4.13×10^{-3}	3.20×10^{-5}	7.28×10^{-11}			
42	5.29×10^{-3}	3.10×10^{-4}	6.81×10^{-11}			
43	5.63×10^{-3}	1.13×10^{-3}	1.33×10^{-10}			
44	5.16×10^{-3}	2.58×10^{-3}	1.33×10^{-7}			
45	3.70×10^{-3}	4.13×10^{-3}	3.20×10^{-5}	7.28×10^{-11}		
46	2.14×10^{-3}	5.29×10^{-3}	3.10×10^{-4}	6.81×10^{-11}		
47	8.95×10^{-4}	5.63×10^{-3}	1.13×10^{-3}	1.33×10^{-10}		
48	2.20×10^{-4}	5.16×10^{-3}	2.58×10^{-3}	1.33×10^{-7}		
49	1.58×10^{-5}	3.70×10^{-3}	4.13×10^{-3}	3.20×10^{-5}	7.28×10^{-11}	
50	1.33×10^{-7}	2.14×10^{-3}	5.29×10^{-3}	3.10×10^{-4}	6.81×10^{-11}	
51	1.13×10^{-10}	8.95×10^{-4}	5.63×10^{-3}	1.13×10^{-3}	1.33×10^{-10}	
52	6.57×10^{-11}	2.20×10^{-4}	5.16×10^{-3}	2.58×10^{-3}	1.33×10^{-7}	
53	1.24×10^{-10}	1.58×10^{-5}	3.70×10^{-3}	4.13×10^{-3}	3.20×10^{-5}	7.28×10^{-11}
Sum in 53 years	3.12×10^{-2}	3.12×10^{-2}	3.49×10^{-2}	3.54×10^{-2}	3.13×10^{-2}	3.12×10^{-2}
Fatingue life (year)	1697	1696	1518	1499	1696	1697

4. Conclusions and Recommendations for Further Work

This study proposed probabilistic methods to study the pipeline's fatigue life under the changes in span length and position due to the local scour and sand wave migration, respectively. For the uncertainty of the span length caused by the local scour, a mathematical transformation relation between the span length and fatigue life was first established. Then, the explicit solution of the fatigue life's PDF was computed. The proposed method can quickly solve the fatigue life's PDF under the various PDF of span length. Furthermore, the change process of span position due to sand wave migration is simulated. The fatigue damage of each point on the pipeline over a cycle is accumulated, and then the fatigue life of the pipeline is predicted. Conclusions were summarized as follows:

The suspension of a submarine pipeline is a dynamic process, and the span length varies with time. In this study, the PDF of the fatigue life of the pipeline was derived. The fatigue life of the pipeline based on the traditional method of the fixed free span was 177 years. The average fatigue life calculated by the different PDFs was 1837 years, 1766 years, and 1876 years, respectively, about 10 times that of a fixed 120 m long span.

The span position changes with time and space. This study accumulated the fatigue damage of the pipeline under different span positions over 53 years and the fatigue life of the pipeline under the condition that the span position changes were predicted. The fatigue life predicted in this condition was 1437 years, about eight times that of a fixed 120 m long span. The comparison results further verified that the moving span due to sand wave migration is beneficial in terms of the fatigue limit state of free spans.

The methodology can be universally adopted for the assessment of spans laying on sandy seabed. The simulation takes a couple of hours for a single case at a common laptop which is acceptable for span analysis. In engineering applications, the inputs, such as flow velocity, soil stiffness, damping ratio, pipeline properties, should be updated. The PDF of span length distribution should be replaced based on field scans of the as laid pipeline positions, e.g., with the multibeam echosounder (MBES). Moreover, the migration rate of sand waves should be captured based on numerical modeling or field scans within a couple of years.

The limitation is that the present work only considers the variation in free span length and position separately without considering the interaction between them. The geometry of sand waves might change during the sand wave migration, and local scour might happen together with the sand wave migration. Both these two phenomena will lead to the interaction of span length and position variations. In the future work, these two physical procedures need to be further investigated, after which the interaction between span length and position variations could be quantitatively assessed.

Author Contributions: Conceptualization and methodology, D.J., X.H. and G.T.; Software and Validation, D.Z. and H.Y.; formal analysis and investigation, D.J., X.H., D.Z., H.Y. and G.T.; writing—original draft preparation, X.H., H.Y. and G.T.; writing—review and editing, D.J. and G.T. All authors have read and agreed to the published version of the manuscript.

Funding: The author Guoqiang Tang would like to acknowledge the support from the National Nature Science Foundation of China through Grant 52371262.

Data Availability Statement: The data presented in this study are available from the corresponding author on request.

Conflicts of Interest: Daoyu Jiang, Xiaowei Huang and Deping Zhao were employed by EHV Transmission Company of China Southern Power Grid Co., Ltd. The remaining authors declare that the research was conducted in the absence of any commercial or financial relationships that could be construed as a potential conflict of interest.

Abbreviations

The following abbreviations are used in this manuscript:

VIV	vortex-induced vibration
POF	Probability of Failure
PDF	Probability density function
COV	Coefficient of Variation

References

1. Kapuria, S.; Salpekar, V.Y.; Sengupta, S. Fatigue due to vortex-induced crossflow oscillations in free spanning pipelines supported on elastic soil bed. In Proceedings of the 9th International Offshore and Polar Engineering Conference, Brest, France, 30 May–4 June 1999.
2. Fyrileiv, O.; Mørk, K. Structural response of pipeline free spans based on beam theory. In Proceedings of the 21st International Conference on Offshore Mechanics and Arctic Engineering, Oslo, Norway, 23–28 June 2002.
3. Gamino, M. FSI Methodology for Analyzing VIV on Subsea Pipeline Free Spans with Practical Boundary Conditions. Ph.D. Thesis, Department of Mechanical Engineering Technology, University of Houston, Houston, TX, USA, 2013.

4. Ulveseter, J.V.; Sævik, S.; Larsen, C.M. Vortex induced vibrations of pipelines with non-linear seabed contact properties. In Proceedings of the 35th International Conference on Ocean, Offshore and Arctic Engineering (OMAE2016-54424), Busan, South Korea, 19–24 June 2016.
5. Yttervik, R.; Larsen, C.M.; Furnes, G.K. Fatigue from vortex-induced vibrations of free span pipelines using statistics of current speed and direction. In Proceedings of the ASME 2003 22nd International Conference on Offshore Mechanics and Arctic Engineering, Cancun, Mexico, 8–13 June 2003.
6. Esplin, G.D.; Stappenbelt, B. Reducing conservatism in free spanning pipeline vortex-induced vibration fatigue analysis. *Aust. J. Mech. Eng.* **2011**, *8*, 11–20. [CrossRef]
7. Gazis, N. A probabilistic approach for reliability assessment and fatigue analysis of subsea free spanning pipelines. In Proceedings of the Twenty-Second International Offshore and Polar Engineering Conference, Rhodes, Greece, 17–22 June 2012.
8. Dong, J.; Chen, X.D.; Wang, B.; Guan, W.X.; Yang, T.X.; Fan, Z.C. The effect of soil on the structural response and fatigue life of free span for submarine pipeline. In Proceedings of the ASME Pressure Vessels and Piping Conference, Boston, MA, USA, 19–23 July 2015.
9. Palmer, A. 10^{-6} and all that: What do failure probabilities mean? *J. Pipeline Eng.* **2013**, *2*, 109–111.
10. Shabani, M.M.; Taheri, A.; Daghigh, M. Reliability assessment of free spanning subsea pipeline. *Thin-Walled Struct.* **2017**, *120*, 116–123. [CrossRef]
11. Shabani, M.M.; Shabani, H.; Goudarzi, N.; Taravati, R. Probabilistic modelling of free spanning pipelines considering multiple failure modes. *Eng. Fail. Anal.* **2019**, *106*, 104169. [CrossRef]
12. Li, X.H.; Zhang, Y.; Abbassi, R.; Khan, F.; Chen, G.M. Probabilistic fatigue failure assessment of free spanning subsea pipeline using dynamic Bayesian network. *Ocean. Eng.* **2021**, *234*, 109323. [CrossRef]
13. Pinna, R.; Weatherald, A.; Grulich, J.; Ronalds, B.F. Field observations and modelling of the self-burial of a North West Shelf pipeline. In Proceedings of the 22nd International Conference on Offshore Mechanics and Arctic Engineering, Cancun, Mexico, 8–13 June 2003.
14. Pu, J.J.; Xu, J.S.; Li, G.X. Self-burial and potential hazards of a submarine pipeline in the sand wave area in the South China Sea. *J. Pipeline Syst. Eng. Pract.* **2013**, *4*, 124–130. [CrossRef]
15. Worden, K.; Farrar, C.R.; Manson, G.; Park, G. The fundamental axioms of structural health monitoring. *Proc. R. Soc. A Math. Phys. Eng. Sci.* **2007**, *463*, 1639–1664. [CrossRef]
16. West, B.M.; Locke, W.R.; Andrews, T.C.; Scheinker, A.; Farrar, C.R. Applying Concepts of Complexity to Structural Health Monitoring. In *Structural Health Monitoring, Photogrammetry & DIC*; Springer: Cham, Switzerland, 2019; Volume 6, pp. 205–215.
17. Ceravolo, R.; Civera, M.; Lenticchia, E.; Miraglia, G.; Surace, C. Detection and localization of multiple damages through entropy in information theory. *Appl. Sci.* **2021**, *11*, 5773. [CrossRef]
18. Thorsen, M.J.; Sævik, S.; Larsen, C.M. Fatigue damage from time domain simulation of combined in-line and cross-flow vortex-induced vibrations. *Mar. Struct.* **2015**, *41*, 200–222. [CrossRef]
19. Passano, E.; Larsen, C.M.; Lie, H.; Wu, J. *VIVANA Theory Manual. User Manual*; Norwegian University of Science and Technology: Trondheim, Norway, 2014.
20. Sumer, B.M. *Hydrodynamics Around Cylindrical Structures*; World Scientific: Singapore, 2006.
21. Dowling, N.E.; Calhoun, C.A.; Arcari, A. Mean stress effects in stress-life fatigue and the Walker equation. *Fatigue. Fract. Eng. M.* **2009**, *32*, 163–179. [CrossRef]
22. Palmgren, A. Die lebensdauer von Kugellagern. *Verfahrenstechnik Berl.* **1924**, *68*, 339–341.
23. Miner, M.A. Cumulative damage in fatigue. *J. Appl. Mech. Sep.* **1945**, *12*, A159–A164. [CrossRef]
24. Leckie, S.H.F.; Draper, S.; White, D.J.; Cheng, L.; Fogliani, A. Lifelong embedment and spanning of a pipeline on a mobile seabed. *Coast. Eng.* **2015**, *95*, 130–146. [CrossRef]
25. Lin, Y.K.M. *Probabilistic Theory of Structural Dynamics*; McGraw-Hill: New York, NY, USA, 1967.
26. Cao, Q.Y.; Hu, S.L.J.; Li, H.J. Global maximum response of linear systems with uncertain damping to irregular excitation. *J. Eng. Mech.* **2020**, *146*, 04020043. [CrossRef]
27. Det Norske Veritas. *Recommended Practice DNV-RP-F105: Free Spanning Pipelines*; DNV: Oslo, Norway, 2006.
28. He, Z.F.; Wei, Y.; Liu, S.L. Analysis of safe span length and fatigue life of submarine pipelines. *COE* **2020**, *34*, 119–130. [CrossRef]
29. Knaapen, M.A.F. Sandwave migration predictor based on shape information. *J. Geophys. Res. Earth Surf.* **2005**, *110*, 3. [CrossRef]
30. Hulscher, S.J.M.H.; van den Brink, G.M. Comparison between predicted and observed sand waves and sand banks in the North Sea. *J. Geophys. Res. Ocean.* **2001**, *106*, 9327–9338. [CrossRef]
31. Bijker, R.; Wilkens, J.; Hulscher, S. Sandwaves: Where and why. In Proceedings of the Eighth International Offshore and Polar Engineering Conference, Montreal, QC, Canada, 24 May 1998.

Journal of
Marine Science and Engineering

MDPI

Article

Finite Element Modelling of Pressure Armour Unlocking Failure in Unbonded Flexible Risers

Rongzhi Wei [1], Xiaotian Li [2,3,4], Murilo Augusto Vaz [1,*] and Anderson Barata Custódio [5]

1 Ocean Engineering Program, Universidade Federal do Rio de Janeiro, Rio de Janeiro 21941-611, Brazil; rongzhiwei@oceanica.ufrj.br
2 School of Marine Engineering and Technology, Sun Yat-sen University, Zhuhai 519082, China; lixiaot5@mail.sysu.edu.cn
3 Southern Marine Science and Engineering Guangdong Laboratory (Zhuhai), Zhuhai 519082, China
4 Key Laboratory of Comprehensive Observation of Polar Environment (Sun Yat-sen University), Ministry of Education, Zhuhai 519082, China
5 Centro de Pesquisas Leopoldo Américo Miguez de Mello (CENPES,) Petrobras, Rio de Janeiro 21941-915, Brazil; andersonbc@petrobras.com.br
* Correspondence: murilo@oceanica.ufrj.br

Abstract: Flexible pipes can be subjected to extreme bending load during installation and in situ operations, as well as high pressure from oil and gas field production fluids. Although field experience shows that the unlocking of the pressure armour layer is a rare failure mode, its consequences are potentially catastrophic. To investigate the unlocking mechanism of the pressure armour layer, a 3D nonlinear finite element model is developed, which contains three layers: a pressure armour layer with a "Zeta" shape cross-section, a polymer layer, and an analytical cylindrical surface representing the radial support from the carcass. The analysis is carried out with the ABAQUS Dynamic Explicit solver using a semi-automatic mass scaling factor. The study focuses on the possibility of pressure armour unlocking due to bending load, where the effects of geometric features of the pressure armour layer and internal pressure value on unlocking are investigated by varying corresponding parameters. The influence of these variables are discussed at the end. Ultimately, in conjunction with the numerical analysis results, critical reflections on the shortcomings of the model and suggestions for improvement are presented based on the numerical analysis results.

Keywords: flexible riser; pressure armour layer; unlocking failure analysis; finite element modelling

1. Introduction

With its superior mechanical properties and design flexibility, unbonded flexible risers have become pivotal O&G transportation equipment for deep-water field exploitation. Figure 1 presents a comprehensive internal architecture view of these risers. Unbonded flexible risers comprise metallic and polymeric layers. The metallic layers include carcass, pressure armour and tensile armour layers, while the polymetric components are comprised of inner and outer sheaths and anti-wear and high-strength anti-birdcage tapes. The pressure armour layer is manufactured by applying the Zeta-section steel wire winding and placement technique around the outer surface of the inner sheath. It conventionally consists of one continuous wire with a laying angle close to 90 degrees with regards to the generatrix. The fundamental duty of the pressure armour layer is to provide resistance against internal pressure. Furthermore, it also endures external loads such as crushing due to pipe handling, installation, or accidents. As with the interlocked carcass, its shape does not make the pipe watertight, still permitting fluid mobility between layers.

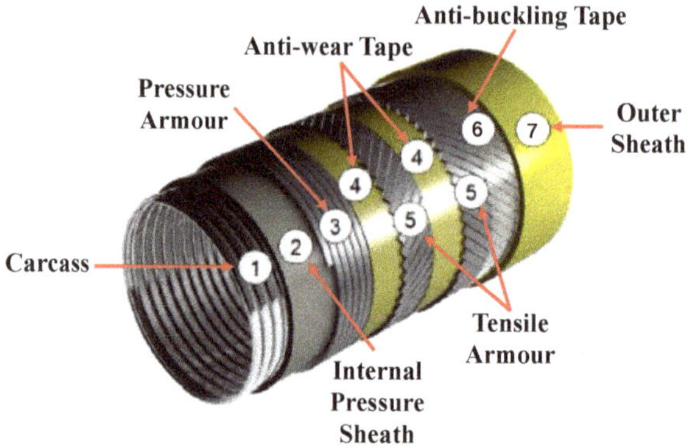

Figure 1. Typical unbonded flexible riser cross-section.

As main international O&G operators turn their attention to deeper marine areas, the harsh operating environment poses even greater challenges for complex offshore production structures with movable components. Special working conditions demand flexible risers endure integrity under high temperature (HT), high pressure (HP), and extreme and alternating dynamic loads. The foregoing statements require that flexible riser manufacturers be more rigorous in the design and verification of products to avoid irreparable damage, as some failure modes, which are considered to have a low probability of occurrence in shallow or deep water, may feasibly be activated in the arduous environment of ultradeep waters.

In 1999, a pressure armour layer unlocking incident that occurred in an eight-inch flexible riser, which transported water, on the Norwegian continental shelf has been officially recorded by the CODAM database [1,2]. Further detail illustrates that the pressure armour was unlocked due to cracks caused by overbending [3]; the riser was then repaired with an entire restoration of the unlocking riser section. The failure mitigation method indicates the severity of the unlocking failure. The creeping and perforation of the inner sheath may occur if the riser is subjected to internal pressure due to the expansive void between the pressure armour wires [4]. One can infer that the inner sheath at the pressure armour wire unlocking location may experience continuous and progressive creep extrusion and perforation under the internal pressure of continuous production conditions. The outcome can be anticipated with the absence of radial restraint of the pressure armour layer, and ultimate failure may manifest, such as burst and leakage. Combining statistical data and relevant reports, the pressure armour layer unlocking can be recognized as a rare but disastrous failure mode, and it is unacceptable because of its threat to riser integrity, expense in terms of a remedy, and environmental contamination risk.

Thereafter, the unlocking failure has been investigated to some extent by the industry sector [4,5]. The possible unlocking initiators are summarized as follows: excessive axial tension, overbending, excessive torsion (at installation), polymer material creeping [6], failure of tensile or backup pressure armour wires, and external impact. It is possible to deduce that unlocking is achievable due to excessive axial tension and overbending. In addition, the loss of support from the neighbouring layers or other forms of pressure armour layer deterioration can facilitate the unlocking.

The present work focuses on the pressure armour layer unlocking caused by overbending, which a flexible riser section can experience after the birdcage incident. The birdcage phenomenon is a unique failure mode for tensile armour layers. Due to the hydrostatic end-cap effect, tensile armour layers can encounter considerable longitudinal compressive load during installation, well shutdown, etc. Flexible pipes are radially dilated as a result of such axial compressive loads. A radial failure mechanism with a substantial localized radial dislocation, namely a birdcage failure, will develop if such radial expansion is not effectively controlled [7]. Based on the preceding discussion, the inference regarding the pressure armour layer unlocking due to overbending is as follows: the tensile armour layer is separated from other inner layers in sections confronted with birdcage failure. Meanwhile, the riser's resistance to bending loads is reduced. Consequently, the excessive bending that takes place in this segmentation will be transmitted to the inner layers. Concisely, the unlocking may become a secondary effect after the "birdcage". It is important to evaluate if the pipe can still maintain its internal fluid tightness after the unlocking. It is deemed valuable to develop a methodology for the risk assessment of pressure armour unlocking after the flexible riser "birdcage" incident.

To date, limited studies are directly related to pressure armour layer unlocking. To simulate the contact behaviour between pressure armour wires, Lu et al. [8] developed a 3D FE model to simulate the loss of interlock of the pressure armour layer. By imposing a pre-calculated displacement on the inner sheath nodes, the bending behaviour is produced in the model. The pressure armour layer acts per the interlayer contact interaction to produce a uniform bending effect. The results showed that the finite element model (FEM) can be used to calculate the Damage Bending Radius (DBR) of the pressure armour layer with a measured cross-section profile. Lu et al. [9] proceeded by introducing the ingression of polymer material and thermo-plasticity into the previous research. Scenarios for the creeping of polymers into the pressure armour wires' gap with different temperatures and different gap size distributions are devised and the resulting reduction in riser flexibility is investigated. The developed FEM is sufficient to simulate the bending of the pressure armour layer with polymer material ingression and gives reasonable stress analysis results according to different pressure armour wire spacing conditions.

Considering the practical function of pressure armour in supporting the inner sheath and protecting the pipe from external pressure or crushing, the mainstream research interest related to the pressure armour layer has been divided into the following two areas: burst and collapse analysis of the flexible riser. For burst analysis, Neto et al. [10,11] proposed three numerical models, including a 2D simplified equivalent ring model, a 3D ring approximation model for helically rolled pressure armour layer, and a full 3D model of the actual spiral structure. The conclusion shows that up to yielding, a 3D FE model is necessary for analysing the stress concentration. Increasing the strain hardening of the pressure armour gives a more significant difference between 2D and 3D models. Subsequently, Cuamatzi-Melendez et al. [12] presented a burst analysis analytically and numerically. The morphology modelling of the pressure armour layer and the polymer layer is of high significance for the unlocking simulation involving the internal pressure effect. The importance of using a 3D model to determine the real stress level in the pressure armour is claimed again, and as compared to a 2D FEM and analytical model, considering the winding pitch and actual profiles, the 3D model produces increased stress results. Liu et al. [13] investigated the internal pressure capacity and burst failure characteristics of an unbonded flexible riser, emphasizing the pressure armour layer. Detailed geometric properties and material elastoplastic behaviour of the pressure armour layer is incorporated into the FEM. In addition to burst, dry and wet annulus collapse failures are becoming more prominent with the increase in deployment depth [14]. Both the pressure armour layer and carcass are involved in the study because they are the key elements that give resistance to external pressure. Firstly, Paumier and Mesnage [15] reported ultradeepwater flexible riser qualification progress with field cases. A local residual stress assessment FEM and an in-house prediction tool of the PSI dynamic wire, namely the pressure armour wire, have been developed per corresponding standards [5,16]. In 2012, Neto et al. [17] developed a four-layer numerical model that considers the actual rolled shape of carcass and pressure armour with a few assumptions. This four-layer complete model was used as a benchmark to validate other simplified models developed by Neto et al. [18,19]. This series of studies is launched to clarify the mechanical behaviour of the pressure armour layer in the wet and dry collapse of straight and curved flexible pipes. Pilot nodes and clamping coupling conditions are used for curvature imposition on the overall flexible riser. For carcass collapse failure, a simplified or equivalent pressure armour layer does not cast significant influence. The text provides a detailed description of the material model section for flexible riser components and contact settings. Meanwhile, in 2017, Cuamatzi-Melendez et al. [20] worked on the FEM and theoretical analysis of two bisymmetric collapses. A helical pressure armour layer with a "Zeta" cross-section was used in an FEM. Simulation results finally proved the lateral resistance and restriction of the pressure armour layer for the carcass ovality before and after the eight-type carcass collapse. Recently, Zhu and Lei [21] analysed how the yield strength of the carcass and pressure armour layer affects flexible pipe collapse failure. Multi-Point Constraint (MPC) and similar finite element modelling methods are used in their study. Progress in simulation algorithms has led to sophisticated numerical models for unbonded flexible risers, capturing intricate geometric details and complex load interactions [22–24]. Structures from flexible risers that were once difficult to simulate, such as the pressure armour layer, can be alleviated to a certain extent. Liu et al. [25–27] explored flexible risers with non-uniform geometries under diverse loads, using explicit algorithms to ensure convergence despite complex contact and geometric profiles. Their work underscores the significant coupling effects of different load types on riser performance.

This study aims to investigate the unlocking mechanism of the pressure armour layer under uniform curvature and internal pressure, leading to unlocking failure. This failure mode is suspected to occur at the Touch Down Zone (TDZ), where elastoplastic deformation causes a loss of interlock between adjacent pressure armour wires, potentially resulting in outer layer failure, such as a birdcage failure of the tensile armour layer. Key factors influencing this process include material properties, structural geometry, and loading conditions. A 3D nonlinear asymmetric finite element model (FEM) is developed, incorporating the pressure armour layer, polymer inner sheath, and a cylindrical surface representing the carcass radial support. The FEM framework is presented in detail, covering the geometric configuration, material properties, meshing strategy, and boundary conditions. A comprehensive sensitivity analysis is performed, examining the effects of critical geometric parameters—such as pitch number, pitch distance, and internal diameter—along with variations in internal pressure. The results are discussed, focusing on the unlocking process, stress distribution, and the influence of various parameters on the unlocking curvature. Key observations and correlations are highlighted. Finally, the key findings are summarized, and future research directions are proposed, including experimental validation and potential improvements for practical design considerations.

2. Finite Element Modelling

This section presents the pressure armour layer unlocking features and FE modelling approach using ABAQUS 2021 software. The FE models are composed of three layers: a rigid surface, inner sheath, and pressure armour layer. Shell and solid elements are used both to ensure the FE simulation can reflect the practical engineering conditions and to reduce computation consumption. Taking into account the contribution of each layer and the extrapolation of the unlocking mentioned in the previous section, several hypotheses are established in advance.

- The structure is assumed to be stress-free in the initial state, and any prestress introduced by fabrication is ignored.
- Geometrical imperfections are not considered. It is assumed that the pressure armour is laid in a perfect helical configuration.
- The inner sheath is modelled as a tubular shell with uniform thickness and its extrusion into the gaps of pressure armour is not considered in the initial unloaded condition.
- It is assumed that the carcass will not be self-contacted and maintain its integrity during the whole simulation. For the sake of simplification, the radial support from the carcass layer is replicated by a frictionless rigid tubular surface.
- Considering that pressure armour layer unlocking may be a secondary effect of a birdcage failure, outer layers are removed to reproduce the birdcage failure and diminish the radially outward displacement constraint these outer layers place on the pressure armour layer.

2.1. Geometry and Material

2.1.1. Pressure Armour Layer

The work by Cuamatzi-Melendez et al. [12] provides the basic geometry and material information for the pressure armour wire and inner sheath. The pressure armour wire profile is claimed to be derived from API RP 17B [16] and matches the dimensions of a two-inch flexible riser. Basic dimensioning parameters and important stress–strain points are displayed in Table 1 and Figure 2, among which ID stands for internal diameter and E, ν, σ_y, σ_u, ε_u, and ρ are, respectively, Young's modulus, Poisson's ratio, yield stress, ultimate tensile stress, and ultimate tensile strain and density. Conclusions drawn from studies of burst failures indicate that a reproduction close to the practical three-dimensional helical structure of the pressure armour layer is important and accurate for stress analysis [12]. Thus, two asymmetry features, armour wire with a true Zeta cross-sectional shape and helical structure, are restored in the modelling. The profile revolved 90 degrees around the rotation axis with a 6.5 mm pitch distance, which equates to the 88.4 degrees helix angle of the pressure armour layer. Multiple discrete quarter armour wire turns are merged head to tail into a pressure armour layer instance with spiral conformation.

Table 1. Geometry and material properties [12].

Layer	ID (mm)	Thickness (mm)	E (GPa)	ν	σ_y (MPa)	σ_u (MPa)	ε_u	Material	ρ (Kg/m^3)	Pitch (mm)
Inner sheath	57.8	6	0.345	0.3	N/A	35	1.4	Nylon PA11	1040	N/A
Pressure armour	69.8	4.8	207	0.3	650	1080	0.25	Carbon Steel	7850	6.5

2.1.2. Inner Sheath

By incorporating the work from Cuamatzi-Melendez et al. [12] and Merino et al. [28], Polymer PA11 is used as the inner sheath material. The inner sheath is not regarded as the primary resisting layer due to its lower stiffness compared to other metallic layers. Nonetheless, transferring the internal pressure of the delivered fluid to the inner surface of the pressure armour layer is its primary function. Since the unlocking failure analysis only includes the external loading implementation until unlocking occurs, and excludes the unloading process, the stress–strain curve is shown in Figure 2 as a linear-hardening form. In FE models, the inner sheath presents a wall with equal thickness and a cylindrical shape generated by revolution with an inner surface distance from the central axis equal to the polymer layer's inner radius.

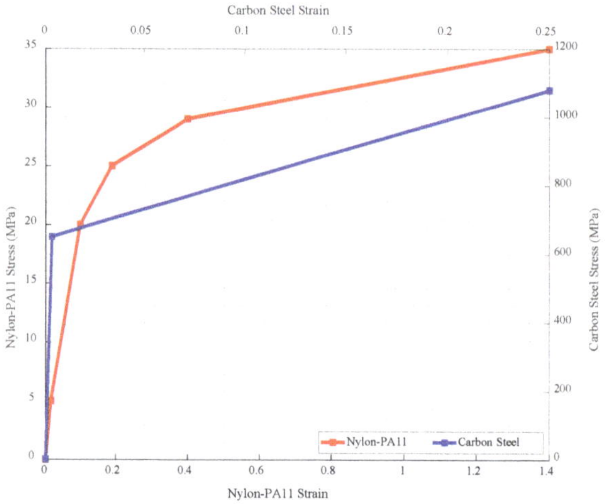

Figure 2. Stress–strain curve of nylon PA11 and carbon steel material.

2.1.3. Carcass

Essentially, the carcass layer is made of a thin steel strip that is spiralled into an S shape. The purpose of the unlocking investigation does not cover modelling complex carcass structures, which is a difficult task for both modelling and computation. During the overbending simulation, it is anticipated that the carcass layer will help maintain the global curvature's regularity. Comparatively to a cylindrical thin-walled shell that does not take thickness into account, the carcass layer using a real cross-section necessitates more complicated operations to maintain the consistency of the overall model's curvature. Additionally, a real cross-sectioned carcass's complex contact behaviour would demand huge processing power during bending. Therefore, the simplification of the carcass as a thin-walled deformable cylinder surface, whose radius is equal to the inner diameter of the polymer layer, serves the goals of lowering computation costs and facilitating simple control of global curvature. (Note: in the following section, "carcass" will be replaced with "simplified surface" textually.)

2.2. Overbending Model

A simplified surface is generated along the positive *X* axis as the carcass layer. The inner sheath and pressure armour layer are spliced by the quarter inner sheath and pressure armour wire turns using the linear and radial pattern and joint head to tail into a whole. The centre of the left cross-section of the simplified surface coincides with the origin of the global coordinate system, and the longitudinal centres of all three layers are aligned in the middle.

Figure 3a shows the three-layer structure of the FE model. One feature of the pressure armour layer is its non-integer pitch number. The pitch number is always an integer plus half of one coil pressure armour wire. Figure 3b depicts the entire assembly of a 14.5-pitch model as well as the unbent and bent shapes of the pressure armour layer to aid in comprehension. It can be observed that half of one coil pressure armour wire at the bending extrados contributes to the bending balance and uniformity, which is important for the overbending-induced unlocking simulation.

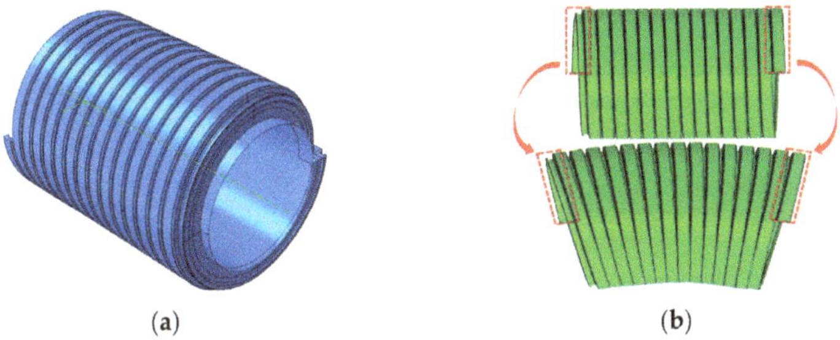

Figure 3. FEM conformation: (a) the FEM's three-layer structure; (b) unbent and bent shapes of the pressure armour layer.

The validity and processability of the simulation results primarily depend on the bending uniformity. The two extremity half coils, which are indicated by the red dashed square at the bending extrados side, where the structure is stretched, ensure the balance of displacement and rotational loading during the bending process.

2.2.1. Contact

The self-contact pairs of the pressure armour layer, interlayer contact pairs, and corresponding contact properties are specified as general contacts. The defined contact pairs' domain and properties are shown in Table 2, and appropriate surface smoothing methods are assigned to eliminate the unwanted disturbance generated by coarse contact surfaces in the numerical solution. According to the works of Halloran et al. [29] and Dong et al. [30], the value of the friction coefficient for three contact pairs is provided. The friction between the simplified shell and polymer layer is ignored given that the former is merely designed to provide internal support and uniformly transfer the bending curvature to other external layers. The sectional drawing with magnified contact pair details is shown in Figures 4 and 5. Within the figures, different contact domains are highlighted by annotation, and for the I-C contact pair, red and yellow lines are used to indicate master and slave surfaces. To prevent penetrations, "hard" contact is employed, which also brings the model closer to actual working conditions. The "Penalty" technique is used to improve convergence efficiency and reduce computation. Specially, the polymer layer is portioned into inner and outer parts for the meshes with different densities. The inner and outer polymer layers (see Figure 6) are unified as a consolidatory polymer layer with TIE constraints, which binds two meshes of different densities jointly and allows the inner and outer polymer layers to behave consistently to imitate a continuous undivided entity.

Table 2. Defined contact pairs and contact properties.

Contact Pair	Master–Slave Assignment	Surface Domain	COF	Smoothing Option
P-P	Master Slave	2–12 14–24	0.3	Spherical Spherical
P-I	Master Slave	12–14 Outer surface of inner sheath	0.1	Spherical Revolution
I-C	Master Slave	Outer surface of carcass Inner surface of inner sheath	0	Revolution Revolution

P, I, C, and COF stand for pressure armour, inner sheath, carcass, and coefficient of friction. P-P (tension) and (compression) stand for self-contact pairs for pressure armour in tension and compression models; P-I and I-C, respectively, denote contact pairs between the pressure armour layer and inner sheath and inner sheath and rigid shell surface.

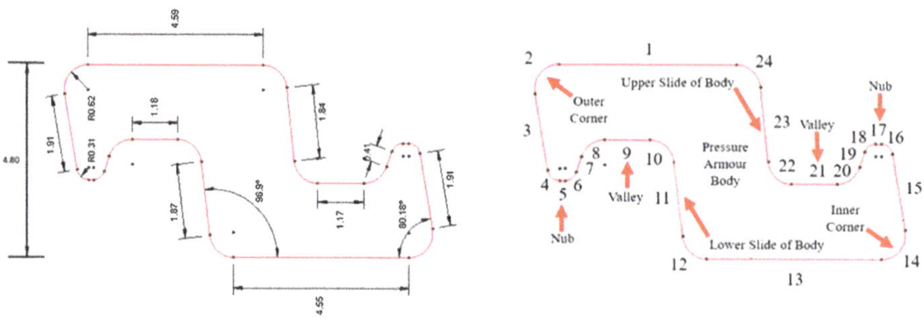

Figure 4. Pressure armour profile and defined surfaces for contact pair definition (dimension in mm).

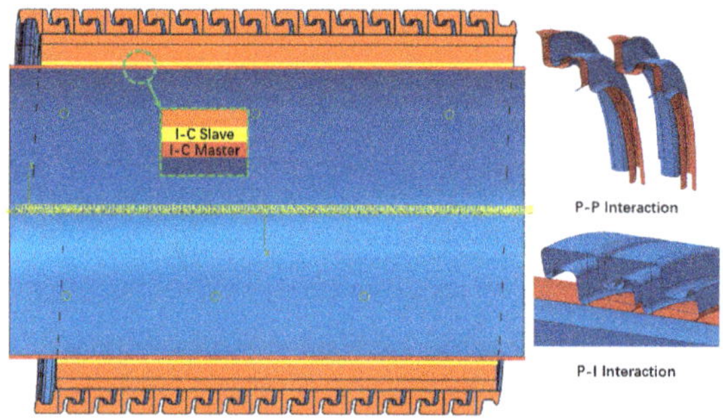

Figure 5. Interaction settings and contact pairs.

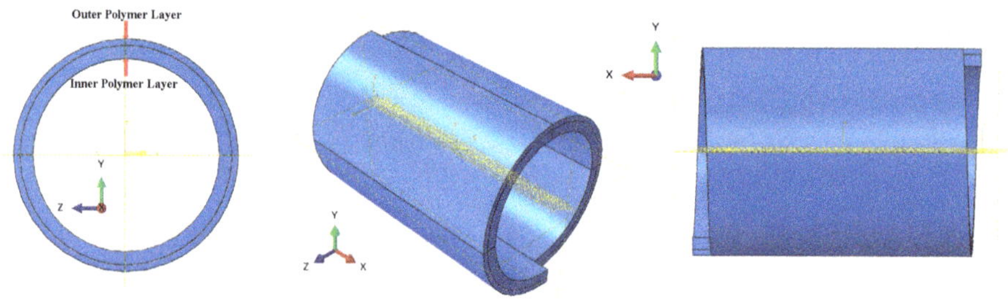

Figure 6. Configuration of the inner sheath of the overbending model.

2.2.2. Mesh Topology

The challenge in meshing is to distinguish clearly between the mesh density of the master and slave surfaces while avoiding an excessive number of FE model elements. To provide a more thorough explanation of the meshing strategy, the description is divided into three points that correspond to the three layers.

- Simplified surface: For the sake of cross-section properties, the mesh of the simplified shell is generated by SFM3D4R element type, which is a 3D linear four-node quadrilateral surface element with reduced integration. The global seed size is fixed at 2 mm while the total length of the simplified shell in a 14.5-pitch model equals 104 mm. The linear eight-node hexahedral brick element with enhanced hourglass and distortion control is selected for both the polymer layer of the structured mesh and the pressure armour layer of the swept mesh.
- Inner sheath: The polymer layer is split into two sub-layers to reduce the element count, as shown in Figure 6. The specific reasons are as follows: (1) the primary

function of the polymer layer is to transmit the internal pressure effect to the pressure armour layer, and a mesh density difference along the wall thickness direction does not affect the transfer of pressure loads; (2) the impacts of polymer deformation and ingression under internal pressure should be considered during simulation. The inner coarse mesh transfers the internal pressure primarily, while the outer polymer layer with a finer mesh can better represent the polymer's deformation and ingression effect on the pressure armour layer. This strategy of different mesh densities between two sub-layers of inner sheath (see Figure 7) enables the model to be functional and reduces the element. The inner layer possesses a 4 mm thickness and is assigned a 1 mm mesh size, and the outer layer is 2 mm thick with a 0.5 mm mesh size. The number of elements in the inner and outer layers is 71,448 and 312,200.

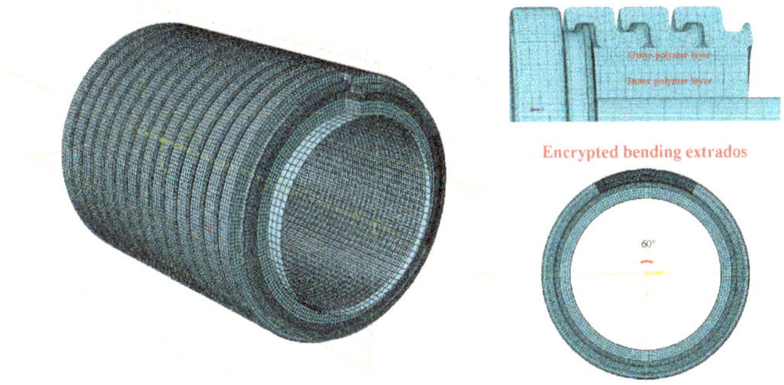

Figure 7. Global view of FEM meshing with local detail.

- Pressure armour layer: Under the premise of maintaining the global mesh size at 1 mm, the pressure armour layer is encrypted emphatically at the nub region and the bending extrados region, where the unlocking is expected to happen. The bending extrados side of the pressure armour layer is segmented along the ring direction by $\pi/3$ radians (see the low right corner of Figure 7). The mesh sizes of the master and slave self-contact surfaces of the pressure armour layer are 0.1 mm to 0.2 mm, respectively. For a 14.5-pitch model, the problem size is determined to contain 982,580 elements with 4,369,692 DOFs.

Both the inner sheath and pressure armour wire are meshed with C3D8R elements with distortion and enhanced hourglass control. Structured mesh is used in the inner sheath. The schematic diagram of the mesh settings, the details of the bending extrados mesh partition, and the different mesh sizes of the inner and outer polymer layers are all shown in Figures 6 and 7.

2.2.3. Boundary and Loading Conditions

The overbending model that includes internal pressure has two simulation steps, while the model without internal pressure just has one step. For comprehension, the overbending model with internal pressure is described as follows:

- First step—pressurization: The first step focuses on applying internal pressure. The pressurization step allows the inner sheath to expand radially, transferring the pressure load to the pressure armour layer, which is essential for the replication of realistic interlayer interactions during riser production. A coupling constraint is applied at two extremities to limit pressure armour layer displacement during pressurization. Internal pressure is imposed on the internal surface of the sheath layer and increased linearly until the target value mimics the gradual pressurization in real-world scenarios.

- Second step—bending: This step replicates the overbending conditions that a flexible riser may experience during installation or after a birdcage failure. By applying a uniform curvature, the model captures the stress and deformation patterns in the pressure armour layer, which are critical for unlocking analysis. Within this step, a global uniform curvature is applied while maintaining a constant internal pressure. Predetermined displacements and rotations, which are calculated based on distances between specific reference points and the model's midpoint, are imposed at specified reference points to ensure bending uniformity. The curvature is applied incrementally until the unlocking condition is achieved, as determined by the loss of interlock between adjacent armour wires.

By implementing predetermined displacements and rotations at specified reference points, uniform curvature is guaranteed during the simulation of unlocking failures due to excessive bending. The information about coupling and boundary condition settings is presented sequentially in the following two subsections. The first subsection fully describes the coupling and boundary conditions setting, and the second subsection introduces the calculation procedure for the pre-defined displacements and rotations on the reference points for the uniform bending application.

2.2.4. Coupling and Boundary Condition Settings

The specifics of coupling and boundary condition settings consider the simulation steps and structural features of different components. A 14.5-pitch overbending model with internal pressure is taken as an example.

- Simplified surface: Figure 8 depicts the schematic of the coupling constraint on the simplified surface. Two cross-sectional centres at the left and right ends of the simplified surface serve as the first and last reference points controlling the simplified surface. The remaining reference points are spaced out evenly at 2 mm intervals along the central longitudinal axis, matching the simplified surface mesh density. "Kinematic" coupling is adopted to constrain translational and rotational DOFs of the coupled nodes in the Cartesian coordinate system. Therefore, the simplified surface transforms into a deformation-controllable instance to achieve a uniformly bent configuration and transmit the curvature to other outer layers.

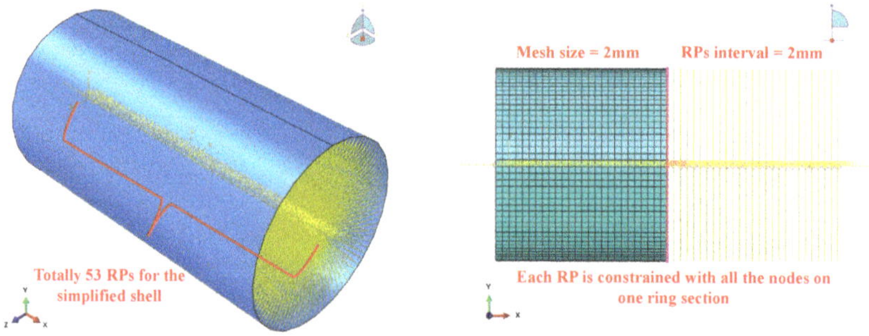

Figure 8. Schematic of the coupling constraint on the simplified surface.

- Inner sheath: Figure 9 depicts the schematic of the coupling constraint on the inner sheath. Except for radial displacement U1, which permits the boundary of the inner sheath layer to radially expand under internal pressure, other DOFs are constrained with the reference points located at two extremities of the polymer layer.

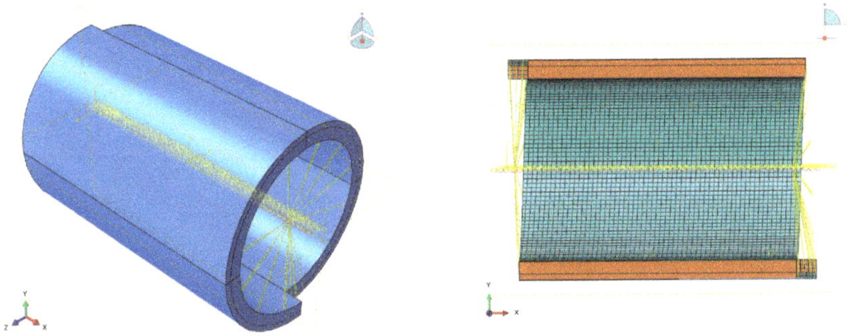

Figure 9. Schematic of the coupling constraint on the inner sheath.

- Pressure armour layer: Figure 10 depicts the coupling constraint on the pressure armour wire. Pressure armour wire is coupled with two reference points at ends of the pressure armour layer. The half sectional coupling and purpose of the partition are shown in Figure 10. The cross-sectional profile is divided into two sections, as indicated by the yellow dot line. Throughout the entire simulation, the "coupled" portion in the armour wire will not contact the neighbouring armour wire and generate deformation. During the bending process, there will be significant contact and distortion effects in the "uncoupled" portion. Therefore, this method of splitting the profile into "coupled" and "uncoupled" sections will reduce the unexpected boundary effects brought on by the coupling constraints.

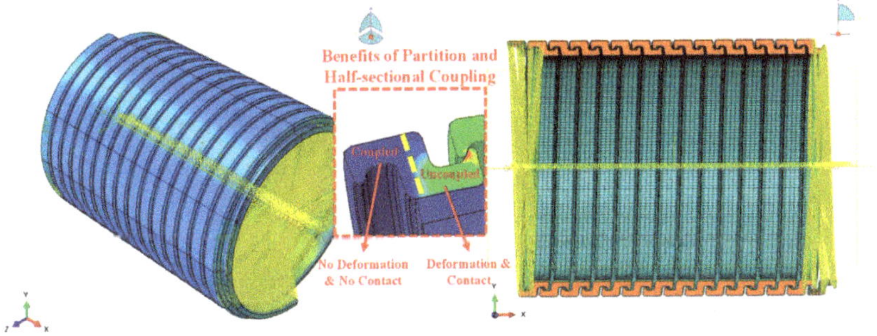

Figure 10. Schematic of the coupling constraint on the pressure armour layer.

Uniform Bending Displacement Rotation Calculation Method

The global rotation angle is linearly loaded with a target minimum curvature radius of 0.1 m. The reference points' displacement and rotation amplitudes over time are obtained based on distances between their location and the model's midpoint. The midpoint of all the reference points for the simplified surface has been marked as RP^1_{mid} in Figure 11.

Only the right half of the model is depicted in Figure 11 to demonstrate the pre-defined displacement and rotation angle calculation process. The undeformed and deformed neutral axes, where all RPs are located, are represented by the green and red curves, respectively. There are a total of N reference points, numbered from RP^1_{mid} to RP^N, with RP^N standing for the Nth reference point. The distance between RP^1_{mid} and RP^N on the undeformed neutral axis is $1/2L_s$, where L_s is the overall length of the simplified surface. Assuming that the neutral axis is neither pulled nor compressed during the overbending process, the translational displacement of any RP^n along the X and Z axes, and the rotation around the Y axis, can be written as Equations (1)–(3).

$$\Delta\theta = \frac{t}{T} \cdot \frac{L_1}{R_{tg}} \cdot \frac{n-1}{N-1} \tag{1}$$

$$\Delta L = \frac{T}{t} \cdot R_{tg} \cdot \sin(\Delta\theta) - L_1 \tag{2}$$

$$\Delta h = \frac{T}{t} \cdot R_{tg} \cdot [\cos(\Delta\theta) - 1] \tag{3}$$

In the above equations, t is the current simulation time, T stands for the duration of the overbending analysis step, R_{tg} is the target minimum curvature radius, and L_1 is the distance between RP^n and RP^1_{mid}.

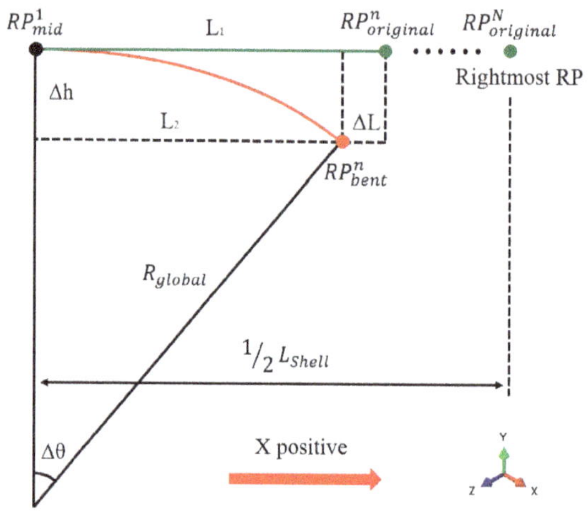

Figure 11. Schematic of the translational and rotational displacement of the reference points during bending.

2.2.5. Unlocking Judgement Methodology

Bending moment and curvature (or curvature radius) are important characterization indicators for determining and studying the overbending-induced pressure armour layer unlocking failure. The bending moment characterizes the magnitude of the flexural load to which the riser section is subjected, while the curvature straightforwardly characterizes the riser's bending degree geometrically. The determination and results with both indicators are introduced later, and the method for determining when the unlocking occurs is described in Figure 12. As an illustration, consider one nub pair at the bending extrados. The difference in distances between the apexes of the nubs and the centre of the bending circle is calculated according to Equation (4). The distance difference would be negative before the loss of interlock. As the bending proceeds, the apexes of nub1 and nub2 overlap, and the distance difference ΔL_{nb} will equal 0, which can be viewed as the unlocking occurrence with a nub-ride-nub configuration. Following the apexes' separation, the distance would turn negative. To sum up, the judgement method can be expressed as Equation (5), where the conversion of the positive and negative signs of the distance difference value allows for the detection of the unlocking.

$$\Delta L_{nb} = L_{cn1} - L_{cn2} \tag{4}$$

$$\begin{cases} \Delta L_{nb} < 0 \text{ Before Unlocking} \\ \Delta L_{nb} = 0 \text{ Loss of Interlock} \\ \Delta L_{nb} > 0 \text{ After Unlocking} \end{cases} \tag{5}$$

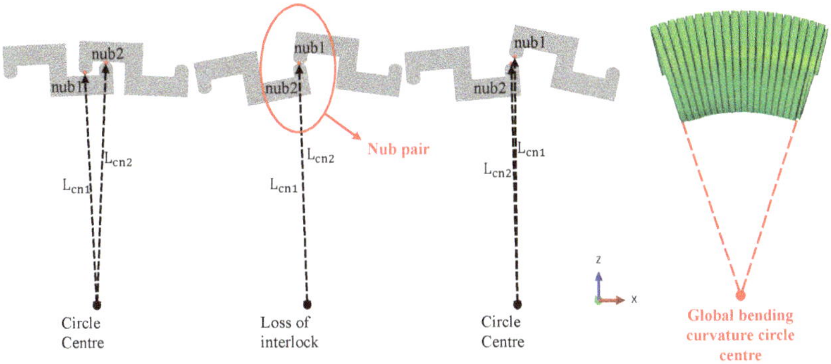

Figure 12. Unlocking moment judgement diagram.

3. Pitch Number Sensitivity

The purpose of conducting a study on the number of pitches is to define an overbending FEM with a predetermined length that can yield results for a convergent unlocking radius or allowable error in comparison to the astringent value while using minimal computational resources. Therefore, this analysis can serve as the basis for the later investigation of other parameters. Figure 13 displays the relationship between the number of pitches and the unlocking curvature radius.

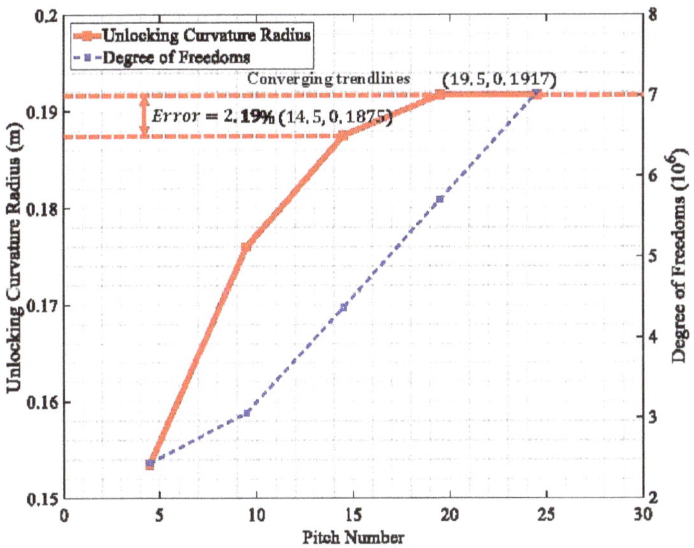

Figure 13. Number of pitches vs. the unlocking curvature radius.

In this figure, 19.5-pitch and 24.5-pitch models produce the same unlocking curvature radius, and the converging value of the unlocking radius is 0.1917 m. Both earlier models, however, are limited by computational resources and are deemed uneconomical when compared with the 14.5-pitch model, which has a tolerable 2.19% error. The unlocking configurations for models of different lengths are shown in Figure 14. The red-dotted circle indicates specific unlocking nub pairs.

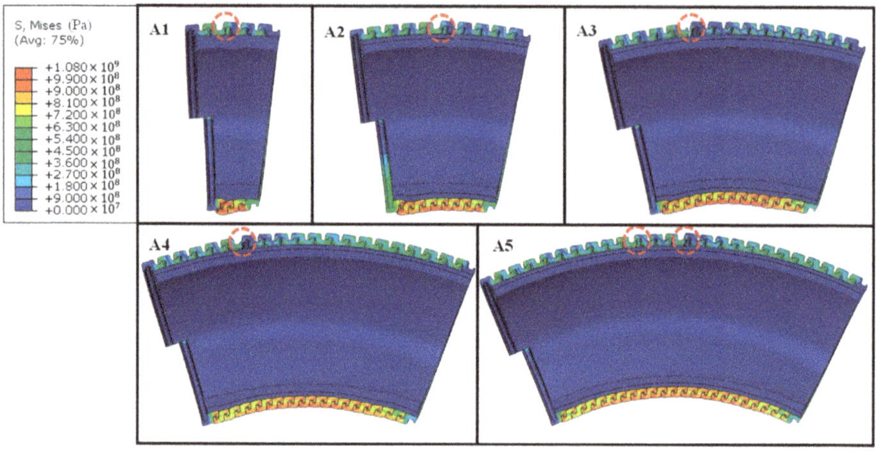

Figure 14. Unlocking configurations for all cases in the pitch number study.

4. Results and Discussion

Parameter studies are performed by varying the model length and internal pressure. Associated unlocking processes and manifestations are described. According to the cross-section profile and parameters used in the FEM, the results are discussed and evaluated.

Four simulation groups are paralleled to study the unlocking process caused by overbending. The examined arguments include the number of pitches (model length), pitch distance, the internal diameter of the pressure armour layer, and the internal pressure. The case study for the number of pitches starts with a model length of 4.5 pitches and gradually increases to a model with a length of 24.5 pitches, with a common difference of 5 pitches. The pitch distance group includes three FEMs with pitch distances equal to 6, 6.5, and 7 mm, respectively. The case study for the pressure armour layer internal diameter includes three FEMs with internal diameters equal to 57.1, 69.8, and 82.5 mm. The inner surface of the flexible riser is applied with a uniform hydrostatic pressure of 0, 5, 10, and 15 MPa to study the influence of internal pressure. The general idea of overbending-induced unlocking is shown in Figure 15, before the presentation of the results and discussion.

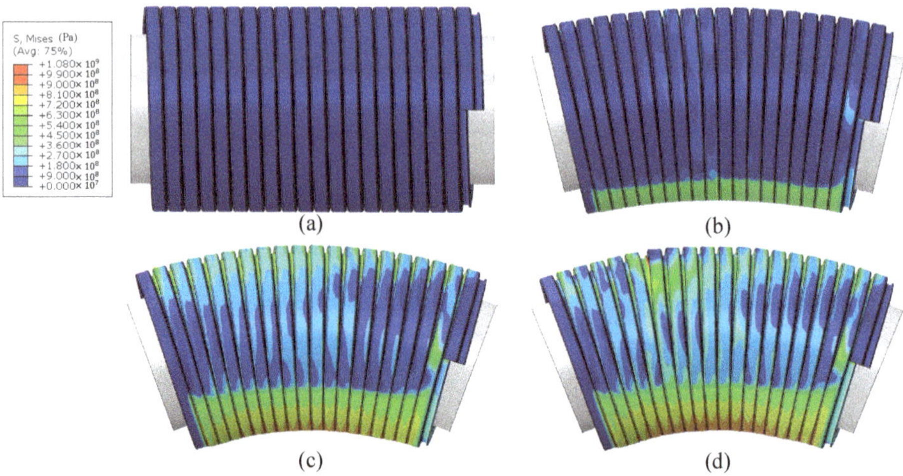

Figure 15. Schematic diagram of the overbending-induced unlocking process.

Figure 15a demonstrates a 19.5-pitch model's unbent conformation at the beginning of the simulation; Figure 15b depicts the pressure armour layer's intrados as the first area to start becoming compressed; and Figure 15c depicts the extrados as the next area to reach the tensioned state. The order in which the intrados and extrados enter stressed conditions is unfixed and depends largely on the pitch distance. Stress develops as soon as contact is established and locking starts. With a greater pitch distance, the extrados can make contact first and transition into the tensioned state earlier. Eventually, the extrados region armour wire unlocks, as shown in Figure 15d. There is a transient nub-nub sliding effect immediately before the unlocking. Figure 16 illustrates the location where the sliding begins to occur. Profiles ① and ② are pulled out in opposite directions. Profile ①'s right nub twists due to elastoplastic deformation. After dividing the force produced by Profiles ① and ② on the middle contact pair into parallel and perpendicular component directions, it is found that both profiles present a clockwise rotating trend, which facilitates unlocking.

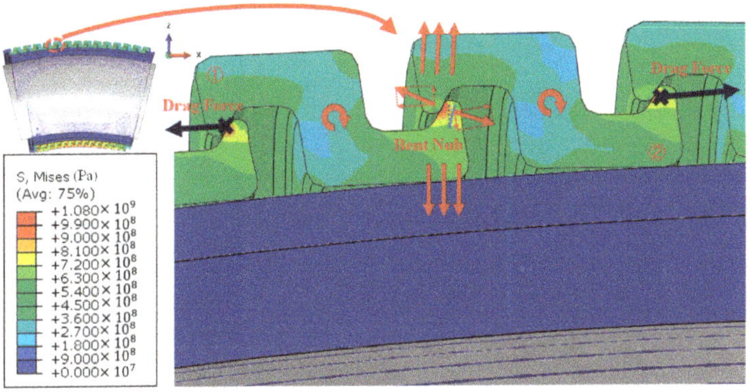

Figure 16. Nub-nub sliding effect before overbending-induced unlocking.

As shown in Figure 17, an indication of the rotating trend is the angle formed between the inner surface of the pressure armour layer and the outer surface of the polymer layer. The rotating trend facilitates unlocking because two of the nubs in the middle nub pair move in opposite directions, apart from one another. As the curvature is further applied, static friction will eventually lose its ability to prevent the nub pair from separating and unlocking. Figure 17 represents the local stress distribution on the armour wire cross-section at the extrados. The stress contour demonstrates that most of the cross-section is still in an elastic state. Only a small portion of the nub region is susceptible to plastic deformation. The intrados of the pressure armour layer will be compacted while the extrados is tensioned during the bending. Despite the tremendous compressive force, it is unlikely that compression induces the unlocking. It is anticipated that the tensioned extrados will unlock.

4.1. Influence of Pitch Distance

For the perfect flexible riser, the pressure armour wire should be distributed evenly. However, under external load, the pressure armour layer in a tube section where a birdcage failure has occurred may be stretched or contracted, causing the wire pitch of the failed section to differ from that of the adjacent section. It is noteworthy how this uneven distribution of wire pitch affects unlocking failure. Figure 18 shows the results of the study on pitch distance. The unlocking radius rises significantly with the pitch distance being elongated. It implies that an armour layer with a long pitch distance can make the unlocking failure easier. The pitch distance effect is schematically illustrated in Figure 19.

From the view of deformation, overbending-induced unlocking can be separated into two phases: (1) armour wire is circumferentially tensioned due to internal pressure, and the two nubs move toward one another until they make contact; (2) two nubs undergo elastoplastic deformation until unlocking. Thus, the unlocking process can be expressed as Equation (6). ΔL_{ext} represents the traction displacement required for the extrados unlocking. The traction displacement ΔL_{ext} consists of two components: the relative distance L_{sd} between two nubs at the initial state, which will be eliminated by the sliding of two nubs, as shown on the right side of Figure 19; and δ_{ts}, which is the elastoplastic deformation contributed by each nub pair after being stretched and making contact for the unlocking. It is assumed that there are n nub pairs at the bending extrados side and the δ_{ts} contribution of each nub pair is constant. It can be inferred from Equation (6) that the larger the pitch distance, the smaller L_{sd} sliding space between a nub pair, and then the less extrados stretching displacement ΔL_{ext} that will be required for unlocking.

Figure 17. Stress contour on the cross-section of the overbending model without internal pressure (Curvature radius = 0.2 m).

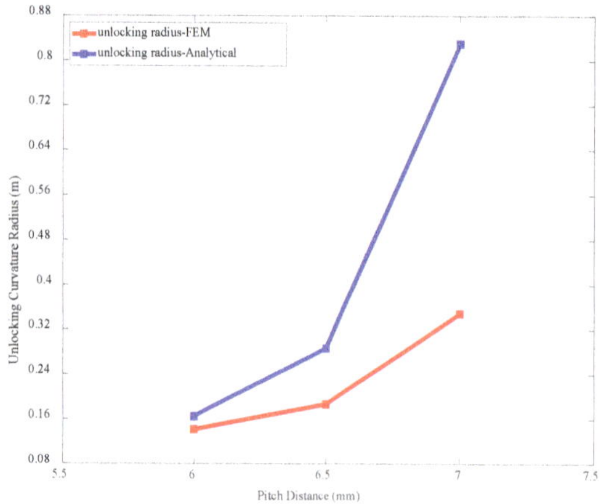

Figure 18. Pitch distance versus unlocking radius.

From the view of bending moment and curvature imposition, the relationship between extrados curvature radius R_{ext} and extrados stretch ΔL_{ext} due to bending can be expressed as Equations (7) and (8). Since pressure armour layer internal diameter ID_{pa} and pressure armour layer thickness t_{pa} are constants, extrados stretching ΔL_{ext} and global bending radians θ_{global} are linearly related. Combining Equations (6)–(8) furnishes Equation (9). It implies that the model will have a smaller global curvature radius as L_{sd} decreases. Consequently, less bending load is required for unlocking if the riser section possesses a longer pitch distance.

$$\Delta L_{ext} = \sum_{i=1}^{n} L_{sd}^{i} + \sum_{i=1}^{n} \delta_{ts}^{i} \tag{6}$$

$$R_{ext} = \frac{L_{neu}}{\theta_{global}} + \frac{1}{2} ID_{pa} + \frac{1}{2} t_{pa} \tag{7}$$

$$\Delta L_{ext} = R_{ext} \cdot \theta_{global} - L_{neu} = \frac{1}{2}\left(ID_{pa} + t_{pa}\right) \cdot \theta_{global} \tag{8}$$

$$\frac{1}{2}\left(ID_{pa} + t_{pa}\right) \cdot \theta_{global} = \sum_{i=1}^{n} L_{sd}^{i} + \sum_{i=1}^{n} \delta_{ts}^{i} \tag{9}$$

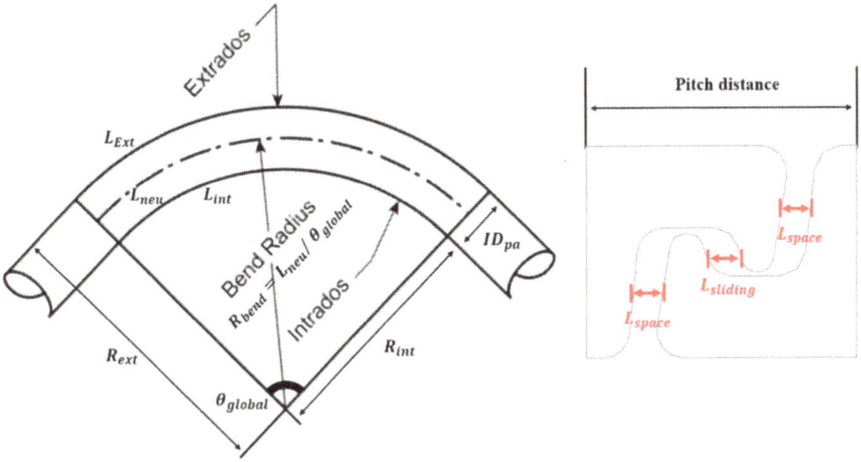

Figure 19. Schematic of the pitch distance effect.

Theoretically, elastoplastic deformation δ_{ts} generated by the nub pairs is relatively small when compared with sliding distance L_{sd}, and computationally hard to calculate. Therefore, the elastoplastic deformation term δ_{ts} is omitted for simplification. An approximate analytical solution for the bending curvature radius that triggers the unlocking failure can be written as Equation (10):

$$R_{bcr} = \frac{L_{neu} \cdot \left(ID_{pa} + t_{pa}\right)}{2\sum_{i=1}^{n} L_{sd}^{i}} \tag{10}$$

The theoretical solution for the bending curvature radius is presented with the FEM result in Figure 18. The results show that both the analytical and numerical solutions increase as the pitch distance increases, suggesting that a large pitch distance can facilitate the unlocking of the riser at low curvature. The analytical solution is larger than the theoretical solution because the elastoplastic deformation of nub pair δ_{ts} after making contact is neglected. The error becomes progressively larger as the pitch distance increases, indicating that the elastoplastic deformation of nub pair δ_{ts} cannot be omitted for large pitch distance cases. As the pitch distance increases, the influence of elastoplastic deformation δ_{ts} on the unlocking becomes even more significant. This is consistent with the FE simulation results, as the nub pair needs to undergo significant elastoplastic deformation δ_{ts} before unlocking as L_{sd} decreases.

Figure 20 depicts the stress contour for models with different pitch distances. For small pitch distance and normal pitch distance models, the intrados have already undergone continuous compression and severe deformation before the extrados engage contact. For the large pitch distance model, one nub pair at the extrados has unlocked, but the intrados region still has not been fully compacted. Moreover, in comparison to the first two models, the large pitch distance model generally presents a lower stress level and a less curved conformation, which suggests that a large pitch distance is a significant unlocking facilitation factor.

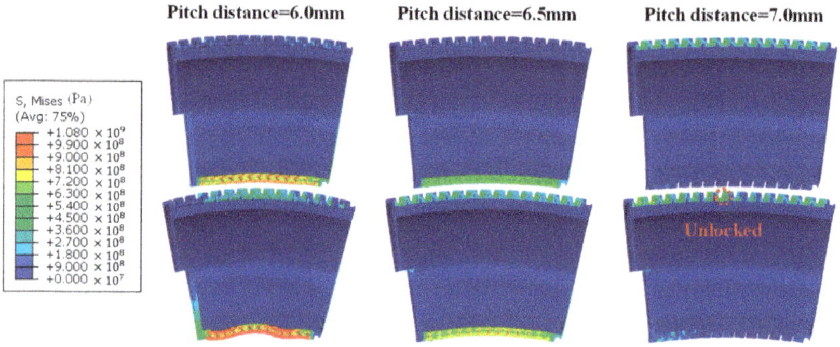

Figure 20. Stress contour of different pitch distance models when the extrados or intrados engage fully in contact.

4.2. Influence of Pressure Armour Layer Internal Diameter

The internal diameter of the riser is a geometrically crucial parameter for its bending performance. Therefore, models with original thickness, pitch distance, no internal pressure, and different internal diameters ID_{pa} are used to study its influence. Figure 21 demonstrates the unlocking curvature radius and γ, which is defined in Equation (11) as the bending degree of the riser at the unlocking moment.

$$\gamma = \frac{ID_{ul}}{ID_{pa}} \quad (11)$$

where γ is the ratio of unlocking curvature diameter ID_{ul} to pressure armour layer internal diameter ID_{pa}. According to the trend in unlocking curvature radius, a model with a higher ID_{pa} has a greater unlocking curvature radius, making the unlocking easier. This tendency is proven and supported by Equation (9). L_{sd} and δ_{ts} are constant values in terms of the FEM studying the pitch distance effect, where θ_{global} drops as ID_{pa} increases. Moreover, the model length is constant, and the unlocking curvature radius increases as θ_{global} decreases. Nevertheless, the γ curve in Figure 21 indicates that the model with the original ID_{pa} demands greater curvature to unlock. There are two possible explanations: although it contradicts the previous inference, it may naturally exist in a non-monotonic relationship between ID_{pa} and the unlocking curvature radius; (2) since ID_{pa} is a crucial parameter, length sensitivity studies must be performed for cases with various ID_{pa} to pinpoint the precise minimum pitch number for numerical simulation.

Figure 22 shows the stress contour for the case study of ID_{pa}. The bending curvature presents a significant decreasing tendency with the increase in ID_{pa}. Additionally, as ID_{pa} increases, the stress caused by intrados compression decreases as ID_{pa} rises. However, the study only offers a cursory analysis of the impact of ID_{pa}. Further investigation is required, along with a more precise and thorough explanation, to perform multiple case studies using FEMs.

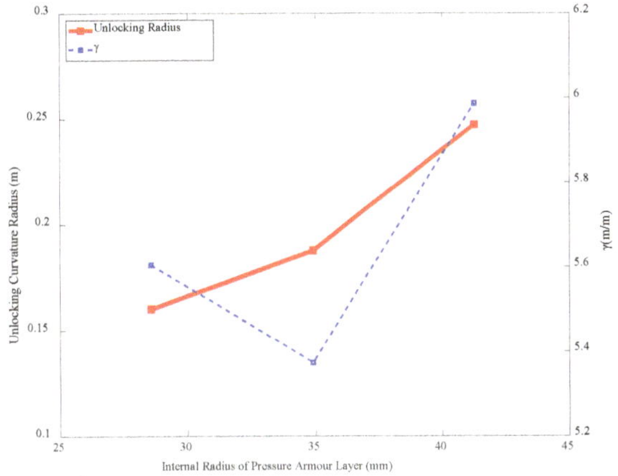

Figure 21. Pressure armour layer internal radius versus unlocking radius.

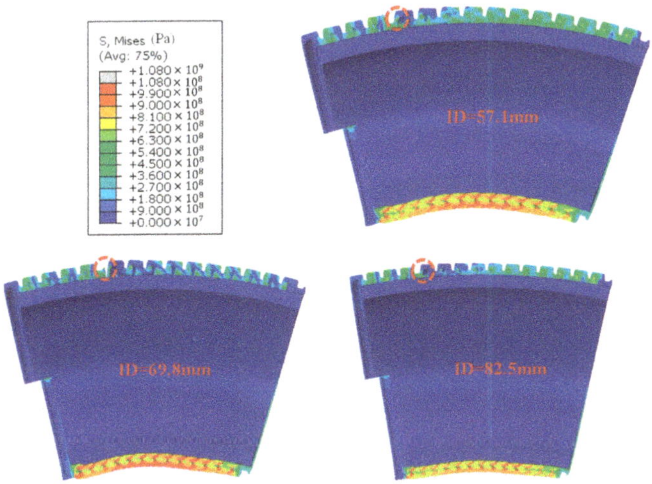

Figure 22. Schematic of the unlocking configuration for models with different ID_{pa}.

4.3. Influence of Internal Pressure

The flexible riser is subjected to internal pressure from the transported fluid during operation. If the axial load is not considered, namely the plane stress assumption, internal pressure can increase the circumferential stress level of the pressure armour wire, which is intuitively thought to be a favourable condition for unlocking. Consequently, simulation is conducted using models with a range of internal pressure values. Internal pressure loads of 5 MPa, 10 MPa, and 15 MPa are implemented in FEMs separately. The numerical results show that internal pressure increases the stress level of the armour wire section. The stress maxima are usually found in the inner corner region, as defined in Figure 4. Figure 23 depicts the stress contour within the armour wire section subjected to various internal pressures. The maximum von Mises stress levels in the wire are approximately 99.8 MPa, 200.9 MPa, and 300.2 MPa as the internal pressure rises from 5 MPa to 15 MPa.

The local stress contour of the model's armour wire cross-section with 10 MPa internal pressure is shown in Figure 24. Three important phenomena can be observed: (1) the inner sheath is pressurized and pushed away from the simplified surface (carcass); (2) the inner sheath has been extruded into the gap between two adjacent armour wires. The pressurized extrusion effect is not pronounced because the polymer material is assumed without initial creeping ingression; (3) the inner sheath, which is radially outward expanded by internal pressure, inhibits the armour wire rotation effect illustrated in Figures 16 and 17.

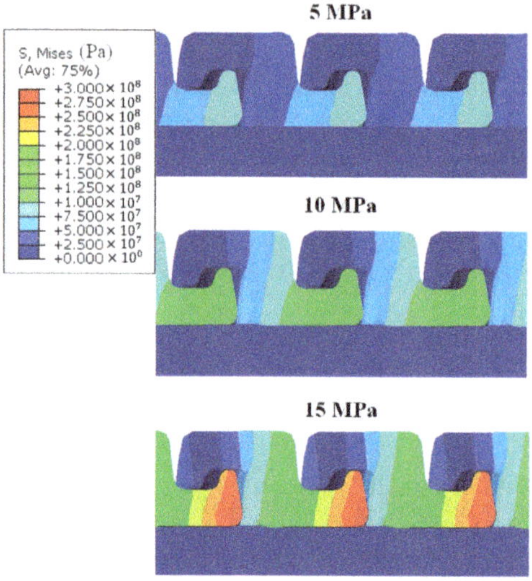

Figure 23. Stress levels within the armour wire cross-section under different internal pressures.

Figure 24. Local cross-section stress contour of the overbending model with internal pressure (10 MPa, curvature radius = 0.2 m).

Figure 25 shows that as the internal pressure rises, the unlocking curvature radius grows. This can be explained by the push-out effect cast by the internal pressure as well as the elevated stress level in the armour wire caused by pressurization, which facilitates easier deformation once the nub pair makes contact. But as the internal pressure increases from 0 to 5 MPa, the riser shows a decrease in unlocking curvature radius. An explanation for this is that the pressure armour layer exhibits a loose spiral conformation in the absence of internal pressure. Therefore, being slightly internally pressurized may stabilize the loose spiral armour wire structure and reduce the unlocking risk to some extent. A mild internal pressure can transform the armour wire into a circumferentially tensioned state, which inhibits the rotation effect demonstrated in Figures 16 and 17.

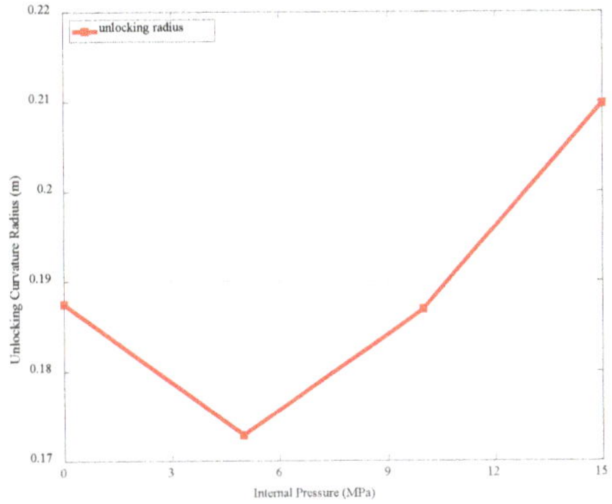

Figure 25. Internal pressure versus the unlocking radius.

5. Conclusions and Perspective

To investigate the overbending-induced unlocking of the pressure armour layer, this paper presents a three-layer finite element model implemented with uniform bending by pre-defined displacement and rotation loads and boundary constraints. Nevertheless, it remains of interest and to be confirmed whether the numerical overbending simulation is compatible with the uniform bending conformation of a true 3D helical pressure armour layer structure and whether the boundary effects of the asymmetric FEM can be further attenuated when internal pressure and other external loads are applied. A number of pitch-sensitivity assessments are necessary because such assessments not only ensure the accuracy of subsequent numerical simulations but also computational efficiency. A three-layer, 14.5-pitch pressure armour layer FEM is confirmed after balancing computation time and relative error. Firstly, the unlocking curvature radius tends to increase as the internal diameter of the riser expands. A pressure armour layer with a larger internal diameter would produce larger displacements in both the extrados and intrados regions compared to models with a small internal diameter. The principle is analogous to the idea that long pitch distance facilitates unlocking. This leads to the conclusion that the armour wire profile thickness and internal diameter may be crucial variables for the analytical model of unlocking curvature radius. To determine the representativeness of the FEM, more case studies are required. Internal pressurization reduces the difficulty of unlocking to a limited extent. The unlocking curvature radius of the 15 MPa model is approximately 0.04 m greater than the unlocking curvature of the 5 MPa model. If extremely high internal pressure is applied, there is a chance that a pronounced result with a steep lift of the unlocking radius will be produced. However, there is a risk that extremely high internal pressure can cast significant boundary effects on the FEM, necessitating advanced measures to counteract them.

It is suggested that future work include laboratory experiments simulating overbending-induced pressure armour unlocking. A comparison of numerical accuracy could be established based on the experimental results and the FEM methodology could be revisited and calibrated. Both the effect of fabrication-induced prestress in pressure armour wire and the effect of polymeric layer extrusion into armour wire gaps also merit further investigation.

Author Contributions: R.W.: conceptualization, methodology, software, investigation, formal analysis, and writing—original draft; X.L.: conceptualization, investigation, formal analysis, and review; M.A.V.: conceptualization, formal analysis, writing—review and editing, and supervision; A.B.C.: conceptualization, formal analysis, and review. All authors have read and agreed to the published version of the manuscript.

Funding: This research received no external funding.

Institutional Review Board Statement: Not applicable.

Informed Consent Statement: Not applicable.

Data Availability Statement: The original contributions presented in this study are included in the article. Further inquiries can be directed to the corresponding author.

Acknowledgments: The authors acknowledge the support for the support from PETROBRAS, Coordination for the Improvement of Higher Education Personnel (CAPES), CNPq (National Council of Scientific and Technological Development), National Science Foundation of China (No. 52201348), and Guangdong Basic and Applied Basic Research Foundation, China (No. 2024A1515011027).

Conflicts of Interest: Anderson Barata Custódio was employed by Petrobras. The remaining authors declare that they have no known competing financial interests or personal relationships that could have appeared to influence the work reported in this paper.

Nomenclature

E	Young's modulus
ID	internal diameter
ID_{pa}	pressure armour layer internal diameter
ID_{ul}	unlocking curvature diameter
L_1	distance between RP^n and RP^1_{mid}
L_{cn1}	distance between nub1 and the bending curvature centre
L_{cn2}	distance between nub2 and the bending curvature centre
L_{neu}	hoop length of the FEM along the neutral axis
L_s	overall length of the simplified surface
L_{sd}	sliding distance between two adjacent nubs at the initial unbent state
n	number of reference points, with N as the total number of reference points on one side of the FEM
R_{bcr}	critical bending curvature that will trigger the unlocking
R_{ext}	extrados curvature radius
R_{tg}	minimum curvature radius applied in the FEM
RP	reference point, with RP^1_{mid} indicating the reference point at the middle of the FEM and RP^N standing for the reference point at the model extremities
t	real simulation time
t_{pa}	pressure armour layer thickness
T	total simulation time of the overbending analysis step
v	Poisson ratio
σ_y	yielding stress
σ_u	ultimate tensile strength
ε_u	ultimate tensile strain
ρ	material density
δ_{ts}	elastoplastic deformation contributed by each nub pair after being stretched and making contact for the unlocking
θ_{global}	global bending radians
γ	ratios of unlocking curvature diameter ID_{ul} to pressure armour layer internal diameter ID_{pa}
$\Delta\theta$	bending angle between RP^1_{mid} and RP^N relative to the curvature centre at time t
ΔL	distance between RP^1_{mid} and RP^N in the x direction at time t
ΔL_{cn}	difference between distances of the two adjacent nubs to be unlocked with the bending curvature centre
ΔL_{ext}	traction displacement required for the extrados unlocking
Δh	distance angle between RP^1_{mid} and RP^N in the y direction at time t

References

1. Stigerørskader-Skader Og Hendelser Fra Petroleumstilsynets CODAM Database. Available online: https://www.havtil.no/contentassets/a13ec8ed3fc94e3a97b76aebb01bf6e3/stigeroersskader-2021-okt.pdf (accessed on 28 July 2021).
2. Muren, J. PSA-NORWAY Flexible Pipes Failure Modes, Inspection, Testing and Monitoring. 2007. Available online: https://www.ptil.no/contentassets/a4c8365164094826a24499ef9f22742b/p5996rpt01rev02cseaflex_janmuren.pdf (accessed on 1 December 2007).
3. Simonsen, A. Inspection and Monitoring Techniques for Un-Bonded Flexible Risers and Pipelines. Master's Thesis, University of Stavanger, Stavanger, Norway, 2014.
4. Fergestad, D.; Løtveit, S.A. *Handbook on Design and Operation of Flexible Pipes*; SINTEF Ocean: Trondheim, Norway, 2014; Volume 1, ISBN 9788271742652.
5. API. *17J Specification for Unbonded Flexible Pipe*; American Petroleum Institute: Washington, DC, USA, 2002.
6. Ted Miller, FPSO Past Issues and Typical Claims 2012. Available online: https://slideplayer.com/slide/1567408/ (accessed on 28 July 2021).
7. Li, X.; Vaz, M.A.; Custódio, A.B. Analytical Study of a Repair Methodology for Flexible Pipes Anti-Birdcage Tapes. *Mar. Struct.* **2019**, *63*, 289–303. [CrossRef]
8. Lu, J.; Tan, Z.; Ma, F.; Sheldrake, T. Simulating Contact Behavior Between the Flexlok Wires. In *International ANSYS Conference*; Wellstream International Limited: Newcastle upon Tyne, UK, 2008.
9. Lu, J.; Ma, F.; Tan, Z.; Sheldrake, T. The Impact of Flexbarrier Ingress on Flexlok Stress. In Proceedings of the ASME 2009 28th International Conference on Ocean, Offshore and Arctic Engineering, Honolulu, HI, USA, 31 May–5 June 2009; Volume 3, pp. 533–538.
10. Neto, A.G.; Martins, C.D.A.; Pesce, C.P.; Meirelles, C.O.C.; Malta, E.R.; Neto, T.F.B.; Godinho, C.A.F. Prediction of Burst in Flexible Pipes. *J. Offshore Mech. Arct. Eng.* **2013**, *135*, 011401. [CrossRef]
11. Neto, A.G.; Martins, C.D.A.; Pesce, C.P.; Meirelles, C.O.C.; Malta, E.R.; Barbosa Neto, T.F.; Godinho, C.A.F. Burst Prediction of Flexible Pipes. In Proceedings of the 29th International Conference on Ocean, Offshore and Arctic Engineering, Shanghai, China, 6–11 June 2010; Volume 5, pp. 511–520.
12. Cuamatzi-Melendez, R.; Castillo-Hernández, O.; Vázquez-Hernández, A.O.; Albiter, A.; Vaz, M. Finite Element Modeling of Burst Failure in Unbonded Flexible Risers. *Eng. Struct.* **2015**, *87*, 58–69. [CrossRef]
13. Liu, X.; Qu, Z.; Liu, Y.; He, J.; Si, G.; Wang, S.; Liu, Q. Numerical Simulation of Burst Failure in 2.5-Inch Unbonded Flexible Riser Pressure Armor Layers. *Metals* **2024**, *14*, 762. [CrossRef]
14. Paumier, L.; Averbuch, D.; Felix-Henry, A. Flexible Pipe Curved Collapse Resistance Calculation. In Proceedings of the ASME 2009 28th International Conference on Ocean, Offshore and Arctic Engineering, Honolulu, HI, USA, 31 May–5 June 2009; Volume 3, pp. 55–61.
15. Paumier, L.; Mesnage, O. PSI Armour Wire for High Collapse Performance of Flexible Pipe. In Proceedings of the ASME 2011 30th International Conference on Ocean, Offshore and Arctic Engineering, Rotterdam, The Netherlands, 19–24 June 2011; Volume 4, pp. 239–246.
16. API. *17B Recommended Practice for Flexible Pipe*; American Petroleum Institute: Washington, DC, USA, 2008.
17. Neto, A.G.; de Arruda Martins, C.; Malta, E.R.; Godinho, C.A.; Neto, T.F.; de Lima, E.A. Wet and Dry Collapse of Straight and Curved Flexible Pipes: A 3D FEM Modeling. In Proceedings of the Twenty-Second International Offshore and Polar Engineering Conference, Rhodes, Greece, 17–22 June 2012.
18. Neto, A.G.; Martins, C.D.A.; Malta, E.R.; Tanaka, R.L.; Godinho, C.A.F. Simplified Finite Element Models to Study the Dry Collapse of Straight and Curved Flexible Pipes. *J. Offshore Mech. Arct. Eng.* **2016**, *138*, 021701. [CrossRef]
19. Neto, A.G.; Martins, C.D.A.; Malta, E.R.; Tanaka, R.L.; Godinho, C.A.F. Simplified Finite Element Models to Study the Wet Collapse of Straight and Curved Flexible Pipes. *J. Offshore Mech. Arct. Eng.* **2017**, *139*, 061701. [CrossRef]
20. Cuamatzi-Melendez, R.; Castillo-Hernández, O.; Vázquez-Hernández, A.O.; Vaz, M.A. Finite Element and Theoretical Analyses of Bisymmetric Collapses in Flexible Risers for Deepwaters Developments. *Ocean. Eng.* **2017**, *140*, 195–208. [CrossRef]
21. Zhu, X.; Lei, Q. Collapse Characteristics of Flexible Pipes under External Pressure. *J. Braz. Soc. Mech. Sci. Eng.* **2021**, *43*, 312. [CrossRef]
22. Yun, R.-H.; Jang, B.-S.; Kim, J. Du Improvement of the Bending Behavior of a Flexible Riser: Part I—Nonlinear Bending Behavior Considering the Shear Deformation of Polymer Layers. *Appl. Ocean Res.* **2020**, *101*, 102204. [CrossRef]
23. Ruan, W.; Chen, M.; Nie, Q.; Xu, P.; Li, J.; Wang, X. Dynamic Response of Steel Lazy Wave Riser Considering the Excitation of Internal Solitary Wave and Ocean Currents. *Ocean Eng.* **2024**, *294*, 116708. [CrossRef]
24. Ruan, W.; Nie, Q.; Han, X.; Li, J.; Bai, Y.; Fu, X. Accurate Modeling and Safety Evaluation of Dented Pipeline with Internal Pressure Based on Reverse Modeling Technique. *Ships Offshore Struct.* **2024**, 1–11. [CrossRef]
25. Liu, Q.; Xue, H.; Tang, W. Behavior of Unbonded Flexible Riser with Composite Armor Layers under Coupling Loads. *Ocean Eng.* **2021**, *239*, 109907. [CrossRef]

26. Liu, Q.; Xue, H.; Tang, W.; Yuan, Y. Theoretical and Numerical Methods to Predict the Behaviour of Unbonded Flexible Riser with Composite Armour Layers Subjected to Axial Tension. *Ocean Eng.* **2020**, *199*, 107038. [CrossRef]
27. Liu, Q.-S.; Qu, Z.-Y.; Wang, G.; He, J.-W.; Ni, J.-F. Advanced Analysis of Cross-Sectional Properties in Unbonded Flexible Risers with Composite Layers. *Ships Offshore Struct.* **2025**, 1–14. [CrossRef]
28. Merino, H.E.M.; de Sousa, J.R.M.; Magluta, C.; Roitman, N. Numerical and Experimental Study of a Flexible Pipe Under Torsion. In Proceedings of the 29th International Conference on Ocean, Offshore and Arctic Engineering, Shanghai, China, 6–11 June 2010; Volume 5, pp. 911–922.
29. O'Halloran, S.M.; Connaire, A.D.; Harte, A.M.; Leen, S.B. Modelling of Fretting in the Pressure Armour Layer of Flexible Marine Risers. *Tribol. Int* **2016**, *100*, 306–316. [CrossRef]
30. Dong, L.; Huang, Y.; Zhang, Q.; Liu, G. An Analytical Model to Predict the Bending Behavior of Unbonded Flexible Pipes. *J. Ship Res.* **2013**, *57*, 171–177. [CrossRef]

Journal of
Marine Science and Engineering

MDPI

Article

Numerical Permeation Models to Predict the Annulus Composition of Flexible Pipes

João Marcos B. Vieira and José Renato M. de Sousa *

LACEO—Laboratory of Analysis and Reliability of Offshore Structures, Civil Engineering Department, COPPE/UFRJ—Post Graduate Institute of the Federal University of Rio de Janeiro, Avenida Pedro Calmon, S/N, Cidade Universitária, Ilha do Fundão, Rio de Janeiro 21941-596, RJ, Brazil; joao.vieira@coc.ufrj.br
* Correspondence: jose.sousa@coc.ufrj.br

Abstract: The migration of acid gases through the pressure sheath of flexible pipes may induce a corrosive environment that can lead to steel armors' failure by SCC (stress corrosion cracking). This permeation process depends on temperature, partial pressures, materials, and the pipe's geometry. However, there are few works related to permeation modeling in flexible pipes, and these works usually contain significant simplification in pipes' geometry. Hence, this work proposes two finite element (FE) permeation models and discusses the effects of the pipe's characteristics. The models were developed in Ansys®, considering two- (2DFE) and three-dimensional (3DFE) approaches, and rely on gas fugacities instead of concentrations to describe the mass transport phenomenon. A radial temperature gradient is also considered, and the heat transfer is uncoupled from the mass transfer. Dry and flooded annulus analyses were conducted with the proposed models. In dry conditions, the results obtained with the 2DFE and the 3DFE approaches showed no significant differences, demonstrating that 3D effects are irrelevant. Hence, the permeation phenomenon is ruled by the permeation properties of the polymeric layers (pressure and outer sheaths) and possible shield effects promoted by the metallic armors. In contrast, the flooded annulus analyses resulted in a non-uniform fugacity distribution in the annulus with significant differences between the 2DFE and the 3DFE approaches, showing the importance of modeling the helical geometries of the metallic armors in this condition. Finally, a conservative 2DFE approach, which neglects the contribution of the pressure sheath, is proposed to analyze the flooded annulus condition, aiming to overcome the high computational cost demanded by the 3DFE approach.

Keywords: diffusion; stress corrosion cracking; finite element method; flexible pipes

1. Introduction

Offshore oil and gas exploitation frequently involves using flexible pipes [1] (Figure 1) to connect subsea wellheads to platforms or floating production units. These pipes are composite multi-layered structures, each layer having a specific function, resulting in a structure with a low bending stiffness, combined with high strength and stiffness in the axial and radial directions.

In a typical unbonded flexible pipe, the metallic layers, such as the interlocked carcass, the pressure armor, and the tensile armors, resist the acting mechanical loads. The interlocked carcass is made from profiled stainless-steel strips wound at angles close to 90°, mainly resisting radial inward forces. The pressure armor is usually made from Z-shaped high-carbon steel wires (one or two) wound at angles close to 90°, supporting the system's internal pressure and radial inward forces. On the other hand, the tensile armors comprise several rectangular high-carbon steel wires laid in two or four layers, cross-wound at angles varying between 20° and 55° to resist tension, torque, and pressure end-cap effects. During operation, the relative movement of the metallic layers may induce friction forces and wear in the metallic layers, which are minimized by antiwear polymeric tapes. Moreover, as the metallic layers are not leakproof, inner and outer polymeric sheaths seal the pipe. Finally, other layers, such as insulation or high-strength tapes, may also be present, depending on the application, as described in API RP 17B [1].

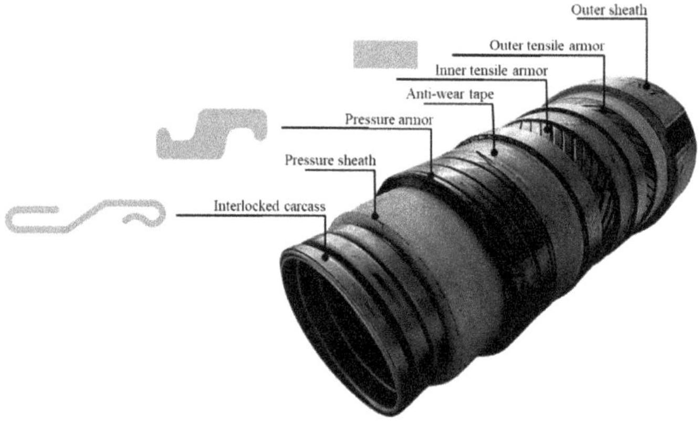

Figure 1. Unbonded flexible pipe.

Flexible pipes are subjected to several multi-physical phenomena during operation, such as fluid flow, pressure, and temperature changes, which affect their performance and integrity. Notably, one critical challenge related to the operation of flexible pipes is stress corrosion cracking (SCC), a material fracture in a favorable environment when neither the tension nor the corrosion, acting singly, would induce the fracture [2]. Revie and Uhlig [2] indicate that the first studies on chemical corrosion and tension were conducted at the beginning of the 20th century. However, although the phenomenon has been known for almost a century, just in 2017, with the progress in the pre-salt exploitation of hydrocarbons in the Brazilian shelf, the Brazilian National Petroleum Agency (ANP) noted the first occurrence of SCC due to CO_2 (SCC-CO_2) in flexible pipes [3].

The SCC mechanism is a significant concern in the offshore oil and gas industry because of its potential to cause sudden and catastrophic failure, leading to severe losses and environmental damage. After its first occurrence in 2017, this failure mode has been noted in several flexible pipes during line recoveries for investigation [4]. Although SCC in flexible pipes has several aspects to be investigated, it is known [4,5] that the phenomenon initiation depends on the simultaneous action of three key factors: tensile stresses, susceptible material, and a favorable environment. The tensile stresses in the tensile armors' wires arise from the acting mechanical loads, such as tension, torsion, internal and external pressures, and bending. As shown by mass transport and structural mechanics coupling [6], the tensile zones concentrate more corrosive species and further the SCC. Also, these wires are composed of high-strength steel, which is highly susceptible to SCC [4,5], and, finally, the water ingress and the presence of acid gases develop a favorable environment for corrosion.

In flexible pipes, acid gases, such as CO_2 or H_2S, may permeate the pressure sheath and access the pipe annulus (space between the inner and outer sheaths), which can be under flooded or dry annulus conditions. The flooded condition means water presence in the annulus due to external sheath damage [7,8] that allows seawater ingress. Instead, the dry condition implies an intact outer sheath, i.e., the pipe is externally sealed. However, even if the outer sheath is intact, water may be present in the annulus due to the diffusion of vapor water through the pressure sheath (following a process like the one related to the acid gases) and/or due to direct condensation [7,8]. Hence, a favorable environment for SCC development can be found in the annulus of flexible pipes in both conditions, dry or flooded.

Therefore, it is essential to understand species diffusion in flexible pipes to manage the risk associated with SCC effectively. Diffusion, the movement of matter from high to low chemical potential, plays a vital role in the SCC-induced unexpected failure of flexible pipes. As indicated, the transport of acid species, especially CO_2, from the bore to the annulus by diffusion may lead to favorable conditions for initiating the SCC mechanism.

Since the early 2000s, numerical models have been proposed to predict the annulus composition of flexible pipes during operation. However, with the rise of SCC incidents in offshore structures, more theoretical models have been developed to predict and understand diffusion in flexible pipes. These models use numerical techniques such as the finite element (FE) method or the finite difference (FD) method to analyze the molecules' permeation from the bore into the annulus of the pipe.

Kristensen [9] developed one of the first works on diffusion in flexible pipes to understand the mechanisms related to gases' permeation through the pressure sheath, including potential damage, such as armor wires' corrosion or the burst of the outer sheath due to increased annulus pressure. Kristensen [9] assumed some hypotheses to model the diffusion mechanism, i.e., a Fickian diffusion process through the sheaths and an ideal gas behavior to calculate the annulus pressure. Then, one- and two-dimensional models were constructed employing the FD method to solve the continuity differential equation. The one-dimensional model assumed a concentration gradient just along the radial direction. In contrast, the two-dimensional model simulated the pipe layers as concentric rings, allowing radial and axial concentration gradients. The model also accounted for the possibility of a region over the pressure sheath with no gas flux in the radial axis, simulating the presence of a steel wire (shield effect). The study concluded that the two-dimensional premise reduces the gas concentration in the annulus space and amplifies the necessary time to reach the stationary state.

Benjelloun-Danaghi et al. [10] proposed a two-dimensional axisymmetric model based on the FE method. The model assumed that the metallic layers could be simulated as rings, like Kristensen's [9] model. Additionally, some simplifying assumptions were adopted, such as using Arrhenius' law to estimate temperature dependence in the diffusion parameters, Fick's law for diffusion through the pipe, Henry's law for modeling the interface between different layers, and the Peng–Robinson cubic equation to calculate the thermodynamic equilibrium in the annulus space. The model also considered the presence of gas-release valves. Through this methodology, the authors presented a simulation to consider the diffusion of CO_2, CH_4, H_2S, and H_2O. The model calculates the temperature gradient along the polymeric layers, the partial pressure of each species in the annulus, and the volume of condensed water per structure length over time. Despite its primary use in analyzing pipes under dry annulus conditions, flooded conditions may also be considered. Finally, the model was verified with experimental tests presented by Taravel-Condat et al. [11] and was implemented in a computer program called MOLDI™.

Last et al. [12] conducted a comparative study in partnership with various offshore oil and gas corporations. The study aimed to evaluate the results and characteristics of eight permeation models from different companies, but no information regarding the models' development or theoretical basis was presented. The models were compared against mid-scale, full-scale, and field tests. The mid-scale test measured CH_4 permeation through a polyethylene (PE) sheath. Full-scale tests considered an actual pipe segment under different bore conditions, while the field tests were performed on three structures: two gas injection risers and a production riser. The authors compared the models with the experimental results and found that models based on fugacity had better agreement with the experimental measurements than those based on partial pressures. Also, the study highlighted a high model sensitivity to the permeation data and an effective default shielding factor of 0.5, addressing the blockage promoted by the pressure armor contact with the pressure sheath (shield effect) in the diffusion process.

In a recent study, Lefbrev et al. [13] improved the model proposed by Benjelloun-Danaghi et al. [10], highlighting several key parameters that affect the permeation phenomenon in flexible pipes. The main improvements were the possibility of a flooded annulus condition, thermodynamic equilibrium through a cubic plus association equation (CPA), and a new approach considering a three-dimensional simplified pipe. Moreover, a simplified three-dimensional FE model was proposed, and its predictions were compared to those obtained with the model developed by Benjelloun-Danaghi et al. [10]. Ultimately, the authors pointed out that the flooded condition led to higher fugacity levels than the dry condition. Furthermore, the difference in fugacity between the flooded and dry annulus conditions highly depended on the permeability of the pressure and outer sheaths.

As perceived, there are a few works published concerning the permeation in flexible pipes, and these studies still leave many questions about the phenomenon, such as the modeling of the polymeric tapes and the impact of some geometric assumptions, such as axisymmetry. Thus, this work uses two models based on the FE method constructed in Ansys® (Release 2023 R1) [14] software to understand and investigate some aspects of the diffusion phenomenon in flexible pipes. The models are divided into two- and fully three-dimensional, aiming to analyze the geometric effects of diffusion, especially in flooded annulus conditions.

The methodology for deriving the proposed diffusion models is detailed in the following sections, including mass transfer modeling and the related physical parameters, thermodynamic equilibrium modeling, and heat transfer modeling. Then, each numerical model is introduced, and the relevant characteristics are described. Finally, a case study is conducted, and the results and conclusions obtained are presented.

2. Modeling Principles

Mass transport in flexible pipes occurs from the bore into the annulus space. Initially, the molecules of the acid gases in the bore adhere to the polymeric surface, a phenomenon known as adsorption. Subsequently, these species migrate through the polymeric matrix and reach the annulus space, creating a corrosive environment. Hence, this section presents the fundamental modeling of permeation in flexible pipes, divided into the following subsections: mass transport, heat transfer (as temperature affects permeation), the estimation of physical properties in mass transport, and thermodynamic equilibrium.

2.1. Mass Transport

Mass transport in flexible pipes begins with the fluid's adsorption on the pressure sheath. Adsorption is an interfacial phenomenon characterized by physical or chemical adherence to liquid or solid surfaces. After adsorption equilibrium, the molecules migrate by diffusion through the pressure sheath. Diffusion is a transport phenomenon governed by a driving force called chemical potential. Therefore, mass transport is divided into adsorption on the interface and diffusion through the body.

The mass transport literature contains some theories about the adsorption of gases and liquids on polymeric membranes. Based on Frisch's [15], this work employs Henry's law to evaluate the adsorption equilibrium between interfaces. According to Frisch [15], for polymers above their glass temperature and species above their critical temperature, the adsorption is governed by Henry's law. These conditions are usually present in flexible pipes' bores. Henry's law evaluates the amounts of adsorbed species from a linear relation with their partial pressures. However, Last et al. [12] demonstrated that changing the partial pressure reference for the species fugacity improves the results compared to experimental data. Fugacity is a thermodynamic property related to the partial pressure of a species in a system, accounting for its non-ideal behavior. This parameter often substitutes the pressure in problems involving real gases or liquids, mainly where deviations from the ideal behavior are significant. Therefore, this work uses Henry's law in terms of fugacity to address possible non-ideal behaviors, i.e., Equation (1).

$$c_i = S_{ij} f_i \tag{1}$$

where c_i is the adsorbed concentration of species i on an interface, f is the fugacity of species i, and S_{ij} is the solubility coefficient of species i on surface j.

The fluid in the bore of a flexible pipe is usually a mixture of chemical species. The Maxwell–Stefan theory describes the simultaneous transport of multiple species. However, according to Frisch [15], mass transport follows Fick's law under specific conditions, such as a polymer above its glass temperature and a species above its critical temperature. Furthermore, Fick's law admits that each species' flux is independent, describing the mass diffusion of a single solute through a homogeneous solvent barrier [6]:

$$\boldsymbol{h_i} = -m_{ij}\nabla\mu_i \tag{2}$$

where $\boldsymbol{h_i}$ is the flux vector of species i, μ_i is species' i chemical potential, and m_{ij} is the mobility parameter of the i solute in the j solvent.

The mobility term in Equation (2), assuming isobaric and isothermal conditions, is given by [16]

$$m_{ij} = -\frac{c_i\, D_{ij}}{R\,T} \tag{3}$$

where D_{ij} is the diffusivity coefficient of species i in j medium, T is the considered temperature, and R is the universal gas constant.

Equation (3) describes the mass flux through the chemical potential driving force. However, this work adopts a change in driving force reference to simplify the equation solution and represents the transport phenomenon in a single variable: the species fugacity. The chemical potential μ_i relates to the fugacity of one species in a mixture [17]:

$$\mu_i = \mu_i^0 + RT\ln\left(\frac{f_i}{f_i^0}\right) \tag{4}$$

where μ_i and μ_i^0 are, respectively, the chemical potential and the fugacity of species i in a reference state, typically an ideal gas state. In this case, the fugacity becomes equal to the partial pressure of the species. Now, applying Equations (3) and (4) in Equation (2), Fick's law is obtained in another reference system:

$$\boldsymbol{h_i} = \frac{-c_i\, D_{ij}}{f_i}\nabla f_i \tag{5}$$

Hemond and Fechner [18] suggest a linear relationship between the species' fugacity and its concentration in a medium. The solubility coefficient S_{ij} is the material property that links the species (i) concentration and fugacity in a medium (j). Also, this statement is analogous to Henry's law for sorption phenomena at different phases' interfaces. This coefficient measures the solvent's ability to dissolve a solute. Therefore, Equation (6) rewrites Fick's law as a function of the solubility coefficient:

$$\boldsymbol{h_i} = -S_{ij}\, D_{ij}\nabla f_i \tag{6}$$

The accumulation of species over time is obtained by applying the mass balance law to Equation (6). Moreover, considering S_{ij} as time-independent, Equation (6) becomes

$$S_{ij}\dot{f_i} = \nabla\cdot\left[S_{ij}\, D_{ij}\nabla(f_i)\right] \tag{7}$$

The boundary conditions regarding fugacity can be determined once the mass transport law has been written. In contrast to the law in concentration reference, the fugacity is continuous at the interface between different materials. Moreover, adsorption equilibrium is assumed at the pressure sheath's interface, and the fugacity equals the fluid's fugacity. On the other hand, the fugacity on the outer sheath's external surface equals zero, as it assumes that the gas permeation does not impact the vast amount of seawater that washes the gas concentration. Finally, when the fluid's bore consists of many substances, Equation (7) must be solved for each component, as Fick's law admits that the permeation of each element is independent.

2.2. Physical Parameters in Mass Transport

The diffusive properties, i.e., the diffusivity coefficient D and the solubility coefficient S, highly depend on the permeation medium. Moreover, these parameters vary with temperature and may depend on the species concentration. As a result, this section presents information regarding the methods considered for determining these properties. This section is divided into three sub-sections: dry annulus, flooded annulus, and polymers, according to the medium permeation. In all models developed in this work, the metallic layers are assumed to be impermeable, i.e., no diffusion occurs in these layers.

2.2.1. Dry Annulus

Fuller et al. [19] propose an equation to evaluate the diffusivity coefficient D (cm^2/s) of a gas component i in a gas medium j. The equation depends on the mean pressure of the medium P (atm), the diffusion volume of the components V, the temperature T (K), and the molar mass of each species $\overline{m}$ (g/mol):

$$D_i = \frac{10^{-3}\, T^{1.75}\left(\frac{1}{\overline{m}_i} + \frac{1}{\overline{m}_j}\right)^{\frac{1}{2}}}{P\left[\left(\sum V_i\right)^{\frac{1}{3}} + \left(\sum V_j\right)^{\frac{1}{3}}\right]}. \tag{8}$$

The diffusion volume is an empirical parameter, and Fuller et al. [19] present its value for several gases. Hence, Equation (8) is initially employed assuming the mean pressure P equals 1 atm. After that, the diffusivity coefficient can be updated considering the new pressure in the annulus.

Furthermore, the ideal gas law can be applied to determine the solubility coefficient within the annulus space. Given the absence of the bore's species concentration within the annulus and the relatively low-pressure condition, this approach is deemed appropriate at the process's inception. It follows that, from ideal behavior, the fugacity is equivalent to the partial pressure of the gas, leading to

$$S_{ij}^{t=0} = \frac{1}{RT}. \tag{9}$$

From Equation (9), two approaches may be established. The first approach assumes that, in dry annulus conditions, the solubility remains constant and equivalent for all species, following the methodology proposed by Kristensen [9]. The second approach involves updating the solubility over time. Bengelloun-Dabaghi et al. [10] admit a parameter correction by calculating fugacity through an equation of state (EOS), using the pressure obtained from the concentration and the annulus' volume. However, this study suggests a temporal properties' correction through the species concentration determined by the EOS, as the FE models proposed in this work directly calculate the fugacity, as described in Section 3.

The approach rectifies the parameter through the concentration of each species obtained by an equation of state (EOS). For example, assuming that the one substance occupies all the annulus' volume, it is possible to obtain the annulus concentration of each component by an EOS, and this property is recalculated by

$$S_{ij}^{\mathrm{t_0+\Delta t}} = \frac{{c_i}^{p\overline{v}T}}{f_i^{\mathrm{t_0}}}. \tag{10}$$

where ${c_i}^{p\overline{v}T}$ is the concentration obtained through the EOS.

2.2.2. Flooded Annulus

The diffusivity coefficient in wet annulus conditions measures the capacity of a gas to permeate through the water. Chang and Wilke [20] present an empirical expression to evaluate this property for any gas substance percolating inside an aqueous medium:

$$D_{ij} = \frac{5.06 \cdot 10^{-10} T}{\eta \, \overline{v}_i^{0.6}} \tag{11}$$

where diffusivity is D (cm^2/s), temperature is T (K), η (poise) is the water viscosity, and $\overline{v}$ (in cm^3/mol) is the molar volume of the solute.

Regarding the fugacity capacity, Lefbrev et al. [13] use the Kontogeorgis et al. [21] model to obtain this property for different gases in water. The model employs a type of CPA EOS with the disadvantage of requiring several experimental data and a complex mathematical solution. Instead, Diamond and Akinfiev's [22] provide a more straightforward closed-form equation to calculate Henry's coefficient, i.e., the inverse of the solubility coefficient. The expression proposed by Diamond and Akinfiev [22] needs three experimental parameters, which, for several components, are given by the authors. However, a second study published by Diamond and Akinfiev [23] showed excellent accuracy for CO_2, the main acid gas in SCC phenomena in flexible pipes, even under extreme pressure and temperature conditions:

$$S_{ij} = \exp\left\{ -\left\{ (1-\xi)\ln(f^*) + \xi\ln\left(\frac{RT}{\overline{m}^*}\rho^*\right) + 2\rho^*\left[a' + b'\left(\frac{1000}{T}\right)^{0.5}\right]\right\}\right\}^{-1} \tag{12}$$

where a'(cm^3/g), b' (cm^3 K$^{0.5}$/g), and ξ are the experimental parameters of the evaluated species and ρ^*(g/cm^3), f^* (MPa), and $\overline{m}^*$ (g/mol) are the water's density, fugacity, and molar mass under prescribed pressure and temperature conditions.

Figure 2 compares the model proposed by Kontogeorgis et al. [21] and the equation from Diamond and Akinfiev [23], Equation (12), with experimental measurements presented by Lefbrev et al. [13]. The results related to Diamond and Akinfiev [23] were obtained considering Henry's law, Equation (1), combined with Equation (12). The fugacity capacities were evaluated under 25 °C and 50 °C, typical temperatures in flexible pipes' annuli. In the end, comparing the equations with the experimental results, Equation (12) better agrees with the experimental results at 25 °C than the Kontogeorgis model. On the other hand, the Kontogeorgis model showed better accuracy at 50 °C, but, at pressures greater than 10 MPa, both converge to similar results. Hence, this work considers Equation (12) to calculate the solubility of the species in water due to its simplicity and good agreement with the tested conditions. However, it is recognized that further investigations to verify the applicability of this equation to other gases and conditions are required.

2.2.3. Polymeric Barriers

Arrhenius's law describes the temperature dependence for solubility and diffusivity coefficients in polymeric materials:

$$D_{ij} = D_{ij,0}\exp\left(-\frac{E_{ij,D}}{R\,T}\right) \tag{13}$$

$$S_{ij} = S_{ij,0}\exp\left(-\frac{E_{ij,S}}{R\,T}\right) \tag{14}$$

where $D_{ij,0}$ and $S_{ij,0}$ are, respectively, the diffusion coefficient and the fugacity capacity when the temperature tends to infinity and $E_{ij,D}$ and $E_{ij,S}$ are the activation energies.

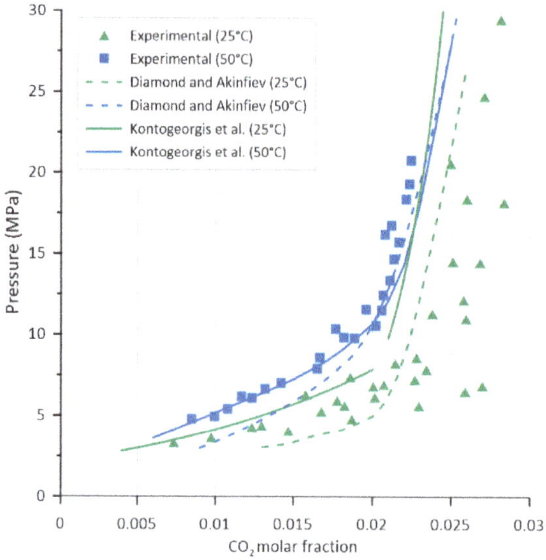

Figure 2. Comparison between capacity fugacity of CO_2 in water from Kontogeorgis et al. [21], Diamond and Akinfiev [23], and experimental data [13].

These parameters are experimentally measured for each pair of gas/polymer. Flaconnèche et al. [24] present experimental data from permeation tests on the diffusivity and solubility coefficients for gases and polymers in flexible pipes' industry under different temperatures. From these data, it is possible to estimate an Arrhenius equation for both properties for a pair of gas/polymer.

2.3. Thermodynamic Equilibrium in the Annulus

The species' fugacity and pressure in the annulus increase with the permeation of the gases from the bore either in dry or flooded annulus conditions, implying a variation in the diffusivity and solubility coefficients with time, as can be perceived in Equations (8)–(12). Therefore, the fugacity and the pressure in the annulus should be updated considering its actual composition, requiring using an equation of state (EOS) to evaluate the thermodynamic equilibrium. This work assumes the Peng–Robinson–Stryjek–Vera (PRSV) EOS [25], as Appendix A describes.

The annulus pressure is calculated assuming that only gaseous substances are present. In such cases, the global system's molar composition becomes the phase molar composition, and the algorithm shown in Figure 3 requires only species' fugacities as input to estimate the molar composition and pressure. This algorithm is a back substitution to evaluate the fugacity through the PRSV equation until it matches the calculated annulus fugacity.

2.4. Heat Transfer in Flexible Pipes

The temperature significantly affects mass transport, so it is necessary to evaluate the pipe's temperature profile to model the mass transport correctly. Fourier's law describes the heat transport in stiff and isotropic materials under a temperature gradient [6]:

$$\rho\, c_s\, \dot{T} = k\, \nabla^2 T \tag{15}$$

where ρ, c_s, and k are, respectively, the material's density, specific heat, and thermal conductivity.

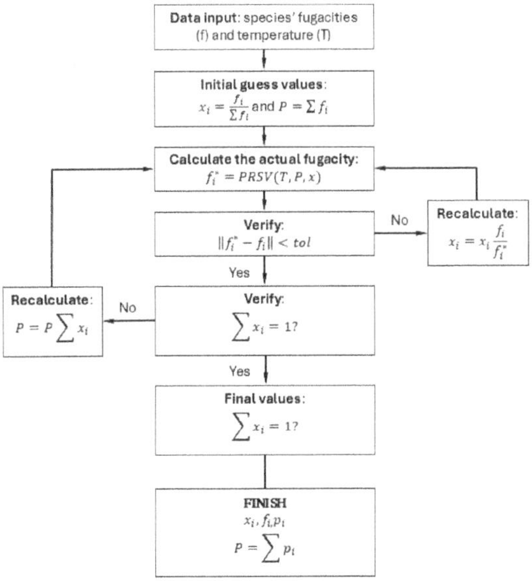

Figure 3. Algorithm to estimate the annulus pressure P.

This work applies simplified hypotheses to simulate heat transport in flexible pipes, such as considering just heat transport by conduction and neglecting gaps in the annulus. Hence, Equation (8) is simplified to a one-dimensional cylindrical equation. Other hypotheses concern the stationary time of the phenomenon, i.e., while mass transport may take years to reach the permanent regime, heat transfer takes hours. Thus, the heat and mass transfer in flexible pipes are uncoupled phenomena, and Equation (8) can be reduced to the one-dimensional steady-state form:

$$\frac{1}{r}\frac{\partial}{\partial r}\left(kr\frac{\partial T}{\partial r}\right) = 0 \tag{16}$$

where r is the radial position of a point in a pipe layer.

This approach's advantage is the possibility of solving Equation (16) analytically. The boundary conditions in Equation (16) are the bore's fluid and seawater temperatures, assuming perfect contact between each layer. The method employed in this study is based on the work of Benjelloun-Dabaghi et al. [10], where further details can be obtained.

In the solution of Equation (16), the polymeric layers present a temperature gradient along their thicknesses due to the low thermal conductivity. At the same time, the steel armors exhibit a uniform temperature due to their high conductivity. The temperature has a direct impact on the materials' diffusive properties. Nonetheless, a study conducted as part of this research [26] indicates that considering mean temperatures in the pipe's polymeric layers leads to acceptable results compared to studies that evaluate the temperature gradient along the layer.

3. Numerical Model

3.1. Assumptions

This section describes the characteristics of permeation modeling in Ansys® software [14], divided into a primary section describing the main hypotheses and three other sub-sections referring to the modeling domain. The domain can be two- or three-dimensional. The two-dimensional model (2DFE) proposes an axisymmetric permeation domain, assuming the armor layers and intermediary tapes as rings. In contrast, the three-dimensional approach (3DFE) aims to detail the flexible pipe annulus space.

All models treat the mass transport problem through a fugacity reference, as it offers some advantages, such as the following:

- Non-ideal representation of the species behavior, as discussed in Section 2;
- Continuity in the annulus of the pipe, including at the layers' interfaces. In contrast, the concentration of the species has a discontinuous distribution;
- Better agreement with experimental measurements, according to Last et al. [10].

However, the model was developed in Ansys®, where no elements with fugacity as a degree of freedom are available. The software has specific finite elements to solve the mass transfer phenomenon, but it is developed in terms of concentration. Therefore, to avoid the development of a new finite element using fugacity as a degree of freedom, this work proposes an analogy between mass and heat transport.

Fourier's law and Fick's law are analogous parabolic equations, so it is possible to calibrate Fourier's law parameters for amounting to Fick's law. Therefore, this work assumes a thermal element with the temperature as the degree of freedom to evaluate the mass transport in fugacity reference. The analogy relies on the following:

- Equaling temperature T and fugacity f;
- Equaling thermal conductivity k and permeability $P_e = S \cdot D$;
- Matching the specific heat capacity c_s with the solubility coefficient S;
- Assuming density ρ as unitary.

Figure 4 indicates typical two- and three-dimensional FE meshes. The meshes are generated for the gaps between the discontinuous layers (tapes and armor layers) and the polymeric layers, which are assumed to be the porous medium. Their main characteristics are described next.

3.2. Two-Dimensional FE Model (2DFE)

The two-dimensional model (2DFE) employs an isoparametric element with one degree of freedom, temperature, at each node to describe the permeation domain (polymeric sheaths and annulus), as seen in Figure 4. This element, named PLANE177 in Ansys® [14], has eight nodes in its quadrangular shape (non-degenerated form) and considers three potential behavior possibilities: plane state, plane with thickness, and axisymmetric. Also, it has a quadratic interpolation. Moreover, the element requires three material properties, i.e., thermal conductivity, density, and specific heat, which are transformed to deal with the mass transport problem following the analogy described in the introduction of this section.

Calibration has been conducted to ensure equivalence between rings and layers in the case of modeling tapes and wires, which possess a helicoidal geometry. This equivalence ensures that the ring's volume, surface area, and length are equivalent to the helicoidal geometry, similar to Benjelloun-Dabaghi et al. [10]. Equation (17) indicates the width s (Figure 5) of the space between two profiles:

$$s = \frac{2\pi R_m}{\tan\varphi} - \frac{n\,w}{\sin\varphi} \tag{17}$$

where n is the number of wires or tapes, R_m is the mean radius, w is the width profile, and, finally, φ is the laying angle.

Nevertheless, this equivalence is not straightforward for the pressure armor layers. Although all discontinuous layers (tapes and wires) are described as rectangular profiles, this assumption does not hold for the pressure armor, which usually has a Z cross-section (Figure 5). Hence, the intralayer gap length s is calculated through the wires' width to ensure an equivalent shielding effect, as illustrated in Figure 5. This approach guarantees that the blocking surface over the pressure sheath is the same for both profiles.

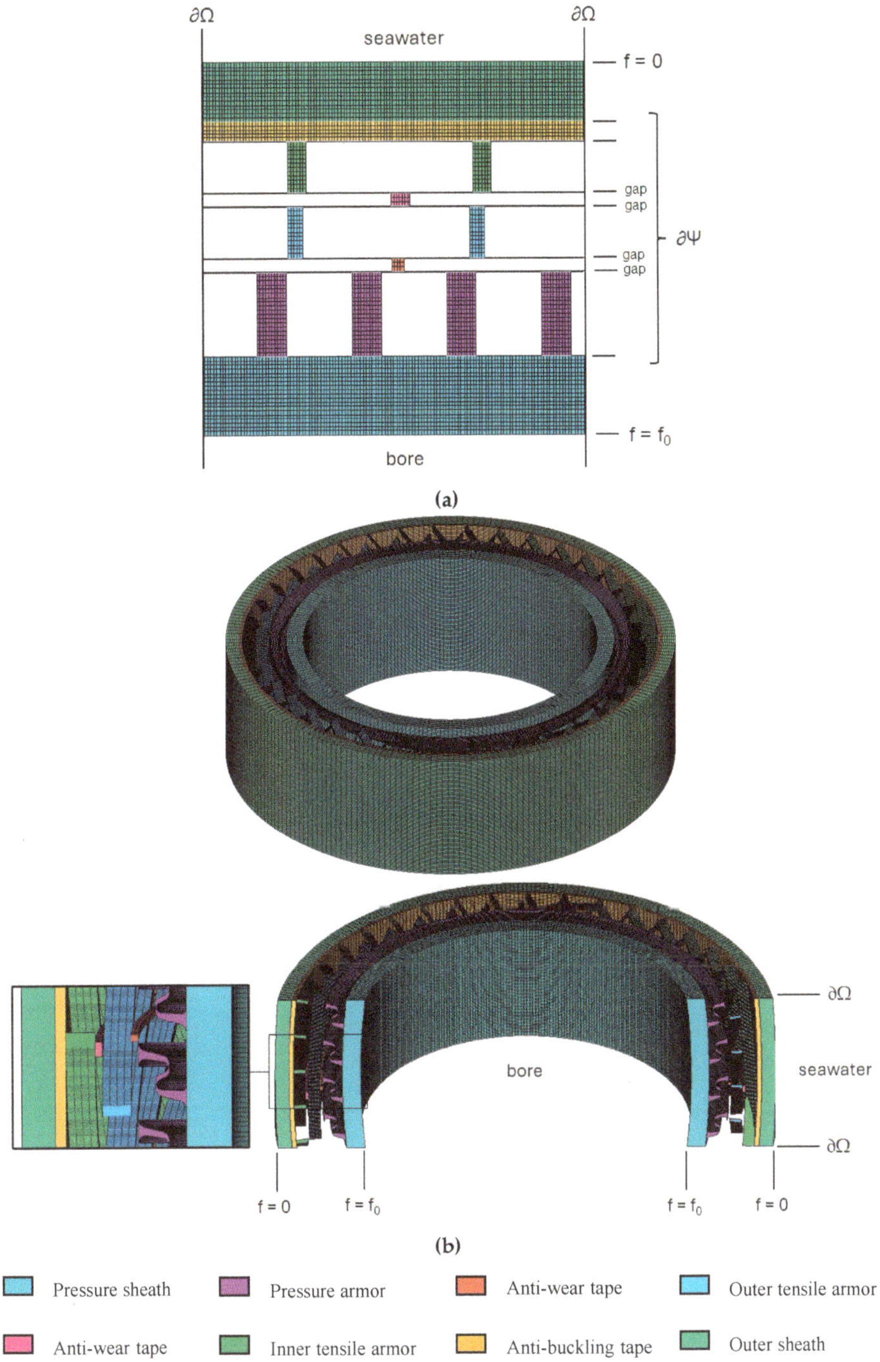

Figure 4. Mesh and boundary conditions: (**a**) 2DFE model and (**b**) 3DFE model.

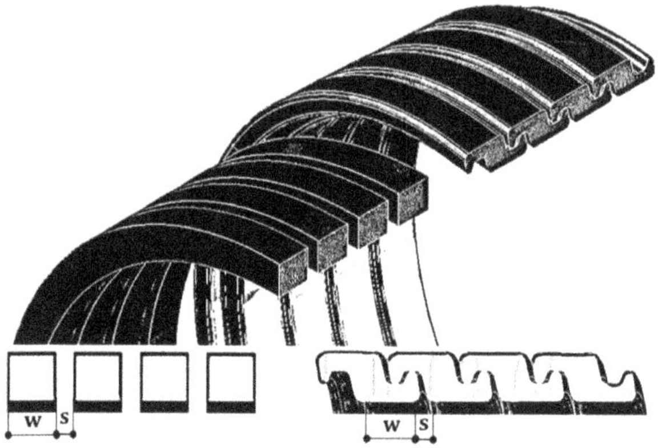

Figure 5. Equivalence for non-rectangular cross-sections.

Furthermore, one fundamental assumption in the modeling process pertains to the tapes, which can be classified as antiwear or anti-buckling. Unfortunately, the existing literature lacks information about the modeling of these tapes. Hence, this study proposes a hypothesis that considers the antiwear layers to be impermeable. This hypothesis assumes that the flux through the polymeric material can be neglected due to its low permeability. Therefore, the gas flows only through the gaps between the tapes. On the other hand, the anti-buckling tape, composed of aramid tissue, is intrinsically permeable due to its voids. Hence, this layer is treated as free volume, and a penalty factor is applied to simulate the actual free volume since gas cannot occupy spaces occupied by aramid.

Moreover, contact elements guarantee the interaction between layers, assuming no fugacity dissipation at the model's interfaces (perfect contact). These contact elements are defined in pairs and are named CONTA174 and TARGE170 in Ansys®. In the proposed model, CONTA174 elements are placed on the annulus interfaces, while TARGE170 elements are placed on the opposite contact surface. If the contact is established between a polymeric sheath or a tape or armor layer, the CONTA174 elements are placed on the sheath surface, while TARGE170 elements are in the tape or armor layer interface.

The metallic layers and the polymeric tapes are admitted as impermeable (white regions in the two-dimensional model, Figure 4). Hence, following Benjelloun-Dabaghi et al. [10]'s remark, a "modeling" gap between the polymeric tape–armor layer interface was considered to avoid gas trapping, leading to an artificial pressure build-up. This gap was assumed to be 1% of the armor layer thickness in each interface and was modeled as an annulus region.

Regarding the boundary conditions, it is assumed that there is no mass flux through the $\partial\Omega$ borders (Figure 4). The inner surface of the pressure sheath is in phase equilibrium with the bore's fluid, meaning that the fugacity of the bore's fluid is equal to the absorbed gas fugacity on the pressure sheath. Conversely, the outer sheath is in equilibrium with seawater, which is admitted to flush out any gas escaping through the sheath, resulting in a fugacity equal to zero over the outer polymeric sheath's external surface. Instead, the fugacity in the internal surface of the pressure sheath equals the fugacity of the gas in the bore. Additionally, equilibrium is maintained at each $\partial\Psi$ interface between different layers, ensuring that the fugacity is equal on the contact regions.

The proposed model is suitable for both steady-state and transient analysis. However, in this work, only transient analyses were conducted. These analyses required a time step equivalent to 3 h to achieve suitable results. The FE meshes were constructed with a maximum element length of 0.5 mm, and a total length higher than 1.5 linear pitches of the pressure armor is sufficient to represent the phenomena.

3.3. Three-Dimensional FE Model (3DFE)

The three-dimensional model represents, in detail, the geometry of the pipe. Hence, isoparametric three-dimensional solid elements describe the permeation domain, as seen in Figure 4. This element is named SOLID238 in Ansys® and has 20 nodes in its hexahedral non-degenerated form and 10 nodes in a degenerated tetrahedral form, suitable to model irregular geometries.

Compared to the two-dimensional model, the actual modeling of the layers' cross-section and lay angle makes the geometric equivalences and the interface gaps unnecessary. The detailed modeling of the cross-sections is achieved by using structured and unstructured FE meshes to capture the irregular geometry of the metallic layers and the annulus space. Moreover, the helical geometry of the tapes and the armor wires avoid trapping [13].

The contact between layers is established with the same contact elements employed in the two-dimensional model. However, the trustworthy representation of the three-dimensional domain increases the computational costs due to the layers' detailed modeling and the necessity of many elements to couple continuous and discontinuous layers properly. In addition, Ansys® requires a minimal coincidence between contact surfaces to generate coupling equations, as establishing contact over the entire polymer and the wire surface would produce equations that link excessively distant nodes and produce incorrect results. The continuous layers are generated by rotating a plane around a Cartesian axis. In contrast, discontinuous layers are developed along the layer's lay angle (Figure 6). Thus, achieving surface proximity between nodes from different layers requires high discretization, and the contact regions were discretized using elements having a maximum dimension of 0.2 mm after a few FE mesh tests. In all other areas, the same maximum element dimension assumed in the two-dimensional model was considered (0.5 mm). The length of the model should be higher than 1.5 linear pitches of the pressure armor.

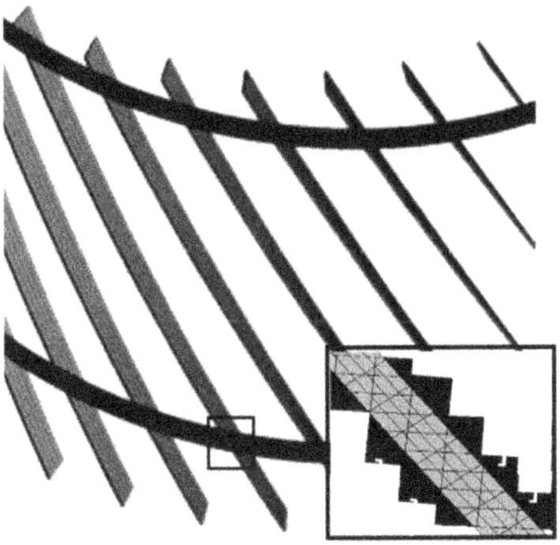

Figure 6. FE mesh in the interface between continuous and discontinuous layers.

Finally, the same time step as the 2DFE approach (3 h) was employed in the analyses with the three-dimensional model.

3.4. Implementation

The 2DFE and the 3DFE models are automatically generated with an APDL (Ansys Parametric Design Language) macro. Moreover, the macro controls the problem solution, updating the annulus diffusivity and solubility coefficients, relying on a user subroutine that calculates the annulus pressure with the algorithm presented in Figure 3 and Equations (8)–(10). Finally, the macro controls the solution of the problem, setting the sparse direct solver as the solution procedure.

4. Case Study

4.1. Description

A typical eight-layer 6″ flexible pipe layout was analyzed to evaluate the permeation phenomena. Table 1 describes its main characteristics. Despite an inner carcass in the pipe, this layer is not considered in the numerical analyses. The carcass profile allows fluid passage due to its interlocked cross-section, as seen in Figure 1. The fluid reaches the internal surface of the pressure sheath, and the imposed internal pressure typically opens a gap between the carcass and the pressure sheath [27]. Hence, a possible shield effect can also be conservatively neglected.

Table 1. The 6″ (ID = 152.4 mm) flexible pipe characteristics.

No.	Layer (Material)	Properties
1	Inner carcass [1] (stainless steel)	Thickness = 7.5 mm; Width = -; No of wires = 1; Lay angle = 87.8°; intralayer gap, s = -
2	Pressure sheath (PA11)	Thickness = 9.3 mm
3	Pressure armor (carbon steel)	Thickness = 10.0 mm; Width = 10.5 mm; No. of wires = 2; Lay angle = 87.6°; intralayer gap, s = 3.9 mm
4	Antiwear tape (polymeric tape)	Thickness = 1.5 mm; Width = 60.0 mm; No. of tapes = 1; Lay angle = 79.7°; intralayer gap, s = 3.0 mm
5	Inner tensile armor (carbon steel)	Thickness = 6.0 mm; Width = 14.0 mm; No. of wires = 2; Lay angle = 35.0°; intralayer gap, s = 1.3 mm
6	Antiwear tape (polymeric tape)	Thickness = 1.5 mm; Width = 60.0 mm; No. of tapes = 1; Lay angle = 79.7°; intralayer gap, s = 3.0 mm
7	Outer tensile armor (carbon steel)	Thickness = 6.0 mm; Width = 14.0 mm; No. of wires = 2; Lay angle = −35.0°; intralayer gap, s = 1.7 mm
8	Anti-buckling tape (aramid fiber)	Thickness = 2.3 mm
9	Outer sheath (HDPE)	Thickness = 9.3 mm

[1] This layer is disregarded in the model, but the models maintain the ID of the pressure sheath. Hence, the width and the intralayer gap are not required.

Furthermore, Table 2 presents the thermal conductivity in each layer and the Arrhenius parameters related to CO_2 and CH_4 for the PA11 and HDPE materials. The tapes and the metallic layers are assumed to be impermeable by hypothesis, as discussed in Section 2.2. The diffusivity and solubility of the annulus in dry and wet annulus conditions are calculated with the approach described in Sections 2.2.1 and 2.2.2.

Table 2. Thermal conduction coefficient, diffusivity, and solubility parameters for Arrhenius equations [24].

Material	k (W/mK)	Gas	D_0 (cm^2/s)	E_D (J/mol)	S_0 (cm^3(STP)/cm^3)	E_S (J/mol)
Stainless steel	15.0	CO_2	-	-	-	-
		CH_4	-	-	-	-
Carbon steel	45.0	CO_2	-	-	-	-
		CH_4	-	-	-	-
Polymeric tape	0.24	CO_2	-	-	-	-
		CH_4	-	-	-	-
Aramid fiber	0.20	CO_2	-	-	-	-
		CH_4	-	-	-	-
PA11	0.24	CO_2	0.40	36,000	5.40×10^{-3}	−12,000
		CH_4	4.50	45,000	7.50×10^{-2}	0
HDPE	0.54	CO_2	0.20	32,000	3.00×10^{-3}	−15,000
		CH_4	0.50	36,000	2.50×10^{-2}	−7000

Finally, as the main objective of this work is to compare the two- and three-dimensional FE models' responses when focusing on the SCC-CO_2 phenomena, the following conditions were considered:

- Internal and external temperatures of 60 °C and 5 °C, respectively;
- Internal and external pressures of 500 bar and 200 bar, respectively;
- CO_2 and CH_4 fugacities of 100 bar and 190 bar, respectively. These fugacities correspond to a bore composition of 60% CH_4 and 40% CO_2;
- Both wet and dry conditions;
- Total permeation analysis time of 2 years.

Next, a thermal analysis was conducted to determine the temperature in each layer, allowing the computation of the diffusive properties (diffusivity and solubility). Then, the permeation analyses' results are presented and discussed.

4.2. Thermal Analysis

The first step of a permeation analysis is to evaluate the thermal equilibrium through the material properties and the external and internal temperatures. Figure 7 shows the pipe's temperature profile along the radial direction.

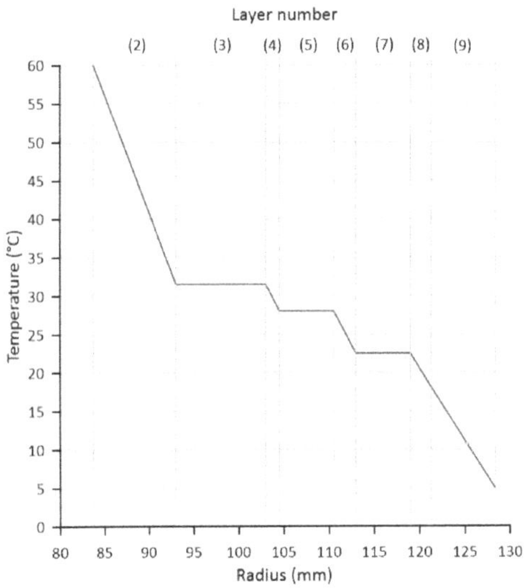

Figure 7. Radial temperature distribution in the analyzed 6″ flexible pipe.

As expected, the analysis indicates that the polymeric sheaths are responsible for the pipe's thermal insulation, while the metallic layers freely conduct the heat. The temperature linearly increases from 5 °C at the external surface of the outer sheath to about 23 °C at its internal surface. Analogously, a linear increase from 32 °C to the bore temperature (60 °C) is observed along the pressure sheath, evidencing the reasonable capacity of the sheaths to insulate the transported fluid. Moreover, it should be highlighted that these values are valid both for dry and wet annulus conditions, as pointed out in [26].

Once the temperature profile over the layers is evaluated, the diffusivity and the solubility of the CO_2 and the CH_4 in the permeation media are calculated with Equations (8)–(14) and Table 2. This work admits the mean temperature in each layer to obtain uniform permeabilities along the same layer.

4.3. Permeation Analyses

First, Figure 8 depicts the CO_2 fugacity distribution across the pipe domain, demonstrating the homogeneous annulus fugacity distribution under dry annulus conditions after an operation time of 2 years. A similar distribution was obtained for CH_4.

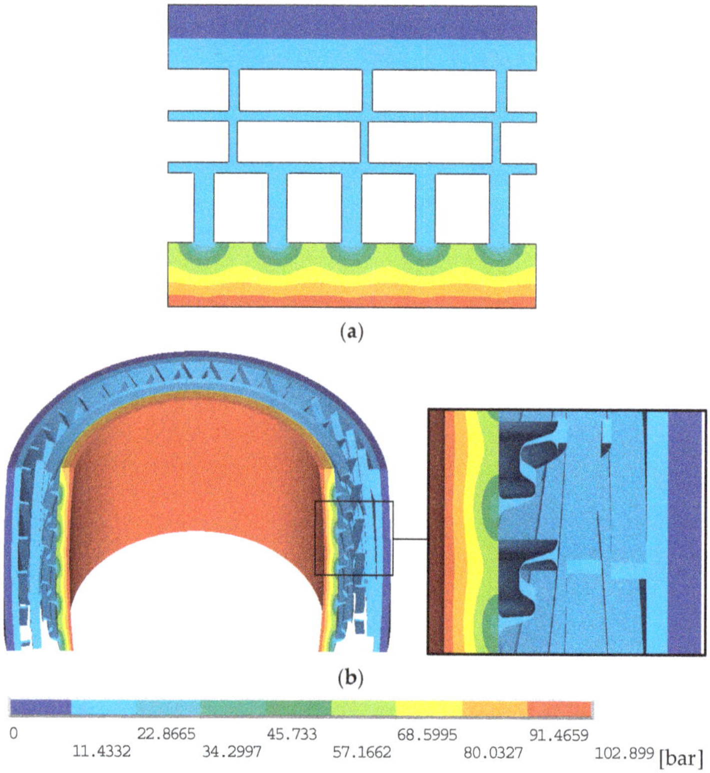

Figure 8. CO_2 fugacity distribution in the annulus of the 6″ pipe with dry conditions after 2 years of operation: (**a**) 2DFE and (**b**) 3DFE.

This uniformity arises from the high diffusivity in the annulus space, which offers no resistance to gas permeation. Consequently, when the fluid enters the annulus space, it fills all available volume. Therefore, as long as the structure remains sealed, the geometric effects occur only at the surfaces between a sheath and a discontinuous metallic sheath (pressure armor). The molecules become trapped beneath the impermeable profiles at these interfaces because the fluid cannot freely permeate through the polymeric matrix, resulting in the shielding effect (Figure 8). Thus, as the annulus internal geometry has minimal resistance and does not affect permeation, the 2DFE and 3DFE approaches are almost equivalent, as the annulus volume and the surface between the barrier and the metallic layer are equal in both models.

However, Figure 8 only qualitatively characterizes the approaches, highlighting the annulus scenario under a CO_2 stationary flux regime and a dry condition. Subsequently, the models were separately evaluated to access the CO_2 and CH_4 fugacities' variation over time, which are presented in Figure 9. This figure indicates that the two- and three-dimensional models agreed well, evidencing the significant potential of the two-dimensional model in accurately representing the permeation phenomenon in dry annulus conditions, in addition to considerably reducing the computational costs. The total time to simulate 2 years of operation was 20 min and 3 h for the 2DFE and 3DFE models, respectively. Moreover, the maximum CO_2 fugacity is about 17.0 bar, while the maximum CH_4 fugacity is 13.5 bar. These values correspond to 17% and 7% of the CO_2 and CH_4 fugacities in the bore and were reached after about 0.2 years and 0.6 years, respectively.

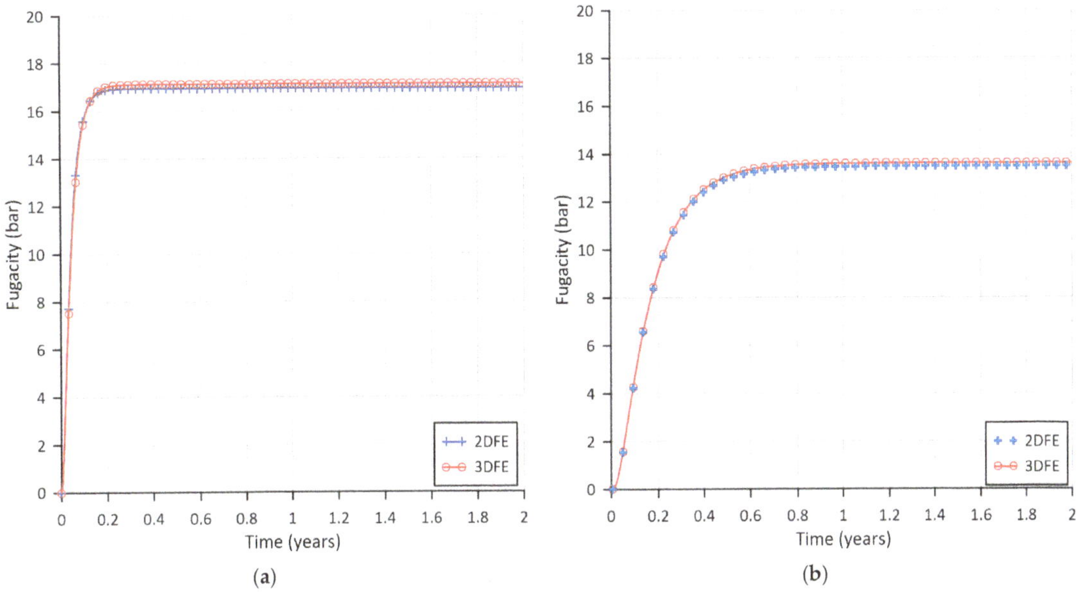

Figure 9. Fugacities' build-up in the annulus space of the 6″ flexible pipe with dry conditions: (**a**) CO_2 and (**b**) CH_4.

On the other hand, when the annulus is flooded, the water induces considerable resistance to the CO_2 flow, and the annulus geometry plays a significant role in permeation. Hence, the annulus fugacity becomes non-homogeneous along the pipe length, presenting higher values close to the pressure sheath than in regions close to the outer sheath (Figures 10 and 11). Observing only the 2DFE model, Figure 10 highlights some interesting permeation aspects. First, the geometric arrangement affects the annulus fugacity along the pipes' length once the fugacity is low in locations with narrower escape paths. This effect results from the resistance induced by water that traps the molecules inside the complex annulus locations, allowing the molecules to flow only through the interconnected gaps, consequently reducing the fugacity in these zones.

Now, looking at the geometric aspects resulting from a three-dimensional analysis (Figure 11), other relevant effects arise. Figure 11 shows a significant difference in fugacity levels along the pipe's circumference. Cross-section AA shows a fugacity profile that is different from the one in cross-section BB. This variation results from the non-axisymmetry of the free spaces, as the gaps between the wires are helicoidal in this model. Consequently, the molecules escape only at the intersection of these gaps.

Figure 12 depicts the contact between the inner tensile armor and the antiwear tape intralayer gaps. In the 3DFE model, the tapes are assumed to be impermeable, allowing molecules to only move from the tape gap to the tensile armor gaps. It is worth remarking that other layers are also impermeable, reducing the possible movement of molecules. In Figure 12, when the gas reaches the flooded gap of the antiwear tape, CO_2 remains trapped until it can escape through the nearest opening, i.e., a contact region with a gap in the inner tensile armor. Hence, the diffusion restriction caused by water presence associated with the restrictions imposed by the impermeable layers results in high CO_2 fugacities, which are reduced in regions where the molecules can freely permeate, as shown in Figure 12. This phenomenon is consistently observed at the interfaces of discontinuous layers, such as pressure and tensile armors and antiwear tapes. Finally, it is worth remarking that all these remarks regarding the flooded condition are also valid for the CH_4 fugacity distribution.

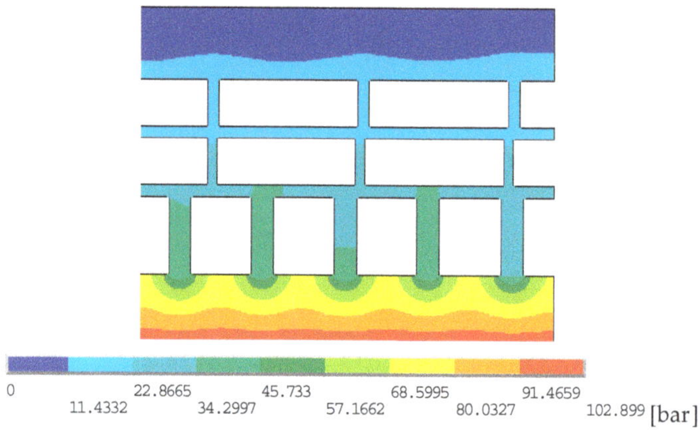

Figure 10. CO_2 fugacity distribution in the annulus of the 6″ pipe with flooded conditions after 2 years of operation: 2DFE model.

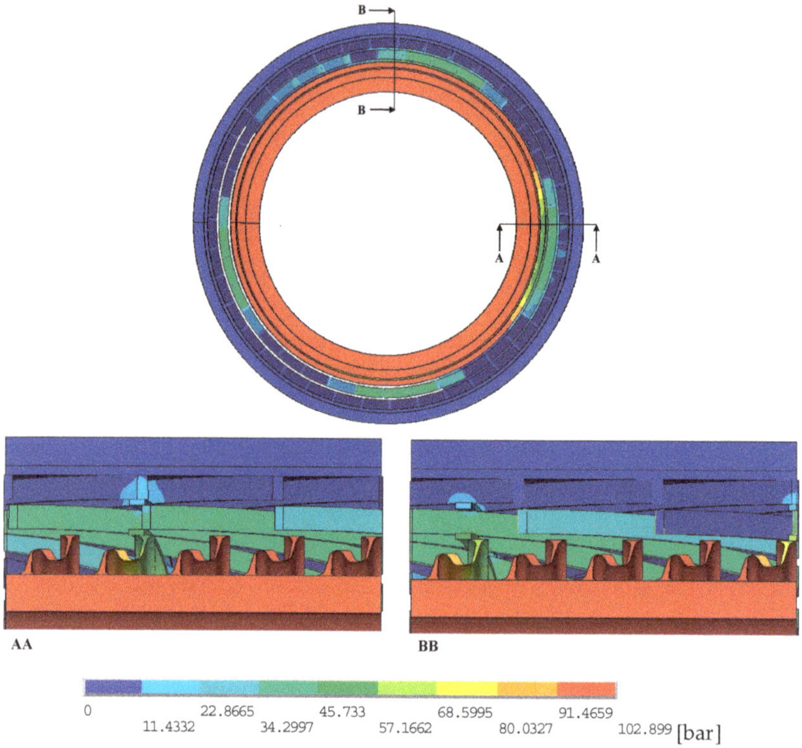

Figure 11. CO_2 fugacity distribution in the annulus of the 6″ pipe with flooded conditions after 2 years of operation: 3DFE model.

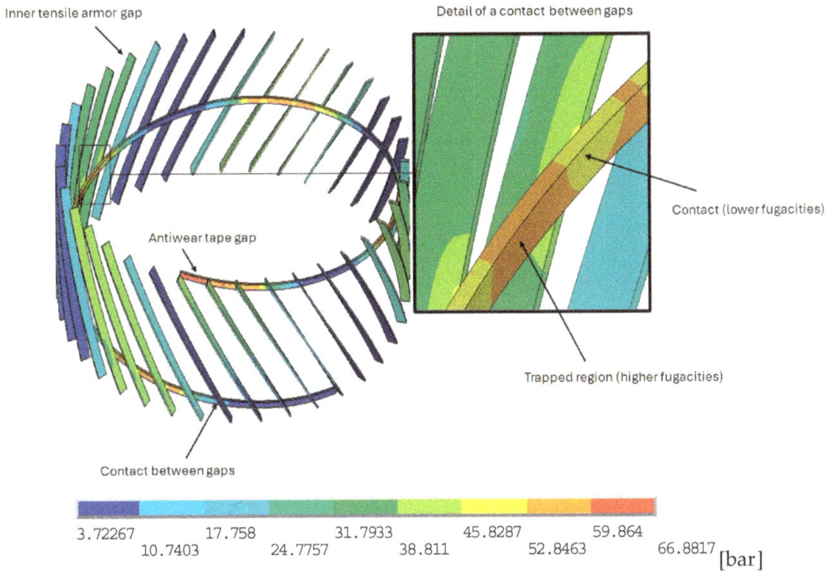

Figure 12. CO_2 fugacity distribution in the gaps between the antiwear tape and the inner tensile armor of the 6″ pipe with flooded conditions after 2 years of operation.

Next, the CO_2 and the CH_4 annulus fugacities' build-up were analyzed with the 2DFE and 3DFE approaches. As illustrated by Figures 10–12, the annulus fugacity distribution is non-homogeneous. Hence, Figure 13 indicates the minimum and maximum CO_2 and CH_4 fugacities obtained with different approaches in SCC critical zones, i.e., in the pressure and inner tensile armors of the analyzed flexible pipe.

Comparing the 2DFE and 3DFE models, Figure 13 demonstrates that non-axisymmetry plays a critical role in predicting the composition of the annulus. In the 2DFE model, a "modeling" gap (represented as a continuous layer) is required at the interface between two discontinuous layers to prevent fluid trapping, allowing the gases to permeate freely from one discontinuous layer to another. On the other hand, in the 3DFE model, the fluid must find a connected gap (as shown in Figures 6 and 12) to escape. Regarding the CO_2, which is critical to the SCC mechanism in flexible pipes, these permeation characteristics result in maximum fugacity levels in the 3DFE model, being 2.7 and 1.8 times higher in the pressure and tensile armor spaces, respectively. Conversely, due to the continuous permeation path, the 2DFE model exhibits a higher minimum CO_2 fugacity in the tensile armor. In contrast, the 3DFE approach has regions where molecules may become trapped or fail to occupy certain areas due to the dependence on the points of intersection between successive gaps. Hence, the 3DFE model shows higher maximum fugacities than the 2DFE model, where the CO_2 may freely permeate. Analogous conclusions can be stated for the CH_4.

Moreover, assessing the impact of a flooded annulus compared to a dry condition in the 2DFE and 3DFE approaches is crucial to the SCC-CO_2 phenomenon. Therefore, Figure 14 illustrates the stationary CO_2 fugacities along the radius of the pipe for the different proposed models and annulus conditions. However, as the geometry of the annulus induces a longitudinal gradient of the CO_2 fugacity, this work proposes an envelope of fugacities. The envelope assumes the maximum and minimum fugacity values in each layer's inner and outer radius. Through these results, a linear variation of the fugacity is considered in the layer.

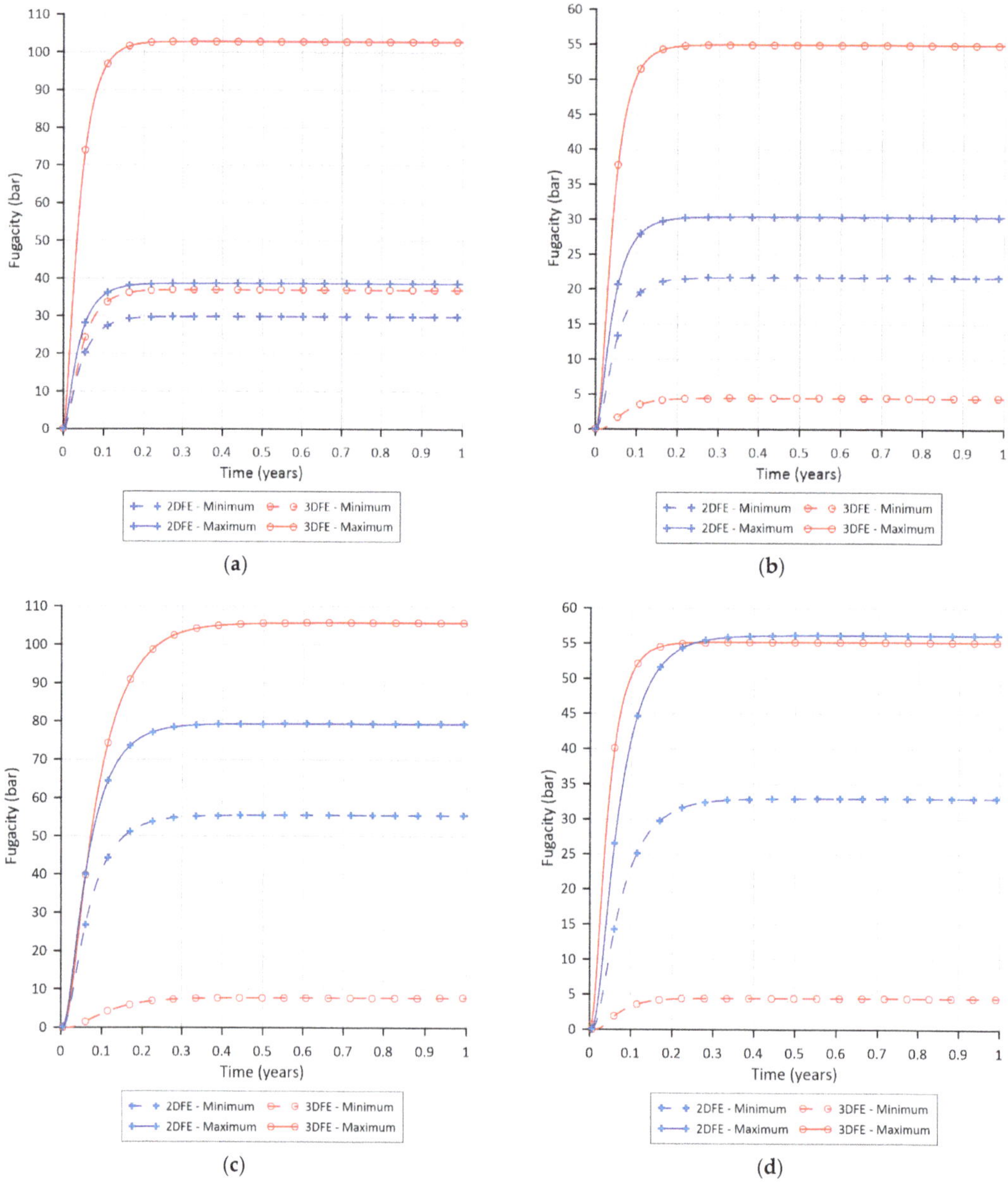

Figure 13. CO_2 and CH_4 fugacities' build-up in the 6″ flexible pipe with flooded annulus conditions: (**a**) pressure armor, CO_2; (**b**) inner tensile armor, CO_2; (**c**) pressure armor, CH_4; (**d**) inner tensile armor, CH_4.

Figure 14 shows that water significantly elevates CO_2 levels within the annulus. In the pressure armor layer, the 3DFE model demonstrates a fugacity 6.1 times higher than the dry condition, while the 2DFE model indicates an increase of 4.2 times. Similarly, in the inner tensile armors, the 3DFE and 2DFE models exhibit CO_2 levels 3.2 and 2.1 times higher, respectively, compared to the dry annulus. However, the dry annulus condition exhibited fugacity levels higher than the flooded condition in the top region due to the absence of resistance after passing the armors, making it easier for the gas to reach this region.

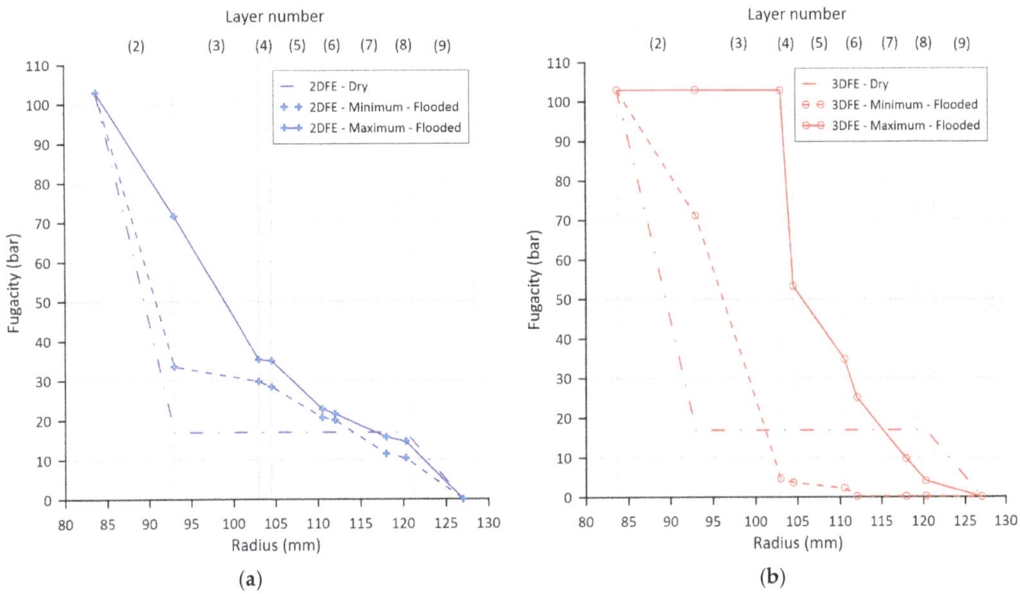

Figure 14. CO_2 fugacity envelopes considering dry and flooded annulus conditions: (**a**) 2DFE and (**b**) 3DFE.

Additionally, when comparing the envelopes of the 2DFE and 3DFE models, Figure 14 indicates that the 3DFE model, constrained by its geometric limitations, increases fugacity levels throughout most radius sections of the pipes, except for the uppermost regions. The lower CO_2 concentration in the upper region of the pipe can be attributed to the confined annulus geometry employed by the 3DFE model, which traps fluid in the lower annulus region, hindering CO_2 from reaching the uppermost areas. However, the maximum fugacity value in the pressure sheath and the pressure armor obtained with the 3DFE model is close to the CO_2 fugacity in the bore, while much lower values are predicted by the 2DFE model.

Nevertheless, as discussed in the dry annulus analyses, the 2DFE model requires much lower computational effort compared to the 3DFE model. Hence, a modified 2DFE model is proposed, aiming to approximate the maximum annulus fugacity values with those obtained with the most complex geometry represented in the 3DFE model.

A 2DFE analysis was conducted considering the same boundary conditions employed in the previously discussed flooded analysis but imposing the bore fugacities at the interface between the pressure sheath and the pressure armor, i.e., the diffusion through the pressure sheath was neglected. Figure 15 compares the 3DFE fugacity envelopes with those obtained with this modified 2DFE approach.

The results revealed that considering the bore's fugacity in the pressure armor layers influences the 2DFE fugacity envelope, leading to a more conservative scenario across most of the free annulus space, except for the top of the pressure armor layer. These elevated values stem from the confinement geometry of the pressure armor's Z profile in the 3DFE approach, which allows gas permeation only in specific zones, thereby increasing the fugacity levels in those trapping zones. Additionally, due to the confinement geometry, the 3DFE model consistently exhibits a significantly smaller minimum fugacity than the 2DFE model. This discrepancy arises because the continuous geometry of the 2DFE strategy facilitates a more uniform distribution of gas concentration, as shown in Figure 8. Nevertheless, apart from the higher levels at the top of the pressure armor layer, the 2DFE model, with the new boundary conditions, yields more conservative values in the more critical areas, the tensile armors. Specifically, the CO_2 levels in the inner and outer tensile armors are 1.8 and 2.5 times higher in the 2DFE model compared to the 3DFE model.

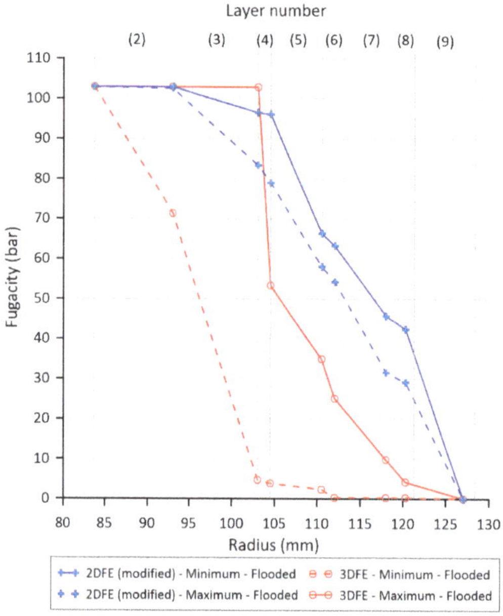

Figure 15. CO_2 fugacity envelopes considering flooded annulus conditions: 2DFE with modified boundary conditions and 3DFE.

5. Conclusions

This research aimed to develop and indicate numerical strategies for predicting the gas permeation from the bore into the flexible pipes' annulus, focusing on CO_2 permeation. Two FE models were developed that differ in the geometric domain. The first assumed an axisymmetric geometry of the annulus space, which reduced the three-dimensional domain to a two-dimensional domain, diminishing the number of degrees of freedom and the computational cost. On the other hand, a full three-dimensional model was admitted to evaluate the consequences of the geometry simplification.

Initially, both models underwent simulation and comparison under a dry annulus condition. The results obtained during this phase demonstrate a remarkable agreement between the 2DFE and 3DFE approaches. The fact that both models yielded similar annulus fugacity values underscores the negligible resistance offered by the dry condition to gas permeation, rendering the annulus geometry insignificant in this phenomenon. As a result, the study confirmed that the 2DFE approach is an excellent method for evaluating annulus composition under dry conditions, as its simplified geometry accurately represents the complete geometry without compromising precision while significantly reducing the computational time by approximately 115 times.

Sequentially, both models were evaluated and compared under a flooded annulus condition. The first observed consequence of a flooded annulus was the nonhomogeneity of fugacity values within the annulus. Consequently, at this stage, the fugacity results along the pipes' radial axis were treated as an envelope of fugacities, which involved evaluating the maximum and minimum values within specific layer zones.

Therefore, the initial study compared the fugacity of a dry annulus under stationary conditions with the fugacity envelope obtained for a wet annulus condition. The findings revealed a significant increase in fugacity levels within the annulus. Specifically, the 2DFE exhibited 2.1 times higher fugacity under flooded conditions in the tensile armors' region compared to the dry condition. Conversely, the 3DFE model demonstrated an even higher CO_2 fugacity than the dry conditions, approximately 3.2 times higher. Hence, significant differences were observed between the 2DFE and the 3DFE models, making the 2DFE model less conservative.

Aiming to approximate the fugacities predicted by the 2DFE model to those evaluated in the 3DFE model under wet annulus conditions, a modified 2DFE model was constructed disregarding the pressure sheath. By doing that, the pressure armor's fugacities equal the bore's fugacity. The results obtained with this approach showed higher fugacities compared to the 3DFE model. Hence, this modified model may be an alternative to a more conservative design regarding SCC-CO_2, demanding much lower computational effort.

Finally, this article dealt with different numerical approaches to predict flexible pipes' annulus compositions during operation, assuming either dry or flooded conditions. However, several other aspects must be studied to adequately address crucial phenomena, such as SCC-CO_2, e.g., the impact of the polymeric tapes in the diffusion phenomena; the presence of other acid gases and, more critically, water vapor in the diffusion process; the quantification of the corrosion phenomenon under different acid gases' fugacities; and comparisons with experimental tests, which are critical to validate the proposed models.

Author Contributions: J.M.B.V.: Conceptualization, Methodology, Formal Analysis; Writing—original draft. J.R.M.d.S.: Conceptualization, Methodology, Supervision, Writing—review and editing. All authors have read and agreed to the published version of the manuscript.

Funding: The first author would like to acknowledge the financial support from ANP (the Brazilian National Agency for Petroleum, Natural Gas and Biofuels) implemented through its Program of Human Resources (code PRH 9.1), during his post-graduate studies at COPPE/UFRJ. Morover, this work was partially supported by the Brazilian funding agencies CNPq (Conselho Nacional de Desenvolvimento Científico e Tecnológico), grant number 308625/2022-6, and CAPES (Coordenação de Aperfeiçoamento de Pessoal de Nível Superior), finance code 001.

Data Availability Statement: The input and output data from the results of the analyses are available from the corresponding author upon reasonable request.

Conflicts of Interest: The authors declare no conflicts of interest.

Appendix A

Using cubic equations of state, such as the Peng–Robinson EOS, associated with the van der Waals mixing rule, is frequently employed to solve the problem of thermodynamic equilibrium in mixtures. However, this equation was used in this work but considering the modification proposed by Stryjek and Vera [25] (PRSV). The Peng–Robinson equation is given by:

$$p = \frac{R \cdot T}{v - \bar{b}} - \frac{\bar{a}}{v^2 + 2 \cdot \bar{b} \cdot v - \bar{b}^2} \tag{A1}$$

where v is the molar volume of the system and a and b are parameters of the EOS. In a mixture, these parameters are obtained by combining the values of the pure components (van der Waals mixing rule):

$$\bar{a} = \sum_{i=1}^{n_f} \sum_{j=1}^{n_f} x_i \cdot x_j \cdot a_{i,j} \tag{A2}$$

$$\bar{b} = \sum_{i=1}^{n_f} x_i \cdot b_i \tag{A3}$$

where x is the molar fraction of each of the n_f components in the mixture. In this work, n_f equals 2 ($i, j = 1$ to 2), as only CO_2 and CH_4 are considered. Moreover,

$$a_{i,j} = \begin{cases} \frac{\alpha_i \cdot p}{(R \cdot T)^2}, & i = j \\ \sqrt{a_{i,i} \cdot a_{j,j}} \cdot (1 - k_{i,j}), & i \neq j \end{cases} \tag{A4}$$

$$\alpha_i = \frac{0.457235 \cdot R^2 \cdot (T_{C_i})^2}{p_{C_i}} \cdot \left[1 + \kappa_i \cdot \left(1 - \sqrt{\frac{T}{T_{C_i}}}\right)\right]^2 \tag{A5}$$

$$b_i = \frac{\beta_i \cdot p}{R \cdot T} \tag{A6}$$

$$\beta_i = \frac{0.077796 \cdot R^2 \cdot (T_{C_i})^2}{p_{C_i}} \tag{A7}$$

where p_C and T_C are the critical pressure and temperature of species i, and k is an interaction parameter, which, for the interaction between CO_2 and CH_4, equals 0.102. The parameter κ is given, according to Stryjek and Vera [25], by

$$\kappa_i = \kappa_i^0 + \kappa_i^1 \cdot \left(1 + \sqrt{\frac{T}{T_{Ci}}}\right) \cdot \left(0.7 - \frac{T}{T_{Ci}}\right) \tag{A8}$$

where κ_i^0 equals 0.397 and 0.673 for CH_4 and CO_2, respectively, while κ_i^1 is -0.0225 and -0.0679 for CH_4 and CO_2, in this order.

Equation (A1) may be written in the cubic polynomial form:

$$Z^3 - \left(1 - \bar{b}\right)Z^2 + \left(\bar{a} - 2\bar{b} - 3\bar{b}^2\right)Z - \left(\bar{a}\bar{b} - \bar{b}^2 - \bar{b}^3\right) = 0 \tag{A9}$$

where Z is the compressibility factor, which is stated as

$$Z = \frac{p \cdot v}{R \cdot T} \tag{A10}$$

Solving Equation (A1) by considering the parameters given by Equations (A2)–(A9), the compressibility factor Z is calculated. After that, the fugacity coefficient ϕ_i may be obtained with

$$\ln(\phi_i) = -\ln\left(Z - \bar{b}\right) + \frac{b_i}{\bar{b}} \cdot (Z - 1) - \frac{\bar{a}}{2\sqrt{2}\bar{b}} \cdot \left(\frac{2 \cdot \sum_{j=1}^{n_f} x_i \cdot a_{i,j}}{\bar{a}} - \frac{b_i}{\bar{b}}\right) \cdot \ln\left[\frac{Z + \left(1 + \sqrt{2}\right) \cdot \bar{b}}{Z + \left(1 - \sqrt{2}\right) \cdot \bar{b}}\right] \tag{A11}$$

Finally, the fugacity of species i in the analyzed conditions is given by

$$f_i = \phi_i \cdot x_i \cdot P \tag{A12}$$

where P is the system's internal pressure.

References

1. API. *API RP 17B: Recommended Practice for Flexible Pipe*, 6th ed.; American Petroleum Institute: Washington, DC, USA, 2024.
2. Revie, R.W.; Uhlig, H.H. *Corrosion and Corrosion Control: An Introduction to Corrosion Science and Engineering*, 4th ed.; Wiley-Interscience: Hoboken, NJ, USA, 2008; pp. 150–166.
3. ANP. Stress Corrosion Due to CO_2 (SCC-CO_2). Security Alert 001. Agência Nacional do Petróleo. 2017. Available online: https://www.gov.br/anp/pt-br/assuntos/exploracao-e-producao-de-oleo-e-gas/seguranca-operacional/incidentes/arquivos-alertas-de-seguranca/alerta-01/alerta-de-seguranca_001_ssm_scc-co2_pt.pdf (accessed on 19 November 2024). (In Portuguese)

4. Fonseca, D.; Tagliari, M.R.; Guaglianoni, W.C.; Tamborim, S.M.; Borges, M.F. Carbon Dioxide Corrosion Mechanisms: Historical Development and Key Parameters of CO_2-H_2O Systems. *Int. J. Corros.* **2024**, *2024*, 5537767. [CrossRef]
5. de Motte, R.; Joshi, G.; Chehuan, T.; Legent, R.; Kittel, J.; Désamais, N. CO_2-SCC in Flexible Pipe Carbon Steel Armour Wires. *Corrosion* **2022**, *78*, 547–562. [CrossRef]
6. Gurtin, M.E.; Fried, E.; Anand, L. *The Mechanics and Thermodynamics of Continua*; Cambridge University Press: Cambridge, UK, 2010.
7. Eide, J.T.W.; Muren, J. Lifetime assessment of flexible pipes. In Proceedings of the ASME 2014 33rd International Conference on Offshore Mechanics and Arctic Engineering (OMAE2014), San Francisco, CA, USA, 8–13 June 2014.
8. Eriksen, M.; Engelbreth, K.I. Outer cover damages on flexible pipes—Corrosion and integrity challenges. In Proceedings of the ASME 2014 33rd International Conference on Offshore Mechanics and Arctic Engineering (OMAE2014), San Francisco, CA, USA, 8–13 June 2014.
9. Kristensen, A.B. Diffusion in Flexible Pipes. Master's Dissertation, Technical University of Denmark, Lingby, Denmark, 2000.
10. Benjelloun-Dabaghi, Z.; Hemptinne, J.-C.; Jarrin, J.; Leroy, J.-M.; Aubry, J.-C.; Saas, J.N.; Condat, C.T. MOLDI^{TM}: A Fluid Permeation Model to Calculate the Annulus Composition in Flexible Pipes. *Oil Gas Sci. Technol.* **2002**, *57*, 177–192.
11. Taravel-Condat, C.; Guichard, M.; Martin, J. MOLDI^{TM}: A Fluid Permeation Model to Calculate the Annulus Composition in Flexible Pipes—Validation With Medium Scale Tests, Full Scale Tests and Field Cases. In Proceedings of the OMAE03 22nd International Conference on Offshore Mechanics and Arctic Engineering, Cancun, Mexico, 8–13 June 2003.
12. Last, S.; Groves, S.; Rigaud, J.; Taravel-Condat, C.; Wedel-Heinen, J.; Clements, R.; Buchner, S. Comparison of Models to Predict the Annulus Conditions of Flexible Pipe. In Proceedings of the Offshore Technology Conference, Houston, TX, USA, 6–9 May 2002.
13. Lefebvre, X.; Khvoenkova, N.; de Hemptinne, J.C.; Lefrançois, L.; Radenac, B.; Pignoc-Chicheportiche, S.; Plennevaux, C. Prediction of Flexible Pipe Annulus Composition by Numerical Modeling: Identification of Key Parameters. *Sci. Technol. Energy Transit.* **2022**, *77*, 1–11. [CrossRef]
14. Ansys, Inc. *Ansys® Mechanical APDL (Release 2023 R1)*; ANSYS, Inc.: Canonsburg, PA, USA, 2023.
15. Frisch, H.L. Sorption and Transport of Penetrants in Glassy Polymers. *Polym. Eng. Sci.* **1980**, *20*, 2–13. [CrossRef]
16. Prausnitz, J.M.; Poling, B.E.; O'Connell, J.P. *Properties of Gases and Liquids*, 5th ed.; McGraw-Hill Education: New York, NY, USA, 2001.
17. Lewis, G.N.; Randal, M. *Thermodynamics and the Free Energy of Chemical Substances*, 1st ed.; McGraw-Hill Book Company: New York, NY, USA, 1923.
18. Hemond, H.F.; Fechner, E.J. *Chemical Fate and Transport in the Environment*, 4th ed.; Elsevier: San Diego, CA, USA, 2022.
19. Fuller, E.N.; Schettler, P.D.; Giddings, C.J. New method for prediction of binary gas-phase diffusion coefficients. *Ind. Eng. Chem.* **1966**, *58*, 19–27. [CrossRef]
20. Chang, P.; Wilke, C.R. Some Measurements of Diffusion in Liquids. *J. Phys. Chem.* **1955**, *59*, 592–596. [CrossRef]
21. Kontogeorgis, G.; Voutsas, E.; Yakoumis, I.; Tassios, D. An Equation of State for Associating Fluids. *Ind. Eng. Chem. Res.* **1996**, *45*, 4310–4318. [CrossRef]
22. Diamond, L.W.; Akinfiev, N.N. Thermodynamic description of aqueous nonelectrolytes at infinite dilution over a wide range of state parameters. *Geochim. Cosmochim. Acta* **2003**, *67*, 613–629.
23. Diamond, L.W.; Akinfiev, N.N. Solubility of CO_2 in water from −1.5 to 100 °C and from 0.1 to 100 MPa: Evaluation of literature data and thermodynamic modeling. *J. Chem. Thermodyn.* **2003**, *208*, 265–290. [CrossRef]
24. Flaconnèche, B.; Martin, J.; Klopffer, M.H. Permeability, Diffusion and Solubility of Gases in Polyethylene, Polyamide 11 and Poly (Vinylidene Fluoride). *Oil Gas Sci. Technol.* **2001**, *56*, 261–278. [CrossRef]
25. Stryjek, R.; Vera, J. PRSV—An Improved Peng-Robinson Equation of State for Pure Compounds and Mixtures. *Can. J. Chem. Eng.* **1986**, *64*, 323–333. [CrossRef]
26. Vieira, J.M.B. Numerical Models to Predict Gases Diffusion in the Annulus of Flexible Pipes. Master's Dissertation, Federal University of Rio de Janeiro, Rio de Janeiro, Brazil, 2023. Available online: http://www.coc.ufrj.br/pt/dissertacoes-de-mestrado/661-2022-5/10176-joao-marcos-bastos-vieira-3 (accessed on 19 November 2023). (In Portuguese)
27. de Sousa, J.R.M.; de Sousa, F.J.M.; de Siqueira, M.Q.; Sagrilo, L.V.S.; de Lemos, C.A.D. A Theoretical Approach to Predict the Fatigue Life of Flexible Pipes. *J. Appl. Math.* **2012**, *2012*, 983819. [CrossRef]

Journal of
Marine Science and Engineering

MDPI

Article

Decoupled Analysis of a Multi-Layer Flexible Pipeline Buried in Clay Subjected to Large Lateral Soil Displacement

Eduardo Ribeiro Malta , Xiaoyu Dong and Hodjat Shiri *

Civil Engineering Department, Faculty of Engineering and Applied Science, Memorial University of Newfoundland, St. John's, NL A1C 5S7, Canada; emalta@mun.ca (E.R.M.); xiaoyu.dong@mun.ca (X.D.)
* Correspondence: hshiri@mun.ca

Abstract: Multilayered flexible subsea pipelines may experience significant lateral movements due to manmade and environmental geohazards. These pipelines incorporate several structural and protective layers to resist different loads, and may require additional protection such as trenching, rock placement, or burial. In practice, simplifications are considered due to the complexities and uncertainties involved in the multi-layer pipe structure and the surrounding soil, compromising the pipe structure or the soil behavior. These simplifications are applied either on the pipe by assuming a rigid section or on the soil by representing it as elastic springs, which may result in inaccuracies. This study proposes a decoupled methodology combining the Coupled Eulerian–Lagrangian (CEL) model for soil displacement with a small-strain finite element analysis of the flexible pipe. This approach aims to accurately capture cross-sectional deformations and local stresses due to soil movement while maintaining reasonable computational effort. A parametric analysis was conducted to assess the impact of several variables on failure risk. The deformed cross-section was then used for a collapse analysis to determine critical loads at maximum operational depth. The study showed that modeling parameters such as soil strength and internal diameter might significantly influence pipe failure and the risk of collapse.

Keywords: finite element analysis; multi-layer flexible pipeline; CEL model; pipe–soil interaction

1. Introduction

Multi-layer flexible pipelines are extensively used in the offshore oil and gas industry. While these pipelines may be buried in the soil or trenched for physical protection [1], they remain vulnerable to significant ground deformations due to landslides, fault crossing, ice gouging in Arctic regions (Figure 1a), and other natural events. When subjected to large soil displacements, flexible pipes are also prone to large deformations. Failures can occur under bending situations, leading to extreme tension and compression within their structure, putting the pipeline's integrity at risk. In contrast to rigid pipes, multi-layer flexible pipes have a low bending stiffness, allowing them to deform considerably and gain strength from the surrounding remolded soil and adjacent undisturbed soil. Thus, it is essential to incorporate the lateral nonlinear soil interaction effects into the performance analysis of buried multilayer flexible pipes subjected to significant lateral soil displacements. However, a coupled analysis incorporating the complexities of both the flexible pipe structure, involving contacts between several different layers, and the advanced constitutive soil models make the analysis computationally costly and inefficient. Therefore, in practice, depending on the area of concern, the accurate modeling of one of these two elements, i.e., the pipe or soil, cannot be achieved at a reasonable computational cost. The current study addressed this practical need by developing a decoupled approach to incorporate the uncertainties of both elements at the same time, while maintaining a reasonable computational effort.

Theoretical and experimental models exist to predict the ultimate lateral resistance or force–displacement responses of embedded moving structures [2–8], with some specifically concentrating on the interactions between trenched/backfilled pipelines and the surrounding soil [9–13]. All these research works assume the pipeline is rigid, with the emphasis placed on the soil properties and the extent of soil resistance against the moving pipe.

The pipe's cross-sectional deformation was considered in the numerical models presented in a few studies [14–16]. The majority of these studies focus on onshore water transportation systems, mainly concrete pipes. In [14], the concrete pipeline is considered static and a continuum FE model is used to examine the impact of external pressure on the pipeline section. The behavior of fiber-reinforced pipelines while under the effect of surface loads was studied in [15]. Although the pipelines considered are onshore water transportation pipes, the soil modeling and the centrifuge tests used to validate the models are quite like the ones used in the offshore industry. Regarding composite pipes, more specifically, very flexible pipes, a numerical study can be found in [16]. The authors developed a more sophisticated FE model using ABAQUS. However, the pipeline was modeled assuming the layers would be perfectly bonded together. The soil itself was modeled using the native ABAQUS soil elements, and the study was able to determine the section deformation and make comparisons between steel and polymeric pipelines.

An FE continuum model was developed by [17] to simulate the impact of fault crossings on a buried pipeline. The authors considered the cross-sectional pipe deformation using the ABAQUS submodeling technique. The study determined the local buckling deformations in the area subjected to bending.

An elastic pipe section was used by [18] to analyze the wave loads' effects on buried offshore pipelines. The authors developed a finite element model for porous soil media. An explicit continuum FE model was used in [19] to replicate ice gouging conditions. The sub-gouge soil deformations were translated into pressures and were directly exerted on the exterior of a buried subsea pipeline. This model used a single-layer shell deformable pipe, making it possible to simulate the ovalization of the cross-section. However, the authors focused further on the global response of the pipeline compared with a simplified Winkler beam-spring model. A continuum finite element model was also employed by [20] to simulate a buried pipeline exposed to the fault displacement caused by an earthquake. Both the pipeline and the seabed soil were represented as continuum Lagrangian domains in the model. Nevertheless, there was no discussion regarding section ovalization, given that the paper's goal was to compare the continuum model with the Winkler models.

Continuum soil models with conventional Lagrangian meshes for large deformation suffer from mesh distortion and instability with large displacements. Thus, the current study employs the Coupled Eulerian–Lagrangian (CEL) method for a large deformation analysis of the seabed soil that allows for the material to flow through the mesh. Also, the pipe layers were set to be detachable to capture the local instabilities and separations.

Typically, the multi-layer flexible pipe assembly comprises multiple concentric layers. The operational requirements of each pipe determine the dimensions and quantities of these layers. Conversely, these loads directly result from the environment in which the pipeline is deployed. This study concentrates on the section of the tube that bears radial loads, referred to as the "structural nucleus" or "annulus" (see Figure 1b).

The CEL model for lateral pipeline–soil interaction analysis was developed in ABAQUS, and the structural model of the multi-layer flexible pipe was constructed in ANSYS. This configuration allowed for a cost-effective analysis of the interaction between flexible pipe and soil, incorporating both the soil and pipe models' complexities. Further details of the analysis sequence will be presented in Section 2.

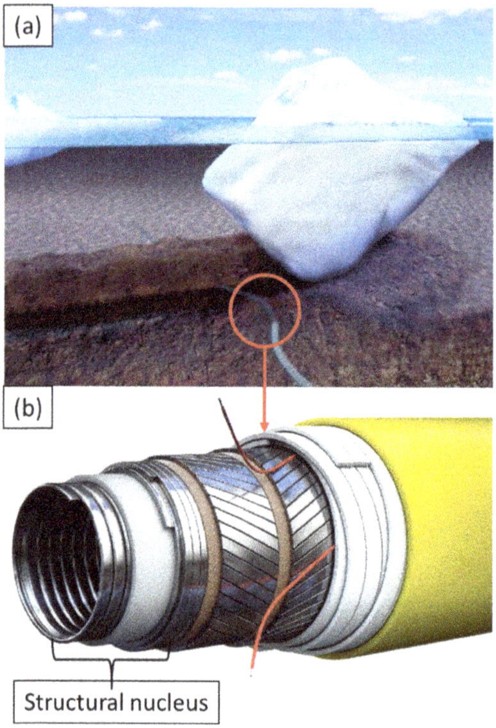

Figure 1. (**a**) Pipeline submitted to ice gouging (adapted from [21]); (**b**) a multi-layered flexible pipe and the layers that comprise the structural nucleus (adapted from [22]).

2. Proposed Methodology and Simulation Flowchart

Overall, the CEL model was developed to simulate the pressure distribution around the rigid pipe section during substantial lateral displacement in clay. Strain-softening effects were also incorporated by programming a user subroutine in ABAQUS. Then, the obtained soil pressure was fed into the flexible pipe model in ANSYS to simulate the structural response of the pipe and the failure analysis.

As per API and DNV regulations [23,24], deformation of the cross-section is not inherently problematic, but it significantly raises the risk of failure due to collapse. Therefore, the suggested approach involves determining the cross-section deformation using soil loads obtained from a CEL model and then employing this deformed cross-section in collapse analyses. Further details of the sequential analysis are illustrated in the flowchart shown in Figure 2, which utilizes four different numerical models:

- Coupled Eulerian–Lagrangian (CEL) pipeline–soil interaction model;
- Transient structural cross-section deformation analysis;
- Wet collapse quasi-static structural analysis;
- Dry collapse quasi-static structural analysis.

The first step is computationally more expensive: the CEL model employs a hybrid mesh and advanced soil formulations to calculate the acting pressure on a pipe section resulting from the pipeline's interaction with the seabed. A simple representation of the flexible pipe is used at this stage as a rigid pipe section, assuming that the cross-section deformation is too small to influence the soil resistance. During this analysis, the strain-softening effects are incorporated into the soil by programming a user subroutine in ABAQUS.

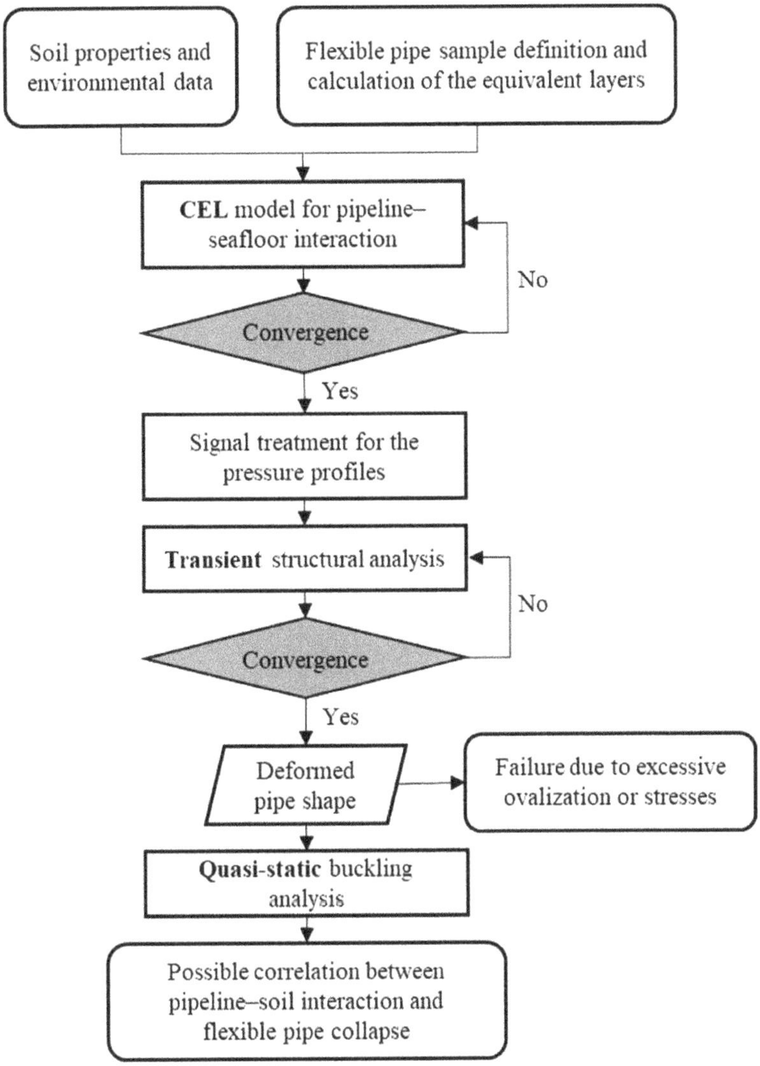

Figure 2. Simulation flow chart.

Once the pressure profiles are obtained from the CEL analysis, the output is fed into the ANSYS as pressure loads on the flexible pipe model. An advanced pipe cross-section analysis is then conducted by ANSYS, allowing for the separation of the pipe layers and by incorporation of elastic–plastic materials and contacts. The equivalent thickness is a consecrated simplification for radial load analysis in flexible pipes [25–27]. The typical transient analysis timeframe is about twenty minutes, considerably reducing the analysis time and making it extremely practical for design purposes. This simplification, associated with the separation between the cross-section analysis and the pipe–soil interaction, allowed for several flexible pipe configurations to be tested with different materials or internal layer geometries. Also, aside from the external diameter, every parameter of the flexible pipe may be altered and quickly simulated.

The transient structural analysis provides the deformed cross-sections during the time frame of the soil interaction process. It is possible to calculate the cross-section ovalization with radial deformations, which can be defined according to two different regulations.

As stated in API 17B [28], Equation (1) can be used to define pipe ovalization. The out-of-roundness of the pipe's cross-section is an essential factor in collapse due to external pressure. The higher the ovalization, the higher the pipe failure risk [27]. Therefore, limits of 0.2% and 3% ovalization were set in the regulation for the unloaded loaded conditions, respectively. This criterion is employed in the crushing analysis during the flexible pipe's launching stage as well.

$$API_{oval} = \frac{D_{max} - D_{min}}{D_{max} + D_{min}} \tag{1}$$

In Equation (1), D_{max} represents the cross-section's largest external diameter, and D_{min} is the smallest. The same regulation also notes that ovalization might be a consequence of the excessive bending of a flowline. The section dedicated to bending tests of tensioned lines specifies that the allowed variations in external diameter (as a consequence of bending) are 3% for the loaded condition and 1% for the unloaded condition. One may notice that the word "ovalization" is not mentioned here; instead, the term direct diameter variation is used. Nevertheless, this will end up causing ovalization.

According to the DNV regulations [24], however, the bending is not the cause of failure itself, but the cross-section deformation caused by it. The regulation defines ovalization, calling it "ovality" (see Equation (2)). It is noticeable that a diameter variation, as per DNV's definition, can result in greater ovalization compared to the API's definition. Consequently, the DNV regulation is more stringent.

$$DNV_{oval} = \frac{D_{max} - D_{min}}{D} \tag{2}$$

In Equation (2), D_{max} and D_{min} remain consistent with API's equation, while D represents the nominal diameter of the cross-section. Regarding failure criteria, the initial ovality for fabrication (unloaded condition) must not exceed 0.5%. The ovalization due to bending should be kept under 3% to prevent any buckling. It is worth mentioning that the DNV regulation mainly applies to steel pipelines; nevertheless, it represents a similar concept to API's equation and it will be used in this work as a comparatively conservative alternative.

Regardless of the definition, the ovalization obtained by the transient analysis may become an input for a collapse analysis. The model employed at this stage was the symmetric ring model, proposed by [29]. The model is quasi-static and uses the Arc-Length (Riks) method for the simulation, enabling the capture of buckling effects such as snap-through and determining the critical loading of the section with more accuracy than linear buckling techniques. This study conducted the analysis until ovalization was obtained, and the results were verified against the regulation limits.

3. Developed Numerical Models

3.1. CEL Model for Lateral Pipeline–Soil Interaction Analysis

A Coupled Eulerian–Lagrangian (CEL) model was created in ABAQUS to investigate the soil resistance mobilized against the pipe section during the large lateral displacement. The numerical model comprises a rigid pipe section embedded in a uniform clay seabed under plane strain conditions. Figure 3 depicts the model configuration and boundary conditions. The coupled Eulerian–Lagrangian model takes advantage of both Eulerian and Lagrangian meshing techniques. The Eulerian mesh is highly efficient in dealing with excessive deformation where there is discontinuity in the material surface across the neighboring elements, while the Lagrangian mesh is good at solving structural problems.

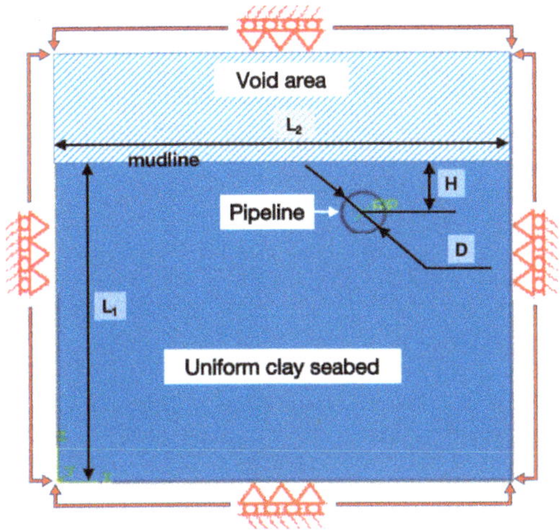

Figure 3. Configuration of the CEL model.

Eight-node linear Eulerian brick elements (EC3D8R) featuring reduced integration and hourglass control were used to represent the soil domain. Although unlikely, hourglass may occur in Eulerian meshes. Besides the hourglass control, the extreme refinement used for the soil mesh compensates for the reduced number of Gauss points in the elements. The contour plots also show no sign of discontinuities or elements with zero-energy modes.

For the material assignment in the soil domain, a discrete field was created to define the space where the soil material should be located, and the Volume Fraction Tool (VFT) was used to assign the material. The mesh size was tested to ensure that the thickness of the propagated shear bands contains at least three layers of elements in order to effectively simulate the failure mechanism, which requires a mesh size of approximately less than 1/30 of the pipe diameter in this research.

A general-purpose linear brick element with reduced integration (C3D8R) was utilized to model the pipeline in the Lagrangian domain, with the pipe section being constrained to a reference point located halfway through of the section to create a rigid body. In that case, the rigid pipeline has a basic carbon steel placeholder material, with 200 GPa of elastic modulus and a Poisson's ratio of 0.3. Due to the multi-point constraints impeding any relative deformation, there was no deformation or stress in the pipe during the CEL phase. The clay seabed was modeled as a Eulerian domain, and the strain-softening effects were incorporated into the seabed model using a modified Tresca model, which is a simple constitutive model for clays in relation to the corresponding models for cohesive soils [30–33], programmed into a VUSDFLD subroutine in ABAQUS (Section 1.2.24 of [34]). More details can be found in [35,36]. Within the plastic region, the undrained shear strength (s), taking into account the strain-softening effect, was determined using the following equation:

$$s = s_u + (s_{u,r} - s_u)\frac{\gamma^p}{\gamma_r^p} \quad (3)$$

In Equation (3), s_u indicates the maximum undrained shear strength, $s_{u,r}$ is the residual undrained shear strength ($s_{u,r} = s_u/S_t$, and S_t represents the sensitivity of soil), γ^p denotes the accumulated plastic shear strain, and γ_r^p represents the γ^p corresponding to the reduction in shear strength from its maximum value to its residual value. The sensitivity of the seabed soil was taken as 1.3, consistent with the research assumptions in [8]. The seabed soil properties in [8] mainly derived from two centrifuge tests by [9] using a mixture of Speswhite Kaolin clay (50% of total weight) and Sil-Co-Silt (50%) and was pre-consolidated to 400 kPa to reach a target undrained shear strength of around 40 kPa.

A void domain above the seabed soil was also included to account for the geometrical variation in the seabed surface and the formation of heave (see Figure 3). To prevent material from escaping the domain, all external surfaces of the Eulerian domain were constrained in their normal directions. The pipeline was extended beyond the front and rear edges of the Eulerian domain to eliminate artificial friction between the soil and the pipeline's end faces. General contact was applied throughout the entire CEL model for the Eulerian domain. However, the pipe–soil interface was specifically defined using Coulomb's friction law, with the interface strength limited to half the soil's undrained shear strength at the pipe's springline, and an overall friction coefficient of 0.5 [8]. Separation is allowed during the contact between the pipe and seabed soil to avoid unrealistic soil resistance being induced during pipe displacement. The refined mesh used in the CEL model allows for a good convergence and captures the shear bands developed in the soil during significant deformations.

The developed CEL model was benchmarked using the solutions proposed by [7] regarding the bearing capacity factors (see Figure 4 [8]). The bearing capacity factors calculated from the CEL model analysis results fall within the range defined by the smooth and rough interface solutions proposed by [7].

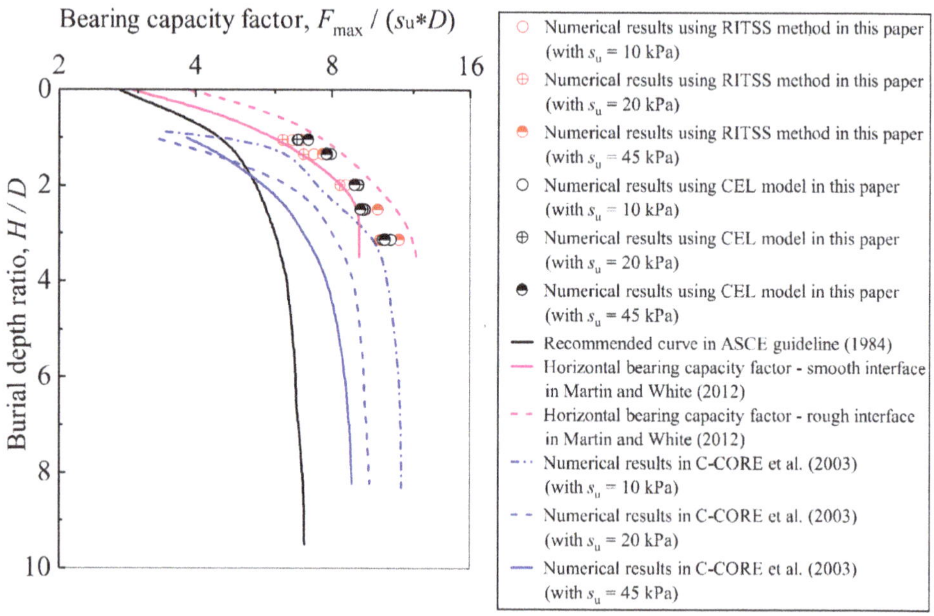

Figure 4. Verification of the developed CEL model on the bearing capacity factor [8,10,37].

ABAQUS/Explicit was used to carry out a two-step analysis. Initially, a geostatic step was conducted to achieve the desired gravity stress level in the soil domain. Subsequently, the pipe section, constrained at the center of the reference point, was laterally displaced at a constant velocity of 0.15 m/s to simulate undrained conditions. This velocity was selected based on the studies by [9,13]. The pipeline's vertical degree of freedom was released during the lateral displacement process.

It is worth mentioning that, due to the bilateral nature of the contact between the pipe and soil, this model works with either the soil or the pipe as the moving component. For this particular study, the pipeline was chosen as the moving part due to the fact that it is easier to constrain using MPCs. However, soil pressure could also increase if an acceleration was applied to the soil while the pipeline was kept in place.

3.2. Structural Model of the Multi-Layer Flexible Pipe

The structural model was constructed in ANSYS, initially proposed and validated by [38], and improved and applied by [29]. The model's premise is to simplify the complex geometries of the interlocked carcass and pressure armor using nonlinear material characteristics and contacts between the layers. Even though flexible pipes are composed of many layers, the model is composed of three concentric arcs, which are, in fact, semi-elliptical shapes (Figure 5). Each arc represents one layer of the structural nucleus, the pressure-resistant part of a flexible pipe. The tensile armor layers and the high-strength tape were deliberately removed since they are designed to withstand tension and protect the armors from local buckling, respectively, without contributing significantly to the radial resistance of the pipe. The model was built using the plane strain model to optimize the simulation times.

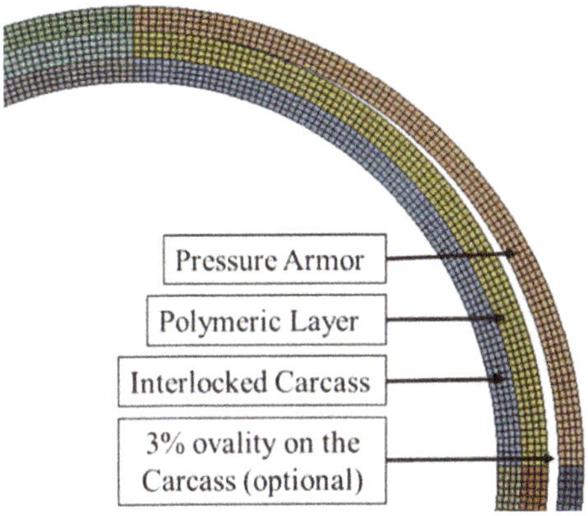

Figure 5. General aspect of the FE model. The surface areas of each concentric ring were broken into circular sectors to facilitate the structured mesh creation, thus appearing in different colors.

The model incorporates material plasticity for the metallic and polymeric layers, frictional contact between the pressure layer and the liner, and bonded contact between the liner and the carcass. The liner was extruded directly on top of the carcass due to the flexible pipe's manufacturing process.

3.2.1. Thickness of Layers

The complex geometry of the flexible pipe's annulus was adapted into a series of equivalent rings in a structural model. To calculate the equivalent thickness, a methodology suggested by [25] was used for both the carcass and pressure armor layers. First, the bending stiffness per unit length of the carcass was calculated by the following equation:

$$ei = \frac{E}{(1-\nu^2)}\frac{2I_{Gmin}}{b} \tag{4}$$

where E denotes the Young's modulus, ν represents the Poisson's ratio, I_{Gmin} represents the minimum inertia moment of the profile cross-section, and b denotes the profile length. The stiffness per unit of length of the equivalent tube (of thickness t_{eq}) is calculated as follows:

$$ei = \frac{E}{(1-\nu^2)}\frac{t_{eq}^3}{12} \tag{5}$$

Notably, the factor of 2 in the numerator of Equation (4) represents the interlocked superposition. This factor could vary based on the profile's superposition from 1 (no superposition) to 2 (total superposition):

$$ei = \frac{E}{(1-\nu^2)} \frac{(1+\psi)I_{Gmin}}{b} \tag{6}$$

in which ψ is the superposition factor, given by:

$$\psi = \frac{sup1 + sup2}{pitch} \tag{7}$$

A generic profile of the superposed parts and the pitch is presented in Figure 6.

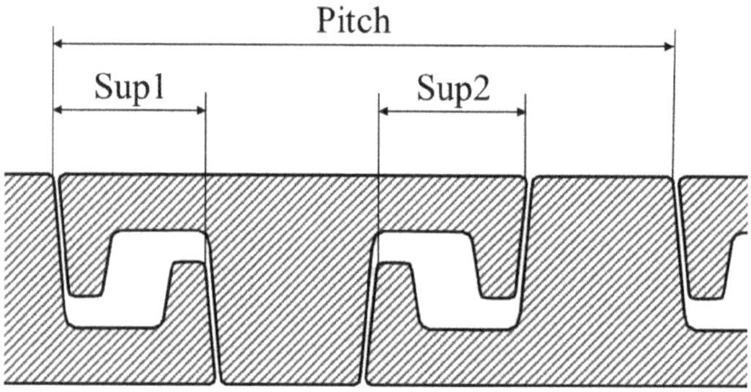

Figure 6. Superposition between pressure armor profiles.

Combining Equations (6) and (7) and rearranging for the thickness, the expression for the equivalent thickness is obtained:

$$t_{eq} = \sqrt[3]{\frac{12(1+\psi)I_{Gmin}}{b}} \tag{8}$$

Equation (8) was employed to calculate the equivalent thickness to replace the interlocked carcass and pressure armor in the structural model.

3.2.2. Boundary Conditions

Different boundary conditions with distinct configurations were considered to facilitate different models, as follows:

- Configuration I: On the symmetry plane, a frictionless support was used, and one node of each layer was horizontally restricted to avoid rigid body motion (see Figure 7);
- Configuration II: Deformable Multi-Point Constraints (MPCs) were employed, connecting the carcass's internal nodes to a master node in the center. This condition allows for the representation of the whole circumference of the flexible pipe section to match the CEL model's geometry.

Configuration I was adopted for both wet and dry collapse quasi-static models, since they use symmetry. The transient model, on the other hand, required a more profound analysis. Figure 8 shows a graphical comparison between the ovalization and stress results (maximum von Mises for the entire model, averaged in the element) for the two versions of the transient model; the only difference between them is the boundary condition.

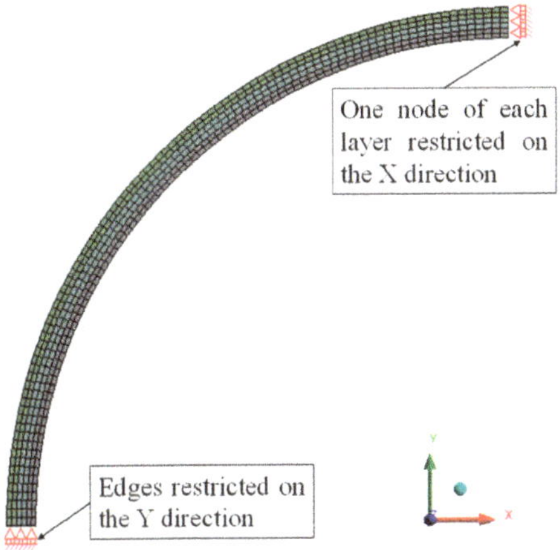

Figure 7. Boundary conditions for Configuration I.

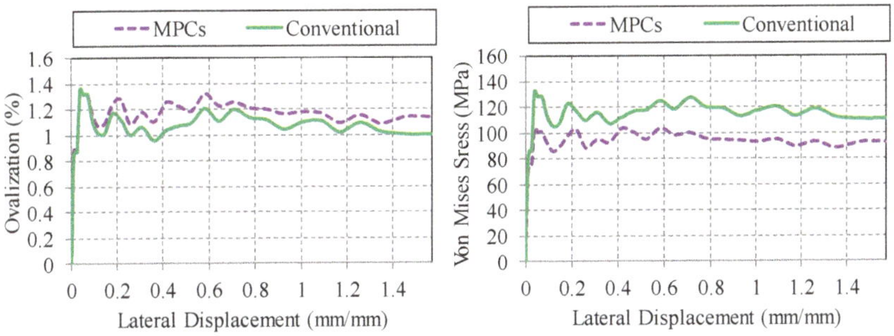

Figure 8. Comparison between the two boundary conditions.

As seen in Figure 8, there is an observable difference in the results shown by both models. The "conventional" curve represents the restrained nodes and edges that correspond to "configuration I", whereas the MPC curve represents "configuration II". With the MPC configuration, the ovalization tends to be larger and the stresses lower than the alternative conventional boundary conditions. This is expected, since the MPC is less restrictive and generates fewer concentrated stress points than nodal restrictions. Another point to be considered is that the CEL model uses a kinematic coupling between the inner edge of the pipe and a central reference node. The MPC reproduces that condition more accurately, with the advantage of being deformable and allowing relative movements between the nodes. For these reasons, configuration II was selected for the transient model.

3.2.3. Material Properties

The multi-layer flexible pipe's cross-section was selected as per the API 17B guidelines [28]. Table 1 shows the calculated thicknesses of various layers and the corresponding material selection. For metallic layers, a typical carbon steel material available in the ANSYS library was used. Beyond the elastic portion of the curve, the material was represented as bilinear with kinematic hardening, where a tangent modulus was also defined, in addition to the Young's modulus. This simple model was extracted from the ANSYS material library, and it is intended to represent a medium carbon steel. The kinematic hardening means that the yield surface is displaced, without increasing in size, accounting for the Bauschinger effect where the properties of material alter as a result of the repeated cycles of loading and unloading. Throughout the transient analysis, a change in loading conditions is expected. A hardening model must therefore be considered. The values used to define the material models are presented in Table 1. Figures 9 and 10 illustrate the stress–strain curves of the carbon steel material and the HDPE, respectively.

Table 1. Material properties of the flexible pipe layers.

Layer	Material	Equivalent Thickness (mm)	Young Modulus (GPa)	Poisson's Ratio	Yield Strength (MPa)	Tangent Modulus (MPa)
Interlocked Carcass	Carbon Steel	6.32	200	0.3	450	1450
Internal Polymeric Layer	HDPE	7	0.571	0.45	20.74	-
Pressure Layer	Carbon Steel	6.52	200	0.3	450	1450

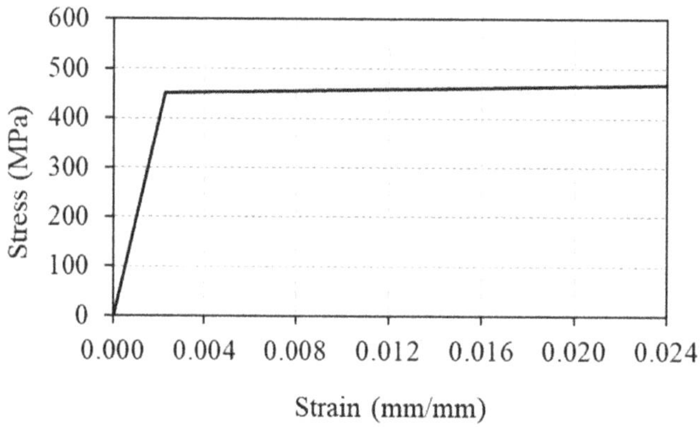

Figure 9. Stress vs. strain curve for the carbon steel.

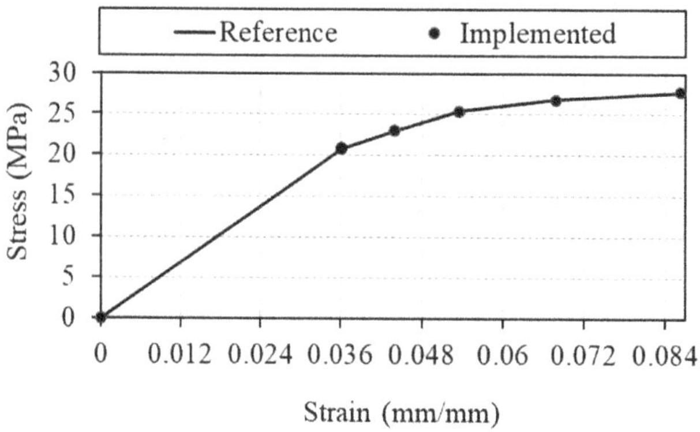

Figure 10. Stress vs. strain curve for the HDPE [26].

3.3. Discussion of the Ovalization Effect on Soil Resistance

The lateral resistance of soil is expected to increase as the ovalization occurs in the cross-section of the pipe. It is intuitive to imagine that an increase in the lateral area due to ovalization under the same lateral pressure would increase the overall force acting on the pipe section. Since a two-way analysis with the CEL model and a deformable cross-section is computationally expensive, a numerical experiment was conducted to verify the effect of a section deformation on the CEL model.

For that purpose, two 2.5″ flexible pipe sections, one circular and another with a 10% API ovalization, were submitted to the same lateral movement while buried in the same soil. Figure 11 shows the difference in the lateral soil resistance exerted on the pipe, where the deformed section increases the lateral soil resistance against the moving section of the pipe.

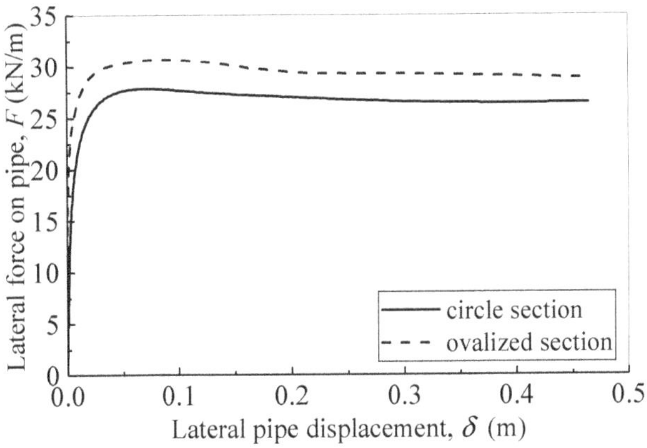

Figure 11. Lateral force on pipe.

Figure 12 shows the bearing capacity, which is the lateral force normalized by the soil stiffness (S_u) and the pipe diameter. One may notice that the effect is significantly reduced when the force results are normalized.

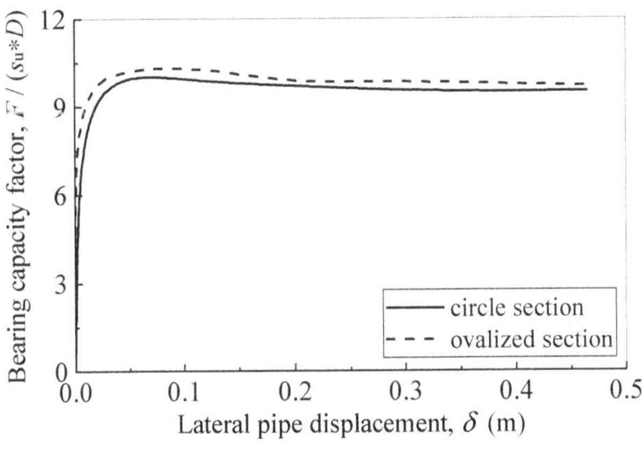

Figure 12. Bearing capacity factor.

While the ovalization certainly influences the p–y curves, in quantitative terms, a relatively large ovalization would be necessary to cause a significant force increase. With an ovalization factor of around 10%, an increase of 11% was observed for the maximum force in the load–displacement curve in comparison with the original circle section, as shown in Figure 11. The influence was significantly decreased by normalizing the resistance with soil properties and the diameter, due to the fact that the ovalization alters the diameter, increasing it in the direction perpendicular to the lateral movement. For both cases, the largest difference occurs near a lateral pipe displacement of 0.13 m; it is close to 8.3% for

Figure 11, and for Figure 12 the difference is about 5.3%. Further analyses are required for future studies to accurately simulate the ovalization effect on lateral soil resistance. However, this brief study showed that the assumption on which the current methodology is based is reasonable.

4. Parametric Study

As an immediate application of the methodology, a parametric analysis was performed to examine the effects of different parameters, namely the flexible pipe's internal diameter, seabed soil strength, and burial depth. As shown in Table 2, these parameters were analyzed individually and compared with the base case (CS-01) using the cross-section ovalization and stress on the metallic layers as output parameters.

Table 2. Different cases for the parametric study.

Case Name	Internal Diameter (inches)	Seabed Soil Strength (kPa)	Burial Depth Ratio (H/D)
CS-01	22	50	1.92
CS-02	22	25	1.92
CS-03	22	5	1.92
CS-04	6	50	1.92
CS-05	16	50	1.92
CS-06	22	50	2.92
CS-07	22	50	3.92

Since the two models are independent, only the external diameter of the flexible pipe must remain fixed to guarantee a geometrical match between the models. Every parameter of the structural model can be easily changed. If the external diameter of the flexible must be changed, then multiple CEL analyses will be required.

The CEL analyses were conducted on a workstation with 24 Intel Xeon Cores, 128 GB of RAM, and a Solid State Drive (SSD), taking around six days to complete. The structural FE analyses, on the other hand, were conducted on a notebook with an i7 Intel processor, 8GB of RAM, and an SSD hard drive and took around 15 min for a 32 s transient analysis.

5. Results of CEL Analysis of the Lateral Pipe Displacement

In the primary case, CS-01, the soil stiffness of 50 kPa could represent overconsolidated clay seabed soil. With the significant lateral movement of the pipe section, the shear bands gradually form in front of the pipe and at the rear side to the seabed soil surface, causing the soil resistance on the pipe section to decrease (see Figure 13).

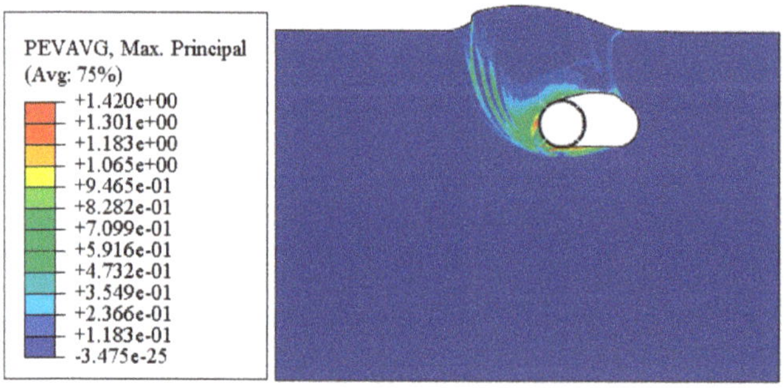

Figure 13. Plastic strain in soil (at $\delta/D = 1.20$). PEVAVG is the average plastic strain in the element, computed as a volume fraction.

Figure 14 illustrates the extracted soil pressure profiles on the pipe's cross-section. Each curve represents a snapshot in time during the lateral displacement, where δ is the displacement normalized by the diameter D. The horizontal axis shows the point on the pipe's surface on which the pressure is acting, where the range from 0 to 0.5 corresponds to the leading edge of the pipeline, and 0.5 to 1 is the trailing edge.

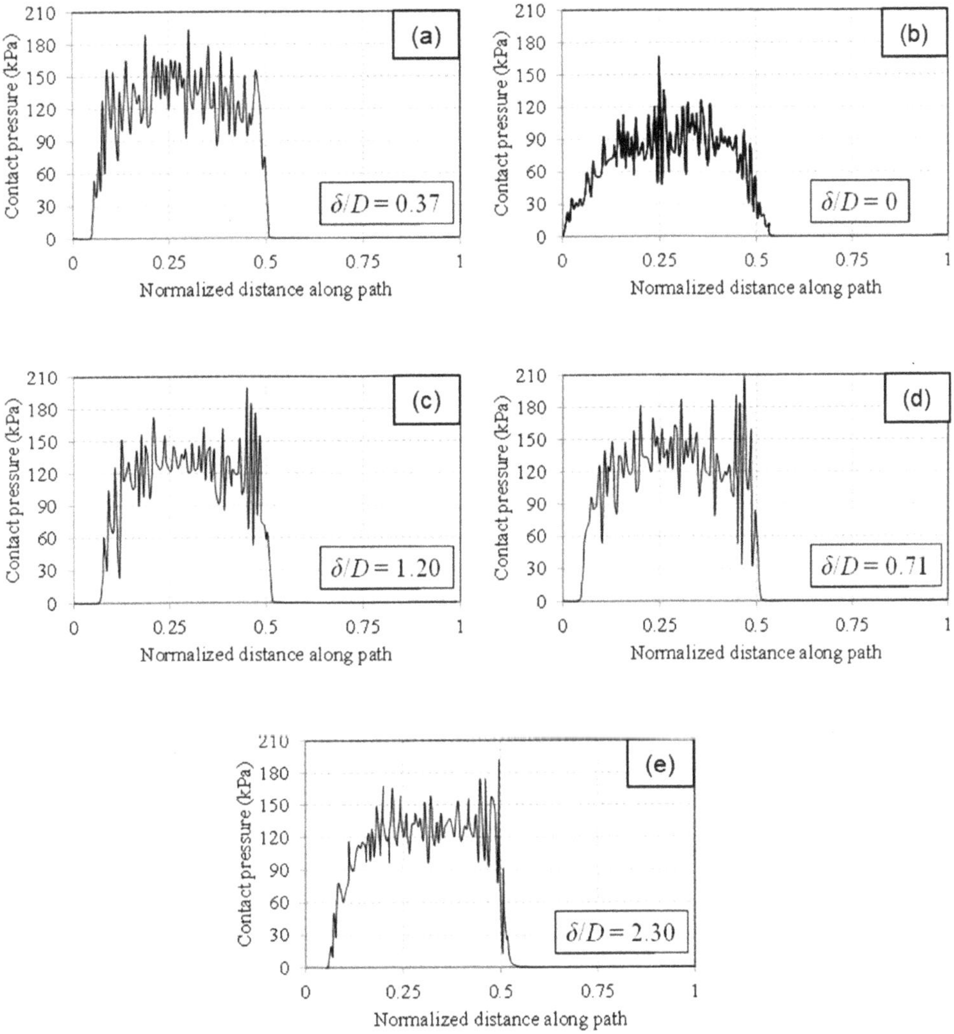

Figure 14. Contact pressure profiles introduced into FE analysis (starting from the top of the pipe and moving in a counterclockwise direction). Figure (**a–e**) show different burial depths normalized by the diameter of the pipeline (δ/D).

As the pipe section is laterally displaced, the soil pressure on the pipe increases in comparison to the initial starting stage (see Figure 14a,b). As displacement continued, a cavity gradually developed in the rear side of the pipe, and strain-softening enhanced the formation of shear bands in the soil, extending from the front bottom of the pipe to the soil surface. This, in turn, reduces contact pressure on the pipe front upon passing the maximum resistance (see Figure 14b,c). At a later stage, the soil pressure remains at a relatively constant magnitude (see Figure 14d,e).

The soil's undrained shear strength (s_u = 50 kPa in CS-01, 25 kPa in CS-02, and 5 kPa in CS-03), the pipe section diameter (D = 22 inches in CS-01, 6 inches in CS-04, 7 inches in CS-05), and the burial depth ratio of the pipe (H/D = 1.92 in CS-01, 2.92 in CS-06, 3.92 in CS-07) can substantially impact the formation of shear bands in the soil during deformation. Figure 15 presents a comparison of the shear bands formed in the soil across different case studies.

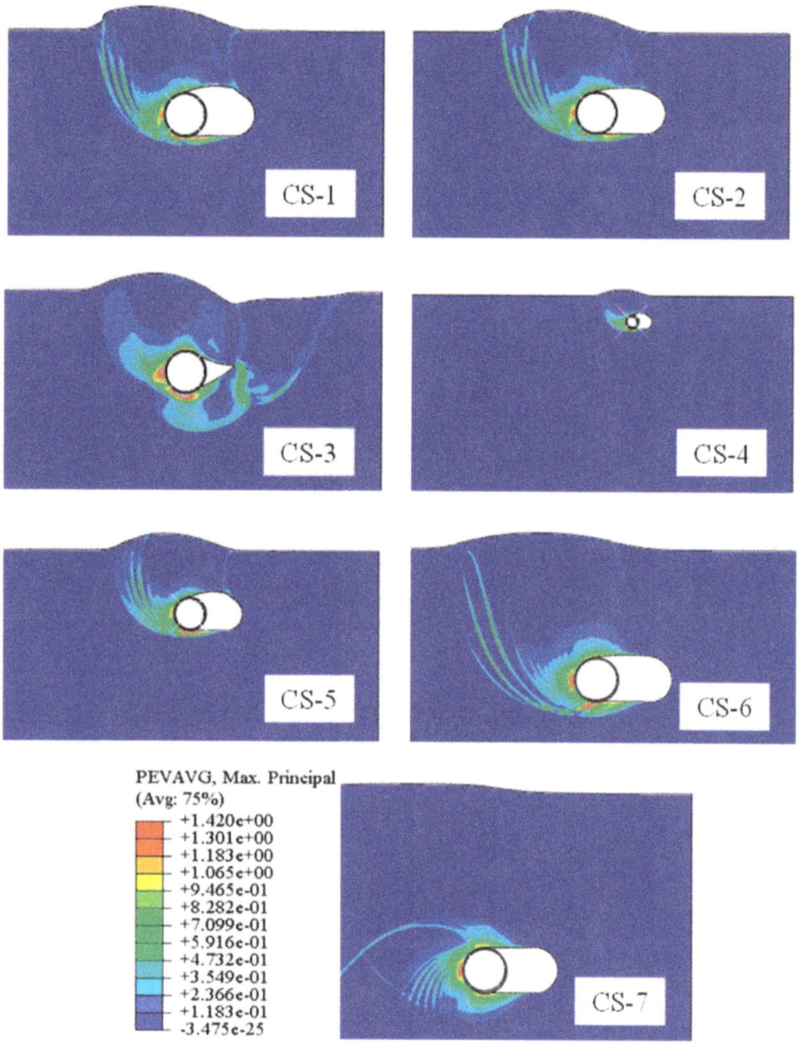

Figure 15. Plastic strain in study cases (at δ/D = 1.20).

Comparing Figure 15a,b, the general distribution of shear bands is similar, but a different number of shear bands could be observed in front of the pipe (more shear bands in CS-02). The shear bands in CS-03 differed a lot from those in CS-01 and CS-02, with extended spiral shear bands initiated from the rear bottom of the pipe and reaching the soil surface at the rear side of the pipe. Comparing Figure 15a,d,e, with different pipe diameters, the distribution pattern of the plastic strain is quite similar. Comparing Figure 15a,f,g, which have an increasing burial depth ratio, significant differences can be observed in both the number and distribution of shear bands.

The pipe section's diameter, the ratio of burial depth, and the soil's undrained shear strength can also significantly influence the stress profiles around the pipe section. The comparison of stress profiles around the pipe among different study cases is shown in Figure 16.

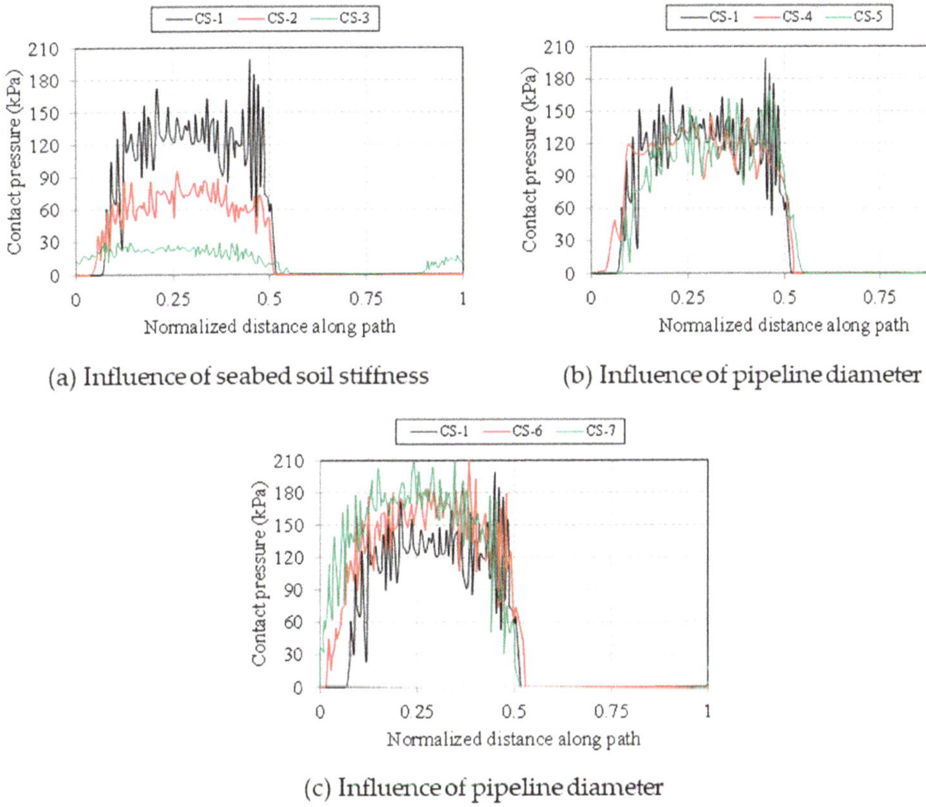

Figure 16. Contact pressure profiles introduced into FE analysis in different study cases (at $\delta/D = 1.20$, starting from the top of the pipe and moving in a counter-clockwise direction).

As indicated in Figure 16a, with a decreased seabed soil stiffness, the magnitude of soil pressure on the front side of the pipe significantly decreases. In the softest seabed (CS-3), the soil surrounded the pipe and a positive soil pressure was retained over a larger area (i.e., regarding the normalized path ratio) on the pipe. The influence of pipe diameter on soil pressure shows scattered results in Figure 16b. Significant increases could be observed in soil pressure magnitude on the pipe with an increase in the burial depth ratio (see Figure 16c).

6. Structural Analysis Results of the Pipe's Cross-Section

6.1. Dynamic Effects and Time Frame

The soil pressures obtained in the previous sections were applied to the structural transient model of the pipeline cross-section. Each analysis was broken down into several time steps and further split into substeps for load application. At the end of each substep, a specific load increment was added and balanced by the FE program's solving routine. Each simulation ran from 0 to 60 s, with frames captured at intervals of one-tenth of a second from 0 to 1 s, one frame per second from 1 to 40 s, and every 2 s in the last 20 s.

Since each change in the load condition required multiple substeps to converge, the results were available for hundreds of instants of time. However, the data were filtered to present the results in this article, providing smoother curves. Figure 17 demonstrates this process, with the raw data shown as a thin red curve and the filtered results depicted as a dashed black curve. Additionally, instead of time, the horizontal axis displays the lateral displacement imposed on the flexible pipe, normalized by the pipe diameter.

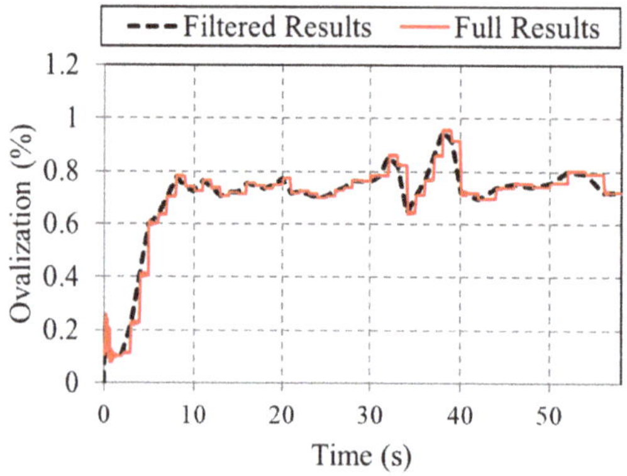

Figure 17. Results of the processing of the numerical results for the case CS-01.

Due to the use of normalized lateral displacement, the first stage of the simulation (from zero to one second) will be compacted in the subsequent results. This stage includes the geostatic step and the beginning of the pipe's lateral displacement, which is a relatively unstable process. These initial stages are illustrated in Figure 17. Since the goal of this section is to compare the results for each case, and every case starts similarly, no significant information will be lost. The upcoming sections will present the outcomes from the parameter variations listed in Table 3.

Table 3. Collapse analysis results.

Soil Strength (kPa)	API Ovalization	DNV Ovalization	Wet Collapse Pressure (MPa)	Dry Collapse Pressure (MPa)
5	0.0539	0.1079	1.2175	3.6800
25	0.4749	0.9497	1.1669	2.2384
50	0.9538	1.9076	1.1191	1.4455
Diameter (inches)	**API Ovalization**	**DNV Ovalization**	**Wet Collapse Pressure (MPa)**	**Dry Collapse Pressure (MPa)**
6	0.0226	0.0452	23.0325	39.4190
16	0.3709	0.7419	2.8420	6.3686
22	0.9538	1.9076	1.1191	1.4455
Burial Depth Ratio (h/D)	**API Ovalization**	**DNV Ovalization**	**Wet Collapse Pressure (MPa)**	**Dry Collapse Pressure (MPa)**
1.92	0.9538	1.9076	1.1191	1.4455
2.92	0.9536	1.9069	1.1194	1.3891
3.92	0.8537	1.7070	1.1301	1.4716

6.2. Dynamic Effects and Time Frame

The first parametric analysis involved cases CS-01, CS-02, and CS-03. The soil stiffness of the seafloor increased from 5 kPa, representing pre-excavated remolded soil as a backfilling material used in a sufficiently wide trench, to 50 kPa, representing over-consolidated clay. To simplify the problem in this research, it was assumed that the flexible pipe is buried in uniform soil. As a suggestion for future studies, it would be interesting to incorporate the trenching and backfilling effects along with the pipeline–backfill–trench wall interaction mechanisms.

Three different response parameters, namely ovalization, stress in the metallic layers, and reaction force on the boundary condition area were used as the basis for comparing the results. Figure 18 compares the results during the entire simulation time frame. It is important to note that the results were normalized for clarity and potential generalization. Instead of displaying the time, the horizontal axis represents the lateral displacement of the cross-section, divided by the flexible pipe's external diameter, which is the normalized pipeline displacement.

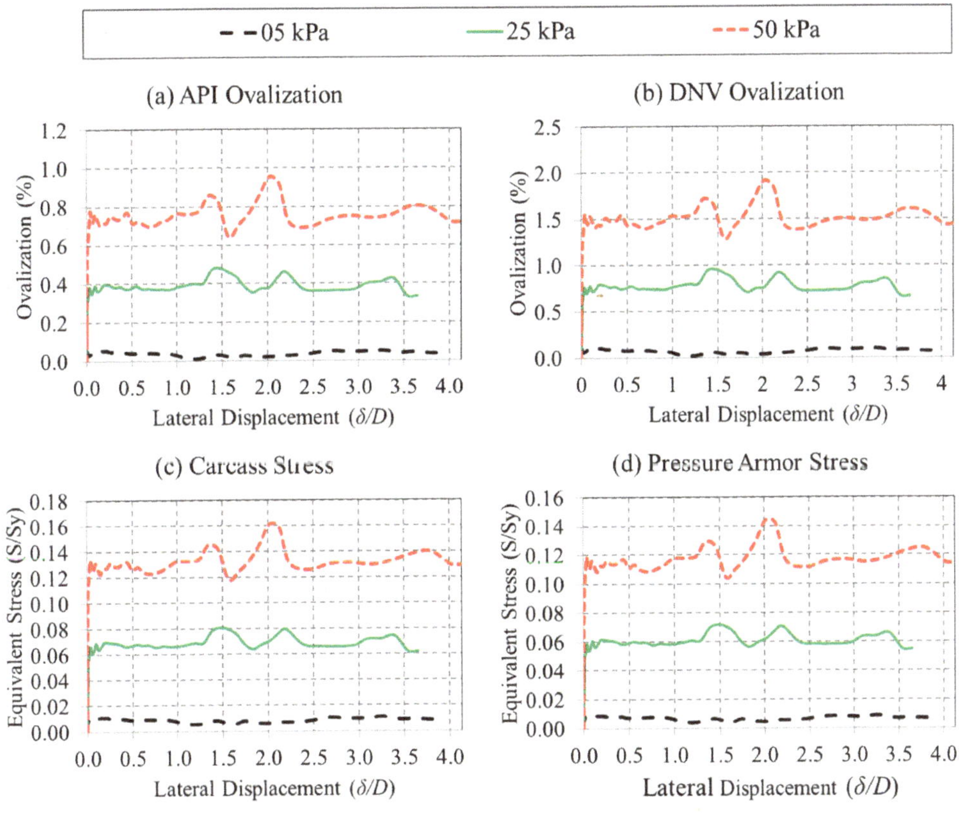

Figure 18. Comparison of soil stiffness results over time.

The type of result variable determines the scale of the vertical axis. The ovalization (API and DNV) discussed at the beginning of this article is a normalized parameter shown as a percentage. At any specific moment, the maximum amount of von Mises equivalent stress is divided by the metallic material's yield stress, which is 450 MPa. Finally, the reaction force is normalized using the highest value observed across all cases, and is shown separately in Figure 19.

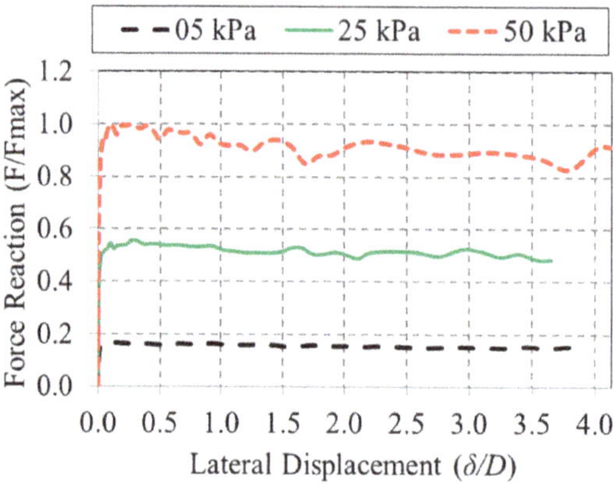

Figure 19. Force reaction results for the soil stiffness cases over time.

The initial noticeable aspect of the analysis of the ovalization graphs is that all cases complied with the regulation limits. The API (Figure 18a) and DNV (Figure 18b) ovalization limits were 3% for the loaded normal service condition. The maximum value for API was 0.95 and that of DNV was 1.91 for the CS-01 (50 kPa) case. Another key observation is that as soil stiffness increases, overall ovalization also increases. The spacing between the curves becomes more pronounced, and the oscillations are more significant. The amplitude of these oscillations is higher with elevated soil stiffness, which is almost unnoticeable for the 5 kPa case.

The equivalent stress curves in the metallic layers exhibit a similar trend, with values remaining quite low, reaching up to 0.16, i.e., 16% of the yielding limit. This indicates the elastic and fully recoverable nature of all strains. As the soil stiffness increases, the stresses also increase, which is an expected outcome since a higher soil stiffness leads to greater resistance to displacement, resulting in increased loads on the pressure-resistant layers. Figure 18c displays the stresses in the interlocked carcass, while Figure 18d illustrates stresses in the pressure armor, which are comparatively lower than those in the carcass.

The graphs in Figure 19 confirm the trends observed in the previous result parameters. The reaction forces are high with a higher soil stiffness. A particular difference between this and the previous curves is that the oscillation is small, with the 5 kPa curve seeming almost constant.

All cases start with zero ovalization or stress, which is visible at the beginning of the curves. This initial slope is more perceivable in the graph of Figure 17, while in Figure 18a,b, the slope is compacted on the left side of the plot area (as a direct result of the normalization process). As the simulation progresses, the values tend to stabilize and oscillate around a straight line. This stabilization can be attributed to the soil accommodations during the lateral displacement. One of these oscillations calls for attention, the "camel hump" in the middle of the displacement, with two peaks, one at $\delta/D = 1.462$ and the other at $\delta/D = 2.297$.

To determine the cause of that phenomenon, the CEL model results for the soil were revisited, and those peaks were likely to be caused by a combination of shear bands and soil berm formations. Looking at Figure 20, Point 1 marks $\delta/D = 1.462$, and it is possible to notice an increase in the height of the soil berm compared with $\delta/D = 1.358$. This is even more noticeable for the second peak at $\delta/D = 1.985$. The soil berm looks higher than that at $\delta/D = 1.771$ (see zone "A" in Figure 20). On the other hand, there was the rapid formation and propagation of new shear bands (see zone "B" in Figure 19). New shear bands can be seen at $\delta/D = 1.567$ and $\delta/D = 2.297$ for the first and second peaks, respectively. The propagation of new shear bands can be associated with the sudden drop in soil stress in the region immediately in front of the pipe.

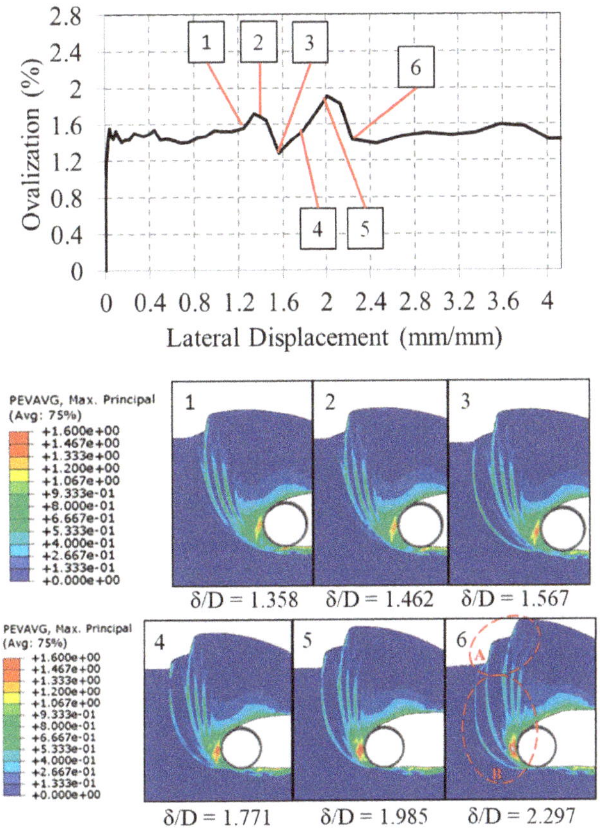

Figure 20. Soil berm and shear band formation.

The previously mentioned combined effects of the increase in the height of the soil berm and formation of new shear bands is likely to be influenced by the allowance for the separation in the pipe–soil interface (since the cavity formed behind the pipe can influence the deformation of soil). The cavity formation is a feature of the CEL model, and in the real scenario, the cavity space would be occupied by collapsing soil material. As observed in experimental tests [9], the cracks tend to show on the soil surface at the front top of the pipe, and the soil blocks would naturally fall to fill the cavity, which seems exceptionally challenging to simulate with finite element analysis. Furthermore, the cavity formed in the model could be due to the native soil's high strength and the inaccuracy introduced by the fraction volume computation regime in the CEL model (fraction of material filled in by the Eulerian element).

6.3. Effect of the Flexible Pipe Diameter

The second parametric analysis verified the influence of the flexible pipe diameter. The flexible pipes were manufactured onshore and stored in large reels. Their main characteristic is that they can withstand high axial and pressure loads with a relatively small bending stiffness. Those features directly result from a complex, multilayered structure built under strict tolerances. Consequently, flexible pipes do not have extremely large diameters like rigid pipelines. The commercial brochure for a major manufacturer [39] states that their current range for flexible pipes is between 2 and 22″. Therefore, the diameters selected for this study were 6, 16, and 22″. Figure 21 shows a comparison of the results.

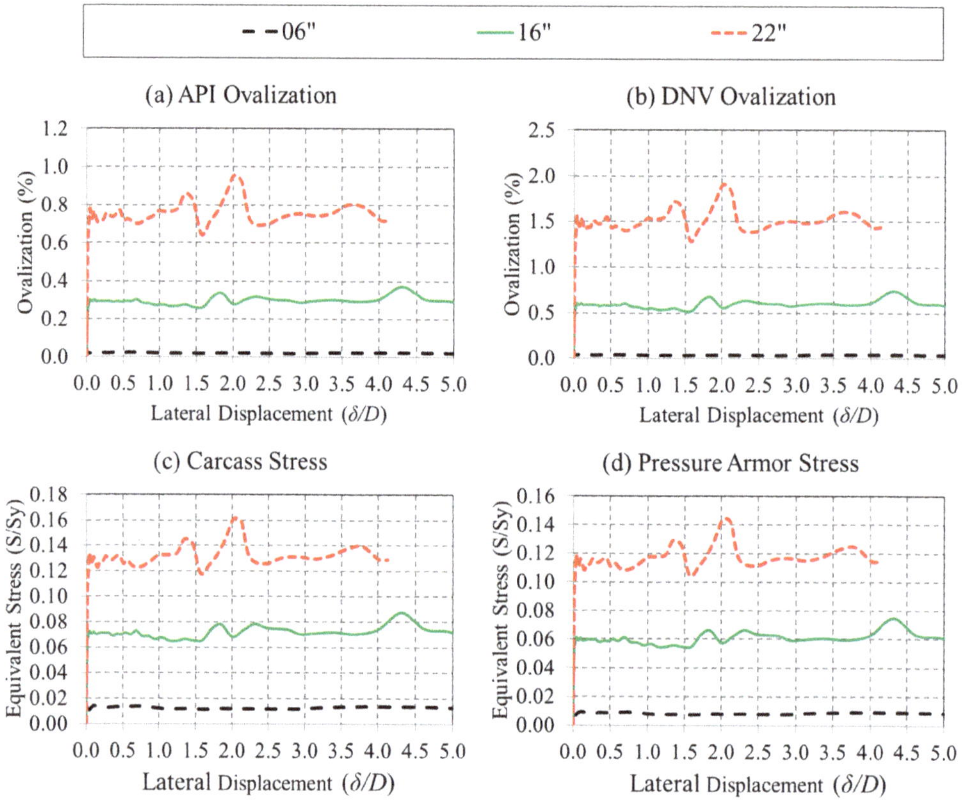

Figure 21. Comparison of flexible pipe internal diameter results over time.

From the curves shown in Figure 21, the first noticeable aspect is that as the internal diameter increases, the ovalization and equivalent stress increase as well. A larger ovalization will lead to larger strains, which will cause higher stresses. The ovalization (both API and DNV) is smaller than the regulation limit of 3% (Figure 21a,b) and a significant difference is observed between the results for medium and large diameters in comparison with small-diameter cases. For the 6″ flexible pipe, the ovalization is almost nonexistent, with a maximum value of 0.0226% for API and 0.1079 for DNV. Similarly to larger diameters, the 6″ flexible pipe ovalization also oscillates with time. Still, it becomes imperceptible when put on the same scale as the larger-diameter results. For the 22″ flexible pipe, the "camel hump" observed in the previous analysis is recurrent. For both ovalization and stresses, it shows up with its distinctive shape, as shown in Figure 20, and the adjacent paragraphs provide a more detailed discussion.

Regarding the stresses in the metallic layers (Figure 21c,d), the soil displacement does not cause any significant stresses on the interlocked carcass or the pressure armor. The maximum values were approximately 16 and 14% of the yielding stress for the carcass and the pressure armor, respectively. Higher stresses on the carcass are typical for rough bore reinforced flexible pipes, with the carcass and the pressure armor sharing the radial loads and generating stresses. The slenderness of the carcass, associated with its complex geometry, is the cause of the higher stresses in comparison with the pressure armor [27].

The stresses in the 6″ flexible pipe were almost negligible, with maximums between 1 and 2% of the yielding stress.

The final comparison between the reaction forces in the different cases is shown in Figure 22. With an increase in the internal diameter, a larger pipe surface is exposed to the soil pressure, resulting in a higher lateral reaction force. There is also a vertical reaction force whose value is relatively small compared to the lateral one. The results are given in the form of a total equivalent force, considering all directions.

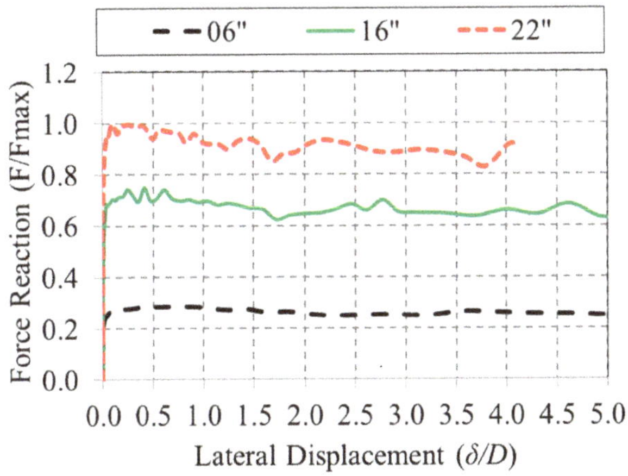

Figure 22. Force reaction results for the internal diameter cases over time.

From these results, it is reasonable to assume that the diameter plays a fundamental role in flexible pipe damage due to soil displacements. The smaller diameter of the pipe suffered from a small ovalization, as well as almost negligible stresses. Although small, these stresses might be relevant in the long term when considering repetitive loads. During its operational cycle, the flexible pipe is submitted to various cyclic loads, namely internal pressure variations, slugging, and vessel movements. These loads might induce lateral displacement on the buried sections, leading to these small stresses, which may eventually contribute to fatigue damage.

6.4. *Effect of the Burial Depth*

The third parametric analysis involved a verification of the burial depth's impact on the flexible pipe's ovalization, reaction forces, and stresses. Burial depth is crucial in pipeline trenching as it helps mitigate issues such as upheaval buckling and lateral movement, and protects the pipeline against impacts from sea anchors, icebergs, and similar hazards. Since a deeper burial increases the installation costs, the objective is to optimize the burial depth, ensuring the flexible pipe is buried as shallowly as possible while still benefiting from the trench's protective features.

As shown in Figure 23, the trench depth is presented as a factor of pipe diameter, with three values selected as a basis for comparison: 1.92D, 2.92D, and 3.92D. There are some significant movements between 1.5 and 3 δ/D, with the "camel hump" oscillation (from the CS-01 case) and a single lump or "dromedary hump" on the CS-06 case. As explained in previous sections, this is due to the soil berm formation and the new shear bands propagation (Figure 20).

Unlike the previous parametric analyses, there is no significant spacing between the curves, and they all seem to oscillate around the same straight line. In terms of the maximums, the 1.92D and 2.92D cases practically share the same value of 0.95% API and 1.91% DNV ovalization, whereas the 3.92D case shows a maximum of 0.85% API and 1.7% DNV ovalization (Figure 23a,b). In terms of stresses, the same pattern can be observed. The

1.92D and 2.92D cases have a maximum of 0.16% for the interlocked carcass yield stress and 0.14% for the pressure armor yield stress, whereas the 3.92D case showed 0.15% and 0.13% for the carcass and pressure armor, respectively (Figure 23c,d). So far, these are the most significant values for stress and ovalization observed in the entire study, but they are still under the allowable limits.

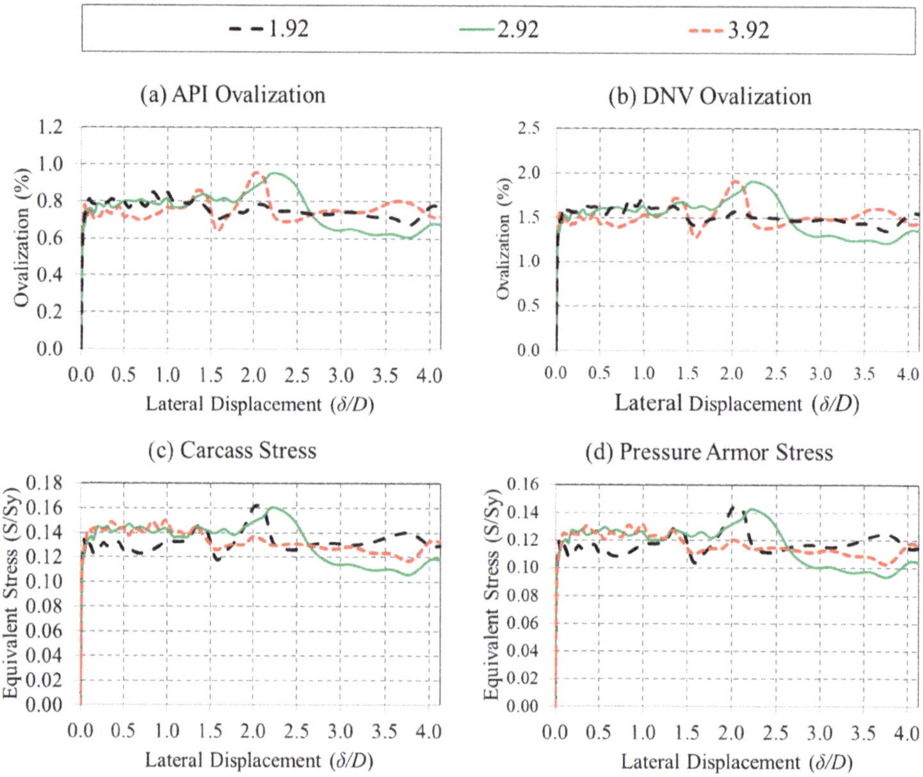

Figure 23. Comparison of burial depth results over time.

The curves in Figure 23 show similar responses during certain intervals of time, allowing for a pattern to be identified and broken into four stages of lateral flexible pipe responses to soil interaction:

- Stage 1 (between 0 and 0.1 δ/D)—Mount from zero, when the initial resistance of the soil is broken and the flexible pipe begins the lateral displacement.
- Stage 2 (between 0.1 and 1.3 δ/D)—The fist soil berms start forming and the sear bands remain the same in number but increase in value, causing minor oscillations.
- Stage 3 (between 1.3 and 2.7 δ/D)—The soil berms reach a maximum height and new shear bands start to propagate, which is seen as the cause of the "camel hump" and "dromedary hump" oscillations.
- Stage 4 (beyond 2.7 δ/D)—Response stabilization, with smoother oscillations.

Provided that there is a long simulation time, it is possible to speculate that stages 2 to 4 will repeat until the movement ceases, simply because the formed soil berms will be left downwind of the moving pipeline and new ones will be formed upwind.

Figure 24 illustrates the curves of the total force reaction for the three cases, with an increasing force reaction for deeper burial depths. The response is also considerably smoother, not showing the oscillations reported in the ovalization and stress graphs, with a slight tendency to decrease.

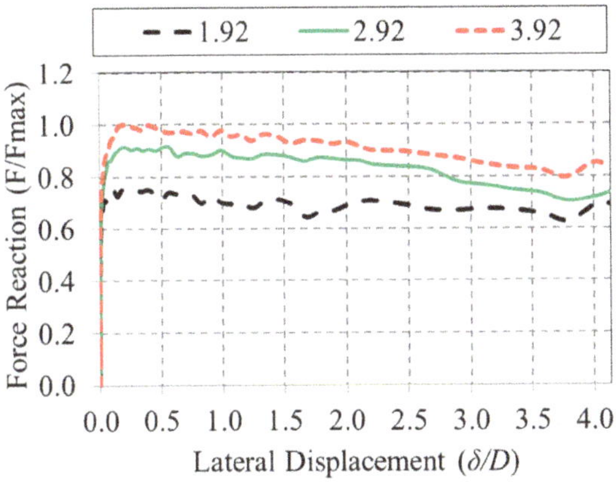

Figure 24. Force reaction results for the burial depth cases over time.

Regarding the cross-section flexible pipe deformation and stresses (not expenditure-wise), the burial depth does not play a significant role. The only variable that shows a visible change with burial depth is the reaction force; however, this does not appear to result in a notable difference in stress levels within the interlocked carcass and pressure armor.

6.5. Extreme Values Comparison

In terms of the design, the extreme values, including the maximum ovalization, force reaction, and stresses for each case, are particularly important. As shown in Figure 25, each row of plots corresponds to one of the three parametric analyses performed in this study, as distinguished by the labels on the horizontal axes. It should be noted that the curves are not fitted to the results; they merely connect the data points.

With the change in the soil strength between 5 kPa and 50 kPa, the API ovalization maximums increased from 0.05 to around 0.95%, showing a 17 times' increase in magnitude. Due to DNV's more conservative approach, the ovalization increased significantly from 0.1 to 1.9% (19 times larger). The stresses also exhibit a tendency to rise with increasing soil stiffness. In the interlocked carcass, the stresses are more significant and increase from 1.14 to nearly 16% of the yield stress, representing a 14 times' increase. At the same time, the pressure armor showed a similar response, with an overall increase of 14.7 times. Finally, the maximum value of the total reaction force increased sixfold with the rise in soil stiffness.

The flexible pipe internal diameter is also shown to significantly impact the result parameters. From the second row of graphs in Figure 25, the API ovalization goes from 0.02 to 0.95% and DNV goes from 0.04 to 1.9% with a diameter increase from 6 to 22 inches. This represents an increase of 42 times for both API and DNV ovalization. The stresses showed an increase of 11.5 times for the interlocked carcass and an increase of nearly 15 times for the pressure armor. Despite the relatively large difference, none of the values surpassed the yield limit of 450 MPa or the maximum allowed ovalization of 3%. The force reaction value increases by about 3.5 times, with a diameter increase of 3.7 times.

In the previous section, alterations in the burial depth did not appear to significantly affect the selected parameters of the results for this analysis. At first glance, a comparison of the extreme values might suggest otherwise, as the curves in the third row of Figure 25 indicate some variation with increased burial depth. However, this impression is misleading and is the result of the graphs' scale. The ovalization parameter shows a 12% overall decrease (API and DNV), which is quite small compared to the effect of the internal diameter, for instance. The stresses show a decrease of 7.2 and 8.7% for the interlocked carcass and pressure armor, respectively, which is also small compared to the 11.5 and 15 times increase observed in the internal diameter study. The force reaction is the parameter that altered the most with increased burial depth. With a depth increase of almost two times, the maximum total reaction force increased by 32%. This is the only parameter that shows some alteration compared to the ovalization and the stresses. However, the scale of the changes is too small for any conclusion regarding the effect of the burial depth on the first two parameters.

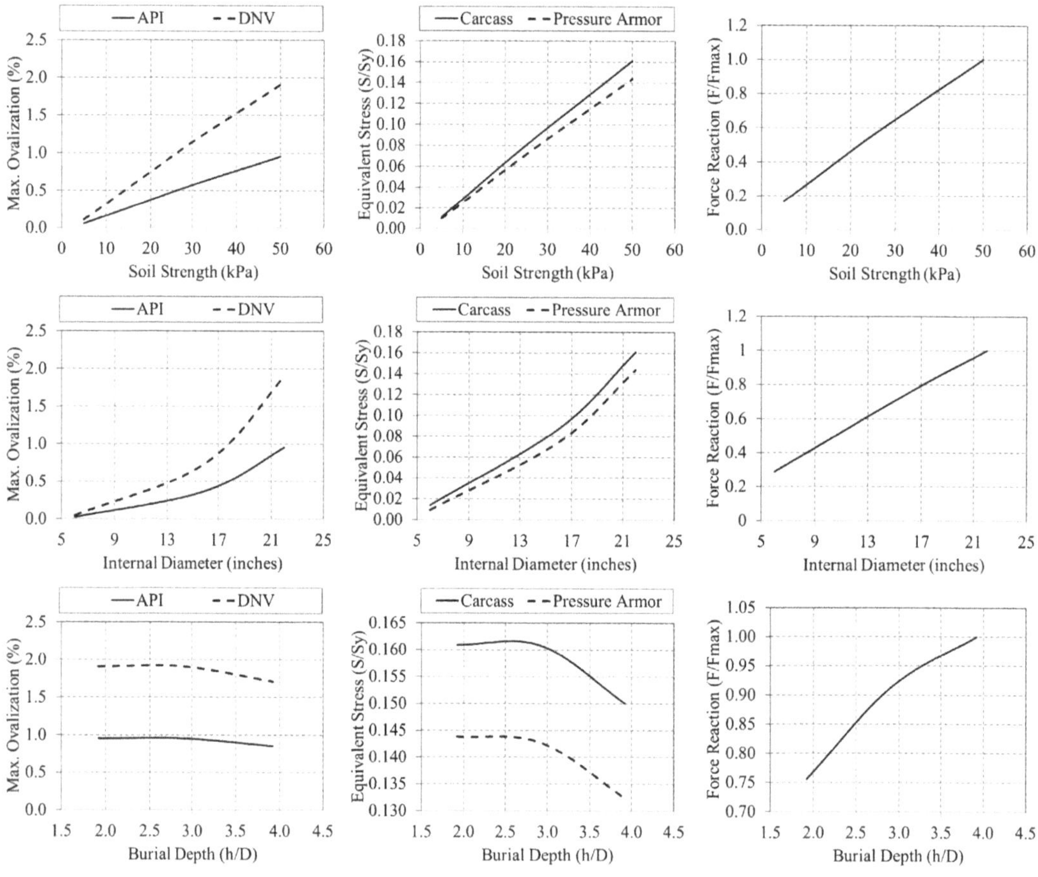

Figure 25. Extreme value comparisons for all the parametric analyses.

6.6. Collapse Analysis

The ovalization results obtained from the previous steps may be used, with the structural model converted to configuration 01, for a collapse analysis (Figure 7). Depending on the type of collapse, the pressure load is exerted either on the surface of the pressure armor or the internal liner. Since the Newton–Raphson method is limited by the requirement of a monotonically increasing force application, the solver is manually switched to the Arc-length method. Despite its being more unstable, this solver allows for negative load increments, which usually appear during the instability analyses. An example of the final results is shown in Figure 26.

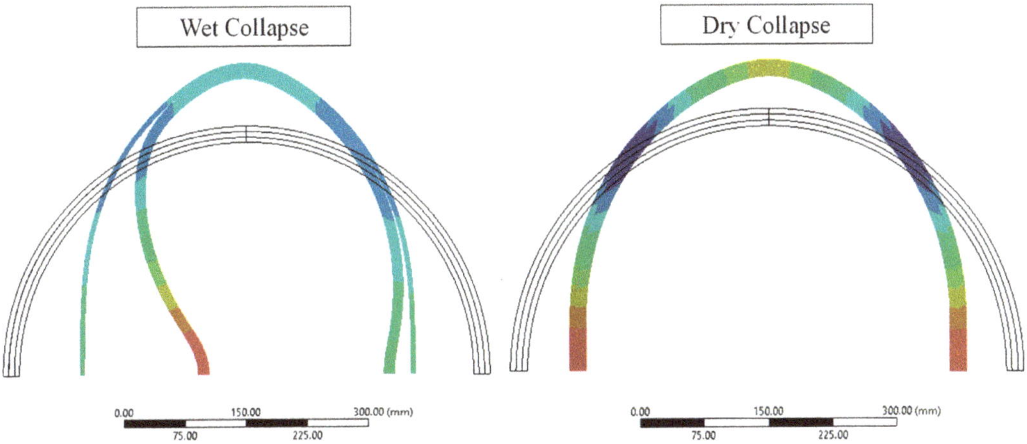

Figure 26. Wet and dry collapse of the cross-section for case CS-03. Red areas show the largest displacements.

Besides the visualization of the flexible pipe's collapsed shape, the primary objective of the analysis is to investigate the critical load, since this marks the moment of structural failure. In the finite element analysis, this event is related to a sudden decrease in the stiffness of the structure. The critical load is the moment at which a stable equilibrium state becomes unstable. A perturbation or an initial imperfection of any kind triggers a large, sudden displacement of the structure, usually catastrophic. Table 4 shows the critical load results, which is also called "collapse pressure" in this article.

Table 4. Collapse analysis results.

Soil Strength (kPa)	API Ovalization	DNV Ovalization	Wet Collapse Pressure (MPa)	Dry Collapse Pressure (MPa)
5	0.0539	0.1079	1.2175	3.6800
25	0.4749	0.9497	1.1669	2.2384
50	0.9538	1.9076	1.1191	1.4455
Diameter (inches)	**API Ovalization**	**DNV Ovalization**	**Wet Collapse Pressure (MPa)**	**Dry Collapse Pressure (MPa)**
6	0.0226	0.0452	23.0325	39.4190
16	0.3709	0.7419	2.8420	6.3686
22	0.9538	1.9076	1.1191	1.4455
Burial Depth Ratio (h/D)	**API Ovalization**	**DNV Ovalization**	**Wet Collapse Pressure (MPa)**	**Dry Collapse Pressure (MPa)**
1.92	0.9538	1.9076	1.1191	1.4455
2.92	0.9536	1.9069	1.1194	1.3891
3.92	0.8537	1.7070	1.1301	1.4716

The columns labeled API ovalization and DNV ovalization show the results of the previous analyses with the CEL and structural transient models. These are the initial imperfections imposed on the collapse models. First, an increase in soil stiffness leads to higher levels of ovalization, reducing the critical load of the structure. This means the flexible pipe will collapse with a smaller pressure load after a lateral displacement on high-strength soil. A similar response can be observed from the diameter results: larger diameters suffer from a more significant deformation during the lateral displacement, reducing the collapse pressure value. The burial depth, however, had minimal influence on the overall deformation of the flexible pipe cross-section. This also reflects the small change observed in the wet and dry collapse pressure values.

The wet and dry collapse of flexible pipes is widely covered in the literature. The results presented in Table 3 are in accordance with the findings of [27,29], whose results showed that increasing the initial imperfection results in a reduction in the collapse pressure. Using the present work, it is possible to directly associate the lateral movement of buried flexible pipes with subsequent collapse failure.

7. Conclusions

A methodology was developed to determine the level of damage inflicted on a buried multilayer flexible pipe due to the large lateral displacements. The methodology employs a CEL model for lateral soil resistance and a transient structural model for the flexible pipe. The CEL pipe–soil interaction model was created, utilizing the modified Tresca model [30,31] to account for the strain-softening effects of the seabed soil. Subsequently, the pressure profiles around the pipe section, calculated in the CEL model, were integrated into the FE model featuring simplified metallic interlocked layers. Both models were individually verified against analytical solutions [40], fully tridimensional numerical models [27], and centrifuge tests [8,13].

The decoupled numerical models address the challenges associated with simulating the large lateral displacements of buried flexible pipes with complex multiple layers, and the use of the equivalent ring for the flexible pipe's complex layers significantly reduces the simulation time. The CEL model provides a more precise modeling of the soil material properties compared to conventional Lagrangian models and is more suitable for large displacement analyses. Soil stiffness significantly affects the flexible pipe's ovalization, stresses, and reaction forces, with stiffer soils leading to larger fluctuations due to the soil berm formations and new shear band propagation. The pipe diameter has a profound impact on response to lateral movements, with maximum ovalization values differing by a factor of 42 and stresses being 15 times higher in larger diameters. Burial depth, however, has a less significant impact, showing minimal differences in ovalization and stresses across various depths, although the reaction forces increase significantly with burial depth. None of the studied cases exceeded the limits for ovalization and stresses according to API and DNV regulations, reducing the risk of collapse due to high external pressure. A collapse analysis corroborates the ovalization and stress findings, showing that soil strength and pipe diameter substantially influence the collapse pressure, while burial depth has little effect. Although validated, the methodology requires further verification to confirm the validity of the decoupled model assumption, which may include testing different levels of ovalization or employing a two-way soil–structure interaction model despite the higher computational costs. The methodology does not account for voids left by pipe movement, although the boundary conditions mitigate the impacts on ovalization and stress outcomes. Future advancements could improve the safeguarding of submarine communication cables, which share structural similarities with offshore multilayered flexible pipes.

Even though both models employed in this study were previously validated, further work is needed to verify the range of validity of the primary assumption of this model, which is the decoupling between the pipe–soil interaction analysis. This can be achieved in different manners, including using cross-sections with different levels of ovalization on the pipe–soil analysis or using a two-way soil–structure interaction model, with the downsides of a significantly higher computational cost and convergence issues. Another limitation of the model is that the void left in the wake of the pipe movement is left unoccupied by the soil particles. However, the boundary conditions impede the pipe from going in the opposite direction, minimizing the effects on the ovalization or stress outcomes. Finally, submarine communication cables share many structural features with offshore multi-layered flexible pipes. Further developments in this methodology and models might be of strategic value, improving the safeguarding of vital communication lines.

Author Contributions: Conceptualization, E.R.M., X.D. and H.S.; methodology, E.R.M. and X.D.; software, E.R.M. and X.D.; validation, E.R.M. and X.D.; formal analysis, E.R.M. and X.D.; investigation, E.R.M. and X.D.; resources, H.S.; data curation, E.R.M. and X.D.; writing—original draft preparation, E.R.M. and X.D.; writing—review and editing, E.R.M., X.D. and H.S.; visualization, E.R.M. and X.D.; supervision, H.S.; project administration, H.S.; funding acquisition, H.S. All authors have read and agreed to the published version of the manuscript.

Funding: The authors sincerely acknowledge the financial support of Atterix Technologies Inc. and Mitacs (214775) through Accelerate Entrepreneurs program (Project IT23317), Wood Group PLC (211474), and NSERC (212534).

Institutional Review Board Statement: Not applicable.

Informed Consent Statement: Not applicable.

Data Availability Statement: Data are contained within the article.

Acknowledgments: Special thanks are also extended to Memorial University for providing excellent resources to conduct this research.

Conflicts of Interest: The authors declare no conflicts of interest. The funders had no role in the design of the study; in the collection, analyses, or interpretation of data; in the writing of the manuscript; or in the decision to publish the results.

References

1. Fergestad, D.; Løtveit, S.A. *Handbook on Design and Operation of Flexible Pipes*; SINTEF Ocean AS: Trondheim, Norway, 2017. Available online: https://sintef.brage.unit.no/sintef-xmlui/handle/11250/3123266 (accessed on 17 July 2024).
2. Tschebotarioff, G.P. *Foundations, Retaining and Earth Structures*; McGraw-Hill Book Company: New York, NY, USA, 1973; 642p.
3. Merifield, R.S.; Sloan, S.W.; Yu, H.S. Stability of plate anchors in undrained clay. *Géotechnique* **2001**, *51*, 141–153. [CrossRef]
4. American Lifelines Alliance. *Guidelines for the Design of Buried Steel Pipe*; APA: Washington, DC, USA, 2001.
5. Guo, P. Numerical Modeling of Pipe–Soil Interaction under Oblique Loading. *J. Geotech. Geoenviron. Eng.* **2005**, *131*, 260–268. [CrossRef]
6. PRCI. *Guidelines for Constructing Natural Gas and Liquid Hydrocarbon Pipelines through Areas Prone to Landslide and Subsidence Hazards*; Pipeline Research Council International: Houston, TX, USA, 2009.
7. Martin, C.M.; White, D.J. Limit analysis of the undrained bearing capacity of offshore pipelines. *Géotechnique* **2012**, *62*, 847–863. [CrossRef]
8. Dong, X.; Shiri, H.; Zhang, W.; Randolph, M.F. The influence of pipeline-backfill-trench interaction on the lateral soil resistance: A numerical investigation. *Comput. Geotech.* **2021**, *137*, 104307. [CrossRef]
9. Paulin, M. An Investigation into Pipelines Subjected to Lateral Soil Loading. Ph.D. Thesis, Memorial University of Newfoundland, St. John's, NL, Canada, 1998.
10. C-CORE; Honegger, D.G.; Honegger Consulting. *Extended Model for Pipe Soil Interaction*; Technical Toolboxes: Houston, TX, USA, 2003.
11. Chaloulos, Y.K.; Bouckovalas, G.D.; Zervos, S.D.; Zampas, A.L. Lateral soil–pipeline interaction in sand backfill: Effect of trench dimensions. *Comput. Geotech.* **2015**, *69*, 442–451. [CrossRef]
12. Kianian, M.; Shiri, H. Experimental study of trench effect on lateral failure mechanisms around the pipeline buried in clay. *J. Pipeline Sci. Eng.* **2021**, *1*, 198–211. [CrossRef]
13. Dong, X.; Zhang, W.; Shiri, H.; Randolph, M.F. Large deformation coupled analysis of embedded pipeline–Soil lateral interaction. *Mar. Struct.* **2021**, *78*, 102971. [CrossRef]
14. Lee, Y.; Feng, M.; Lee, E.T. Deflection of buried prestressed concrete cylinder pipe with soil-pipe interaction. *KSCE J. Civ. Eng.* **2014**, *18*, 2191–2195. [CrossRef]
15. Bryden, P.; El Naggar, H.; Valsangkar, A. Soil-structure interaction of very flexible pipes: Centrifuge and numerical investigations. *Int. J. Geomech.* **2014**, *15*, 04014091. [CrossRef]
16. Soveiti, S.; Mosalmani, R. Mechanical behavior of buried composite pipelines subjected to strike-slip fault movement. *Soil Dyn. Earthq. Eng.* **2020**, *135*, 106195. [CrossRef]
17. Banushi, G.; Squeglia, N.; Thiele, K. Innovative analysis of a buried operating pipeline subjected to strike-slip fault movement. *Soil Dyn. Earthq. Eng.* **2018**, *107*, 234–249. [CrossRef]
18. Jeng, D.S.; Postma, P.F.; Lin, Y.S. Stresses and deformation of buried pipeline under wave loading. *J. Transp. Eng.* **2001**, *127*, 398–407. [CrossRef]

19. Nobahar, A.; Kenny, S.; Phillips, R. Buried pipelines subject to subgauge deformations. *Int. J. Geomech.* **2007**, *7*, 206–216. [CrossRef]
20. Odina, L.; Tan, R. Seismic fault displacement of buried pipelines using continuum finite elements methods. In Proceedings of the ASME 2009 28th International Conference on Ocean, Offshore and Arctic Engineering, Honolulu, HI, USA, 31 May–5 June 2009.
21. Journal of Petroleum Technology. Design Tools for Arctic Subsea Pipelines. (17 July 2017). Available online: https://jpt.spe.org/design-tools-arctic-subsea-pipelines (accessed on 14 May 2022).
22. Journal of Petroleum Technology. Direct Electrical Heating of a Flexible Pipe. (9 October 2016). Available online: https://jpt.spe.org/direct-electrical-heating-flexible-pipe (accessed on 13 May 2022).
23. *API-1104*; Welding of Pipelines and Related Facilities. American Petroleum Institute: Washington, DC, USA, 2007.
24. *DNV-OS-F101*; Submarine Pipeline Systems. Det Norske Veritas AS: Høvik, Norway, 2007.
25. Martins, C.A.; Pesce, C.P.; Aranha, J.P. Structural Behavior of Flexible Pipe Carcass during Launching. In Proceedings of the 22th International Conference on Offshore Mechanics and Arctic Engineering, Cancun, Mexico, 8–13 June 2003.
26. Pesce, C.P.; Ramos, R., Jr.; Silveira, L.Y.; Tanaka, R.L.; Martins, C.A.; Takafuji, F.M.; Godinho, C.F. Structural behavior of umbilicals—Part I: Mathematical modeling. In Proceedings of the ASME 2010 29th International Conference on Ocean, Offshore and Arctic Engineering, Shanghai, China, 6–11 June 2010; American Society of Mechanical Engineers: New York, NY, USA, 2010.
27. Gay Neto, A.; Martins, C.A.; Malta, E.R.; Godinho, C.F.; Barbosa Neto, T.F.; Lima, E.A. Wet and dry collapse of straight and curved flexible pipes: A 3D FEM modeling approach. In Proceedings of the 22nd International Offshore and Polar Engineering Conference, Rhodes, Greece, 17–22 June 2012.
28. American Petroleum Institute. *Recommended Pratice for Flexible Pipe (API Recommended 17B)*; American Petroleum Institute: Washington, DC, USA, 2002.
29. Malta, E.R.; Martins, C.A.; Gay Neto, A.; Toni, F.G. An investigation about the shape of the collapse mode of flexible pipes. In Proceedings of the Twenty-Second International Offshore and Polar Engineering Conference, Rhodes, Greece, 17–22 June 2012.
30. Zhang, W.; Wang, D.; Randolph, M.F.; Puzrin, A. Catastrophic failure in planar landslides with a fully softened weak zone. *Géotechnique* **2015**, *65*, 755–769. [CrossRef]
31. Zhang, W.; Randolph, M.F.; Puzrin, A.; Wang, D. Transition from shear band propagation to global slab failure in submarine landslides. *Can. Geotech. J.* **2019**, *56*, 554–569. [CrossRef]
32. Kavvadas, M.; Amorosi, A. A constitutive model for structured soils. *Géotechnique* **2000**, *50*, 263–273. [CrossRef]
33. Savvides, A.A.; Papadrakakis, M.A. Computational study on the uncertainty quantification of failure of clays with a modified Cam-Clay yield criterion. *SN Appl. Sci.* **2021**, *3*, 659. [CrossRef]
34. Simulia. *Abaqus User Subroutines Reference Guide*; Dassault Systemes Simulia Corp.: Providence, RI, USA, 2016.
35. Guo, X.; Liu, Z.; Zheng, J.; Luo, Q.; Liu, X. Bearing capacity factors of T-bar from surficial to stable penetration into deep-sea sediments. *Soil Dyn. Earthq. Eng.* **2023**, *165*, 107671. [CrossRef]
36. Guo, X.; Stoesser, T.; Zheng, D.; Luo, Q.; Liu, X.; Nian, T. A methodology to predict the run-out distance of submarine landslides. *Comput. Geotech.* **2023**, *153*, 105073. [CrossRef]
37. American Society of Civil Engineers; Committee on Gas, and Liquid Fuel Lifelines. *Guidelines for the Seismic Design of Oil and Gas Pipeline Systems*; American Society of Civil Engineers: Reston, VA, USA, 1984.
38. Gay Neto, A.; Martins, C.A. Flexible Pipes: Influence of the Pressure Layer in the Wet Collapse Resistance. In Proceedings of the 30th International Conference on Ocean, Offshore and Arctic Engineering, Rotterdam, The Netherlands, 19–24 June 2011.
39. TechnipFMC. *Flexible Pipe. A Broad Range of Flexible Pipe Systems with Advanced Integrated Solutions*; TechnipFMC plc: London, UK, 2019.
40. Ramos, R.; Pesce, C.P., Jr. A consistent analytical model to predict the structural behavior of flexible risers subjected to combined loads. *J. Offshore Mech. Arct. Eng.* **2004**, *126*, 141–146. [CrossRef]

Journal of
*Marine Science
and Engineering*

MDPI

Review

Key Parameters for Design Analysis and Optimization of Dynamic Inter-Array Power Cable Configurations in Floating Offshore Wind Farms

Anja Schnepf [1] and Ove Tobias Gudmestad [1,2,*]

[1] Department of Mechanical, Civil and Material Science, University of Stavanger, 4036 Stavanger, Norway; anjaschnepf@web.de
[2] Department of Marine Science, Western Norway University of Applied Science, Haugesund, 5020 Bergen, Norway
* Correspondence: otgudmestad@gmail.com

Abstract: Power cables transmit electricity from offshore wind turbines (OWTs) to consumers. The configuration, design, and optimization of power cables for floating offshore wind turbines (FOWTs) are subject to various objectives, variables, and constraints. These components are outlined, and existing research gaps that need to be addressed for a more sustainable and robust design for future developments are highlighted. The main aim of power cable configuration design is to optimize performance and minimize costs. This can be achieved through thoughtful consideration of variables like power cable selection, configuration design, and the integration of specialized components and ancillaries. An extensive overview of constraints for power cable configuration design is provided, and the most important ones are identified. The local conditions determine which constraints are of key importance in optimization. Environmental factors like wind, waves, and especially currents significantly influence design processes, introducing uncertainties when comprehensive data are lacking. Marine growth posed a significant challenge in prior FOWT projects and must be considered carefully. Considering the potential impact of climate change is crucial, especially for extreme weather events. Early integration of environmental considerations and assessment of effects on socio-economic impacts is essential for a successful project. The power cable response is also influenced by its inherent limitations, including tension and compression thresholds, curvature constraints, and the necessary minimum fatigue life. A flowchart is provided to aid in choosing variables and constraints in the design and optimization processes.

Keywords: floating offshore wind turbines; power cables; umbilicals; wind farms; optimization; constraints; risers

1. Introduction

According to the forecast by Det Norske Veritas (DNV) [1], floating offshore wind technologies are expected to account for 15% of the total offshore wind capacity by 2050. To achieve this, around 20,000 OWTs must be mounted on floating platforms [1]. However, these technologies are in the early stages of development and require commercialization for broader adoption. Cost reduction is crucial for economic viability, with standardization, larger OWTs, and industrialization being key factors [1]. Further, larger offshore wind farms and optimized mooring and cable systems are essential to lower costs [1].

Over 20% of total project capital costs account for offshore transmission, inter-array cabling, and offshore substations, as reported by the International Energy Agency (IEA) [2].

According to GCube [3], subsea cables, both inter-array and export, accounted for a majority (30%) of all insurance claims in the offshore wind industry from 2010 to 2020 and exceeded 50% of the total claims expenditure during this period. Dynamic power cables are relatively new to the offshore wind industry and incur higher costs for floating than bottom-fixed offshore wind technologies, which do not require them. Therefore, thorough design analysis and optimization of power cable configurations are essential for successful FOWT farms. This is particularly critical with the expected growth of FOWT farm sizes and the need for a significant amount of inter-array electricity transmission. Therefore, the present study discusses the different parameters that influence the design and optimization of dynamic inter-array power cable configurations. Further, the uncertainties of the design parameters are discussed.

The present paper is organized as follows. Section 2 presents the state-of-the-art dynamic power cable configuration design processes. Section 3 comprises the rules and regulations. The design objectives are discussed in Section 4, the variables in Section 5, and the constraints in Section 6. Section 7 shows an overview of published optimization algorithms. A flowchart to determine the most relevant optimization parameters is presented in Section 7. Conclusions are drawn in Section 8. A comprehensive list of references is also presented.

2. The State of the Art of Dynamic Power Cable Configuration Design Processes

OWTs convert wind kinetic energy into electrical energy, which is transmitted to consumers through a network of power cables. The inter-array cables connect OWTs within offshore wind farms, forming feeders that link to collectors or substations [4]. From there, an export cable transmits electrical energy to consumers, such as land-based grids or offshore platforms. In offshore wind farms with bottom-fixed OWTs, inter-array cables are either statically laid on the seabed or buried. In floating offshore wind farms, the power cables connect to FOWTs. Therefore, they must be designed to account for dynamic responses caused by FOWT motions and environmental loads.

In the FOWT farms in operation today, one FOWT acts as the collector or substation of all feeders. This can be seen in commercial FOWT farms currently in operation: Hywind Scotland [5], Kincardine Offshore Wind Farm [6], Hywind Tampen [7], WindFloat Atlantic [8], and the METCentre test site [9]. Larger FOWT farms may require substations with a transformer connecting the inter-array and export cables. The test site Fukushima FORWARD is the first to have a floating substation in operation [10]. The transformer contains a significant amount of toxic oil and is an environmental threat in the case of substation failure [11]. Therefore, the substation must have a resilient mooring system with high safety levels [12]. The dynamic section of the export cable from the FOWT or the substation faces constraints like those of the inter-array cables. Nevertheless, its configuration might be designed differently than for inter-array cables in an FOWT farm, as they usually have a larger diameter and other stiffness values.

The design of a dynamic power cable configuration is usually a time-consuming iterative process performed by an experienced engineer. First, a hypothesis is made on the configuration setup that will be feasible for the given problem. Simulations are then performed, and the results are evaluated against the design constraints. Afterwards, the hypothesis is updated, and the process is repeated until it converges to a solution. Power cable configuration design is often considered after the FOWT floater and mooring system are designed. It is commonly assumed that power cable motions have a negligible influence on FOWT behavior and can be considered separate.

The design of oil and gas (O&G) industry risers and mooring systems has been performed together numerous times [13]. Ja'e et al. [14] provided an overview of such studied and implemented systems. The combined consideration is crucial for the functionality and safety of riser systems, as damage can have fatal repercussions. Similarly, the safety levels for offloading lines must be high to prevent severe incidents. Mooring systems and power cables are not as closely reliant on each other in offshore wind farms as mooring systems and O&G lines containing highly flammable fluids. That is because damaged power cables have fewer consequences on human lives and the marine environment than a damaged riser or an offloading line can have. Even though this means that lower safety levels may be applied to power cables [15], they are nevertheless a crucial part of the functionality of offshore wind farms. A damaged power cable disrupts the connection between the electrical energy-generating wind turbine and the consumer, resulting in zero revenue and possibly high repair costs for the wind farm owner and operator.

Offshore wind farm power cables are also referred to as "power umbilicals" as they are a specific type of umbilical. Umbilicals focus on providing various services to offshore equipment, whereas power cables concentrate on transmitting electrical energy over long distances from FOWTs to consumers. OWTs have already exceeded the number of facilities installed in the O&G industry by tenfold, and the demand is projected to rise [15]. That means a much larger quantity of power cables is necessary for an offshore wind farm project compared with the umbilicals needed in an O&G project. In addition, FOWTs often have response characteristics or failure modes that differ from O&G platforms [15]. FOWTs are mainly without crew and do not contain explosive hydrocarbon fluids that can threaten humans and the environment. The overall configuration design can be like the umbilical design in the O&G industry. However, caution is needed when applying umbilical configuration design practices to dynamic power cable configurations due to differences in diameters, stiffness, components, function, applications, and system-level reliability.

3. Rules and Regulations

Several rules and regulations have already been implemented for power cable configuration design. Nevertheless, limited regulations on dynamic power cable configurations are currently being established. For example, the standard DNV-ST-0119 [16] states that this is the responsibility of the supplier. The following list shows relevant codes, standards, and recommended practices:

- ISO 19900 "Petroleum and natural gas industries—General requirements for offshore structures" [17]
- ISO 19901-1 "Petroleum and natural gas industries—Specific requirements for offshore structures—Part 1: Metocean design and operating considerations" [18]
- ISO 13628-5 "Petroleum and natural gas industries—Design and operation of subsea production systems—Part 5: Subsea umbilicals" [19]
- ISO 13628-11 "Petroleum and natural gas industries—Design and operation of subsea production systems—Part 11: Flexible pipe systems for subsea and marine applications" [20]
- DNV-ST-0119 "Floating wind turbine structures" [16]
- DNV-ST-0359 "Subsea power cables for wind power plants" [21]
- DNV-OS-E301 "Position mooring" [22]
- DNV-RP-0360 "Subsea power cables in shallow water" [23]
- DNV-RP-F401 "Electrical power cables in subsea applications" [24]
- DNV-RP-C205 "Environmental conditions and environmental loads" [25]
- NORSOK-N003 "Action and action effects" [26]

- API RP 2A-LRFD "Recommended practice for planning, designing and constructing fixed offshore platforms—load and resistance factor design" [27]
- API RP 2A-WSD "Recommended practice for planning, designing and constructing fixed offshore platforms—working stress design" [28]
- API SPEC 17E "Specifications for subsea umbilicals" [29]

Other regulations that the configuration must comply with are defined by the government responsible for the site. This often includes environmental restrictions.

In the pursuit of cost competitiveness, there have been considerable cost reductions in the planning phase of offshore wind projects over the last decade, and thus also in production, building, and operation. However, this emphasis on cost efficiency has negatively impacted safe design concepts. Consequently, there has been a rise in both the frequency and seriousness of insurance claims within the sector [3]. Proper insurance is essential to protect the public from paying for a malfunctioning offshore wind farm project [11].

4. Design Objectives

The design objective refers to the specific goal or outcome that the design process of a project aims to achieve. For power cable configurations, they are mostly related to economic and performance indicators, which are explained in more detail in this section. Overall, the most important objective should always be to create a safe design that will work.

4.1. Economic Indicators

Most design and optimization objectives are directly or indirectly related to minimizing costs. For power cable configurations, several economic indicators can be relevant. Some key indicators considered early in the design process can help minimize long-term costs and are presented here:

- Levelized Cost of Energy (LCOE): It is a widely used metric to assess cost-effectiveness by considering total lifetime costs divided by total lifetime energy output, with lower values indicating efficiency [30]. The LCOE of floating offshore wind projects is about five times as high as for bottom-fixed offshore wind projects [1]. Predictions for the future vary, including higher, equal, and lower LCOE values by 2050 for floating compared to bottom-fixed offshore wind projects [1]. Nevertheless, they all predict a decrease in the LCOE for floating offshore wind projects due to technological advancement [31].
- Capital Expenditure (CapEx): This metric refers to the initial investment required to design, manufacture, and install the floating wind turbine and its associated power cables. For offshore wind projects, most lifetime costs occur for the development and construction of the project as CapEx [31].
- Installation and decommissioning costs are considered parts of capital expenditure: A design can have low costs considering the price of the components involved, but it might be costly during transport, installation, and decommissioning. Improper consideration of these costs has been a major problem in past renewable energy projects, such as when removing wave energy converters (WECs). It can quickly lead to bankruptcy [32]. Further, the available equipment for installation and removal must be considered in the design process to avoid using unusual or hardly available equipment, which can be very costly. Also, the choice of ancillaries should be made considering the need for potentially new molding and their installation time, which can increase and raise costs.
- Operating Expenditures (OpEx): These are the day-to-day costs, covering expenses like rent, utilities, salaries, maintenance, and other essential operational needs. Maintenance costs include regular inspections and repairs throughout the lifetime of the

power cable configuration. These costs can vary based on cable design, materials used, and accessibility for maintenance crews. Critical components, such as the hang-off section and the touch-down point (TDP), are crucial for the integrity of the system. Addressing these early in the design process can prevent future repair work. Regularly removing marine growth from the power cables and ancillaries can significantly raise maintenance costs. If considered correctly, the need for removals or buoyancy additions throughout the lifetime might be limited. Rinaldi et al. [30], Garcia-Teruel et al. [33], and Yang et al. [34] present analyses including operation and maintenance costs for FOWTs.

- Regulations and permits: Compliance with regulations and permitting requirements can influence design choices and costs. For example, factors such as seabed intervention (e.g., trenching or similar activities), marine life entanglement, trawling entanglement, and ghost net entanglement must be carefully assessed, as these can be regulated and require specific permits. Further, it may be necessary to obtain an initial technical qualification for a new power cable system.
- Risk assessment: Economic indicators should also incorporate risk analysis at every design stage. Identifying potential failure points, downtime risks, and their financial consequences is crucial for estimating overall project costs accurately. It helps to account for unforeseen events such as supply chain disruptions or regulatory changes.
- Future expansions and upgrades: Designing for future expansion and upgrades helps minimize project costs by avoiding costly modifications later. Anticipating potential expansions and technology improvements in FOWT farms, leading to increased power generation, is crucial for long-term viability.

Further economic indicators that influence the design objectives include, for instance, the selection of the discount rate for the project, the optimization of the Net Present Value (NPV) of the project, given the selected discount rate, Return on Investment (ROI), and Weighted Average Cost of Capital (WACC) [35,36].

In academic research and industrial projects, specific cost data are often unavailable. Therefore, different objectives related to cost reduction may be used, particularly for feasibility research studies and preliminary designs in early project stages. The following presents some of these objectives:

- Minimize cable lengths: Shorter cable lengths reduce both electrical losses and material costs associated with cable manufacturing and installation. Power transmission efficiency can be improved by strategically planning cable configurations and routes and minimizing unnecessary detours, ensuring the electricity reaches its destination with minimal loss. However, a difference in cable length by a few percentage points may not significantly affect capital and operational costs, depending on the project.
- Minimize the use of ancillaries: Ancillaries represent an additional cost to the power cable. Using as few ancillaries as possible can reduce costs as well as installation and maintenance times.

4.2. Performance Indicators

Proper design objectives are crucial for long-term project feasibility, ensuring reliability, safety, and regulatory compliance. Performance-focused goals are essential for power cable configurations, alongside economic indicators, to guarantee project viability and safety. Several possible objectives are the following:

Maximize performance: The performance of power cables relates to their ability to handle the required data transmission rates, power levels, and signal quality without significant losses or degradation. Finding the right balance between performance and cost is essential for ensuring the economic viability of the project.

Maximize reliability: For example, at some FOWT farm locations, such as off the eastern coast of Japan, the cable cannot be repaired for half a year due to the harsh environmental conditions during certain seasons [37]. Design efforts should focus on minimizing the risk of cable failures, downtime, and maintenance disruptions.

Safety and compliance: Design efforts should aim to facilitate the design to comply with relevant codes and standards.

Minimize environmental impact: The environmental impact should be considered at every project stage [33]. Offshore cable designs must minimize the environmental impact by choosing routes away from sensitive ecological areas and using installation, maintenance, and decommissioning techniques to reduce impact.

Minimize footprint: The footprint of an offshore structure refers to the area on the seabed or ocean floor that the subsea structure occupies or covers. Its size can be minimized to possibly decrease the amount of material needed, reduce the environmental impact, keep the equipment in the regulatory zones, or keep clear of fishing routes. The placement of the power cable configuration heavily depends on the environmental conditions and the mooring system, which all need to be considered together.

Extend the lifetime of the cable configuration: Offshore projects often exceed their initial timelines. To extend the life of a power cable configuration, the current system is assessed, and analyses are conducted to determine its expected lifetime. Necessary measures may be taken to make the extension feasible, such as adding buoyancy modules or redesigning hang-off sections.

5. Design Variables

A design variable is an adjustable parameter or factor that influences the design, behavior, or performance of a system. For the design and optimization of power cable configurations, key variables are the material choice of the actual cable, overall configuration design, and the design of special components and ancillaries.

5.1. Power Cable Selection

The design of the power cable is usually heavily site- and project-dependent. First, the choice of the voltage rating and between alternating current (AC) or direct current (DC) technology must be made. Usually, the inter-array cables use medium- or high-voltage AC due to the short distances covered. They are rated up to 66 kV in current projects, depending on the wind turbines and the cable length [5,38–40]. With OWT sizes increasing compared to OWTs that have been installed in the past [41,42] the power cable sizes increase as well, and are expected to rise to a voltage rating of 132 kV [43] and beyond [44]. Larger cable diameters lead to new challenges, especially regarding cable buckling and conductor creep [38].

The power transmission capacity within a feeder varies as the cables transmit energy from nearby and distant OWTs. A cost-effective approach is using a single cable design sized for the largest load, applicable to all inter-array cables [45]. The different feeders are combined at the collector or substation, and an export cable with a high-voltage AC or an ultra-high-voltage AC connects the FOWT farm to consumers [5,38]. With increasing FOWT farm sizes and further distances from shore or consumers, DC technology can be used for the export cable, as has been installed in fixed-bottom offshore wind farms [46,47]. Note that the electrical energy loss in DC cables is less than that in AC cables. However, the costs of a substation transforming the current from DC to AC at the farm location and back to DC at a substation near the consumers can increase capital expenditures.

Figure 1 shows an exemplary cross-section of a power cable, including the conductors, optical fibers, insulation, and any shielding or protective layers. The conductor area determines the current capacity, with larger areas allowing higher currents. Power cables commonly have several layers of armor to define axial, bending, and torsional stiffness. Smaller diameter cables of lower current rating do not necessarily have to be equipped with armor, reducing costs and facilitating the installation and maintenance; see Ringsberg et al. [48].

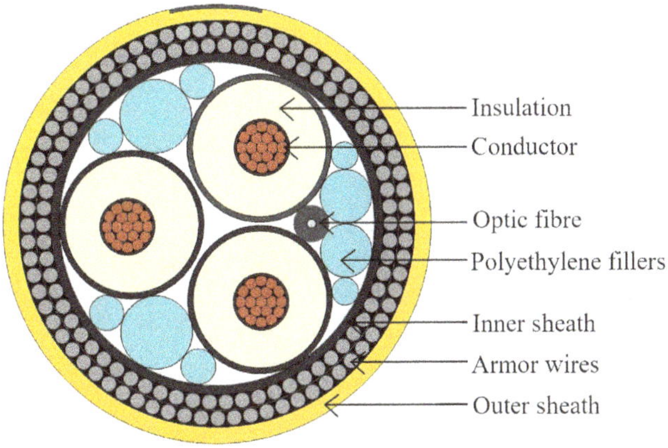

Figure 1. Exemplary power cable cross-section with copper cores.

The primary objective of various cross-sectional components is to determine the behavioral properties of the power cable and to protect the conductor core. Copper and aluminum are commonly used for cable conductor cores, with copper being the prevailing choice today. Aluminum cables were studied by Thies et al. [49], Schnepf et al. [50], and Ahmad et al. [51]. An actual aluminum cable was first implemented for the TetraSpar FOWT in the Northern North Sea, which is currently operating [52]. In cables with identical voltage ratings, the cross-sectional area of the copper conductor core is typically smaller than that of its aluminum counterparts. Additionally, copper cables exhibit a more significant weight per unit length when compared to aluminum cables of equal conductivity [53]. This difference is due to copper being more conductive than aluminum and the low density of aluminum [53]. Therefore, the submerged-weight-to-diameter ratio is usually lower for aluminum cables than for copper cables of the same conductivity [50].

5.2. Configuration Design

Ikhennicheu et al. [45] and Clausen and D'Souza [54] provided overviews of overall configurations successfully implemented for O&G risers, offloading hoses, and umbilicals. Figure 2 shows several configurations that can be used for dynamic power cable configurations of FOWTs.

A vastly researched configuration is the lazy-wave configuration shown in Figure 2a (left), which was the study subject of, for example, Rentschler et al. [4,55], Thies et al. [56], and Zhao et al. [57]. It has been installed in the FOWT farm WindFloat Atlantic [8]. The lazy-wave configuration has a section of buoyancy attached to the cable between the FOWT and the TDP on the seabed. This way, the motions of the floater can be decoupled from the lower hanging cable sections and the TDP to a large extent. It is a simple configuration that only needs buoyancy modules attached to the cable, a bend restrictor at the connection to the FOWT, and possibly cable protection at the seabed.

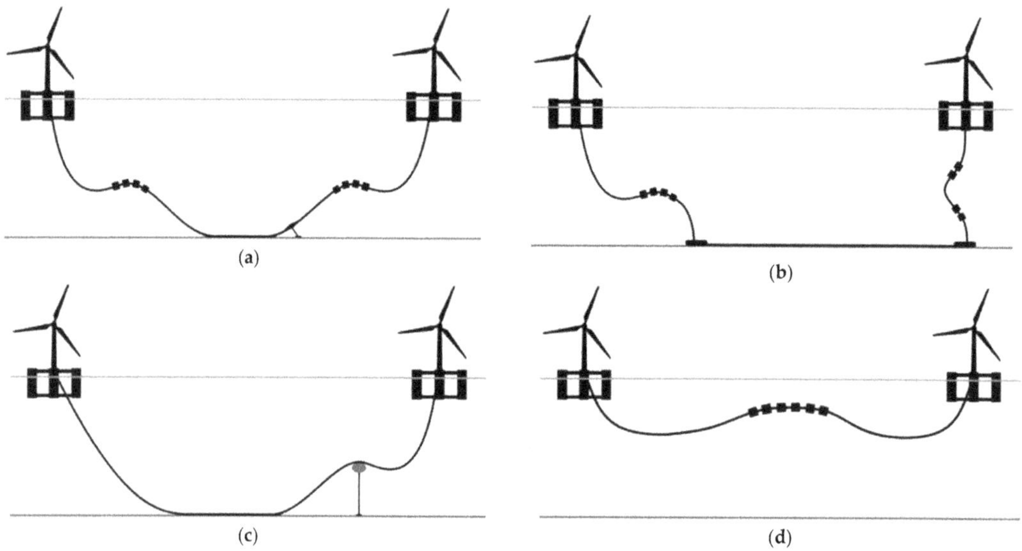

Figure 2. Different power cable configuration types. (**a**) Lazy-wave (left) and tethered-wave (right) configurations. (**b**) Steep-wave (left) and Chinese lantern (right) configurations. (**c**) Catenary (left) and lazy-S (right) configurations. (**d**) Suspended configuration.

Double lazy-wave power cable configurations are usually considered in shallower water depths, having two buoyancy sections instead of one [57–59]. Such a system has been installed in the China Three Gorges (CTG) Yangxi Shapa III wind farm, where an FOWT is connected to a fixed-bottom monopile wind turbine in water depths from 29 m to 35 m [60,61].

The tethered-wave configuration is very similar to the lazy-wave configuration, and is also named the pliant-wave configuration. In addition to the lazy-wave setup, it has a tether restraining the motions of the power cable at the TDP, as shown in Figure 2a (right). This effectively limits motions at the TDP when the wave configuration decouples them from the floater. Further, it controls the TDP location throughout its lifetime, preventing damage and alterations caused by factors such as extreme weather and marine growth. On the other hand, the tethered-wave configuration needs more equipment than the lazy-wave configuration, making installation more complex. The commercial floating offshore wind farms Hywind Scotland [5] and Hywind Tampen [7] use tethered-wave configurations.

The steep-wave configuration in Figure 2b (left) is a wave configuration, with the distinction that the cable is vertically connected to a seabed junction with a bend stiffener. It has the advantage that an excursion of the cable close to the seabed can be prevented. This configuration has often been used in O&G applications to connect subsea equipment with a vessel or platform at the sea surface, for example, using a riser from a subsea wellhead to a floating production facility [62].

Chinese lantern configurations, as shown in Figure 2b (right), have been used primarily with Catenary Anchor Leg Mooring (CALM) buoys [63]. A buoyancy section to decouple the motions of the floater from the bottom location keeps the cable in a U-shape in a vertical direction. The cable is usually vertically connected to a seabed junction with a bend stiffener. This configuration is usually installed in shallower water depths [63].

The most straightforward configuration is the catenary configuration shown in Figure 2c (left), also often referred to as "free hanging". It does not require additional equipment except for possibly a bend restrictor at the hang-off location and protection at the TDP. Nevertheless, it does not decouple the motions of the floater from the TDP, which can make it challenging to apply [4].

The lazy-S configuration in Figure 2c (right) is also like the lazy-wave configuration. However, instead of buoyancy modules attached to create an arch, it uses fewer and larger buoyancy devices, such as a buoy. That buoy can be tethered or buoyant, controlling the hog bend location. It reduces the excursions of the power cable under the current. An altered lazy-S power cable configuration with six buoys attached to a tether configuration was successfully used for two years at the Haenkaze FOWT while it was installed off Kabashima Goto, Nagasaki prefecture, Japan [64]. Another example is the Wello wave energy converter project off the coast of Scotland, United Kingdom, where the lazy-S configuration had been implemented with a buoyed loop stabilized by a chain [65]. The subsea buoy can also be implemented in a tethered-S configuration with an additional tether to control the TDP and a steep-S configuration with a vertical connection at the bottom.

The suspended power cable configuration is shown in Figure 2d. In this configuration, the cable does not touch the seabed, which can be useful for deepwater applications and reduces costs due to shorter cable lengths and less additional equipment than the other configurations. Suspended configurations are common practice for O&G offloading lines, as described by Araújo et al. [66], He et al. [67], and Kang et al. [68]. They have been researched for application with FOWTs by Beier et al. [69,70], Schnepf et al. [50,71], Ahmad et al. [51], and Li et al. [72]. Their applicability depends on the environmental conditions of the location where the cable will be installed. Buoyancy elements can be designed in one section or divided into several sections distributed over the cable length. Subsea buoys may also replace buoyancy modules to minimize drag.

The presented power cable configuration options do not depict all possible designs. They only summarize tried and tested practices in O&G and renewable energy applications. To name a few, further power cable configurations can be the double-tethered-wave, lazy-camel [54], camel-S [54], and corkscrew [73] configurations. Configurations applied in the O&G industry that have weights added in the configuration are, for example, the weight-added wave [74,75], multi-wave with weights [76], and touchdown chain-added wave [74]. Power cables usually need extra ancillary equipment, such as buoyancy modules and bend stiffeners, to successfully implement configurations with weights attached, not exceeding the maximum allowable tension values. An example of weight chains successfully implemented in an umbilical configuration is the "cage-hose" concept connecting large fish cages in the northern Norwegian Sea [77]. The configuration choice must be made based on the specific environmental conditions and further constraints it will encounter, which are all heavily site-dependent.

5.3. Special Components and Ancillaries

The tension in the power cable throughout the dynamic configuration is usually highest at the hang-off section from the FOWT [49,50,56,78]. This section must be designed to support and suspend the weight of the power cable, restrict it from overbending, and withstand the loads from the environmental conditions. It consists of clamps, connectors, and a bend stiffener or bell mouth. Typically, it is placed beneath the sea surface, although it can also be installed above sea level for ease of access. Dai et al. [79] addressed the behavior of dynamic steel tube umbilicals in a bell-mouth hang-off section. Svensson et al. [80] showed that the bell mouth is the most critical point for fatigue. The conductor of their power cable failed inside the bell mouth in their numerical models and laboratory tests. Further research on the hang-off section and its design variables should be carried out for dynamic power cable configuration design.

The lazy-wave and similar configurations employ numerous buoyancy modules between the hang-off point and the TDP. The size and number of the modules vary; they can all be customized [81]. As casting a mold is usually the most expensive part of the buoyancy module fabrication, standard sizes can be used to save costs. The attached buoyancy modules change the behavior of the flow around the cable, which must be considered in the analysis [82].

The cable section around the TDP can be designed in various ways. One decision is whether a tether should be installed and, if so, in what manner. This influences the overall configuration design significantly. The decision depends on the soil conditions and power cable motions, which can lead to damage. A cable protection device [81] can be attached to the power cable at the TDP to reduce wear and tear.

6. Design Constraints

The power cable configuration must be designed to endure the environmental conditions it is exposed to, as well as adhere to lifecycle, regulations, modeling, environmental, and socio-economic factors. This section discusses these aspects in detail.

Figure 3 shows critical points of the power cable configuration to which special attention must be paid in the design. These points are the hang-off location, sag bend, hog bend, TDP, and, if applicable, the tether clamp close to the TDP with the respective bend stiffeners. Further, the attached buoyancy modules and their attachment points on the power cable need special consideration. These points are especially prone to exceeding the limits of certain constraints detailed in the following.

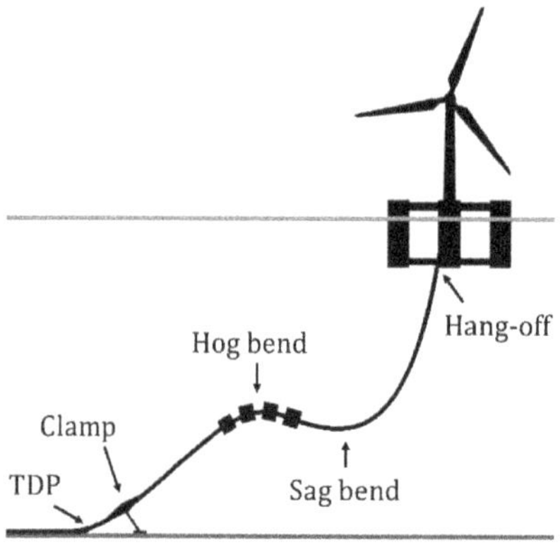

Figure 3. The critical power cable configuration points in the design.

6.1. Environmental Constraints

This section describes how environmental factors impose limitations on power cable configurations. Extensive measurements and appropriate modeling methods are essential for determining statistical descriptions and extreme conditions for analysis [83,84]. Usually, the measurements must be taken directly at the site or nearby.

6.1.1. Wind and Waves

If a part of the power cable configuration extends above the sea surface, it becomes directly susceptible to wind and wave loads. Such a configuration is only chosen in rare instances. Typically, the entire power cable configuration is submerged beneath the sea surface, experiencing an indirect impact from wind and waves. Furthermore, wind and waves significantly determine the FOWT behavior and, consequently, the behavior of the power cable from its hang-off location.

According to DNV-RP-C205 [25] the modeling of waves for short durations is assumed to be stationary, with a duration of 3 h. For analysis regarding stable wind speed, a duration of 1 h is considered as per DNV-RP-C205. This can be disputed, as such stable conditions are somewhat uncommon [85]. Transient conditions, characterized by varying wind speeds and wave conditions, are frequently overlooked in the analyses and might be worth investigating if they can potentially lead to larger responses than initially assumed.

It is crucial to account for the most severe wind and wave conditions, particularly those where the FOWT exerts the greatest influence on the behavior of the power cable. These extreme conditions typically coincide with the generation of currents induced by wind and waves. The most extreme responses of FOWTs can occur during events such as hurricanes and typhoons. Such conditions can generate extreme FOWT motions and currents at the same time, which the power cable configuration must survive.

Furthermore, it is essential to consider special wave conditions, including exceptional occurrences like freak waves and tsunamis. Freak waves, often also referred to as rogue waves, are distinguished by their significantly greater height than the surrounding waves. These phenomena can occur across varying water depths and arise abruptly. Certain regions exhibit a higher statistical occurrence of freak waves, notably the North Atlantic, the North Sea, the waters surrounding Australia, and the Agulhas Current south of Africa [86,87]. Their formation is attributed to the crossing of different sea states characterized by distinct energy levels and frequencies [86], as well as other factors such as interactions between ocean currents and waves [87,88]. On the other hand, tsunamis must be considered where appropriate. Tsunamis typically originate from seismic events and have extremely high water-particle velocities, directly impacting power cable configurations. Unlike freak waves, tsunamis exhibit small wave amplitudes at sea but pose a significant threat as they grow substantially in height when approaching coastal areas.

Another crucial factor to consider is the availability of sufficient wind resources, which determines the feasibility of offshore wind projects. Figure 4 shows the mean wind speeds along the coastlines of the world, with higher speeds generally indicating more favorable wind resources. Additionally, it is essential to consider the local mean wind power density, which indicates the available wind resources at a specific location [89,90].

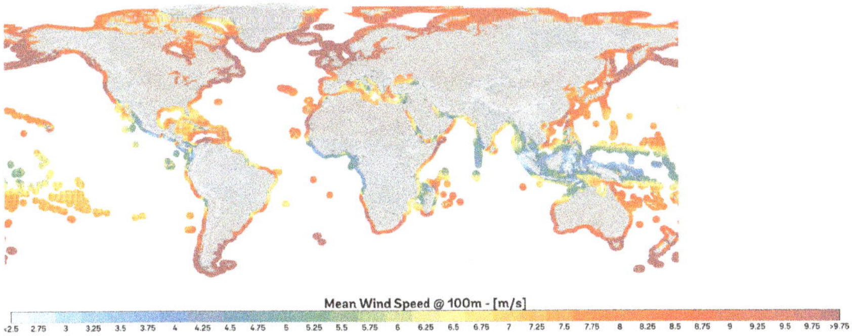

Figure 4. World map of mean wind speed for wind resources (Reference [89]).

6.1.2. Currents

Different currents impacting offshore structures include oceanic currents (like the Gulf Stream), upwelling/downwelling currents (bringing nutrients to the surface/transporting surface water downward), wind-generated currents (influenced by wind speed and direction), tidal currents (due to the gravitational pull of the moon and sun, causing regular patterns of water movement), and coastal currents (influenced by tidal and wind-driven forces, local topography, and storm surges). The local current that must be designed for is usually a combination of the different current types.

Knowing statistics and extreme current events is essential for the design. Most current measurements are averaged over 5 to 20 min, which might undermine "gusty" strong current events [91,92]. This can lead to an underestimation of the prevalent current and its turbulence that must be considered for the design.

Current velocity profiles show the distribution of water particle velocities at different water depths. Often, the fastest velocities occur near the surface and decrease with depth. This is also the case for the general unidirectional current profile suggested in the recommended practice DNV-RP-C205 [25] characterized as a power law function. Therefore, a common design approach for power cable configurations is to avoid strong current regions by installing large cable sections in deeper waters.

Currents can exhibit varying degrees of directionality, ranging from unidirectional to bidirectional or even multidirectional. In unidirectional current profiles, water flows consistently in a single direction, such as in a river or an oceanic current. Bidirectional profiles occur when currents exhibit a reversal in the flow direction, for example, caused by tidal influences. Multidirectional profiles indicate a complex flow pattern, where currents may change direction frequently and vary in intensity. Numerous factors, including geographical features, wind patterns, tidal forces, and the presence of obstructions or constrictions in the water body, influence the directionality of current velocity profiles. Therefore, multidirectional profiles should be considered for analyzing power cable configurations according to the local conditions [93–95].

Extreme current events occur especially at the sea surface and can result from specific local circumstances. An exemplary region with very high surface current velocities is the Northern North Sea, reaching velocities of 2.5 m/s, as shown in Figure 5 [96]. The Norwegian Coastal Current (NCC) causes this. It originates primarily from the freshwater outflow from the Baltic Sea and the freshwater runoff from Norway. It flows northwards along the coast of Norway as a low-salinity current [97]. Figure 6 shows a satellite image of the outflow of the Norwegian Coastal Current from the Baltic Sea and its multidirectional meandering in eddies northwards along the Norwegian Coast. After storm surges from the North, vast amounts of water can flow out from the Baltic Sea. When combined with the tidal current, high current velocities occur along the coast of Norway [11,26,96].

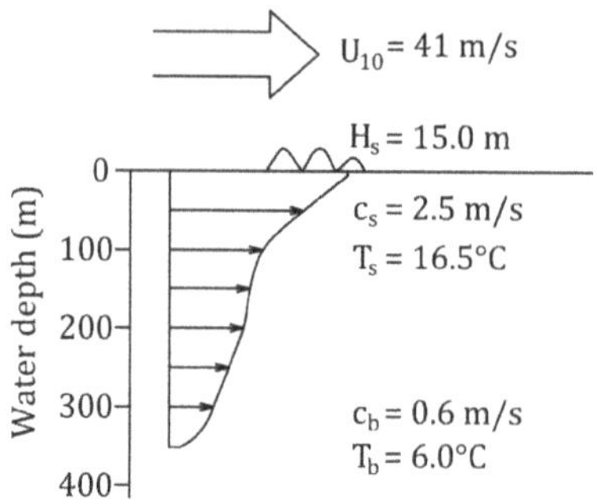

Figure 5. The 100-year metocean design criteria for the oil and gas field Troll location in the Northern North Sea. The figure shows the extreme horizontal current velocity as a function of the water depth. U_{10}—mean wind velocity 10 m above mean sea surface level, H_S—significant wave height, c_S—surface current velocity, c_b—current velocity at the seabed, T_S—water temperature at the surface, and T_b—water temperature at the seabed (see Saetre [96] and Marechal [88]). Adapted from [26] with permission from Standards Norway.

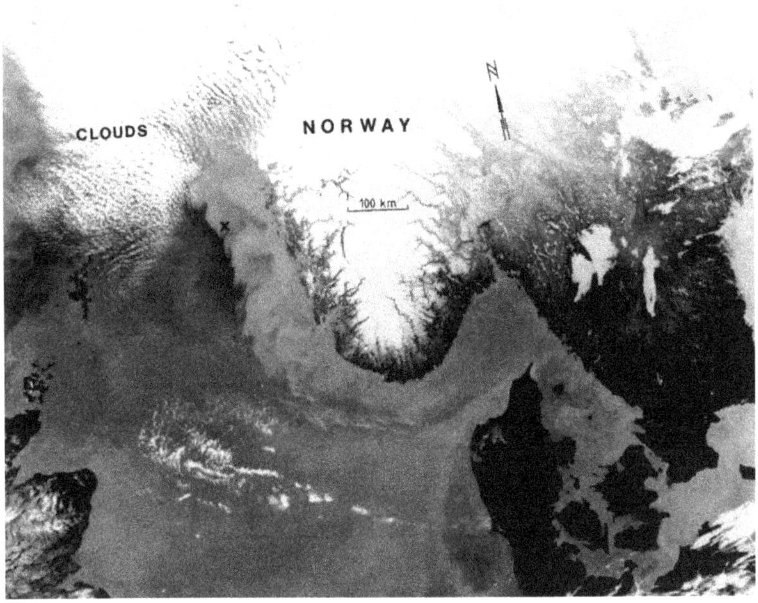

Figure 6. A satellite image showing the Norwegian Coastal Current on 13 April 1981. Courtesy of Tromsø Telemetry Station, Norway.

Another extreme local current example is the loop current in the Gulf of Mexico. It forms from warm subtropical water streaming northwards and through the geographical shape of the Gulf [84]. It leads to large current direction and velocity variabilities throughout the entire water depth, which can also change quickly. Extreme current velocities of up to 2.10 m/s have also been measured at deeper waters, with even larger velocities at the sea surface [84]. Such extreme events can statistically occur every 50 to 100 years, making long-term measurements for currents critical, as well as proper modeling techniques to predict them [98]. Often, measurement campaigns are only carried out for a couple of months to years, which can be insufficient to estimate extreme conditions correctly.

6.1.3. Internal Waves

Internal waves are like waves at the surface of a water body. However, they occur where different water densities meet below the surface, often caused by different water temperatures [84]. The internal density gradients are significantly smaller than those at the water surface, giving the internal waves slightly different characteristics from surface waves [99]. Internal waves can have large amplitudes, wavelengths, and periods. For example, Hyder et al. [99] observed internal waves with amplitudes of about 40 m and periods of 40 min. These waves cause water particles to move both vertically and horizontally with high accelerations [100]. Hyder et al. [99] measured currents up to 0.6 m/s for their particular case of internal waves. A current with this velocity can have a significant impact on power cable configurations.

Internal waves occur in areas all over the world, often near fjords, estuaries, and straits. Some exemplary areas with potential offshore wind farm development with frequent internal wave phenomena are the South China Sea, the Sulu Sea in the southwestern Philippines, the Australian North West Shelf, and the Scotian Shelf southeast of Canada [101]. An overview of possible locations with internal waves of large amplitudes, i.e., internal solitary waves, has been provided by Global Ocean Associates (GOA) [102].

6.1.4. Water Depth

The water depth is critical in determining the dynamic subsea power cable configuration for FOWTs. Shallow waters usually require complex configurations to decouple the cable motions from the floater and the TDP [57]. Suspended configurations are not feasible in shallow water either. The currents imposed on the power cables, as well as the influence of the waves and the FOWT motions, can be significant. Additionally, the space between the seabed and the sea surface is limited, requiring complex configurations that can be prone to damage. Intermediate and deep waters facilitate power cable configuration design compared to shallow waters. However, the weight of the cable itself can pose a problem, as well as large FOWT motions, due to the large flexibility in the mooring systems, especially for deep water configurations. Nevertheless, cables at deeper water depths are less susceptible to marine growth and environmental forces, allowing different power cable setups with space for buoyancy and motion decoupled from the FOWT and the TDP or the next connection point. Furthermore, FOWTs are generally installed at intermediate to deep water depths because bottom-fixed OWTs typically have better economic viability for shallower waters. Therefore, the development of floating renewable energy technologies should focus on areas where bottom-fixed solutions are not feasible.

6.1.5. Marine Growth

Marine growth refers to the accumulation of organisms such as barnacles, mussels, algae, and others on surfaces. As FOWTs are located at the water surface, their power cables span various layers of water where they are subjected to different kinds of marine growth. The attached biomass can drastically increase the size and mass of the cable, strongly impacting its responses.

The presence of marine growth can have several detrimental effects on the power cable. First, it increases the actual diameter and roughness, and thus the hydrodynamic drag, potentially leading to higher structural loads. This can be particularly problematic when exposed to strong currents, waves, and floater motions. Second, the additional weight of marine growth can affect the load distribution on the power cable. It can lead to uneven loading and stress concentrations, potentially compromising the safety of the power cable configuration. Third, marine growth can also cause localized corrosion due to biofouling, accelerating the degradation of protective coatings and metal surfaces.

Several offshore standards, such as ISO 19901-1 [18], DNV-OS-E301 [22], NORSOK N-003 [26], and API RP 2A-LRFD [27] recommend different considerations of the marine growth effects in the design analysis of offshore structures. They agree to increase the diameter to an equivalent diameter corresponding to the roughness of the surface. Further, the drag coefficient is adjusted depending on the roughness expected at the installation location. Generally, higher water temperatures facilitate the accumulation of marine growth compared with lower water temperatures. Deeper waters have less marine growth that will be attached to the power cable than close to the surface due to the temperature, the sunlight availability, and the general flora and fauna in the different regions [26]. Accumulation can, in part, be avoided by configuring the buoyancy arc of configurations in deeper waters.

Nevertheless, the different variables used are the subject of several publications that challenge the standard. For example, Marty et al. [103] conducted experiments on cylinders with mussels and corals attached. When comparing their experimental results to calculations from DNV-OS-E301 [22] and API RP 2A-WSD [28] they found that these standards sufficiently predict drag coefficients under current conditions within the sub-critical regime but differ significantly for oscillating drag and inertia coefficients in wave and current conditions.

Further research suggests that the assumed marine growth in specific regions is insufficient when applied as described in the standards because marine growth is strongly site-specific [104]. Such an underestimation led to an incident in the Pacific Ocean in the test field off the coast of Fukushima, Japan, where a dynamic power cable connected to an FOWT became damaged due to contact with the seabed [37]. The marine growth changed the configuration of the dynamic cable so that it sank and touched the seabed in its sag point, where it was damaged and needed to be repaired. Since then, marine growth must be removed regularly, which adds costs to the research project. Okubo et al. [105] showed that the marine growth on an FOWT in the Sea of Japan (East Sea) also led to its cable sagging significantly more than expected during the design work. Later, additional installations of buoyancy devices were necessary when a sag point of the dynamic power cable configuration came critically close to the seabed. The marine growth measured at the location exceeds the general reference values suggested in DNV-OS-E301 [22]. The DNV reference values were obtained from locations in the North Sea and the Norwegian Sea, which have lower water temperatures and less marine growth than other locations with, for example, usually higher water temperatures, such as around Japan. Measurement campaigns and analyses should be carried out considering the marine growth that can occur at the specific location.

Preventive measures are often employed to manage marine growth. These can include applying antifouling coatings or materials to the surfaces of the structures to deter or minimize attachment. Okubo et al. [105] successfully applied an antifouling coating on the buoyancy modules attached to a power cable for an FOWT. Nevertheless, antifouling should be employed sparingly or ideally avoided altogether, given the significant environmental hazards stemming from its toxicity [106]. Periodic cleaning and maintenance are also carried out to remove accumulated growth and ensure the structural integrity of the power cable in several projects, as reported by Ishii et al. [37]. If unavoidable, the cost of cleaning the cable on a regular basis is usually significantly lower than risking a critical cable failure and replacement.

6.1.6. Temperature

The temperature of the surrounding environment can have significant effects on power cables. Therefore, they must be designed to withstand the different temperatures in different layers of water and the seabed while efficiently transmitting electrical power [107].

The primary concern for subsea power cables is the temperature of the conductor, which carries the electrical current. High temperatures can be generated due to the resistance of the conductor itself and the electrical load it carries. This heat must be managed to prevent cable overheating, which can lead to insulation degradation and, ultimately, cable failure [38]. Excessive heat can also accelerate the aging of the insulation materials, potentially shortening the operational lifespan. It can also affect the electrical properties of the cable, leading to losses and reduced efficiency. Cold water temperatures can reduce the conductor's temperature. While this may help prevent overheating, it can also make the cable more brittle, which may affect its mechanical properties.

Furthermore, biofouling around the power cable can modify the heat transfer and change the temperature in the conductor, affecting its fatigue life [108]. Depending on the age of the mussels attached to the power cable in experiments by Maksassi et al. [108] the global thermal resistance led the conductor to either overheat or cool down.

6.1.7. Cold Climate

The polar regions are characterized by their unique cold climate and extensive glacial ice. These remote areas have recently become more accessible for commercial use due to technological progress and the receding sea ice. However, the polar regions have delicate ecosystems that can take considerable time to recover from even minor disruptions [109, 110]. Therefore, all activities in these regions must undergo thorough environmental assessments to guarantee long-term sustainability.

Snow and icing, whether from atmospheric conditions or sea spray, can impact FOWTs, subsequently affecting the motions and loads experienced by the power cable. In colder regions, more immediate hazards to power cable setups also include sea ice and icebergs, where the latter can also be a threat to the entire wind farm. Drifting sea ice ridge keels can extend several meters below the sea surface, exceeding the hang-off location of the power cable from the FOWT. In such situations, the hang-off section and the upper part of the power cable must be equipped to withstand the presence of sea ice adhering to it or potential collisions.

6.1.8. Bathymetry and Geotechnical Conditions

Bathymetry defines the setup of the entire footprint of an FOWT farm. The primary concern when laying power cables on the seabed is the geotechnical condition of the ocean floor. Variations in sediment composition, bedrock, and unstable formations can greatly affect the stability and integrity of the power cables. Extensive geotechnical surveys must be conducted to assess the seabed conditions and make informed decisions about cable placement; these surveys should be conducted in parallel with data collection for the mooring systems. The on-bottom stability of the power cable must be considered accordingly.

In regions with soft seabed soil, there is a concern that power cables may dig themselves into the sediment over time. This phenomenon could lead to a snap load with excessive tension and potential damage when the cable is released from the soil. Burying cables over certain distances can be a solution to prevent damage. Also, more unconventional solutions, such as cable protection systems, have been implemented successfully [111]. Areas with rocky seafloor usually have rock piles or mats on top of the power cables to prevent them from moving. A solution without additional measures to keep the cable in place on the rocky seabed has been successfully tested for a power cable connected to a tidal stream turbine in the UK [112].

Certain areas have a significant risk of underwater avalanches, rock falls, and sediment shifting [113]. They can disconnect cables and create catastrophic failures in the FOWT farm, mainly due to anchor failures. On a smaller scale, the cable sections on the seabed can be suspended over distances, leading to vortex-induced vibrations (VIVs) and fatigue. Identifying and avoiding these zones is essential during the planning process of the inter-array cables and the entire FOWT farm.

6.1.9. Climate Change Impact

Due to climate change, statistics over the last decades might not always be a reliable basis to estimate future environmental conditions accurately [114,115]. The environmental conditions have been changing in the last few decades, with a general trend towards more extreme sea states. In some locations, extreme values may appear more frequently than before; this can, for example, mean that the 100-year condition is the new 50-year condition. For example, Feng et al. [115] obtained a change of 0.82 m in annual H_s and 1.39 m in winter H_s over 30 years of measurements in the Norwegian Sea. According to them, this is a change that never occurred that rapidly before their measurements. As another example, Taiwan experiences typhoons significantly less frequently than in the past, whereas the sea temperature around the island is rising at the same time [116,117]. This influences, for example, the wind, waves, current, and marine growth conditions that FOWTs installed there will experience. Modeling techniques and designs must consider these uncertainties and should refer to a 'dynamic design baseline' in the future, where appropriate.

6.2. Response Constraints

Multiple factors influence the behavior of a dynamic power cable. External factors are loads exerted by the surrounding environment and the motions of the components to which the power cable is connected. The inherent structure of the power cable not only dictates how it responds to these external forces but also constrains the flexibility of the cable. The power cable structure has specific thresholds that must not be exceeded to maintain the integrity of the system. Therefore, several methods of analysis to obtain the failure modes of the power cable components by carrying out global configuration analysis have been proposed, for example, by Ran et al. [118] and Young et al. [119]. This section is about these crucial constraints governing the responses of the power cable in the global configuration.

Each structural component and section of a power cable has a distinctive failure mode linked to its material properties and strain response. For example, when a cable is bent beyond its minimum bend radius, it can suffer from a phenomenon known as 'bird caging' [120]. When a typical power cable is bent too sharply, the outer layers are stretched along the outer curve of the bend while the inner layers are compressed. This unequal stretching and compression can cause the outer insulation and shield to deform and create small gaps or openings resembling a birdcage. This damage often occurs due to sharp bends during the installation of power cables. However, it can also appear during operations when the cable is subjected to excessive compression. Dobson et al. [38] provided an overview of the critical failure modes of power cables.

6.2.1. Tension and Compression Limits

The manufacturer of the power cable provides tension and compression limits. Tension limits denote the maximum tensile stress a cable can bear without undergoing permanent deformation or failure. Compression limits define the maximum compressive stress a cable can endure without buckling or collapsing. Exceeding these limits can lead to elongation, plastic deformation, or even breakage, compromising the structural stability of the entire system. Experiments are conducted to obtain these limits; see Ringsberg et al. [48], Grivas et al. [121], and Nasution et al. [122].

One particularly high-risk area for exceeding the maximum tension limit is the hang-off, where the cable experiences motions from the floater, curvature loads, and static tensile loads from the weight of the hanging cable [50,56,123]. Factors such as cable temperature, service loads, solar radiation, insulation, and mechanical and electrical loads contribute to this risk.

The sag and hog bends of the power cable configuration are usually subjected to cyclic compression and bending, which makes them a significant risk area for exceeding the maximum compression limit [118]. If buoyancy modules, bend stiffeners, and clamps are applied with excessive pressure, they can hinder the lateral movement of helical components in the cable [38]. This restriction might cause compressive overload or amplify fatigue loading in the armor or conductors, posing a severe threat to the integrity of the cable. Proper attachment techniques are essential to prevent unintended stress concentrations that could compromise the reliability of the cable.

6.2.2. Curvature Limits

The ability of a power cable to withstand maximum curvature is directly tied to its capacity for enduring compression forces. In most cases, designs allow for minimal to zero compression stresses, strictly adhering to maximum allowable curvature requirements. The manufacturer of the power cable provides the limit values. Note that the minimum bend radius is often used to ensure the maximum allowable curvature instead of the maximum curvature constraint.

An important point of the power cable configuration where exceeding the maximum curvature becomes critical is typically the TDP location. It is subjected to cyclic bending, compression, and abrasive loading due to contact with the seabed. Preventing failure at this point often involves implementing a cable protection system. Such a system restricts bending and shields the power cable from abrasion. By limiting cable movements with a tether, such as in the tethered-wave configuration, stability at the TDP can be better controlled compared with a lazy-wave configuration.

Additionally, there are other crucial locations in the design where surpassing the maximum curvature constraint can be a concern. These include the hog bend, sag bend, hang-off location, and tether clamp. The sag and hog bends must be equipped with appropriate buoyancy to support the cable and prevent overbending. Furthermore, the hang-off location requires a bend stiffener or bell mouth to prevent the power cable from overbending [79,124]. The hang-off section and the seabed tether connection are commonly custom-designed for each project to meet the specific design requirements and ensure optimal performance.

6.2.3. Fatigue

Due to the dynamic loading of the power cable configuration, the cable itself is subjected to significant fatigue. The primary cause of fatigue damage in dynamic cables is the stress resulting from cable bending rather than tension [119]. The relevant design constraint is the fatigue life, meaning the minimum lifetime the cable can operate before breaking.

In academic literature, cables are often simplified as single homogeneous materials with equivalent cross-sections, assuming infinite friction between layers, as seen in Rentscher et al. [55]. This methodology generally offers a conservative assessment of stress and strain levels, as it does not account for potential slippage between layers that might mitigate stress within the cable. Land et al. [125] supported this assumption, as the obtained strain for their full-stick cable model is significantly larger than for their full-slip model. Nevertheless, they chose to use the full-slip model to be more realistic. That the non-slip assumption is conservative can be challenged, as it might not apply to every case. Beier et al. [69] demonstrated higher stress values in their power cable model, incorporating friction, in contrast to the identical cable configuration with zero friction. This finding suggests that considering interlayer slip could result in more realistic estimations of stress and strain in cables.

Most published fatigue estimation studies primarily focus on assessing the fatigue of copper cable conductors. A more extensive study by Beier et al. [69] supports that approach, as the conductor is the first component to fail due to fatigue. The cores are typically made from electrolytic tough pitch (ETP) copper, which exhibits mechanical properties such as stress relaxation and non-linear stress–strain behaviors [125]. Therefore, when evaluating fatigue in such cables, it is recommended by Karlsen et al. [126] to use strain-cycle (ε-N) curves instead of the more common stress-cycle (σ-N) curves.

A larger transmission capacity, either by increased voltage or conductor cross-sections, increases the physical footprint of cable components or introduces new components in a dynamic cable context. For example, a radial water barrier is needed to protect the insulation system against humidity and prevent the water treeing failure mechanism [127]. Fatigue as a failure mode becomes increasingly relevant and critical where the relevant mechanisms and material selection differentiate from the lower voltage standard [44].

A notable issue encountered is the underprediction of low-frequency floater surge motions in numerical software [128–131]. Holcombe et al. [128] compared the numerical results from three potential-flow-based software to the experimental results of an FOWT exposed to waves. All three software models underpredicted the FOWT surge response in low-frequency wave regions compared to the experimental results. This can result in a considerable underestimation of power cable fatigue damage. Land et al. [125] found that by increasing the low-frequency surge responses of the FOWT by 20%, based on the findings of Robertson et al. [130] there was a notable increase of 2.1% in the curvature of the power cable. Furthermore, this adjustment led to a significant 13% increase in the annual fatigue damage compared to the simulation results that did not incorporate the increased surge responses.

6.2.4. Torsion Limits

In general, umbilicals possess relatively low torsional stiffness, making them vulnerable to damage caused by excessive torque or twisting [132]. The risk of torsion damage is particularly high during installation [132]. During operation, torsional failure is not regarded as a large threat, as it is seldom reported as an incident [133]. Torsion has a minimal impact on the structural response of umbilical cables during the global configuration analysis [134]. When the wires are coiled in a helical pattern, the power cable experiences tension when it is twisted. These loads can lead to the rupture of helical elements if applied excessively [133]. However, the torsion within helical elements is minimal and does not affect the cross-section of the umbilical cable; instead, tension induces the torsion [134]. Therefore, torsion limits are usually neglected in the analysis of power cable configurations, and often even the effect of torsion itself.

6.2.5. Snap Loading

Snap loading refers to a sudden, significantly higher load on a power cable than its rated capacity. The most common occurrence of such an event is during installation, but it can also occur during operation. This excessive load can be caused by factors such as intense drag forces on specific cable parts, like the buoyancy section, and pulling forces induced by the FOWT. A more common scenario involves the cable becoming stuck in seabed mud. If weather conditions cause substantial floater motions, the cable can be strongly pulled at the TDP, leading to a snap load. Design considerations must prevent such occurrences.

6.2.6. Resonance

The oscillations of the FOWT floater can be close to the natural frequencies of the power cable in specific degrees of freedom. Resonance poses a significant risk, potentially causing cable damage and fatigue. The fundamental frequency of the power cable setup can be adjusted, especially by positioning buoyancy at specific distances from the hang-off location. The fundamental frequency of the power cable heavily relies on the catenary between its hang-off and first buoyancy attachment [50].

When the power cable is suspended and the connected FOWTs move asynchronously, as is typical in actual sea conditions, a resonance between the movements of the FOWTs and the power cable can occur. In this context, the bending radius can be a critical factor, potentially leading to failures.

Resonance can also arise from vibrations originating in the FOWT, transmitted to the power cable through the hang-off point. This is important because the FOWT inherently vibrates due to rotor rotation and system elasticity. The configuration must be designed to avoid exposure to its natural frequencies to prevent issues.

The impact of the power cable configuration on the FOWT is usually small. The power cable is usually considerably lighter than the FOWT structure. However, when a heavy cable is used on a lightweight platform, it can substantially affect the responses of the platform. Schnepf et al. [50] showed that their suspended power cable increases the low-frequency sway and roll responses of the spar FOWT to which it is attached. If this occurs, it must be accounted for in the analysis.

6.2.7. Vortex-Induced Vibrations (VIVs)

Vortices are formed behind a structure encountering a current. These vortices impose forces on the structure, resulting in VIVs. They can cause increased drag forces in the flow direction and significant accumulated fatigue damage [101,135]. The impact angle of the current towards the power cable configuration greatly influences the VIV behavior [136]. Discrepancies between different VIV estimation software can occur [136] and should be subject to further investigations. If VIVs are present in the power cable configuration, they should be suppressed with devices such as strakes, or the configuration should be redesigned to avoid VIVs [137,138]. Optimization of the configuration to avoid VIV effects can save costs due to a longer asset lifetime [138,139].

6.3. *Geometrical Constraints*

Geometrical constraints involve seabed and sea surface clearance, as well as ensuring safe placement regarding other infrastructure and ancillaries.

6.3.1. Clearance to the Seabed and Sea Surface

A specified minimum distance from the seabed, i.e., seabed clearance, in specific sections, is a constraint in nearly all published optimization algorithms for power cable configurations. This requirement is crucial to prevent uncontrolled contact between the power cable and the seabed, which can damage the cable. Additionally, it aims to prevent a change in behavior in the power cable due to the near-wall effect. Among all sections of the power cable configuration, the sag bend is the most critical area where unwanted contact with the seabed can occur.

Opposite to the sag bend, the hog bend has a risk of encountering the sea surface. This must be avoided to prevent large loads on the power cable from waves. Depending on the type of optimization algorithm, a constraint on the sea surface clearance should be implemented.

6.3.2. Placement to Other Infrastructure

The power cable must be placed so that it does not cross any other lines when viewed from atop, such as other power cables, moorings, and other infrastructure. This is a primary concern in optimization algorithms for the overall layout of the inter-array cable system. Also, this needs to be considered, especially for the dynamic power cable configuration excursions. Depending on the environmental conditions, the power cable can have large excursions leading to potential interaction with mooring lines, which must be avoided. Crossing pipelines and cables can be necessary and possible with special precautions [21].

6.3.3. Inherent Constraints of Ancillaries

Ancillary components are typically designed to meet specific configuration requirements, but also have inherent limitations. For instance, buoyancy modules are rigid and cannot conform to the curvature of the power cable. Similarly, bend stiffeners are limited in their flexibility due to material constraints [140,141]. These limitations primarily stem from production constraints, such as restricted mold casting and oven sizes, which impose geometric restrictions. Additionally, it is important to guarantee that the design of ancillary components facilitates installation with minimal risk and cost.

6.4. Lifecycle Constraints

The project lifecycle for a power cable configuration can be divided into seven phases: preliminary design, detailed design, manufacturing and testing, load-out, transport and installation, commissioning, in-service, and decommissioning [21]. Each phase imposes limitations on the design, which are discussed in this section.

6.4.1. Installation

According to GCube [3] approximately 44% of their cable insurance claim expenses can be traced back to errors made by contractors during transit and cable laying. Therefore, it is crucial to consider complying with the constraints during load-out, transport, installation, and commissioning. During installation, similar response and geometrical constraints apply as those during operation. Additionally, the torsion damage risk is high, occurring below the final sheave on a laying vessel when it turns. This occurs due to the weight of the cable inducing a twist, leading to friction preventing rotation as it exits the sheave, potentially causing various failures [142]. Preventive measures involve maintaining cable tension and utilizing torque-balanced armor to minimize torsion effects. Other noteworthy risks include elongation damage and crushing, where the cable falls uncontrolled onto the seabed. Okkerstrøm et al. [143,144] provided an overview of risks and factors to consider during the installation.

6.4.2. Operation and Maintenance

The primary operational cost drivers for offshore wind farms are regular inspections and failures that necessitate maintenance and repair of moorings and cables [1]. A durable system design with minimal failure is crucial to reduce these costs. Also, the power cable configuration must be designed to guarantee adequate access for inspection, maintenance, and repair. When major repairs on FOWTs are required, they might be towed to port, as for the Hywind Scotland FOWT farm [145]. The cable will be disconnected from the FOWT and left at its current location. The design must consider how the cable can be placed in this situation. Possibilities include dropping the cable to the seabed or keeping it afloat with buoys. Measures should also be implemented to prevent entanglement with nearby fishing equipment.

6.4.3. Decommissioning

How the power cable configuration will be decommissioned must be considered from the start of the design. How the seabed will look afterwards is specified in governmental regulations. The decommissioning process of the power cable configuration needs to be planned from the initial design stage. Ongoing debates exist about whether all objects added during installation should be removed or if disturbing the seabed during cable extraction is acceptable [146]. Governmental regulations specify the post-removal seabed appearance [147]. Additionally, the recycling possibilities for power cables should influence the initial cable selection [53].

6.5. Modeling and Analysis Limitations

Efficient simulations are crucial for cost savings, sparking ongoing debates in research on minimizing simulation durations [148,149]. Fast and accurate results are crucial, especially in optimization when numerous simulations are required. The equation of motion is solved in many software, such as described by Orcina [150]. Often, a compromise between time efficiency and precision must be found. Various methods are therefore employed:

Frequency domain analyses: Analyzing power cable configurations in the frequency domain shows their responses to various frequencies. Identifying critical resonance frequencies enables adjustments in the system design to prevent resonances. Frequency domain analysis is generally fast. However, it fails to capture the dynamic responses of coupled complex systems, for example, the FOWT floater and rotor.

Static and steady-state analyses: Steadier results can sometimes be obtained through steady-state or static calculations in optimization algorithms. However, they often necessitate the use of safety factors for accuracy.

Dynamic simulations: For the most precise analyses and optimizations, dynamic time-domain simulations are employed. However, these simulations usually demand significantly more time than frequency and static analysis. Optimization typically means a compromise between time efficiency and precision.

6.6. Environmental Impacts

Installing, operating, and decommissioning subsea power cables have significant environmental impacts, as outlined by Taormina et al. [151]. Installation and decommissioning cause underwater noise and risk of chemical pollution, while burying or laying the cable on the seabed leads to sediment resuspension and habitat reworking. During operation, a primary chemical concern involves potentially releasing buried pollutants in the sediment, such as heavy metals and hydrocarbons [151], resulting from sediment resuspension at the TDP. Sediment toxicity assessments in potentially polluted areas can help cable routing planning to prevent pollutant mobilization [152]. Sea electrodes at monopolar DC cables emit harmful electrolysis by-products such as chlorine and bromine, polluting the water [151]. Artificial electromagnetic fields can disrupt the geomagnetic abilities of marine organisms [153]. Low-frequency electromagnetic fields can negatively affect various species, with evidence from studies on cockles [154], crabs [155], sponges [156], and eels [157]. Adequate shielding and cable burial can reduce exposure from sensitive species [152]. The dynamic cable section between the seabed and FOWT poses entanglement risks for animals, especially for large mammals [158,159].

On the other hand, FOWT farms can accommodate a wide variety of marine life. It can be a refuge for fish, as fishing activities cannot be carried out between the FOWTs due to mooring lines and power cables. They can also have a "reef effect" [151,153], which is the creation of artificial reefs due to the increased habitat complexity of the unburied cables. It is commonly regarded as a favorable human-caused influence since artificial reefs have higher fish and decapod biomass densities than nearby seabed areas [151]. Without the fisheries in the area, it can function as a reserve for these populations and stabilize them in the surrounding areas. That was already determined for the European lobster and the edible crab in offshore wind parks [160,161]. All potential effects of the power cables on the environment should be considered in future developments, and negative impacts should be avoided [53,162,163].

6.7. Socio-Economic Aspects

Fishing is the livelihood of many coastal communities. Ensuring that fishing activities remain undisturbed by the FOWT farm is crucial. Finding a compromise that satisfies all parties involved is essential if disturbances occur. Of particular concern are power cables, which pose significant risks to the fishing industry. Nets and crustacean traps can become entangled in the dynamic sections of these cables, while cables on the seabed can obstruct trawling operations. Fishing activities account for 52% of the causes of damage to telecom cables in the Atlantic [53]. It is essential to engage the fishing industries from the beginning of the project, ensuring their consultation and active involvement in the planning process [7].

Like the fishing industry, the placement of the FOWTs must take shipping routes into account. This is crucial to prevent collisions between ships and FOWT farm structures [164] and avoid significant detours for ships [7]. Suppose an FOWT farm is near a port where ships often anchor while waiting for entry; it poses a significant risk to the power cables. This concern should be addressed from the beginning of the project, and early engagement with local harbors and shipping industries is essential.

The primary immediate benefit of an FOWT farm, in contrast to other existing sources of electrical energy for society, lies in its offshore location, often hidden from public view. The only link between the offshore wind farm and the land where the electricity is required is through power cables. The power distribution network is frequently ill-equipped for integrating an FOWT farm, potentially requiring construction activities on land. This situation could also allow coastal communities to create new employment opportunities. For further information, Holmås et al. [165] provide an overview of the societal repercussions of the Hywind Tampen FOWT farm.

7. Optimization

This section presents the most important parameters for design and optimization in a flowchart. Further, published studies on power cable configuration design optimization, as well as for umbilical and riser configurations in the oil and gas industry, are presented and discussed.

7.1. Determining the Most Relevant Optimization Parameters

Due to limited computational capabilities and time, analysis and optimization cannot be performed on all design parameters. Further, variables and constraints on power cable configuration design are heavily site-dependent and determined by the local conditions. Therefore, choosing the most relevant variables and constraints for the specific location is crucial for a safe and efficient design. Figure 7 shows a flowchart as a reference to assist in the process of optimizing parameter selection. It is divided into the following main sections: objectives, variables, and constraints. The subsections outline the critical factors of each parameter.

The key is optimizing with respect to the constraints that influence design most. In the flowchart, each constraint includes the crucial factors to consider and their influence on the design of the power cable configuration. Yet, uncertainties surrounding these constraints necessitate careful consideration and integration into the design and optimization processes. Using this flowchart as a decision-making tool aids in identifying parameters with significant design impact.

The selection of design parameters can also involve considering multiple, often conflicting, factors. For example, it is crucial to design the connection of the power cable to the FOWT properly. This typically involves using a bend stiffener. The multi-criteria assessment at this stage considers factors such as the ability of the bend stiffener to bend and the minimum bend radius requirements of the power cable set by the manufacturer.

The ultimate constraint on selection depends on the experience of the engineer. For instance, if the current is very weak, bringing this constraint into the optimization procedure might not be relevant, as the water wave kinematics is much larger than the current velocity. However, it might be the most important constraint for the design in the case of a relatively strong current; see Figure 5. Choices made in the project, such as whether marine growth will be cleaned regularly or not, will influence whether this is an important parameter to implement in the optimization procedure. These constraints must be carefully considered one by one.

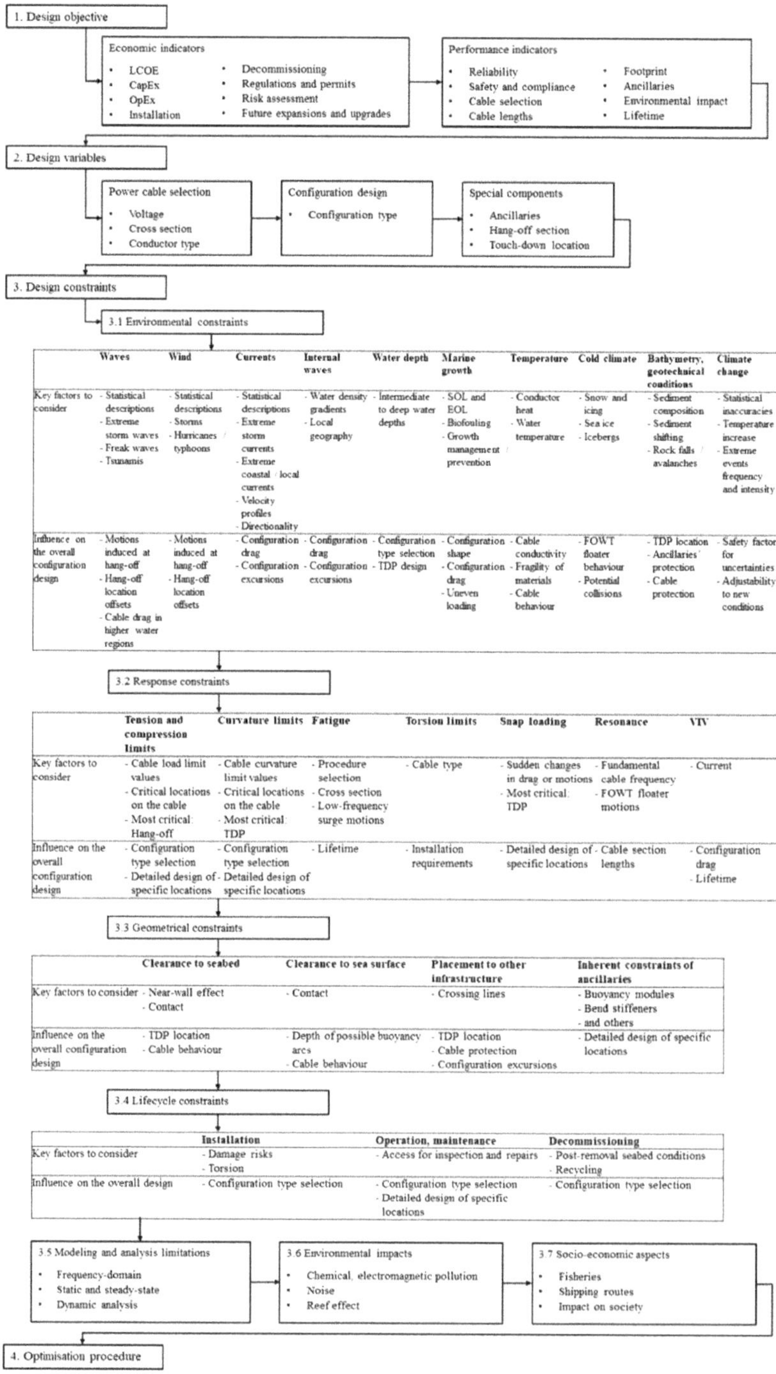

Figure 7. A flowchart outlining a step-by-step approach for identifying key optimization parameters in dynamic cable configuration design.

Appendix A shows the application of the flowchart for a location in the Northern North Sea to demonstrate the selection of the most important design objectives, variables, and constraints for design and optimization.

7.2. Power Cable Configuration Optimization for FOWTs

Kallinger et al. [166] and Song et al. [167] provided reviews on the existing literature relevant to optimizing the layout of the entire electrical system of offshore wind farms. Song et al. [167] and Lerch et al. [168,169] developed optimization methods specifically for the inter-array layout planning of FOWT farms, highlighting the significant impact of dynamic cable configurations, which is absent in bottom-fixed wind farms. Similarly, Giassi et al. [170] concluded that the hydrodynamical interactions significantly influenced the optimal design of their studied WEC farm power cable layout. In contrast, the lengths of the inter-array cables did not substantially impact the WEC farm economy.

Patryniak et al. [171] and Faraggiana et al. [172] reviewed numerous optimization algorithms for FOWT support structures. It is noteworthy that power cable design is integrated into only one algorithm [173] among their listings. However, ensuring the integrity of the power cable should be a fundamental consideration in the platform and mooring system design, limiting FOWT motions accordingly [119,171,174–176].

Table 1 presents an overview of published studies on the optimization of dynamic power cable configurations for FOWTs. Their optimization procedures utilize Sequential Quadratic Programming (SQP), a Genetic algorithm (GA), or a fitness function.

Table 1. Publications on the optimization of power cable configurations for FOWTs.

Reference	Fylling and Berthelsen 2012 [173]	Poirette et al., 2017 [177]	Rentschler et al., 2019 [55]	Ahmad et al., 2023 [51]
Optimization technique	Non-Linear Programming by Quadratic Lagrangian (NLPQL) algorithm	Derivative-free Sequential Quadratic Approximation (SQA) algorithm	Genetic algorithm (GA)	Fitness criteria
Objective function	Minimize the material costs of spar, mooring, and cable	Minimize the material costs of power cable configuration	Minimize cable section lengths, buoyancy section lengths	Fittest configuration considering cable tension, curvature, and length
Design variables	Cable and mooring fairlead vertical position, line direction, pretension, section lengths and diameters; weight of possible buoy; FOWT spar height, diameter, damper plate diameter	Cable and buoyancy section length, cable and buoyancy diameter, length, and diameter of the base of the bend stiffener at hang-off	Cable length, buoyancy section lengths, and positions	Cable length, attached subsea buoy number, and spacing
➢ Response constraints	Cable max. curvature, tension, min. static horizontal pretension; mooring max. tension, min. fatigue life, static horizontal pretension; FOWT max. inclination, max. draught, min. and max. heave and pitch period, max. nacelle acceleration	Cable max. allowable curvature, min. and max. tension, max. elongation	Cable max. accumulated fatigue damage, max. allowable curvature, min. breaking load	Cable max. effective tension, min. bend radius

Table 1. *Cont.*

Reference	**Fylling and Berthelsen 2012 [173]**	**Poirette et al., 2017 [177]**	**Rentschler et al., 2019 [55]**	**Ahmad et al., 2023 [51]**
➢ Geometrical constraints	Cable and mooring max. horizontal offset; mooring max. slope angle at the anchor	Cable min. seabed clearance, min. sea surface clearance; bounded location of the buoyancy section	Cable seabed clearance; hang-off section dependent on simulation setup	Cable min. seabed clearance, max. horizontal excursion
➢ Environmental constraints	Multidirectional conditions with wave spectrum, wind spectrum, wind profile, and constant current (no current profile provided)	Unidirectional ultimate limit states (ULS) with wave spectrum, constant wind speed, wind profile, and constant current throughout depth	Multidirectional conditions with wave spectrum, constant wind speed throughout height	Unidirectional load cases with wave spectrum, wind spectrum, wind profile, and constant current profile
Modeling and analysis method	Quasi-static mooring line analysis: frequency domain	Finite Element Analysis (FEA), dynamic	Finite Element Analysis (FEA), dynamic	Finite Element Analysis (FEA); steady-state (static), dynamic
Cable configuration type	Lazy wave	Lazy wave	Lazy wave	Suspended
Other components included?	Yes: FOWT spar, mooring	No	No	No
Reference	**Schnepf et al., 2023 [178]**		**Schnepf and Giljarhus [179,180]**	
Optimization technique	Sequential Least Squares Programming (SLSQP) algorithm		Efficient Global Optimization (EGO)	
Objective function	Minimize cable section lengths, buoyancy section length		Minimize cable bending	
Design variables	Cable length, buoyancy section length, and position		Cable section lengths, buoyancy section length, and tether location	
Response constraints	Cable min. and max. tension, tether min. and max. tension		Min. and max. tether tension	
Geometrical constraints	Cable min. and max. section lengths, min. and max. curvature, min. and max. hang-off angles, min. seabed clearance		Cable hang-off angle, seabed clearance	
Environmental constraints	Unidirectional load cases with wave spectrum, wind spectrum, wind profile, and constant current profile		Unidirectional constant current profile, hang-off location displacement	
Modeling and analysis method	Finite Element Analysis (FEA), steady-state (static)		Finite Element Analysis (FEA), steady-state (static)	
Cable configuration type	Tethered-wave		Tethered-wave	
Other components included?	No		No	

7.3. *Umbilical and Riser Optimization in Further Industries*

Other floating renewable energy technologies, such as WECs and floating solar plants, similarly have power cables connecting them with consumers for electrical energy transmission. However, no published optimization procedures have been found for their dynamic power cable configurations.

Optimization procedures have been applied in the O&G industry to design umbilical and riser configurations. Typically, umbilicals in O&G applications are larger and heavier than power cables, and they may also include pipes that carry fluids. Power cables are a specific type of offshore umbilicals, often also referred to as 'power umbilicals.' Risers are pipes that transport different fluids throughout their use. Unlike risers or many O&G umbilicals, power cables do not carry any fluids; they have a solid core at their center, which is a key difference. The pressure inside the umbilical pipes and risers means that their design needs can be different from those of power cables, which can have a lot of friction between their internal and external parts. Nevertheless, their similarities in many properties, and especially in shape, are important when looking at optimization procedures that can also be considered for power cable configuration design.

Table 2 summarizes published optimization methods for O&G industry umbilical configurations. Remarkably, there are very few published optimization procedures. The ones published utilize either GA or SQP.

Optimization procedures for riser configurations have been studied extensively in the O&G industry. An overview is provided in Table 3 for configuration types that could also be relevant to dynamic power cable designs. As such, the optimization procedures presented could potentially be considered for application in the design of power cable configurations. This table serves as a resource for selecting and refining optimization procedures for power cable designs in future research. Some modifications might be necessary to adjust the parameters from riser pipes to power cables. For example, von Mises stress limits of the steel pipe are usually critical for risers. However, for power cables, the strain and stress in the conductor carrying the electricity are of high importance.

Furthermore, the oil industry uses suspended pipes to transfer oil from floating ship-type production platforms to tankers. Failures in the pipes and connections, likely caused by fatigue, have been reported, particularly for pipelines located off the coast of West Africa near Angola [181]. In this region, a strong current towards the north is prevailing, and long swell waves propagate northwards from the southern Atlantic. These long swell waves are in resonance with the natural periods of the vessel.

Cunliffe and McCarthy [182], Vieira [183], Vieira [184], and Laddé [185] published further optimization procedures using Genetic algorithms; the authors could not obtain the publications. Optimization procedures with GAs are the most often published studies, with 21 entries. GAs do often have a long runtime, which is considered unfavorable compared to other optimization methods used in several studies, for example, by Cardoso [186] and Vieira et al. [187,188]. Gradient-based methods, like SQP, are usually faster in runtime than GA methods. Nevertheless, they bear a higher risk of finding a local minimum than the best global solution. Converging to a local minimum may be acceptable if it leads to useful improvements in the design. However, if a local minimum is far from the global optimum or does not result in a better solution than the initial setup, it might be preferable to use more effective methods. Even though the combined design of the mooring system and the riser of a floating platform is considered crucial for its functionality [14], the published studies focus solely on riser optimization without the mooring system.

Successful umbilical and riser optimization procedures must be considered for optimizing power cable configurations to ensure that the learnings from the O&G industry are also used in the wind industry to avoid repeating failures. The main learning from the O&G industry is the need to identify the most important design constraints. Focusing on critical constraints makes the optimization process more efficient, leading to better performance and reliability in dynamic power cable designs. This approach is taken forward in the present work. Additionally, integrating robust optimization techniques, or a combination of different methods, can mitigate the limitations inherent to each method when used independently, ensuring a balance between computational efficiency and solution accuracy.

The two most common optimization objectives in Tables 2 and 3 are minimizing cost and line lengths. Line length minimization is often selected to reduce material costs, particularly relevant for risers with significantly larger diameters. However, the length reductions achieved are often not significant enough to decrease costs considerably. Other optimization goals, such as a long design life, should be preferred. Most optimization procedures also apply response, geometrical, and environmental constraints.

Table 2. Published studies on the optimization of umbilical configurations.

Reference	**Wang et al., 2011 [189], Yang et al., 2012 [190]**	**Yang et al., 2021 [191]**	**Yan et al., 2023 [192]**
Optimization technique	Non-dominated Sorting Genetic Algorithm II (NSGA-II)	Non-dominated Sorting Genetic Algorithm II (NSGA-II)	Non-Linear Programming by Quadratic Lagrangian (NLPQL) algorithm
Objective function	Minimize the buoyancy per unit length, length of the buoyancy section	Minimize umbilical max. tension strain, and bending moment	Minimize the umbilical length, cumulative fatigue damage
Design variables	Buoyancy per unit length, end positions of the buoyancy section	Umbilical thickness, helical angle, length, hang-off angle, buoyancy section length, and position	Angles at the hang-off and buoyancy section ends
➢ Response constraints	Umbilical max. top tension, min. bending radius, max. equivalent stress	Umbilical max. tension, max. curvature, min. fatigue life	Min. fatigue life
➢ Geometrical constraints	Min. and max. buoyancy per unit length, min. and max. position of the buoyancy section ends	Umbilical min. and max. thickness, min. and max. helical angle, min. and max. length, buoyancy section min. and max. length, min., and max. position; min. and max. hang-off angle	Min. and max. angles at hang-off and buoyancy section ends, umbilical min. and max. length
➢ Environmental constraints	Unidirectional load cases with regular waves, constant current, and current profile	Multidirectional extreme load cases with regular waves, constant current (no current profile provided)	Multidirectional load cases with wave spectrum, current (no current profile provided)
Modeling and analysis method	Response Surface Model (RSM) and Radial Basis Function (RBF) approximation models	Approximation through a neural network based on Radial Basis Function (RBF)	Approximation through a surrogate model with Radial Basis Function (RBF)
Umbilical configuration type	Lazy wave	Lazy wave	Lazy wave
Other components included?	No	Yes: Cross-section	No

Table 3. Published studies on the optimization of riser configurations.

Reference	**Riggs 1989 [193]**	**Riggs and Leraand 1991 [194]**	**Larsen and Hanson 1999 [195]**	**Vieira et al., 2003 [196]**	**Cunliffe et al., 2004 [197]**
Optimization technique	Sequential Quadratic Programming (SQP)	Adaptive Discretization algorithm	Sequential Quadratic Programming (SQP)	Genetic algorithm (GA)	Genetic algorithm (GA)
Objective function	Minimize riser length	Minimize the riser's overall length	Minimize riser wall thickness linked to material cost	Minimize the acquisition cost of riser sections	Minimize the bill of materials

Table 3. *Cont.*

Reference	**Riggs 1989 [193]**	**Riggs and Leraand 1991 [194]**	**Larsen and Hanson 1999 [195]**	**Vieira et al., 2003 [196]**	**Cunliffe et al., 2004 [197]**
Design variables	Riser section lengths, hang-off height, and buoyancy section weight	Riser section lengths, weight of the second riser section, and the mean location of the vessel	Riser section lengths, wall thickness, and horizontal distance from the lower to upper end (far position)	Riser section lengths, buoyancy section length	Riser section lengths
Response constraints	Riser max. top tension, max. top tension variation, min. service life	Riser min. bend radius, max. top tension variation	Riser sections' max. equivalent stress, critical bending strain, pressure, buckling threshold	Riser max. von Mises stress, max. operating stress	Riser max. wall stress, collapse pressure, collapse propagation
Geometrical constraints	Riser min. bend radius, max. angle at hang-off, min. seabed clearance, max. span, buoyancy max. elevation	Riser min. seabed and sea surface clearance, max. buoyancy elevation, min. first section length, min. and max. bottom angle, max. bottom angle variation, max. top angle variation	Riser min. overall length, min. length on the seafloor in the far position	Riser max. top angle, min. and max. section lengths, buoyancy section min. and max. length	Riser far end is horizontal on the seabed for tie-in, material 30 m from TDP on the seabed is changed from titanium to duplex
Environmental constraints	Varying hang-off positions	Uniform current, hang-off location position	Hang-off location position, current (no profile provided)	None	Near and far ultimate limit states (ULS) (applied parameters not provided)
Modeling and analysis method	Analytic equations; static	Catenary equations, cable elements, static, steady state	Finite Element Analysis (FEA); static	Modeling method not provided; static	Finite Element Analysis (FEA); static, dynamic
Riser configuration type	Steep-wave, lazy-wave, lazy-S	Steep-wave	Catenary	Lazy wave	Catenary
Other components included?	No	No	No	No	No
Reference	**Rodrigues 2004 [198]**	**Roveri et al., 2005 [199]**	**Lima et al., 2005 [200]**	**Tanaka and Martins 2006 [201]**	**Tanaka et al., 2006 [202]**
Optimization technique	Genetic algorithm (GA)	Parametric analysis	Hybrid fuzzy/Genetic algorithm (GA)	Genetic algorithm (GA)	Simulated Annealing (SA)
Objective function	Minimize the acquisition costs of riser components	Find feasible configurations	Minimize riser section lengths and buoyancy section lengths	Minimize acquisition cost and launch cost	Minimize pipe and buoy material volume by price
Design variables	Riser section lengths, tendon length, and buoy net buoyancy	Riser section lengths	Riser section lengths, buoyancy section length	Riser section lengths, outer diameters, and buoyancy section outer diameters	Riser overall length, external diameter, and buoy thickness
➢ Response constraints	Riser top max. tension, max. traction force, tendon top max. traction force	Riser min. fatigue life	Riser max. von Mises stress	Riser min. curvature radius, propagating buckling collapse criteria, max. stress, max. top tension	Riser propagating buckling collapse criteria, max. stress
➢ Geometrical constraints	Tendon vertical max. and min. inclination angles	Riser max. top angle, min. curvature radius, min., and max. section lengths	Riser min. and max. section lengths, max. top angle, buoyancy section min. and max. lengths	Riser min. and max. section lengths, min., and max. outer diameters, buoyancy modules outer min. and max. diameters, max. top angle	Riser max. top angle, min. and max. lengths, min. and max. external diameters, buoy min. and max. thicknesses

Table 3. *Cont.*

Reference	**Rodrigues 2004 [198]**	**Roveri et al., 2005 [199]**	**Lima et al., 2005 [200]**	**Tanaka and Martins 2006 [201]**	**Tanaka et al., 2006 [202]**
➢ Environmental constraints	Variable current velocities, three-dimensional current profile	Current (no profile provided), waves (no specifications provided)	None	Load cases with current at the surface from different directions (no profile provided)	Load cases with different current directions (no profile provided)
Modeling and analysis method	Catenary equations integration, steady-state (static)	Differential ordinary equations with the Runge–Kutta method, static, dynamic	Finite Element Analysis (FEA), static	Modeling method not provided, static	Solving differential equations with the Runge–Kutta method, static
Riser configuration type	Lazy-S	Lazy-wave	Catenary, lazy wave	Catenary, lazy wave	Catenary, lazy wave
Other components included?	No	No	No	No	No
Reference	**Tanaka and Martins 2007 [203]**	**Vieira et al., 2008 [204]**	**Yang et al., 2009 [205]**	**Tanaka 2009 [206]**	**Vieira et al., 2009 [188]**
Optimization technique	Genetic algorithm (GA)	Artificial Immune System (AIS), Genetic algorithm (GA)	Island-based Genetic algorithm (IGA)	Simulated Annealing (SA), Genetic algorithm (GA), Sequential Linear Programming (SLP), Linear Programming, Powell's Conjugate Directions Method, Quasi-Newton, Interval Division, Fletcher-Reeves, Hessian matrix approximation	Artificial Immune System (AIS), Genetic algorithm (GA), Particle Swarm Optimization (PSO)
Objective function	Minimize max. stress amplitude	Minimize riser length and buoy volume by price	Minimize riser material	Minimize volume per price, max. static stress, max. dynamic stress amplitude	Minimize riser length and buoy volume by price
Design variables	Riser section lengths, buoyancy floater diameter	Riser section lengths, buoy diameter, length, spacing	Riser wall thickness, coating density	Riser section material, length; buoyancy module number, material, inner and outer diameter	Riser section lengths, buoy diameter, length, spacing
Response constraints	Riser min. tension, min. curvature radius, max. stress	Riser max. von Mises stress, max. top stress, min. stress	Riser max. von Mises stress, collapse criteria	Riser buckling threshold, max. top traction, min. and max. tension, max. equivalent tension	Riser max. von Mises stress, max. top stress, min. stress
Geometrical constraints	Riser min. and max. section lengths; buoyancy min. and max. floater diameter	Riser min. and max. section lengths, min. and max. top angles, max. top angle variation, buoy min. and max. diameters, min. and max. lengths, min. and max. spacing	Not specified	Riser external diameter > internal diameter, max. length, max. top angle, min. curvature	Riser min. and max. section lengths, min. and max. top angles, max. top angle variation; buoy min. and max. diameters, min. and max. lengths, min. and max. spacing
Environmental constraints	One load case with steady current profile, hang-off offset, and motion	Not specified	Not specified	Multidirectional steady current profiles	Not specified

Table 3. *Cont.*

Reference	**Tanaka and Martins 2007 [203]**	**Vieira et al., 2008 [204]**	**Yang et al., 2009 [205]**	**Tanaka 2009 [206]**	**Vieira et al., 2009 [188]**
Modeling and analysis method	Finite Element Analysis (FEA), frequency domain, static, dynamic	Analytic catenary, static	Finite Element Analysis (FEA), dynamic	Finite Element Analysis (FEA), static, dynamic	Analytic catenary, static
Riser configuration type	Catenary and lazy-wave	Lazy wave	Catenary	Lazy wave	Lazy wave
Reference	**Andrade et al., 2010 [207]**	**Pina et al., 2008, 2011 [208,209]**	**Zhen and Yang 2010 [210], Yang and Zheng 2011 [211]**	**Tanaka and Martins 2011 [212]**	**Vieira et al., 2012 [187]**
Optimization technique	Non-dominated Sorting Genetic Algorithm II (NSGA-II)	Particle Swarm Optimization (PSO)	Single-loop reliability-based design optimization (RBDO)	Genetic algorithm (GA); Simulated Annealing (SA)	Genetic algorithm (GA); Particle Swarm Optimization (PSO); Artificial Immune System (AIS)
Objective function	Minimize the buoyancy module volume, riser top tension	Minimize the cost of material	Minimize the wall thicknesses of riser sections	Minimize max. dynamic stress amplitude along the riser	Minimize the cost and length of each riser section, and the cost and volume of the buoyancy module type applied
Design variables	Riser section lengths, buoyancy section length, buoyancy modules diameters, lengths, and spacing	Riser section lengths, buoyancy section length, buoyancy module length, diameter, spacing	Riser sections' wall thicknesses	Riser section lengths, buoyancy thickness	Riser section lengths, buoyancy modules diameter, length
Response constraints	Installation riser max. tension, stresses, compression, max. VIV fatigue life	Riser max. von Mises stress, max. top tension, min. tension at the bottom	Riser max. dynamic equivalent stress, min. required probability value	Riser propagating buckling collapse criteria, min. curvature, min. tension, max. stress	Riser sections max. von mises Stress, min. tension at seabed sections, max. top tension
Geometrical constraints	Sag and hog bend min. and max. vertical positions; TDP min. and max. horizontal position	Riser top min. and max. angle, max. angle variation, min., and max. riser section lengths; buoyancy section min. and max. length; buoyancy module min. and max. diameter, length, spacing	Riser sections min. and max. wall thicknesses	None	Riser min. and max. top angle, max. 'built-in' angle variation, min., and max. section lengths; buoyancy modules min. and max. diameter, length, spacing
Environ-mental constraints	Multi-directional steady current profiles (profiles not provided)	None	Cosine motion imposed at hang-off	Multidirectional extreme load cases with sea states, steady current profiles	None
Modelling and analysis method	Finite Element Analysis (FEA); static, dynamic	Analytical catenary; static	Design of Experiment (DoE) with Finite Element Analysis (FEA), during optimization, a Kriging metamodel constructed by Latin hypercube sampling (LHS); static, dynamic	Linear model using frame finite elements solved iteratively; frequency domain	Analytical catenary; static
Riser configuration type	Lazy wave	Lazy wave	Catenary	Catenary, lazy-wave	Lazy-wave

Table 3. *Cont.*

Reference	**Andrade et al., 2010 [207]**	**Pina et al., 2008, 2011 [208,209]**	**Zhen and Yang 2010 [210], Yang and Zheng 2011 [211]**	**Tanaka and Martins 2011 [212]**	**Vieira et al., 2012 [187]**
Other components included?	No	No	No	No	No
Other components included?	No	No	No		No
Reference	**Arruda et al., 2012 [213]**	**Martins 2011 [214], Martins et al., 2013 [215]**	**Voie and Sødahl 2013 [216]**	**Nariño et al., 2014 [217]**	**Chen et al., 2016 [218]**
Optimization technique	Genetic algorithm (GA)	Non-dominated Sorting Genetic Algorithm II (NSGA-II)	Sequential Quadratic Programming (SQP)	Non-dominated Sorting Genetic Algorithm II (NSGA-II)	Surrogate model: Kriging, Radial Basis Function (RBF); multi-island genetic algorithm (MIGA); Non-linear Programming by Quadratic Lagrangian (NLPQL) algorithm
Objective function	Minimize the total volume of floaters	Minimize total riser volume, stress utilization factor	Find a design to meet the target fatigue life	Minimize the amplitude variations in the von Mises stresses and the total length of the riser sections with hydrodynamic dampers	Minimize riser curvature
Design variables	The floater section starts and ends on the riser	Riser section lengths, outer diameter, heavy coating wall thickness, overlength fraction, buoyancy module diameter, and well horizontal offset	Weight section length, distance to hang-off, distance to buoyancy section, buoyancy key factor, buoyancy section length, buoyancy key factor	Number, position, and length of the hydrodynamic dampers	Riser section length, buoyancy section length, hang-off to TDP distance
Response constraints	Riser max. service tension, max. traction tension at the floaters	Riser max. von Mises stress, min. effective tension, min. radius of curvature	Riser min. fatigue life	Riser max. stress	Riser max. dynamic tension, max. curvature
Geometrical constraints	Riser min. seabed clearance, floater section min. distance to hang-off, max. length	None	None	Max total length of the hydrodynamic damper sections	Riser min. and max. section lengths, max. overall length, max. hang-off angle, min. and max. buoyancy section length, min., and max. hang-off to TDP distance, seabed clearance to sag bend
Environmental constraints	Load case with current and current profile	Two directions of collinear waves with a spectrum, current	Several unidirectional ultimate and accident limit states with regular waves, hang-off offsets	Wave and current (specifications not provided)	One condition with a regular wave, constant current (no current profile provided)
Modeling and analysis method	Finite Element Analysis (FEA), static	Finite Element Analysis (FEA), static, dynamic	Finite Element Analysis (FEA), dynamic	Finite Element Analysis (FEA), static, dynamic	Finite Element Analysis (FEA), dynamic
Riser configuration type	Lazy-wave	Vertical access with weight between the hang-off and buoyancy section	Lazy-wave with weights between the buoyancy section and the hang-off	Catenary with hydrodynamic dampers	Steep-wave

Table 3. *Cont.*

Reference	**Arruda et al., 2012 [213]**	**Martins 2011 [214], Martins et al., 2013 [215]**	**Voie and Sødahl 2013 [216]**	**Nariño et al., 2014 [217]**	**Chen et al., 2016 [218]**
Other components included?	No	No	No		No
Reference	**Yang et al., 2018 [219]**	**Diakonova et al., 2019 [220]**	**Cardoso 2019 [186,221]**	**Ai et al., 2019 [222]**	**Bhowmik et al., 2019, 2020 [223,224]**
Optimization technique	Particle Swarm Optimization (PSO)	Brute force with limited load case selection	Globalized Bounded Nelder–Mead method (GBNM)	Genetic algorithm (GA)	Random Forest (RF) metamodel coupled with Evolutionary Algorithm (RF-EA), Radial Basis Function (RBF) metamodel coupled with Evolutionary Algorithm (RBF-EA)
Objective function	Maximize the utilization rate of tension, curvature	Minimize the cost of procurement and installation	Minimize total riser length	Minimize the max. stress, fatigue damage	Minimize material cost, fatigue damage
Design variables	Riser section length, buoyancy section length, riser wall thickness	Riser top angle, section lengths, buoyancy section length, buoyancy factor	Riser section lengths, buoyancy module diameter, section length, spacing	Riser section length, buoyancy section length, buoyancy module diameter	Riser top angle, sag bend elevation, hog bend elevation, buoyancy section length, buoyancy module diameter
Response constraints	Riser max. tension, max. curvature, min. fatigue life, min., and max. yield stress	Riser min. and max. interface loads (topside, seabed), min. fatigue life	Riser combined loading criterion (max. bending moment and effective tension)	None	Riser max. tension at hang-off, min. tension at TDP, max. von Mises stress, min. fatigue life
Geometrical constraints	Riser min. and max. section lengths, min. and max. hang-off angles, min. and max. spiral angles at buoyancy section ends, min. and max. buoyancy section lengths; min. and max. wall thicknesses	Riser max. offset	Riser min. and max. section lengths, buoyancy module min. and max. diameters, min. and max. section lengths, min. and max. spacing	Riser min. and max. section lengths, min. and max. vertical offset, max. hang-off angles, buoyancy section min. and max. lengths, buoyancy module min. and max. diameters	Riser max. hang-off angle, min. sag bend seabed clearance
Environmental constraints	Unidirectional extreme load cases with waves (wave type not provided), constant current (no current profile provided)	Extreme load case (no specifications provided)	One imposed riser top-end motion case	Unidirectional extreme load case with irregular waves, steady current profile	One extreme condition with a regular wave, constant current, and current profile
Modeling and analysis method	Bi-scale response surface	Finite Element Analysis (FEA), static, dynamic	Finite Element Analysis (FEA), dynamic	Finite Element Analysis (FEA), dynamic	Finite Element Analysis (FEA), dynamic
Riser configuration type	Lazy wave	Lazy wave	Lazy wave, catenary with hydrodynamic damper	wave	Lazy wave
Other components included?	No	No	No	No	No
Reference	**Lal et al., 2020 [225]**	**Ogbeifun et al., 2021 [226]**	**Lal et al., 2021 [227]**	**Elsas et al., 2021 [228]**	
Optimization technique	Genetic algorithm (GA)	2D tabular optimization method	Genetic algorithm (GA)	Bayesian Optimization (BO) algorithm, derivative-free evolutionary hybrid algorithm MIDACO, Globalized Bounded Nelder–Mead (GBNM) method, Genetic algorithm (GA)	

Table 3. *Cont.*

Reference	**Lal et al., 2020 [225]**	**Ogbeifun et al., 2021 [226]**	**Lal et al., 2021 [227]**	**Elsas et al., 2021 [228]**
Objective function	Minimize bending stress, fatigue	Minimize riser hanging length, hang-off tension, stress utilization, fatigue damage around the bends, smeared buoyancy section length, and thickness	Maximize strength and fatigue response, and minimize the cost of the buoyancy section	Minimize total riser length
Design variables	Riser section lengths, hang-off angle, buoyancy section length, buoyancy factor	Riser hanging length, smeared buoyancy section length, and thickness	Riser length between hang-off and buoyancy section, buoyancy section length	Riser section lengths, buoyancy modules spacing, length, diameter, and hydrodynamic damper section lengths
Response constraints	None	Minimum apparent mass ratio of the buoyant sections	Riser max. von Mises stress	Riser max. effective load
Geometrical constraints	Minimum seabed clearance, sag bend always lower than hog bend	Minimum sag height elevation from the seabed, min. arc height	Riser min. and max. lengths between hang-off and buoyancy section, min. seabed clearance, buoyancy section min. and max. lengths, minimum seabed clearance must be higher than the sag bend	Riser min. and max. section lengths, min. and max. top angles, buoyancy modules' min. and max. spacing, min. and max. lengths, min. and max. diameters, hydrodynamic dampers' min. and max. section lengths, min. seabed clearance, max. total riser length
Environmental constraints	Regular heave motion at hang-off	Two unidirectional regular wave load cases	Several load cases (no specifications provided)	One load case with steady current profile, prescribed hang-off motion
Modeling and analysis method	Finite Element Analysis (FEA), dynamic	Finite Element Analysis (FEA), steady-state (static)	Finite Element Analysis (FEA) and neural network, dynamic	Finite Element Analysis (FEA), steady-state (static)
Riser configuration type	Lazy-wave	Catenary, lazy-wave	Lazy-wave	Lazy-wave, catenary with hydrodynamic dampers
Other components included?	No	No	No	No
Reference	**Abam et al., 2022** [229]		**Abam et al., 2022** [230]	**Lal et al., 2023** [231]
Optimization technique	Combined Genetic Algorithm (GA) and Artificial Neural Network (ANN)		Genetic algorithm (GA)	Particle Swarm Optimization (PSO)
Objective function	Minimize riser steel weight		Minimize total structural weight	minimize bending stress, fatigue, and material cost
Design variables	Riser wall thickness, declination angle, and length		Riser wall thickness, length, declination angle at hang-off	Riser hang-off angle, section lengths, buoyancy section length, buoyancy factor
➢ Response constraints	Bursting criteria, buckling (collapse) criteria, buckling propagation criteria, yielding criteria, and fatigue limit state (FLS) criteria		Bursting criteria, buckling (collapse) criteria, buckling propagation criteria, yielding criteria, fatigue limit state (FLS) criteria	Riser max. von Mises stress
➢ Geometrical constraints	Riser min. and max. wall thickness, min. and max. declination angle, min. and max. length		Riser min. and max. wall thickness, min. and max. declination angle, min. and max. length	Minimum seabed clearance, Sag bend always lower than hog bend

Table 3. *Cont.*

Reference	Abam et al., 2022 [229]	Abam et al., 2022 [230]	Lal et al., 2023 [231]
➢ Environmental constraints	One unidirectional load case with irregular waves, wind spectrum, and steady current profile	One unidirectional load case with irregular waves, wind spectrum, and steady current profile	Harmonic motion at the hang-off
Modelling and analysis method	Finite Element Analysis (FEA), dynamic	Finite Element Analysis (FEA), dynamic	Finite Element Analysis (FEA), dynamic
Riser configuration type	Catenary	Catenary	Lazy-wave
Other components included?	No	No	No

8. Conclusions

The operation of FOWTs relies on power cables to connect them with consumers for economic success. Therefore, a durable power cable configuration is crucial for functionality. The primary goal in designing these configurations is to achieve optimal performance while minimizing costs. The design process considers various variables, predominantly centered around power cable selection, configuration design, and incorporating specialized components and ancillary elements.

Local conditions play a fundamental role in determining the constraints applied in design and optimization processes, with the physical environment conditions at the location being paramount. Factors such as wind and wave conditions, and particularly the influence of currents, are critical considerations. A lack of comprehensive data on local current conditions often introduces significant uncertainties in the design. Additional phenomena, including internal waves and issues such as snow, icing, and sea ice in cold climates, can significantly impact the robustness of the configuration. The water temperature, bathymetry, geotechnical conditions, and water depth further contribute to the overall complexity. Marine growth, identified as a major challenge in previous FOWT projects, requires expanded study. Several long-term measurement campaigns already indicate the effects of climate change, potentially leading to more severe environmental conditions than previously experienced. This aspect should be considered for future designs. Environmental impact considerations must be integrated into the initial stages of the design process, aligning with governmental regulations and necessitating long-term planning well before installation commences. Similarly, assessing the effects on fisheries, shipping routes, and coastal communities should be an integral part of the project from the start.

Consideration must be given to the inherent limitations of the power cable, which impact its response. This includes tension and compression limits, curvature constraints, and the required minimum fatigue life. Additionally, it is crucial to consider responses like snap loading, resonance, and vortex-induced vibrations (VIVs), with the latter having the potential to decrease the fatigue life substantially.

Further constraints arise during installation, decommissioning, and maintenance, predominantly influenced by the available operational equipment carrying out the operation. Moreover, limitations linked to modeling, analysis methodologies, and software should be considered, necessitating the incorporation of adequate redundancy for a robust design.

A flowchart is provided (Figure 7) to facilitate the design and optimization of power cable configurations, guiding the selection of optimal objectives, variables, and constraints. The key is to optimize based on the constraints that have the greatest impact on the design.

Author Contributions: Conceptualization and methodology, A.S. and O.T.G.; investigation, A.S.; writing—original draft preparation, A.S.; writing—review and editing, O.T.G. All authors have read and agreed to the published version of the manuscript.

Funding: The first author received funding from CoreMarine and the Research Council of Norway (grant number 320902). The second author was supported in part by funds provided by "The Program for Maritime Competence (MARKOM II)", https://www.markomii.no/om-markom-ii/, accessed 23 April 2025, project 10031; and funds made available to Western Norway University of Applied Science, Haugesund, Norway.

Acknowledgments: The first author would like to thank Øyvind Johnsen and Carlos Lopez-Pavon from CoreMarine for the insightful discussions on power cable configuration design.

Conflicts of Interest: The authors declare no conflicts of interest.

Appendix A. Application Case: Northern North Sea

Finding key parameters for the design and optimization of inter-array power cable configurations in FOWT farms is demonstrated in this section based on a location in the Northern North Sea. An exemplary location in that region is Utsira Nord, shown in Figure A1. It has good potential for FOWT farm development: wind resources are abundant [232,233], and it is close to cities and industries [233].

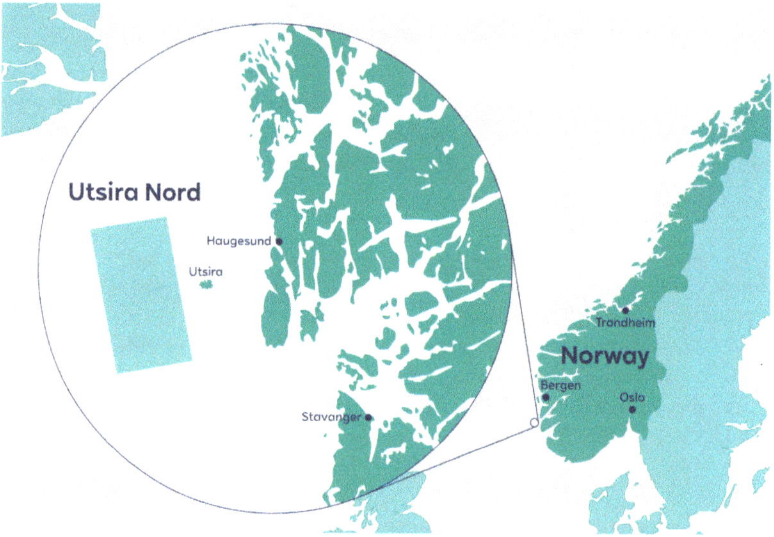

Figure A1. Location in the Northern North Sea with planned FOWT farm development. The figure is availa-ble online: https://utsira-nord.rwe.com/ (accessed on 13 October 2024).

Before designing and optimizing a power cable configuration, or any structure, it is of utmost importance to know the parameters, and especially the constraints, that must be considered. Without a proper design basis, a project can fail easily, and in the worst case, cause huge damage, and cost even human lives. Therefore, the design objectives and variables for power cable configuration design at this location are described in the following section. Afterwards, the design constraints are discussed in more detail.

Design objectives to consider for power cable configuration design are plentiful. Most objectives can be related to costs and performance indicators. Overall, the most important objective should always be a safe design that works. This is important to always keep in mind when choosing more detailed design objectives while designing and optimizing power cable configurations. For example, objectives related to costs are most important for the feasibility of the project. But if the design fails in the long run, the safety might not have been considered correctly.

The variables in a power cable configuration design project can be many. A decision on the type of power cable configuration must be made, a power cable must be chosen, the ancillary equipment must be selected, and the variables inside the chosen configuration must be optimized. A first approach could be to consider the tethered-wave configuration with a copper conductor power cable type like that used in other FOWT farms in the North Sea. Nevertheless, the variables are highly dependent on the constraints of the configuration.

Figure A2 shows constraints that must be considered when designing power cable configurations in the Northern North Sea [234–250]. The environmental conditions are of utmost importance. Wind and waves will significantly influence the behavior of the FOWT floater, which defines the behavior of the power cable hang-off. Storms are frequent in the Northern North Sea, resulting in wind and waves being among the most important environmental loads to consider. The current loads that the power cable configuration will experience are potentially the most important environmental loads to consider. The Norwegian Coastal Current can have very extreme multidirectional current events. They drive changes in the shape of the power cable configuration and might cause the power cable to experience very large tensions that must be considered in the design. Another important factor to consider is the marine growth that will settle on the power cable surface. Even though it might not be as significant in the Northern North Sea as in warmer regions in the world, it nevertheless influences the behavior of the power cable significantly. Other factors, such as internal waves and cold climate, will potentially have a minor effect on the power cable configuration and might therefore need to be considered with less detail than the most important factors.

The responses of the power cable itself must be within the limits the cable can stand. This heavily depends on the chosen power cable. Factors to consider are, for example, tension and compression limits, as well as curvature limits. Some constraints depend on other constraints, such as fatigue, which depends on the load cycles the power cable experiences.

The placement of the power cable configuration might be constrained because of the mooring systems of the FOWTs and other existing infrastructure. Further, it must be possible to install and decommission the power cable configuration, to safely operate it, as well as to maintain and repair it when necessary. Each stage in the project lifecycle poses different constraints on the power cable configuration that must be considered.

Other important aspects to consider are requirements by the issuing government, especially on environmental impacts, which are mainly stemming from socio-economic aspects. For example, several studies must be carried out to show the impact that the power cable configurations as part of the whole FOWT farm can have on the seabed or fish populations.

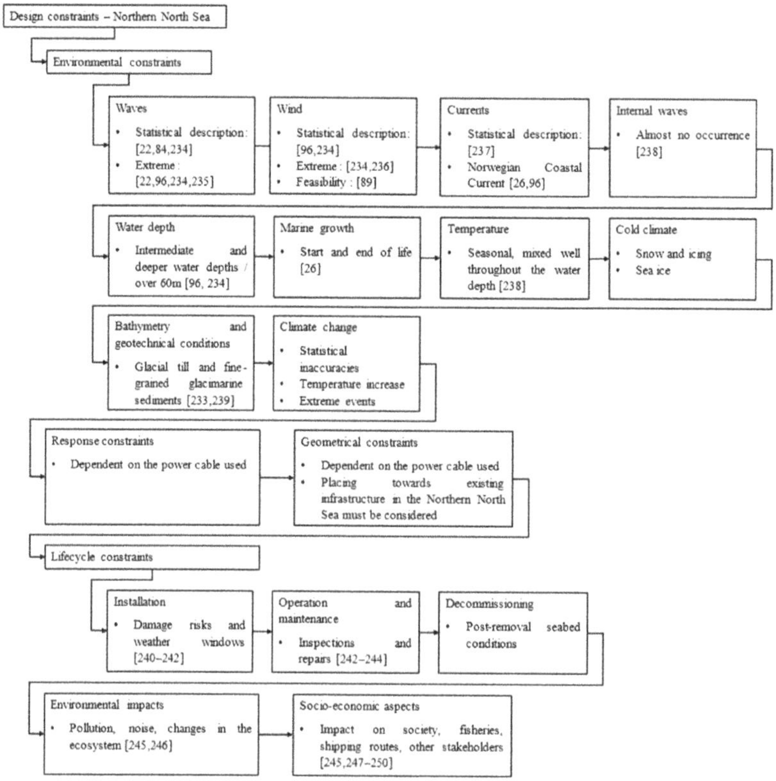

Figure A2. The most relevant optimization constraints for a location in the Northern North Sea [22,26,84,89,96,234–250]

References

1. DNV. *Floating Wind: Turning Ambition into Action*; Det Norske Veritas: Oslo, Norway, 2023.
2. Cozzi, L.; Wanner, B. *Offshore Wind Outlook 2019: World Energy Outlook Special Report*; International Energy Agency (IEA): Paris, France, 2019.
3. WPED. GCube Releases Decade-Long Research Report on Offshore Wind Project Insurance Claims. *Windpower Engineering & Development (WPED)*. 2021. Available online: https://www.windpowerengineering.com/gcube-releases-decade-long-research-report-on-offshore-wind-project-insurance-claims/ (accessed on 9 October 2023).
4. Rentschler, M.U.T.; Adam, F.; Chainho, P.; Krügel, K.; Vicente, P.C. Parametric study of dynamic inter-array cable systems for floating offshore wind turbines. *Mar. Syst. Ocean Technol.* **2020**, *15*, 16–25. [CrossRef]
5. Eldøy, S.; Vold, O.; Graven, H.; Guttormsen, T.; Delp, L. *Hywind Scotland Pilot Park Project Plan for Construction Activities 2017*; Statoil: Stavanger, Norway, 2017.
6. Bowie, M.; Parry, A.; Giles, J.; West, A. *Development Specification and Layout Plan Kincardine Offshore Windfarm Project*; Kincardine Offshore Wind Farm Ltd. (KOWL): Edinburgh, UK, 2018.
7. Equinor. *Hywind Tampen, PL050-PL057-PL089, PUD del II-Konsekvensutredning*; Equinor ASA: Stavanger, Norway, 2019.
8. Duarte, T.; Price, S.; Peiffer, A.; Pinheiro, J.M. WindFloat Atlantic Project: Technology Development Towards Commercial Wind Farms. In Proceedings of the Offshore Technology Conference, Houston, TX, USA, 2–5 May 2022; p. 11. [CrossRef]
9. Metcentre. Test Areas—Metcentre 2023. Available online: https://www.norwegianoffshorewind.no/about/initiatives/met-centre (accessed on 19 November 2023).
10. Yoshimoto, H.; Awashima, Y.; Kitakoji, Y.; Suzuki, H. Development of floating offshore substation and wind turbine for Fukushima FORWARD. In Proceedings of the International Symposium on Marine and Offshore Renewable Energy, Tokyo, Japan, 27–29 October 2013.
11. Gudmestad, O.T.; Schnepf, A. Design Basis Considerations for the Design of Floating Offshore Wind Turbines. *Sustain. Mar. Struct.* **2023**, *5*, 26–34. [CrossRef]

12. *DNV-ST-0145*; Offshore Substations. Det Norske Veritas (DNV): Oslo, Norway, 2021.
13. Do Prado, V.G.; Monteiro, B.D.F.; De Lima, B.S.L.P.; Jacob, B.P. Optimal design of mooring systems for floating production units based on metamodel-assisted Differential Evolution. *Ocean Eng.* **2022**, *266*, 112752. [CrossRef]
14. Ja'e, I.A.; Osman Ahmed Ali, M.; Yenduri, A.; Nizamani, Z.; Nakayama, A. Optimisation of mooring line parameters for offshore floating structures: A review paper. *Ocean Eng.* **2022**, *247*, 110644. [CrossRef]
15. Griffiths, T.; Draper, S.; Cheng, L.; An, H.; Schläppy, M.-L.; Fogliani, A.; White, D.; Noble, S.; Coles, D.; Johnson, F.; et al. The offshore renewables industry may be better served by new bespoke design guidelines than by automatic adoption of recommended practices developed for oil and gas infrastructure: A recommendation illustrated by subsea cable design. *Front. Mar. Sci.* **2023**, *10*, 1030665. [CrossRef]
16. *DNV-ST-0119*; Floating Wind Turbine Structures. Det Norske Veritas (DNV): Oslo, Norway, 2021.
17. *ISO 19900:2019*; Petroleum and Natural Gas Industries. General Requirements for Offshore Structures. International Organization for Standardization: Geneva, Switzerland, 2019.
18. *ISO 19901-1:2015*; Petroleum and Natural Gas Industries. Specific Requirements for Offshore Structures. Part 1: Metocean design and operating considerations (EN ISO 19901-1:2015). Standard Norge: Lysaker, Norway, 2015.
19. *European Standard 13628-5*; Petroleum and Natural Gas Industries. Design and Operation of Subsea Production Systems. Part 5: Subsea Umbilicals. International Organization for Standardization: Geneva, Switzerland, 2009.
20. *European Standard 13628-11*; Petroleum and Natural Gas Industries-Design and Operation of Subsea Production Systems-Part 11: Flexible Pipe Systems for Subsea and Marine Applications. International Organization for Standardization: Geneva, Switzerland, 2007.
21. *DNV-ST-0359*; Subsea Power Cables for Wind Power Plants. Det Norske Veritas: Oslo, Norway, 2021.
22. *DNV-OS-E301*; Position Mooring, Offshore Standard. Det Norske Veritas: Oslo, Norway, 2021.
23. *DNV-RP-0360*; Subsea Power Cables in Shallow Water, Recommended Practice. Det Norske Veritas: Oslo, Norway, 2021.
24. *DNV-RP-F401*; Electrical Power Cables in Subsea Applications, Recommended Practice. Det Norske Veritas: Oslo, Norway, 2021.
25. *DNV-RP-C205*; Environmental Conditions and Environmental Loads, Recommended Practice. Det Norske Veritas: Oslo, Norway, 2021.
26. NORSOK. *N003: Actions and Action Effects*. NORSOK: Oslo, Norway, 2018.
27. API. *Recommended Practice for Planning, Designing, and Constructing Fixed Offshore Platforms-Load and Resistance Factor Design, 2A-LRFD*; Supplemented 1997; American Petroleum Institute (API): Washington, DC, USA, 1993.
28. API. *Recommended Practice for Planning, Designing, and Constructing Fixed Offshore Platforms-Working Stress Design 2A-WSD*; American Petroleum Institute (API): Washington, DC, USA, 2014.
29. API. *Specification for Subsea Umbilicals, API SPEC 17E*; American Petroleum Institute: Washington, DC, USA, 2017.
30. Rinaldi, G.; Garcia-Teruel, A.; Jeffrey, H.; Thies, P.R.; Johanning, L. Incorporating stochastic operation and maintenance models into the techno-economic analysis of floating offshore wind farms. *Appl. Energy* **2021**, *301*, 117420. [CrossRef]
31. Musial, W.; Spitsen, P.; Duffy, P.; Beiter, P.; Shields, M.; Mulas Hernando, D.; Hammond, R.; Marquis, M.; King, J.; Sathish, S. *Offshore Wind Market Report: 2022 Edition*; U.S Department of Energy, Office of Energy Efficiency and Renewable Energy: Washington, DC, USA, 2022.
32. BBC. Jobs Lost as Wave Energy Firm Aquamarine Power Folds. *BBC News 2015*. Available online: https://www.bbc.com/news/uk-scotland-scotland-business-34901133 (accessed on 18 September 2023).
33. Garcia-Teruel, A.; Rinaldi, G.; Thies, P.R.; Johanning, L.; Jeffrey, H. Life cycle assessment of floating offshore wind farms: An evaluation of operation and maintenance. *Appl. Energy* **2022**, *307*, 118067. [CrossRef]
34. Yang, C.; Jia, J.; He, K.; Xue, L.; Jiang, C.; Liu, S.; Zhao, B.; Wu, M.; Cui, H. Comprehensive Analysis and Evaluation of the Operation and Maintenance of Offshore Wind Power Systems: A Survey. *Energies* **2023**, *16*, 5562. [CrossRef]
35. Gudmestad, O.T.; Zolotukhin, A.B.; Jarlsby, E.T. *Petroleum Resources with Emphasis on Offshore Fields*, 1st ed.; WIT Press: Southampton, UK, 2010.
36. Kausche, M.; Adam, F.; Dahlhaus, F.; Großmann, J. Floating offshore wind. Economic and ecological challenges of a TLP solution. *Renew. Energy* **2018**, *126*, 270–280. [CrossRef]
37. Ishii, M.; Ueda, E.; Utsunomiya, T.; Kiyomiya, L.; Harada, B. *Floating Offshore Wind Power Generation Systems off the Coast of Fukushima-Demonstration and Research Project-General Committee Report*; Ministry of Economy, Trade, and Industry; Agency for Natural Resources and Energy: Tokyo, Japan, 2018.
38. Dobson, A. *Challenges Associated with High Voltage Dynamic Inter-Array Cables. Volume 3: Materials Technology; Pipelines, Risers, and Subsea Systems*; American Society of Mechanical Engineers: Melbourne, Australia, 2023; p. V003T04A024. [CrossRef]
39. Equinor. Hywind Tampen: The World's First Renewable Power for Offshore Oil and Gas. *Hywind Tampen 2022*. Available online: https://www.equinor.com/energy/hywind-tampen (accessed on 9 August 2023).
40. Fujii, S.; Tanaka, H. Power transmission system for "Fukushima FORWARD Project"—Power cable system for offshore floating-type wind farm pilot plant. In Proceedings of the Fall ICC Meeting, Washington, DC, USA, 9–10 December 2013.

41. Edwards, E.C.; Holcombe, A.; Brown, S.; Ransley, E.; Hann, M.; Greaves, D. Evolution of floating offshore wind platforms: A review of at-sea devices. *Renew. Sustain. Energy Rev.* **2023**, *183*, 113416. [CrossRef]
42. Hartman, L. Wind Turbines: The Bigger, the Better. Office of Energy Efficiency & Renewable Energy, USA. 2022. Available online: https://www.energy.gov/eere/articles/wind-turbines-bigger-better (accessed on 9 August 2023).
43. The Carbon Trust. *132 kV Array Cable Requirements and The Need for Improved Testing Standards*; The Carbon Trust: London, UK, 2023.
44. Johanson, A.; Gjølmesli, E. Dynamic High Voltage Subsea Cables and Fatigue of their Components. In Proceedings of the Offshore Technology Conference, Houston, TX, USA, 1–4 May 2023; OTC: Houston, TX, USA, 2023. [CrossRef]
45. Ikhennicheu, M.; Lynch, M.; Doole, S.; Borisade, F.; Wendt, F.; Schwarzkopf, M.A.; Matha, D.; Vicente, R.D.; Tim, H.; Ramirez, L.; et al. *Review of the State of the Art of Dynamic Cable System Design*; Corewind: Copenhagen, Denmark, 2020.
46. Srinil, N. Cabling to connect offshore wind turbines to onshore facilities. In *Offshore Wind Farms*; Ng, C., Ran, L., Eds.; Woodhead Publishing: Sawston, UK, 2016; pp. 419–440. [CrossRef]
47. Antunes, T.A.; Castro, R.; Santos, P.J.; Pires, A.J. Power-from-Shore Optioneering for Integration of Offshore Renewable Energy in Oil and Gas Production. *Energies* **2023**, *17*, 151. [CrossRef]
48. Ringsberg, J.W.; Dieng, L.; Li, Z.; Hagman, I. Characterization of the Mechanical Properties of Low Stiffness Marine Power Cables through Tension, Bending, Torsion, and Fatigue Testing. *J. Mar. Sci. Eng.* **2023**, *11*, 1791. [CrossRef]
49. Thies, P.R.; Harrold, M.; Johanning, L.; Grivas, K.; Georgallis, G. Performance Evaluation of Dynamic HV Cables with Al Conductors for Floating Offshore Wind Turbines. In Proceedings of the ASME 2019 2nd International Offshore Wind Technical Conference, St. Julian's, Malta, 3–6 November 2019; American Society of Mechanical Engineers: St. Julian's, Malta, 2019; p. V001T01A039. [CrossRef]
50. Schnepf, A.; Lopez-Pavon, C.; Ong, M.C.; Yin, G.; Johnsen, Ø. Feasibility study on suspended inter-array power cables between two spar-type offshore wind turbines. *Ocean Eng.* **2023**, *277*, 114215. [CrossRef]
51. Ahmad, I.B.; Schnepf, A.; Ong, M.C. An optimisation methodology for suspended inter-array power cable configurations between two floating offshore wind turbines. *Ocean Eng.* **2023**, *278*, 114406. [CrossRef]
52. UNITECH. TetraSpar Floating Wind Turbine Pilot Installed. *GCE Ocean Technology 2021.* Available online: https://www.gceocean.no/news/posts/2021/august/tetraspar-floating-wind-turbine-pilot-installed/ (accessed on 15 August 2023).
53. Worzyk, T. *Submarine Power Cables: Design, Installation, Repair, Environmental Aspects*; Springer: Berlin/Heidelberg, Germany, 2009. [CrossRef]
54. Clausen, T.; D' Souza, R. DEEPWATER E&P: Dynamic Risers' Key Component for Deepwater Drilling, Floating Production. Offshore 2001. Available online: https://www.offshore-mag.com/deepwater/article/16759046/deepwater-ep-dynamic-risers-key-component-for-deepwater-drilling-floating-production (accessed on 10 August 2023).
55. Rentschler, M.U.T.; Adam, F.; Chainho, P. Design optimization of dynamic inter-array cable systems for floating offshore wind turbines. *Renew. Sustain. Energy Rev.* **2019**, *111*, 622–635. [CrossRef]
56. Thies, P.R.; Harrold, M.; Johanning, L.; Grivas, K.; Georgallis, G. Load, and fatigue evaluation for 66 kV floating offshore wind submarine dynamic power cable. In Proceedings of the Jicable'19-10th International Conference on Insulated Power Cables, Paris-Versailles, France, 23–27 June 2019.
57. Zhao, S.; Cheng, Y.; Chen, P.; Nie, Y.; Fan, K. A comparison of two dynamic power cable configurations for a floating offshore wind turbine in shallow water. *AIP Adv.* **2021**, *11*, 035302. [CrossRef]
58. Hung, Y.-C.; Yang, R.-Y. Experimental and Numerical Study on Suitable Configuration of Dynamical Power Cable Connected to Floating Substation in Shallow Water. SSRN 2023. Available online: https://dx.doi.org/10.2139/ssrn.4349135 (accessed on 23 April 2025).
59. Zhao, S.; Wang, Y.; Chen, P. Dynamic Analysis of Multi-Wave Configurations of a Dynamic Export Cable for a Floating Wind Turbine in Shallow Water. In Proceedings of the 32nd International Ocean and Polar Engineering Conference, Shanghai, China, 5–10 June 2022; p. ISOPE-I-22-061.
60. Durakovic, A. First Floating-to-Fixed Wind Turbine Connection Completed Offshore China. Offshore Wind 2021. Available online: https://www.offshorewind.biz/2021/09/02/first-floating-to-fixed-wind-turbine-connection-completed-offshore-china/ (accessed on 14 August 2023).
61. Durakovic, A. Floating-to-Fixed Wind Turbine Connection Comes Online Offshore China-VIDEO. Offshore Wind 2021. Available online: https://www.offshorewind.biz/2021/12/10/floating-to-fixed-wind-turbine-connection-comes-online-offshore-china-video/ (accessed on 14 August 2023).
62. Calvert, S.J.; Balash, C.; Chai, S.; Izarn, C.; Morand, H. *Effects of Buoyancy Modules Geometry on Hydro-Elastic and Hydrodynamic Performance of a Steep Wave Riser in Steady Current. Volume 5A: Pipeline and Riser Technology*; American Society of Mechanical Engineers: St. John's, NL, Canada, 2015; p. V05AT04A037. [CrossRef]
63. Amaechi, C.V.; Chesterton, C.; Butler, H.O.; Wang, F.; Ye, J. Review on the design and mechanics of bonded marine hoses for Catenary Anchor Leg Mooring (CALM) buoys. *Ocean Eng.* **2021**, *242*, 110062. [CrossRef]

64. Taninoki, R.; Abe, K.; Sukegawa, T.; Azuma, D.; Nishikawa, M. Dynamic Cable System for Floating Offshore Wind Power Generation. *SEI Tech. Rev.* **2017**, *190*, 45–50.
65. Thomson, D.; Thomson, A.; Muoniovaara, M.; Long, C.; Scuse, J. *Decommissioning Penguin WEC 1 and Mooring Components-Method Statement*; Orcades Marine Management Consultants Ltd.: Scotland, UK, 2020.
66. Araújo, J.B.; Fernandes, A.C.; Sales, J.S., Jr.; Thurler, A.C.; Vilela, A.M. Innovative Oil Offloading System for Deep Water. In Proceedings of the Offshore Technology Conference, Houston, TX, USA, 6–9 May 2019; OTC: Houston, TX, USA, 2019; p. D022S057R006. [CrossRef]
67. He, N.; Zhang, C.; Kang, Z. Analysis of Coupling Characteristics of the Offloading Buoy System in West Africa Seas. In Proceedings of the Twenty-Eighth (2018) International Ocean and Polar Engineering Conference, Sapporo, Japan, 10–15 June 2018.
68. Kang, Y.; Sun, L.; Kang, Z.; Chai, S. Coupled Analysis of FPSO and CALM Buoy Offloading System in West Africa. In Proceedings of the ASME 2014 33rd International Conference on Ocean, Offshore and Arctic Engineering, San Francisco, CA, USA, 8–13 June 2014; American Society of Mechanical Engineers: New York, NY, USA, 2014. [CrossRef]
69. Beier, D.; Schnepf, A.; Van Steel, S.; Ye, N.; Ong, M.C. Fatigue Analysis of Inter-Array Power Cables between Two Floating Offshore Wind Turbines Including a Simplified Method to Estimate Stress Factors. *J. Mar. Sci. Eng.* **2023**, *11*, 1254. [CrossRef]
70. Beier, D. Dynamic and Fatigue Analyses of Suspended Power Cables for Multiple Floating Offshore Wind Turbines. Master's Thesis, University of Stavanger, Stavanger, Norway, 2023.
71. Schnepf, A.; Devulder, A.; Johnson, Ø.; Ong, M.C.; Lopez-Pavon, C. Numerical Investigations on Suspended Power Cable Configurations for Floating Offshore Wind Turbines in Deep Water Powering an FPSO. *J. Offshore Mech. Arctic. Eng.* **2023**, *145*, 030904. [CrossRef]
72. Li, D.-R.; Su, Y.-S.; Yang, R.-Y. Experimental and Numerical Study of Suspended Inter-Array Cable Configurations for Floating Offshore Wind Farm. *J. Mar. Sci. Eng.* **2024**, *12*, 853. [CrossRef]
73. Bech, A.; Storvoll, E. Marine Flexible Elongate Element and Method of Installation. US10370905B2, 6 August 2019.
74. Amaechi, C.V.; Chesterton, C.; Butler, H.O.; Wang, F.; Ye, J. An Overview on Bonded Marine Hoses for Sustainable Fluid Transfer and (Un)Loading Operations via Floating Offshore Structures (FOS). *J. Mar. Sci. Eng.* **2021**, *9*, 1236. [CrossRef]
75. Tan, Z.; Zhang, J.; Sheldrake, T.H.; Hou, Y.; Hill, T. Weight Added Wave (WAW) System in Production Brazil Offshore Application. In Proceedings of the Offshore Technology Conference (OTC), Houston, TX, USA, 6–9 May 2013; OnePetro: Richardson, TX, USA, 2013. [CrossRef]
76. Hanonge, D.; Luppi, A. Challenges of Flexible Riser Systems in Shallow Waters. In Proceedings of the Offshore Technology Conference (OTC), Houston, TX, USA, 3–6 May 2010; OnePetro: Richardson, TX, USA, 2010; p. 14.
77. Tess. Large Bore Flexible Umbilical for Aquaculture 2021. Available online: https://techtransfer.no/en/fishing-and-aquaculture/tess/ (accessed on 29 November 2023).
78. Martinelli, L.; Lamberti, A.; Ruol, P.; Ricci, P.; Kirrane, P.; Fenton, C.; Johanning, L. Power Umbilical for Ocean Renewable Energy Systems-Feasibility and Dynamic Response Analysis. In Proceedings of the 3rd International Conference on Ocean Energy (ICOE), Bilbao, Spain, 6–9 October 2010; p. 8.
79. Dai, T.; Sævik, S.; Ye, N. Experimental and numerical studies on dynamic stress and curvature in steel tube umbilicals. *Mar. Struct.* **2020**, *72*, 102724. [CrossRef]
80. Svensson, G.; Sævik, S.; Ringsberg, J.W. Fatigue damage analysis of dynamic power cables by laboratory testing and FE analysis. In *Developments in the Analysis and Design of Marine Structures*, 1st ed.; CRC Press: London, UK, 2021; pp. 477–484. [CrossRef]
81. Ovun. *Offshore & Subsea-Buoyancy Systems and Cable Protection Solutions*; Ovun AS: Åndalsnes, Norway, 2024.
82. Shaukat, U.; Schnepf, A.; Giljarhus, K.E.T. Flow over a step cylinder using partially averaged Navier-Stokes equations with application towards dynamic subsea power cables. *IOP Conf. Ser. Mater. Sci. Eng.* **2023**, *1294*, 012002. [CrossRef]
83. Li, L.; Gao, Z.; Moan, T. Joint Distribution of Environmental Conditions at Five European Offshore Sites for Design of Combined Wind and Wave Energy Devices. *J. Offshore Mech. Arctic. Eng.* **2015**, *137*, 031901. [CrossRef]
84. Chakrabarti, S.K. (Ed.) *Handbook of Offshore Engineering*; Elsevier: London, UK, 2005. [CrossRef]
85. Obhrai, C.; Kalvig, S.; Gudmestad, O.T. A Review of Current Guidelines and Research on Wind Modelling for the Design of Offshore Wind Turbines. In Proceedings of the Twenty-Second International Offshore and Polar Engineering Conference, Rhodes, Greece, 17–22 June 2012; OnePetro: Richardson, TX, USA, 2012.
86. Bitner-Gregersen, E.M.; Toffoli, A. Occurrence of rogue sea states and consequences for marine structures. *Ocean Dyn.* **2014**, *64*, 1457–1468. [CrossRef]
87. Ponce De León, S.; Guedes Soares, C. Numerical study of the effect of current on waves in the Agulhas Current Retroflection. *Ocean Eng.* **2022**, *264*, 112333. [CrossRef]
88. Marechal, G.; Ardhuin, F. Surface Currents and Significant Wave Height Gradients: Matching Numerical Models and High-Resolution Altimeter Wave Heights in the Agulhas Current Region. *J. Geophys. Res. Oceans* **2021**, *126*, 1–16. [CrossRef]
89. DTU Wind Energy, VORTEX, The World Bank. Offshore Wind Resource Map-Mean Wind Speed 2019. Available online: https://globalwindatlas.info (accessed on 22 November 2023).

90. DTU Wind Energy, VORTEX, The World Bank. Offshore Wind Resource Map-Wind Power Density Potential 2019. Available online: https://globalwindatlas.info (accessed on 22 November 2023).
91. Olagnon, M.; Quiniou-Ramus, V. Specification of Metocean Design Conditions: Recent Progress and Remaining Gaps. In Proceedings of the Thirteenth International Offshore and Polar Engineering Conference, Honolulu, HI, USA, 25–30 May 2003; OnePetro: Richardson, TX, USA, 2003.
92. Yamaguchi, A.; Taki, S.; Ishihara, T. Metocean Measurement at Fukushima Offshore Site. In Proceedings of the Grand Renewable Energy 2018 Proceedings, Yokohama, Japan, 20–22 June 2018.
93. Kim, S.W.; Sævik, S.; Wu, J.; Leira, B.J. Time domain simulation of marine riser vortex-induced vibrations in three-dimensional currents. *Appl. Ocean Res.* **2022**, *120*, 103057. [CrossRef]
94. Santala, M. Examining Variations in West of Shetlands Current Profile Shapes in Light of Regional Oceanography. In Proceedings of the Offshore Technology Conference, Houston, TX, USA, 30 April–3 May 2012; OnePetro: Richardson, TX, USA, 2012. [CrossRef]
95. Yu, S.; Wu, W.; Naess, A. Current Profile Extreme Prediction in the South China Sea Based on the EOF-ACER Method, by Considering Current Directions. *IJOPE* **2021**, *31*, 275–282. [CrossRef]
96. Saetre, H.J. Acquisition of Critical Metocean Data on the Norwegian Deepwater Frontier. In Proceedings of the Offshore Technology Conference, Houston, TX, USA, 3–6 May 1999; OnePetro: Richardson, TX, USA, 1999. [CrossRef]
97. McClimans, T.A.; Lønseth, L. Oscillations of frontal currents. *Cont. Shelf Res.* **1985**, *4*, 699–707. [CrossRef]
98. Lima, J.A.M.; Scofano, A.M. New Oceanographic Challenges in Brazilian Deepwater Oil Fields. In Proceedings of the Offshore Technology Conference, Houston, TX, USA, 3–6 May 1999; OnePetro: Richardson, TX, USA, 1999. [CrossRef]
99. Hyder, P.; Jeans, D.R.G.; Cauquil, E.; Nerzic, R. Observations and predictability of internal solitons in the northern Andaman Sea. *Appl. Ocean Res.* **2005**, *27*, 1–11. [CrossRef]
100. Umehara, M.; Matsuki, S. Measurements of velocity and trajectory of water particle for internal waves in two density layers: Internal Wave Trajectory. *Geophys. Res. Lett.* **2011**, *38*, 1–7. [CrossRef]
101. Felisita, A. On the Application of Steel Lazy Wave Riser for Deepwater Locations with Harsh Environments (no. 330). Ph.D. Thesis, University of Stavanger, Stavanger, Norway, 2016.
102. GOA-Global Ocean Associates. An Atlas of Internal Solitary-Like Waves and Their Properties. 2004. Available online: http://internalwaveatlas.com/ (accessed on 3 January 2024).
103. Marty, A.; Schoefs, F.; Soulard, T.; Berhault, C.; Facq, J.V.; Gaurier, B.; Germain, G. Effect of Roughness of Mussels on Cylinder Forces from a Realistic Shape Modelling. *J. Mar. Sci. Eng.* **2021**, *9*, 598. [CrossRef]
104. Ameryoun, H. Probabilistic Modeling of Wave Actions on Jacket-Type Offshore Wind Turbines in the Presence of Marine Growth. Ph.D. Thesis, Université de Nantes, Nantes, France, 2015.
105. Okubo, S.; Mitani, S.; Azumaya, O.; Yuto, H.; Niizato, H.; Kosasih, K.M.A.; Takeuchi, K.; Suzuki, H. Survey of Marine Growth on the Dynamic Cable for Floating Offshore Wind Turbine-Comparison of Survey Results and Design Values. In *Proceedings of Lectures of the Japan Society of Shipbuilding and Marine Engineering No. 33*; Japanese Shipbuilding and Marine Engineering Society: Tokyo, Japan, 2021.
106. Tian, L.; Yin, Y.; Bing, W.; Jin, E. Antifouling Technology Trends in Marine Environmental Protection. *J. Bionic Eng.* **2021**, *18*, 239–263. [CrossRef] [PubMed]
107. Gulski, E.; Anders, G.J.; Jongen, R.A.; Parciak, J.; Siemiński, J.; Piesowicz, E.; Paszkiewicz, S.; Irska, I. Discussion of the electrical and thermal aspects of offshore wind farms' power cables' reliability. *Renew. Sustain. Energy Rev.* **2021**, *151*, 111580. [CrossRef]
108. Maksassi, Z.; Garnier, B.; Moctar, A.O.E.; Schoefs, F.; Schaeffer, E. Thermal Characterization and Thermal Effect Assessment of Biofouling around a Dynamic Submarine Electrical Cable. *Energies* **2022**, *15*, 3087. [CrossRef]
109. Peyton, H.R. Arctic engineering. *J. Pet. Technol.* **1970**, *22*, 1076–1082. [CrossRef]
110. Winther, S.N.; Gudmestad, O.T. Impact of and solutions to the effects of climate change for Longyearbyen, Svalbard, Norway. *IOP Conf. Ser. Material Sci. Eng.* **2023**, *1294*, 012036. [CrossRef]
111. Jorgensen, J.; Griffiths, T.; McGrath, N.; Görmüş, D.; Severs, G.; Thurstan, B.; Pistani, F.; Ledoux, B. Application of STABLEpipe Method C Assessment to Determine On-Bottom Stability of a Cable Crossing in North Sea Mobile Seabed Conditions. In Proceedings of the ASME 2023 42nd International Conference on Ocean, Offshore and Arctic Engineering, vol. 8, Melbourne, Australia, 11–16 June 2023; The American Society of Mechanical Engineers (ASME): New York, NY, USA, 2023. [CrossRef]
112. McGrath, N.; Griffiths, T.; Jorgensen, J.; Pistani, F.; Draper, S.; Cheng, L. Improved Reliability Assessment Methods for Subsea Cables on Rocky Seabed Using NPV Calculation. In Proceedings of the ASME 2023 42nd International Conference on Ocean, Offshore and Arctic Engineering, vol. 8, Melbourne, Australia, 11–16 June 2023; The American Society of Mechanical Engineers (ASME): New York, NY, USA, 2023. [CrossRef]
113. Allianz. *A Turning Point for Offshore Wind-Global Opportunities and Risk Trends*; Allianz: Munich, Germany, 2023.
114. Bitner-Gregersen, E.M.; Waseda, T.; Parunov, J.; Yim, S.; Hirdaris, S.; Ma, N.; Soares, C.G. Uncertainties in long-term wave modelling. *Mar. Struct.* **2022**, *84*, 103217. [CrossRef]

115. Feng, X.; Tsimplis, M.N.; Yelland, M.J.; Quartly, G.D. Changes in significant and maximum wave heights in the Norwegian Sea. *Glob. Planet. Chang.* **2014**, *113*, 68–76. [CrossRef]
116. Jensen, S. Taiwan May Face Fewer Typhoons—But Harsher Droughts-as the Planet Warms. Reuters 2020. Available online: https://www.reuters.com/article/us-climate-change-taiwan-typhoon-feature-idUSKBN28W00D (accessed on 24 October 2023).
117. Lee, H.-Y. One-Third of Taiwan's Corals Are Dying Due to Bleaching: Researchers. Focus Taiwan-CNA English News 2021. Available online: https://focustaiwan.tw/society/202101120010 (accessed on 23 November 2023).
118. Ran, X.; Yu, Z.; Lijun, Z.; Zhou, Z.; Jiangyan, G. Reliability analysis of an umbilical under ultimate tensile load based on response surface approach. *Underw. Technol.* **2020**, *37*, 87–93. [CrossRef]
119. Young, D.; Ng, C.; Oterkus, S.; Li, Q.; Johanning, L. Predicting failure of dynamic cables for floating offshore wind. In Proceedings of the RENEW 2018, Lisboa, Portugal, 8–10 October 2018.
120. Rabelo, M.A.; Pesce, C.P.; Santos, C.C.P.; Ramos, R.; Franzini, G.R.; Gay Neto, A. An investigation on flexible pipes bird caging triggering. *Mar. Struct.* **2015**, *40*, 159–182. [CrossRef]
121. Grivas, K.; Moraiti, A.; Georgallis, G.; Bareiro, O. *D3.4-Final 72.5 kV Dynamic Cable Sample*; FLOTANT: Patras, Greece, 2021.
122. Nasution, F.P.; Sævik, S.; Berge, S. Experimental and finite element analysis of fatigue strength for 300 mm^2 copper power conductor. *Mar. Struct.* **2014**, *39*, 225–254. [CrossRef]
123. Yan, Q.; Hong, G.; Lei, Z.; Zhenqin, Y.; Juyue, L. Extreme and Fatigue Analysis of a Dynamic Subsea Power Umbilical. In *Volume 5A: Pipelines, Risers, and Subsea Systems*; American Society of Mechanical Engineers: Glasgow, UK, 2019; p. V05AT04A004. [CrossRef]
124. Balena, R.; Williams, P. Risk-Based Structural Analysis of Frade Umbilical Cables. In Proceedings of the 20th International Congress of Mechanical Engineering COBEM 2009, Gramado, Rio Grande Do Sul, Brazil, 15–20 November 2009; Brazilian Society of Mechanical Sciences and Engineering, ABCM: Parana, Brazil, 2009.
125. Land, E.; Hu, Z.; Haley, N.; Brindley, W.; Ng, C. The effects of surge underprediction on fatigue damage of floating offshore wind turbine dynamic cables. In *Advances in the Analysis and Design of Marine Structures*, 1st ed.; CRC Press: London, UK, 2023; pp. 503–510. [CrossRef]
126. Karlsen, S.; Slora, R.; Heide, K.; Lund, S.; Eggertsen, F.; Osborg, P.A. Dynamic Deep-Water Power Cables. In Proceedings of the RAO/CIS Offshore Conference and Exhibition 2009, St. Petersburg, Russia, 15–18 September 2009; p. 10.
127. Resner, L.; Paszkiewicz, S. Radial Water Barrier in Submarine Cables, Current Solutions, and Innovative Development Directions. *Energies* **2021**, *14*, 2761. [CrossRef]
128. Holcombe, A.; Edwards, E.; Tosdevin, T.; Greaves, D.; Hann, M. A Comparative Study of Potential-Flow-Based Numerical Models to Experimental Tests of a Semi-Submersible Floating Wind Turbine Platform. In Proceedings of the 33rd International Ocean and Polar Engineering Conference, Ottawa, ON, Canada, 19–23 June 2023; International Society of Offshore and Polar Engineers (ISOPE): Cupertino, CA, USA, 2023.
129. Robertson, A.N.; Gueydon, S.; Bachynski, E.; Wang, L.; Jonkman, J.; Alarcon, D.; Amet, E.; Beardsell, A.; Bonnet, P.; Boudet, B.; et al. OC6 Phase I: Investigating the underprediction of low-frequency hydrodynamic loads and responses of a floating wind turbine. *J. Phys. Conf. Ser.* **2020**, *1618*, 032033. [CrossRef]
130. Robertson, A.N.; Wendt, F.; Jonkman, J.M.; Popko, W.; Dagher, H.; Gueydon, S.; Qvist, J.; Vittori, F.; Azcona, J.; Uzunoglu, E.; et al. OC5 Project Phase II: Validation of Global Loads of the DeepCwind Floating Semisubmersible Wind Turbine. *Energy Procedia* **2017**, *137*, 38–57. [CrossRef]
131. Pinkster, J.A. Low-Frequency Phenomena Associated with Vessels Moored at Sea. *Soc. Pet. Eng. J.* **1975**, *15*, 487–494. [CrossRef]
132. Bryant, M.J.; Hickok, D.D. Umbilical Failure-Mechanisms and Avoidance. In Proceedings of the Subsea Control and Data Acquisition: Proceedings of an International Conference, London, UK, 4–5 April 1990; OnePetro: Richardson, TX, USA, 1990.
133. Simonsen, A. Inspection and Monitoring Techniques for Unbonded Flexible Risers and Pipelines. Master's Thesis, University of Stavanger, Stavanger, Norway, 2014.
134. Saneian, M.; Gao, Y.; Bai, Y.; Liu, T. Analysis of Subsea Umbilical Mechanical Behavior Under Simultaneous Bending, Tension and Torsion. In Proceedings of the 38th International Conference on Ocean, Offshore and Arctic Engineering; Volume 5A: Pipelines, Risers, and Subsea Systems, Glasgow, Scotland, UK, 9–14 June 2019; American Society of Mechanical Engineers (ASME): New York, NY, USA, 2019; p. V05AT04A032. [CrossRef]
135. Börü, M.E. VIV Fatigue of Dynamic Power Cables Applied in Offshore Wind Turbines. Master's Thesis, Norwegian University of Science and Technology, Torgarden, Norway, 2021.
136. Elrick, P.; Venugopal, V. Vortex-Induced Vibrations of Dynamic Power Cable for Floating Wind Turbines. In Proceedings of the 42nd International Conference on Ocean, Offshore and Arctic Engineering; Volume 7: CFD & FSI, Melbourne, Australia, 11–16 June 2023; American Society of Mechanical Engineers (ASME): New York, NY, USA, 2023; p. V007T08A026. [CrossRef]
137. Fuglsang, A.; Kusangaya, A.J.; Dillon-Gibbons, C.; Bauer, P.; Marcollo, H. Comparison of Vessel Motion Induced-VIV Response of Lazy-Wave and W-Shaped FOWT Power Cable Configurations. In Proceedings of the OMAE Conference, Melbourne, Australia, 12–15 June 2023; ASME: New York, NY, USA, 2023. [CrossRef]

138. Lal, M.; Sebastian, A.; Rana, Y. Application of Artificial Intelligence for VIV Strake Optimization of Steel Lazy Wave Risers. In *Volume 3: Materials Technology; Pipelines, Risers, and Subsea Systems*; American Society of Mechanical Engineers: Hamburg, Germany, 2022; p. V003T04A035. [CrossRef]
139. Yin, D.; Wu, J.; Lie, H.; Jin, J.; Passano, E.; Saevik, S.; Riemer-Sorenson, S.; Rustad, A.M.; Tognarelli, M.A.; Grytyr, G.; et al. Optimization of Hydrodynamic Coefficients for VIV Prediction. Volume 8: CFD and FSI. Virtual, online. *Am. Soc. Mech. Eng.* **2021**, *85185*, V008T08A025. [CrossRef]
140. Liu, D.; Ahmad, I.; Janocha, M.J.; Nystrøm, P.; Ong, M.C. Optimization of bend stiffeners of a suspended inter-array power cable between two floating offshore wind turbines. *IOP Conf. Ser. Mater. Sci. Eng.* **2023**, *1294*, 012018. [CrossRef]
141. Tang, M.; Yan, J.; Chen, J.; Yang, Z.; Yue, Q. Nonlinear analysis and multi-objective optimization for bend stiffeners of flexible riser. *J. Mar. Sci. Technol.* **2015**, *20*, 591–603. [CrossRef]
142. Mainçon, P.; Longva, V. Torsion Failures in Handling Operations for Power Cables, Umbilicals and Flexible Pipes. In Proceedings of the 40th International Conference on Ocean, Offshore and Arctic Engineering; Volume 4: Pipelines, Risers, and Subsea Systems, Virtual online, 21–30 June 2021; American Society of Mechanical Engineers (ASME): New York, NY, USA, 2021; p. V004T04A003. [CrossRef]
143. Okkerstrøm, L. Sensitivity Analysis of Peak Tension Loads on Subsea Power Cables during Laying Operations. Master's Thesis, Western Norway University of Applied Sciences, Bergen, Norway, 2021.
144. Okkerstrøm, L.; Gudmestad, O.T.; Pedersen, E. Simulation of peak tension loads in Subsea power cables during installation. *IOP Conf. Ser. Mater. Sci. Eng.* **2021**, *1201*, 012010. [CrossRef]
145. Buljan, A. World's First Floating Wind Farm to Undergo First Major Maintenance Campaign, Turbines to Be Towed to Norwegian Port. Offshore Wind 2024. Available online: https://www.offshorewind.biz/2024/01/15/worlds-first-floating-wind-farm-to-undergo-first-major-maintenance-campaign-turbines-to-be-towed-to-norwegian-port/ (accessed on 19 January 2024).
146. Al-Sallami, O. Cables Decommissioning in Offshore Wind Farms: Environmental and Economical Perspective. Master's Thesis, Uppsala Universitet, Uppsala, Sweden, 2021.
147. ARUP, Marine Scotland. *Review of Approaches and Costs of Decommissioning Offshore Wind Installations*; ARUP: Scotland, UK, 2018.
148. Haid, L.; Stewart, G.; Jonkman, J.; Robertson, A.; Lackner, M.; Matha, D. Simulation-Length Requirements in the Loads Analysis of Offshore Floating Wind Turbines. In Proceedings of the 32nd International Conference on Ocean, Offshore and Arctic Engineering; Volume 8: Ocean Renewable Energy, Nantes, France, 9–14 June 2013; American Society of Mechanical Engineers (ASME): New York, NY, USA, 2013; p. V008T09A091. [CrossRef]
149. Pillai, A.C.; Thies, P.R.; Johanning, L. Mooring system design optimization using a surrogate-assisted multi-objective genetic algorithm. *Eng. Optim.* **2019**, *51*, 1370–1392. [CrossRef]
150. Orcina Ltd. OrcaFlex 2024. Available online: https://www.orcina.com/webhelp/OrcaFlex/ (accessed on 29 February 2024).
151. Taormina, B.; Bald, J.; Want, A.; Thouzeau, G.; Lejart, M.; Desroy, N.; Carlier, A. A review of potential impacts of submarine power cables on the marine environment: Knowledge gaps, recommendations, and future directions. *Renew. Sustain. Energy Rev.* **2018**, *96*, 380–391. [CrossRef]
152. Merck, T.; Wasserthal, R. *Assessment of the Environmental Impacts of Cables*; OSPAR Commission: London, UK, 2009.
153. Farr, H.; Ruttenberg, B.; Walter, R.K.; Wang, Y.-H.; White, C. Potential environmental effects of deepwater floating offshore wind energy facilities. *Ocean Coast. Manag.* **2021**, *207*, 105611. [CrossRef]
154. Jakubowska-Lehrmann, M.; Białowąs, M.; Otremba, Z.; Hallmann, A.; Śliwińska-Wilczewska, S.; Urban-Malinga, B. Do magnetic fields related to submarine power cables affect the functioning of a common bivalve? *Mar. Environ. Res.* **2022**, *179*, 105700. [CrossRef]
155. Scott, K.; Harsanyi, P.; Easton, B.A.A.; Piper, A.J.R.; Rochas, C.M.V.; Lyndon, A.R. Exposure to Electromagnetic Fields (EMF) from Submarine Power Cables Can Trigger Strength-Dependent Behavioral and Physiological Responses in Edible Crab, *Cancer pagurus* (L.). *J. Mar. Sci. Eng.* **2021**, *9*, 776. [CrossRef]
156. Dunham, A.; Pegg, J.R.; Carolsfeld, W.; Davies, S.; Murfitt, I.; Boutillier, J. Effects of submarine power transmission cables on a glass sponge reef and associated megafaunal community. *Mar. Environ. Res.* **2015**, *107*, 50–60. [CrossRef]
157. Westerberg, H.; Lagenfelt, I. Sub-Sea power cables and the migration behavior of the European eel. *Fish. Manag. Ecol.* **2008**, *15*, 369–375. [CrossRef]
158. Garavelli, L. 2020 State of the Science Report. In *Chapter 8: Encounters of Marine Animals with Marine Renewable Energy Device Mooring Systems and Subsea Cables*; A report prepared by Pacific Northwest National Laboratory on behalf of the U.S. Department of Energy; U.S. Department of Energy: Washington, DC, USA, 2020. [CrossRef]
159. Lin, T.-H.; Yang, R.-Y. Stability Analysis and Environmental Influence Evaluation on a Hybrid Mooring System for a Floating Offshore Wind Turbine. *J. Mar. Sci. Eng.* **2023**, *11*, 2236. [CrossRef]
160. Hooper, T.; Austen, M. The co-location of offshore windfarms and decapod fisheries in the UK: Constraints and opportunities. *Mar. Policy* **2014**, *43*, 295–300. [CrossRef]

161. Krone, R.; Dederer, G.; Kanstinger, P.; Krämer, P.; Schneider, C.; Schmalenbach, I. Mobile demersal megafauna at common offshore wind turbine foundations in the German Bight (North Sea) two years after deployment-increased production rate of Cancer pagurus. *Mar. Environ. Res.* **2017**, *123*, 53–61. [CrossRef] [PubMed]
162. Hutchison, Z.L.; Gill, A.B.; Sigray, P.; He, H.; King, J.W. A modelling evaluation of electromagnetic fields emitted by buried subsea power cables and encountered by marine animals: Considerations for marine renewable energy development. *Renew. Energy* **2021**, *177*, 72–81. [CrossRef]
163. Maxwell, S.M.; Kershaw, F.; Locke, C.C.; Conners, M.G.; Dawson, C.; Aylesworth, S.; Loomis, R.; Johnson, A.F. Potential impacts of floating wind turbine technology on marine species and habitats. *J. Environ. Manag.* **2022**, *307*, 114577. [CrossRef]
164. Hirokawa, E.; Suzuki, H.; Hirabayashi, S.; Muratake, M. Estimation of Risk of Progressive Drifts in a Wind Farm Caused by Collision of Drift Ship. In Proceedings of the 34th International Conference on Ocean, Offshore and Arctic Engineering; Volume 6: Ocean Space Utilization, St. John's, NL, Canada, 31 May–5 June 2015; American Society of Mechanical Engineers (ASME): New York, NY, USA, 2015; p. V006T05A011. [CrossRef]
165. Holmås, H.E.; Dale, M.S.; Gjerde, S.S.; Holmås, H.E.; Jacobsen, A.K.; Magnus, E.; Mekki, A.; Notkevich, L.; Persen, G.; Vevatne, T.I. *Hywind Tampen-Samfunnsmessige Ringvirkninger*; Equinor: Stavanger, Norway, 2019. (In Norwegian)
166. Kallinger, M.D.; Rapha, J.I.; Trubat Casal, P.; Domínguez-García, J.L. Offshore Electrical Grid Layout Optimization for Floating Wind; A Review. *Clean. Technol.* **2023**, *5*, 791–827. [CrossRef]
167. Song, D.; Yan, J.; Gao, Y.; Wang, L.; Du, X.; Xu, Z.; Zhang, Z.; Yang, J.; Dong, M.; Chen, Y. Optimization of floating wind farm power collection system using a novel two-layer hybrid method. *Appl. Energy* **2023**, *348*, 121546. [CrossRef]
168. Lerch, M.; De-Prada-Gil, M.; Molins, C. A metaheuristic optimization model for the inter-array layout planning of floating offshore wind farms. *Int. J. Electr. Power Energy Syst.* **2021**, *131*, 107128. [CrossRef]
169. Lerch, M.; De-Prada-Gil, M.; Molins, C. Collection Grid Optimization of a Floating Offshore Wind Farm Using Particle Swarm Theory. *J. Phys. Conf. Series.* **2019**, *1356*, 012012. [CrossRef]
170. Giassi, M.; Castellucci, V.; Göteman, M. Economic layout optimization of wave energy parks clustered in electrical subsystems. *Appl. Ocean Res.* **2020**, *101*, 102274. [CrossRef]
171. Patryniak, K.; Collu, M.; Coraddu, A. Multidisciplinary design analysis and optimisation frameworks for floating offshore wind turbines: State of the art. *Ocean Eng.* **2022**, *251*, 111002. [CrossRef]
172. Faraggiana, E.; Giorgi, G.; Sirigu, M.; Ghigo, A.; Bracco, G.; Mattiazzo, G. A review of numerical modelling and optimisation of the floating support structure for offshore wind turbines. *J. Ocean Eng. Mar. Energy* **2022**, *8*, 433–456. [CrossRef]
173. Fylling, I.; Berthelsen, P.A. WINDOPT: An Optimization Tool for Floating Support Structures for Deep Water Wind Turbines. In Proceedings of the 30th International Conference on Ocean, Offshore and Arctic Engineering; Volume 5: Ocean Space Utilization; Ocean Renewable Energy, Rotterdam, The Netherlands, 19–24 June 2011; The American Society of Mechanical Engineers Digital Collection (ASMEDC): New York, NY, USA, 2011; pp. 767–776. [CrossRef]
174. Brommundt, M.; Krause, L.; Merz, K.; Muskulus, M. Mooring System Optimization for Floating Wind Turbines using Frequency Domain Analysis. *Energy Procedia* **2012**, *24*, 289–296. [CrossRef]
175. Hegseth, J.M.; Bachynski, E.E.; Martins, J.R.R.A. Integrated design optimization of spar floating wind turbines. *Mar. Struct.* **2020**, *72*, 102771. [CrossRef]
176. Leimeister, M.; Kolios, A.; Collu, M.; Thomas, P. Design optimization of the OC3 phase IV floating spar-buoy, based on global limit states. *Ocean Eng.* **2020**, *202*, 107186. [CrossRef]
177. Poirette, Y.; Guiton, M.; Huwart, G.; Sinoquet, D.; Leroy, J.M. An Optimization Method for the Configuration of Inter-Array Cables for Floating Offshore Wind Farm. In Proceedings of the 36th International Conference on Ocean, Offshore and Arctic Engineering; Volume 10: Ocean Renewable Energy, Trondheim, Norway, 25–30 June 2017; American Society of Mechanical Engineers: New York, NY, USA, 2017; p. 10. [CrossRef]
178. Schnepf, A.; Giljarhus, K.E.T.; Johnsen, Ø.; Lopez-Pavon, C. Dynamic Power Cable Configuration Design for Floating Offshore Wind Turbines Using Gradient-Based Optimization. In Proceedings of the 42nd International Conference on Ocean, Offshore and Arctic Engineering (OMAE); Volume 8: Ocean Renewable Energy, Melbourne, Australia, 11–16 June 2023; American Society of Mechanical Engineers (ASME): New York, NY, USA, 2023; p. 11. [CrossRef]
179. Schnepf, A. Design Optimisation of Power Cable Configurations for Floating Offshore Wind Turbines. Ph.D. Thesis, University of Stavanger, Stavanger, Norway, 2024.
180. Schnepf, A.; Giljarhus, K.E.T. Efficient Global Optimization of dynamic power cable configurations for floating offshore wind turbines. *Offshore Mech. Arctic Eng.* **2024** *(submitted)*.
181. Nehme, R.; Dethoor, X. Oil Offloading Lines and Fiber Optic Cable Package. Presented at the Offshore Technology Conference, Houston, TX, USA, 4–7 May 2015; 2015; p. OTC-25885-MS. [CrossRef]
182. Cunliffe, N.; McCarthy, T.J. Application of Genetic Algorithms for the Automated Design of Offshore Riser Systems. In Proceedings of the Seventh International Conference on the Application of Artificial Intelligence to Civil and Structural Engineering, Egmond-Aan-Zee, The Netherlands, 2–4 September 2003; p. 35. [CrossRef]

183. Vieira, L.T. Projeto Ótimo de Risers via Algoritmos Genéticos (Optimal Riser Design via Genetic Algorithms). Ph.D. Thesis, COPPE/UFRJ, Rio de Janeiro, Brazil, 2004.
184. Vieira, L.T. Otimização de Sistemas de Risers Para Exploração de Petróleo Offshore Através de Algoritmos Genéticos Paralelos (Optimization of Riser Systems for Offshore Oil Exploration Using Parallel Genetic Algorithms). Ph.D. Thesis, Universidade Federal do Rio de Janeiro, COPPE, Rio de Janeiro, Brazil, 2008.
185. Laddé, S. Multi-Objective Optimization of Steel Lazy Wave Riser Design Through Genetic Algorithms. Master's Thesis, TU Delft, Delft, The Netherlands, 2011.
186. Cardoso, P.H.D.S. Otimização da Configuração Geométrica de Risers Rígidos (Optimization of the Geometric Configuration of Rigid Risers). Master's Thesis, Pontifícia Universidade Católica do Rio de Janeiro, Rio de Janeiro, Brazil, 2019. [CrossRef]
187. Vieira, I.N.; de Lima, B.S.L.P.; Jacob, B.P. Bio-inspired algorithms for the optimization of offshore oil production systems. *Int. J. Numer. Methods Eng.* **2012**, *91*, 1023–1044. [CrossRef]
188. Vieira, I.N.; Silva, A.J.M.; Lima, B.; Jacob, B.P. A Comparative Study Applied to Risers Optimization Using Bio-Inspired Algorithms. *Int. J. Model. Simul. Pet. Ind.* **2009**, *3*, 12.
189. Wang, A.; Yang, H.; Li, H. Multi-objective Optimization for Dynamic Umbilical Installation Using Non-Dominated Sorting Genetic Algorithm. In Proceedings of the 30th International Conference on Ocean, Offshore and Arctic Engineering (OMAE); Volume 4: Pipeline and Riser Technology, Rotterdam, The Netherlands, 19–24 June 2011; American Society of Mechanical Engineers (ASME): New York, NY, USA, 2011; pp. 121–129. [CrossRef]
190. Yang, H.; Wang, A.; Li, H. Multi-objective optimization for deepwater dynamic umbilical installation analysis. *Sci. China Phys. Mech. Astron.* **2012**, *55*, 1445–1453. [CrossRef]
191. Yang, Z.; Yan, J.; Sævik, S.; Lu, Q.; Ye, N.; Chen, J.; Yue, Q. Integrated optimisation design of a dynamic umbilical based on an approximate model. *Mar. Struct.* **2021**, *78*, 102995. [CrossRef]
192. Yan, J.; Su, Q.; Li, R.; Xu, J.; Lu, Q.; Yang, Z. Optimization Design Method of the Umbilical Cable Global Configuration Based on Representative Fatigue Conditions. *IEEE J. Ocean. Eng.* **2023**, *48*, 188–198. [CrossRef]
193. Riggs, H.R. Automated Preliminary Design of Flexible Riser Configurations. In Proceedings of the Eighth International Conference on Offshore Mechanics and Arctic Engineering, The Hague, the Netherlands, 19–23 March 1989; American Society of Mechanical Engineers: New York, NY, USA, 1989; pp. 503–510.
194. Riggs, H.R.; Leraand, T. Efficient Static Analysis and Design of Flexible Risers. *J. Offshore Mech. Arctic Eng.* **1991**, *113*, 235–240. [CrossRef]
195. Larsen, C.M.; Hanson, T. Optimization of Catenary Risers. *J. Offshore Mech. Arctic Eng.* **1999**, *121*, 90–94. [CrossRef]
196. Vieira, L.T.; de Lima Bde, S.L.P.; Evsukoff, A.G.; Jacob, B.P. Application of Genetic Algorithms to the Synthesis of Riser Configurations. In Proceedings of the 22nd International Conference on Offshore Mechanics and Arctic Engineering, Cancun, Mexico, 8–13 June 2003; Offshore Technology; Ocean Space Utilization; American Society of Mechanical Engineers (ASME): New York, NY, USA, 2003; Volume 1, pp. 391–396. [CrossRef]
197. Cunliffe, N.; Baxter, C.; McCarthy, T.; Trim, A. Evolutionary Design of Marine Riser Systems. In Proceedings of the 23rd International Conference on Offshore Mechanics and Arctic Engineering, Parts A and B, Vol. 1, Vancouver, BC, Canada, 20–25 June 2004; American Society of Mechanical Engineers: New York, NY, USA, 2004; pp. 733–739. [CrossRef]
198. de Rodrigues, G.J.O. Ferramentas Computacionais para Otimização e Síntese de Sistemas Híbridos de Risers Baseados no Conceito de Bóia de Subsuperfície (Computational Tools for the Optimisation and Synthesis of Hybrid Risers Based on the Subsurface Buoy Concept). Ph.D. Thesis, Universidade Federal do Rio de Janeiro, COPPE, Rio de Janeiro, Brazil, 2004.
199. Roveri, F.E.; de Arruda Martins, C.; Balena, R. Parametric Analysis of a Lazy-Wave Steel Riser. In Proceedings of the 24th International Conference on Offshore Mechanics and Arctic Engineering: Volume 1, Parts A and B, Halkidiki, Greece, 12–17 June 2005; The American Society of Mechanical Engineers Digital Collection (ASMEDC): New York, NY, USA, 2005; pp. 289–295. [CrossRef]
200. de Lima, B.S.L.P.; Jacob, B.P.; Ebecken, N.F.F. A hybrid fuzzy/genetic algorithm for the design of offshore oil production risers. *Int. J. Numer. Meth. Eng.* **2005**, *64*, 1459–1482. [CrossRef]
201. Tanaka, R.L.; de Martins, C.A. A Genetic Algorithm Approach to Steel Riser Optimization. In Proceedings of the 25th International Conference on Offshore Mechanics and Arctic Engineering; Offshore Technology, Hamburg, Germany, 28 May–2 June 2006; Offshore Wind Energy; Ocean Research Technology; LNG Specialty Symposium; American Society of Mechanical Engineers (ASME): New York, NY, USA, 2006; Volume 1, pp. 271–277. [CrossRef]
202. Tanaka, R.L.; de Martins, C.A.; de Takafuji, F.C.M. Steel riser optimization with simulated annealing. In Proceedings of the 13th Rio Oil & Gas Expo and Conference, Rio de Janeiro, Brazil, 11–15 September 2006.
203. Tanaka, R.L.; de Martins, C.A. Dynamic Optimization of Steel Risers. In Proceedings of the Sixteenth (2007) International Offshore and Polar Engineering Conference, Lisbon, Portugal, 1–6 July 2007; The International Society of Offshore and Polar Engineers (ISOPE): Cupertino, CA, USA, 2007.

204. Vieira, I.N.; de Lima, B.S.L.P.; Jacob, B.P. Optimization of Steel Catenary Risers for Offshore Oil Production Using Artificial Immune System. In *Artificial Immune Systems*; Bentley, P.J., Lee, D., Jung, S., Eds.; Springer: Berlin/Heidelberg, Germany, 2008; pp. 254–265.
205. Yang, H.; Jiang, R.; Li, H. *Optimization Design of Deepwater Steel Catenary Risers Using Genetic Algorithm. Computational Structural Engineering*; Springer: Dordrecht, The Netherlands, 2009. [CrossRef]
206. Tanaka, R.L. Otimização da configuração de risers rígidos (Optimising the Configuration of Rigid Risers). Doutorado em Engenharia de Controle e Automação Mecânica, Universidade de São Paulo, São Paulo, Brazil, 2009. [CrossRef]
207. Andrade, E.Q.d.; Aguiar, L.L.d.; Senra, S.F.; Siqueira, E.F.N.; Torres, A.L.F.L.; Mourelle, M.M. Optimization Procedure of Steel Lazy Wave Riser Configuration for Spread Moored FPSOs in Deepwater Offshore Brazil. In Proceedings of the Offshore Technology Conference (OTC), Houston, TX, USA, 3–6 May 2010; p. OTC-20777-MS. [CrossRef]
208. de Pina, A.A.; Albrecht, C.H.; de Lima, B.S.L.P.; Jacob, B.P. Tailoring the particle swarm optimization algorithm for the design of offshore oil production risers. *Optim. Eng.* **2011**, *12*, 215–235. [CrossRef]
209. de Pina, A.A.; de Lima, B.S.L.P.; Albrecht, C.H.; Jacob, B.P. Parameter Selection and Convergence Analysis of the PSO Algorithm Applied to the Design of Risers. In Proceedings of the EngOpt 2008-International Conference on Engineering Optimization, Rio de Janeiro, Brazil, 1–5 June 2008.
210. Zheng, W.; Yang, H. Dynamic Optimization of Steel Catenary Risers Based on Reliability Using Metamodel; American Society of Mechanical Engineers Digital Collection, New York, NY, USA, 2010; pp. 223–228. [CrossRef]
211. Yang, H.Z.; Zheng, W. Metamodel approach for reliability-based design optimization of a steel catenary riser. *J. Mar. Sci. Technol.* **2011**, *16*, 202–213. [CrossRef]
212. Tanaka, R.L.; Martins, C.d.A. Parallel Dynamic Optimization of Steel Risers. *J. Offshore Mech. Arctic Eng.* **2011**, *133*, 011302. [CrossRef]
213. Arruda, M.D.S.V.; de Martins, M.A.L.; Lages, E.N. Otimização dos flutuadores de risers com catenária em configuração complexa (Optimization of complex configuration of catenary risers buoyancy modules). *Semin. Ciências Exatas E Tecnol.* **2012**, *33*, 149–164. [CrossRef]
214. de Martins, M.A.L. Otimização da Configuração de Risers Verticais Complacentes (Compliant Vertical Access Risers Optimization). Master's Thesis, Universidade Federal de Alagoas, Maceió, Brazil, 2011.
215. de Martins, M.A.L.; Lages, E.N.; Silveira, E.S.S. Compliant vertical access riser assessment: DOE analysis and dynamic response optimization. *Appl. Ocean Res.* **2013**, *41*, 28–40. [CrossRef]
216. Voie, P.E.; Sødahl, N. Optimisation of Steel Catenary Risers. In Proceedings of the 32nd International Conference on Ocean, Offshore and Arctic Engineering; Volume 4A: Pipeline and Riser Technology, Nantes, France, 9–15 June 2013; American Society of Mechanical Engineers: New York, NY, USA, 2013; p. V04AT04A011. [CrossRef]
217. Nariño, G.A.R.; Martha, L.F.; de Menezes, I.F.M. Otimização de Risers em Catenária com Amortecedores Hidrodinâmicos (Optimization of Catenary Risers with Hydrodynamic Dampers). In Proceedings of the XXXV Iberian Latin-American Congress on Computational Methods in Engineering, Fortaleza, CE, Brazil, 23–26 November 2014; Pontifícia Universidade Católica do Rio de Janeiro: Rio de Janeiro, Brazil, 2014. [CrossRef]
218. Chen, J.; Yan, J.; Yang, Z.; Yue, Q.; Tang, M. Flexible Riser Configuration Design for Extremely Shallow Water with Surrogate-Model-Based Optimization. *J. Offshore Mech. Arctic Eng.* **2016**, *138*, 041701. [CrossRef]
219. Yang, Z.; Yan, J.; Sævik, S.; Zhen, L.; Ye, N.; Chen, J.; Yue, Q. Multi-Objective Optimization Design of Flexible Risers Based on Bi-Scale Response Surface Models. In Proceedings of the 37th International Conference on Ocean, Offshore and Arctic Engineering (OMAE); Volume 5: Pipelines, Risers, and Subsea Systems, Madrid, Spain, 17–22 June 2018; American Society of Mechanical Engineers: New York, NY, USA, 2018; p. V005T04A018. [CrossRef]
220. Diakonova, E.; Bonnet, G.; Brouard, Y. SLWR Innovative Design Methodology Supporting Concept Engineering Through to Project Cost Optimisation. In Proceedings of the Offshore Technology Conference, Houston, TX, USA, 6–9 May 2019; p. D021S018R004. [CrossRef]
221. Cardoso, P.H.S.; Casaprima, N.A.G.; Senhora, F.V.; Menezes, I.F.M. Optimization of catenary risers with hydrodynamic dampers. *Ocean Eng.* **2019**, *184*, 134–142. [CrossRef]
222. Ai, S.; Xu, Y.; Kang, Z.; Yan, F. Performance comparison of stress-objective and fatigue-objective optimisation for steel lazy wave risers. *Ships Offshore Struct.* **2019**, *14*, 534–544. [CrossRef]
223. Bhowmik, S.; Noiray, G.; Naik, H. Riser Design Automation with Machine Learning. In Proceedings of the Abu Dhabi International Petroleum Exhibition & Conference, Abu Dhabi, United Arab Emirates, 11–14 November 2019; SPE Society of Petroleum Engineers: Richardson, TX, USA, 2019; p. D041S106R003. [CrossRef]
224. Bhowmik, S.; Naik, H.; Noiray, G. Lazy Wave Steel Catenary Riser Design Optimisation in Digital Field Twin. In Proceedings of the Offshore Technology Conference Asia, Kuala Lumpur, Malaysia, 2–6 November 2020; OnePetro: Richardson, TX, USA, 2020; p. D012S001R074. [CrossRef]

225. Lal, M.; Sebastian, A.; Wang, F.; Lu, X. Steel Lazy Wave Riser Optimization Using Artificial Intelligence Tool. In Proceedings of the 39th International Conference on Ocean, Offshore and Arctic Engineering, Virtual, Online, 3–7 August 2020; American Society of Mechanical Engineers (ASME): New York, NY, USA, 2020; Volume 4. [CrossRef]
226. Ogbeifun, A.M.; Oterkus, S.; Race, J.; Naik, H.; Moorthy, D.; Bhowmik, S.; Ingram, J. A tabular optimisation technique for steel lazy wave riser. *IOP Conf. Ser. Mater. Sci. Eng.* **2021**, *1052*, 012022. [CrossRef]
227. Lal, M.; Sebastian, A.; Rana, Y. Steel Lazy Wave Riser Optimization Using Artificial Neural Networks and Genetic Algorithm. In Proceedings of the International Conference on Offshore Mechanics and Arctic Engineering, Volume 4: Pipelines, Risers, and Subsea Systems, Virtual, Online, 21–30 June 2021; American Society of Mechanical Engineers: New York, NY, USA, 2021; p. V004T04A034. [CrossRef]
228. Elsas, J.H.; Casaprima, N.A.G.; Cardoso, P.H.S.; Menezes, I.F.M. Bayesian optimization of riser configurations. *Ocean Eng.* **2021**, *236*, 109402. [CrossRef]
229. Abam, J.T.; Pu, Y.; Hu, Z. Cost-Effective Optimal Solutions for Steel Catenary Riser Using Artificial Neural Network. In *Volume 3: Materials Technology; Pipelines, Risers, and Subsea Systems*; American Society of Mechanical Engineers: Hamburg, Germany, 2022; p. V003T04A031. [CrossRef]
230. Abam, J.T.; Pu, Y.; Hu, Z. *Weight Optimisation of Steel Catenary Riser Using Genetic Algorithm and Finite Element Analysis*; American Society of Mechanical Engineers Digital Collection: Hamburg, Germany, 2022; p. 9. [CrossRef]
231. Lal, M.; Yenduri, A.; Sebastian, A. Steel Lazy Wave Riser Optimization Using Particle Swarm Optimization Algorithm. In *Volume 3: Materials Technology; Pipelines, Risers, and Subsea Systems*; American Society of Mechanical Engineers: Melbourne, Australia, 2023; p. V003T04A048. [CrossRef]
232. Jakobsen, E.A. Performance Efficiency and Reliability Analysis of Offshore Wind Power Plants: Case Study of Utsira Nord. Master's Thesis, NTNU, Trondheim, Norway, 2023.
233. Petrie, H.E.; Eide, C.H.; Haflidason, H.; Watton, T. A conceptual geological model for offshore wind sites in former ice stream settings: The Utsira Nord site, North Sea. *JGS* **2022**, *179*, jgs2021-163. [CrossRef]
234. Cheynet, E.; Li, L.; Jiang, Z. Metocean Conditions at Two Norwegian Sites for Development of Offshore Wind Farms. *Renew. Energy* **2024**, *224*, 120184. [CrossRef]
235. Reistad, M.; Magnusson, A.K.; Haver, S.; Gudmestad, O.T.; Kvamme, D. How severe wave conditions are possible on the Norwegian Continental Shelf? *Mar. Struct.* **2005**, *18*, 428–450. [CrossRef]
236. Liang, G.; Jiang, Z.; Merz, K. Dynamic analysis of a dual spar floating offshore wind farm with shared moorings in extreme environmental conditions. *Mar. Struct.* **2023**, *90*, 103441. [CrossRef]
237. Asplin, L.; Albretsen, J.; Johnsen, I.A.; Sandvik, A.D. The hydrodynamic foundation for salmon lice dispersion modeling along the Norwegian coast. *Ocean Dyn.* **2020**, *70*, 1151–1167. [CrossRef]
238. Jackson, C.R.; Alpers, W. *North Sea and Skagerrak Strait*; Office of Naval Research: Alexandria, VA, USA, 2004.
239. Petrie, H.E. An Integrated Geological Characterization of Marine Ground Conditions in the North Sea. Ph.D. Thesis, The University of Bergen, Bergen, Norway, 2023.
240. Hansen, T.D. Utsira Nord Floating Wind Farm–Optimalisation of Marine Operations Related to Inter-Array Cable Installation. Master's Thesis, University of Stavanger, Stavanger, Norway, 2023.
241. von Schlanbusch, F. Driving Factors for Levelized Cost of Energy in Floating Wind Farms. Master's Thesis, The University of Bergen, Bergen, Norway, 2022.
242. Jodar Hernandez, P. Optimization of Marine Operations for the Assembly and Installation of a New Floating Wind Concept. Master's Thesis, University of Stavanger, Stavanger, Norway, 2023.
243. Høines, I. Modelling Analysis of Maintenance Logistics Optimization for a Floating Wind Park: A Case Study for Utsira Nord. Master's Thesis, University of Stavanger, Stavanger, Norway, 2022.
244. Austreim, A.H. Hvilke drift og vedlikeholds oppgaver må gjennomføres under vann på den flytende vindparken Utsira Nord, og hva vil være kostnaden? (Which Operation and Maintenance Tasks Must Be Carried Out Underwater at the Utsira Nord Floating Wind Farm, and What Will the Cost Be?). Master's Thesis, University of Stavanger, Stavanger, Norway, 2023.
245. Olsson, E.S. Potential Conflicts in Offshore Wind Development–a Qualitative Study of Utsira Nord; Norwegian University of Life Sciences, Aas, Norway, 2021.
246. Wang, W. Significance of Offshore Wind Farm Sound on Marine Populations. Master's Thesis, NTNU, Trondheim, Norway, 2019.
247. Tjøsvoll, S.E.G. What Socioeconomic Dynamics Will Offshore Wind Affect? A Regional Case Study of Utsira North. Master's Thesis, University of Stavanger, Stavanger, Norway, 2022.
248. Anda, L.M. Communication and Stakeholder Management in Renewable Energy Development; Case Study Utsira Nord Floating Wind Offshore Energy Development. Master's Thesis, University of Stavanger, Stavanger, Norway, 2022.

249. Oliveira, R.V. Whispers in the Wind: Ethical dimensions of social conflict in offshore wind. *J. Phys. Conf. Ser.* **2023**, *2626*, 012070. [CrossRef]
250. Flermoen, P. Motvind eller medvind for norske havvindutbygginger (Headwind or Downwind for Norwegian Offshore Wind Developments). Master's Thesis, NTNU, Trondheim, Norway, 2023. (In Norwegian)

Journal of
*Marine Science
and Engineering*

MDPI

Review

A Review of Research Progress on Cables and Towed Objects of the Ocean Engineering Towing System

Kefu Qi *, Jianing Zhang *, Lei Zhang, Jinlong Zhang and Ruijun Gan

Naval Architecture and Ocean Engineering College, Dalian Maritime University, Dalian 116026, China; zhanglei2020@dlmu.edu.cn (L.Z.); zhangjinlongchn@163.com (J.Z.); gan15072227733@163.com (R.G.)
* Correspondence: userkefuqi@dlmu.edu.cn (K.Q.); zhangjianing@dlmu.edu.cn (J.Z.)

Abstract: Towing operations are widely applied in various fields such as maritime accident rescue, assisting large vessels entering and exiting ports, and transporting large ocean platforms. Tugboats and the towed objects form a complex multi-body system connected by flexible cables, and during operations, they are subjected to the effects of complex marine environmental loads. Current research focuses on using numerical simulations and model tests in water tanks to study the motion response of towed objects and cables under the action of environmental loads. There is a lack of research that combines the mechanical response and structural strength with the load conditions of towing operations. Taking cables as an example, most studies focus on the mechanical properties of cables without considering the impact of towing conditions. After reviewing the literature, this paper summarizes the shortcomings of the existing research and points out several potential research directions in the field of towing: the mechanical response of cables during the initial stage of towing, experiments on towing by multiple tugboats, research on composite fiber cables using experimental and finite element simulation methods, and structural optimization of components related to towing operations.

Keywords: towing system; mechanical response; motion response; cables; towed objects

1. Introduction

In recent years, with the development of large-scale shipbuilding, the towing operation of ships has become a research hotspot. As shown in Figure 1a, this is the scene of the ultra-large container ship "XinFu 102" leaving the port [1]. When such large vessels enter or leave the ports, due to the limitations of narrow waterways, they must rely on tugboats to assist with a series of complex operations such as berthing, turning, and controlling navigational speed. Against the backdrop of rapidly increasing shipping demands, ship collisions, aging fires, groundings, and other maritime accidents are frequent. As shown in Figure 1b, vessels that have lost their self-propulsion capability due to collision accidents urgently require tugboat assistance to leave the incident area [2], otherwise, it may lead to serious consequences such as waterway blockages, fuel leakage, and even secondary accidents. In addition, offshore oil and gas platforms, floating wind turbines, and resource exploration robots, which lack self-propulsion capabilities, also rely on towing technology.

(**a**) (**b**)

Figure 1. Towing operation application scenarios [1,2]. (**a**) The "XinFu 102" towing operation site. (**b**) Damage to the barge collision accident.

Ship towing refers to the process of using tugboats to tow the ship from one location to another; in accordance with the purpose, it can be broadly divided into harbor towing, point-to-point towing, rescue towing, salvage towing, and emergency towing. As shown in Figure 2, in order to meet different operational requirements, it is necessary to choose the appropriate tug arrangement scheme. According to the U.S. Navy Towing Manual [3], the tug arrangement scheme is divided into double towing, catenary towing, tandem towing, and side-by-side towing.

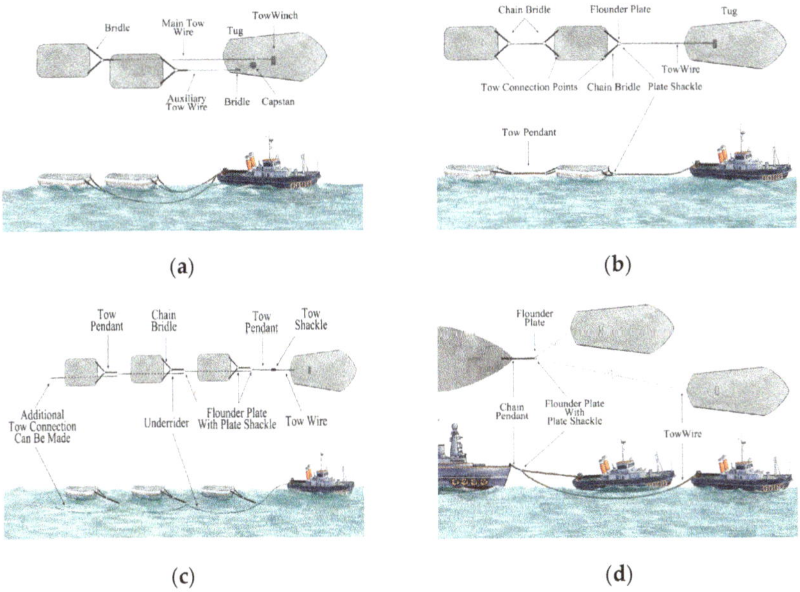

Figure 2. Composite of arrangement scheme. (**a**) Double towing; (**b**) catenary towing; (**c**) tandem towing; (**d**) side-by-side towing.

The towing process involves the tugboat, the towed objects (such as cargo ships, oil drilling platforms, and disabled vessels), and the connecting cable, which together form a complex multi-body system. In the system, the towing force provided by the tugboat is transmitted to the towed objects through the cable, and its motion state (such as speed and direction) directly affects the tension of the cable and the movement of the towed objects. The cable, acting as a medium, has characteristics such as stiffness, length, and mass per unit length that determine the manner and efficiency of force transmission. The size, shape, mass, and hydrodynamic characteristics of the towed object affect its response to the cable tension, thus influencing the motion state of the tugboat. To address the numerous challenges involved in the towing process, researchers have conducted extensive and in-depth research. However, most of the existing review articles focus on the analysis of the overall motion characteristics of the towing system or specific application scenarios, and the systematic review of the mechanical and dynamic responses of the tow cable and the towed object is relatively rare. By leveraging databases such as Web of Science, ScienceDirect, and ResearchGate, this paper systematically reviews the relevant literature in the field of marine engineering towing. This study primarily focuses on research related to towing cables and towed objects within towing systems, with an emphasis on summarizing and discussing the latest research progress on their mechanical and motion responses. The research directions are categorized as follows: the motion response of the ship towing system, the research of cable mechanical performance and dynamic model, and the mechanical response analysis of the towed object. In addition, the current research status in each direction is sorted out and summarized, which will provide references for scholars who will further study this field. The current state of research in each direction is reviewed, providing a reference for scholars who will delve deeper into this field in the future.

2. Motion Response of Towing System

During the towing process, the tugboat, towed object, and cable are subjected to the combined loads of wind, wave, and current, and the special connection relationship among the three makes their motion response interrelated and influences each other, which constitutes an extremely complex dynamic response system. At present, researchers all over the world have carried out a lot of scientific research on the motion response problems involved in the towing process by using different methods, including, but not limited to, numerical simulation methods, theoretical model methods, and experimental methods, which are suitable for solving different towing problems. This section introduces innovative and representative research cases in detail, using research methods as the thread, and elaborates on the advantages and disadvantages of some studies.

2.1. Numerical Simulation Method

Scholars from various countries have conducted extensive theoretical research on the motion response of towing systems from multiple perspectives. Numerical simulation methods have become an indispensable key technology in research across various fields due to their high efficiency, high precision, and low cost. Advanced CFD (Computational Fluid Dynamics) technology and innovative numerical models can be employed to deeply analyze the dynamic interaction effects of towing systems under different conditions [4,5]. These models are not only capable of accurately simulating fluid resistance and motion response on calm water surfaces, but also take into account the complex dynamic interactions under wave conditions [6]. As shown in Figure 3, the use of multi-zone overlapping grid technology achieves a more precise simulation of the dynamic interaction effects between the towing system and the mother ship, better capturing the subtle interactions between the towing systems, and enhancing the accuracy of simulation results. Figure 3 illustrates the variation of the six degrees of freedom of the tugboat over time during towing operations in calm water conditions, with particular attention to the significant vibrations in heave and pitch during the tugboat's startup phase.

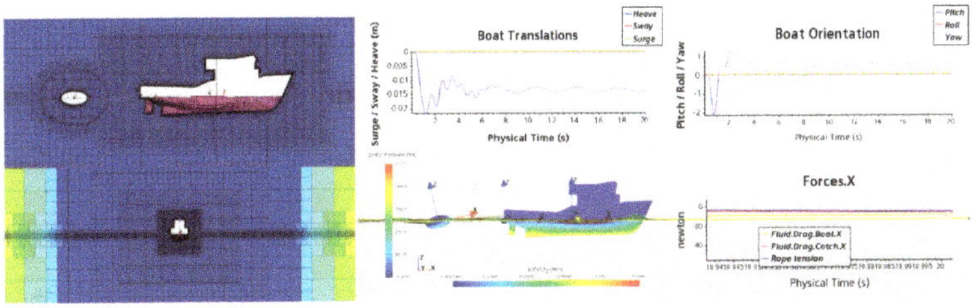

Figure 3. Mesh division and simulation calculation results [6].

CFD technology can reduce reliance on physical model experiments, lowering research costs and time, while improving the flexibility and repeatability of tests. To verify the installation of temporary fins at the stern to improve the steering stability of FPSO (Floating Production Storage and Offloading) during towing, Li used the Vortex Lattice Method to predict the hydrodynamic lift coefficient of fins [7]. A three-degrees-of-freedom numerical towing system model is established, considering the interaction of hydrodynamic forces, the hull, and other appendages, as well as the influence of towing cables. It is noteworthy to combine the fluid dynamic prediction of fins with the dynamic response analysis of the towing system, thereby providing researchers with a comprehensive assessment framework. Furthermore, the impact of fins on the dynamic response and stability of the towing system is assessed in sync through time domain simulation and eigenvalue analysis, which is relatively rare in previous studies. Similarly, Figari presented an all-round design-to-simulation approach for a new Z-drive escort tug class, combining CFD calculations and towing tank tests to optimize the hull design and predict the maneuvering capabilities of the tug in different operational scenarios [8]. Their study highlights the importance of integrating numerical simulations and experimental validations to improve the design and performance of towing vessels.

In the realm of resolving towage issues, Workbench AQWA has garnered widespread recognition for its comprehensive functional platform in frequency domain calculations. OrcaFlex stands out for its convenient and efficient user interface, as well as its high-performance and precise computational capabilities in the time domain analysis of models. Consequently, AQWA and OrcaFlex have gradually become key tools for researchers employing numerical simulation methods to study towing and mooring problems. In the field of ocean engineering towing, the Munk moment is an important concept that significantly affects the hydrodynamic response and motion stability of towing systems. Figure 4 presents a schematic diagram of the ocean towing cable-array system model. Zhang and colleagues used the lumped mass method to discretize the towing cable into a series of concentrated mass points, forming a lumped mass model [9]. This method, based on a series of assumptions, establishes the relationship between the expression of the Munk moment in classical towing body kinematics and that in the hydrodynamic analysis software OrcaFlex. The research results indicate that the maximum effective tension of the towing cable exhibits a periodic fluctuation trend along its length under the influence of a single Munk moment coefficient. Moreover, when the Munk moment coefficient ranges from 0.5 to 0.9, the effective tension of the towing cable demonstrates better synchronization along its length. As the Munk moment coefficient increases, the pitch angle of the towing body in the steady state gradually increases and eventually approaches a stable value. This study reveals the impact of the Munk moment coefficient on the tension and bending moment of the towing cable, as well as its role in the stable state and dynamic response characteristics of the towing body, providing a new method for assessing the performance of towing systems under complex sea conditions.

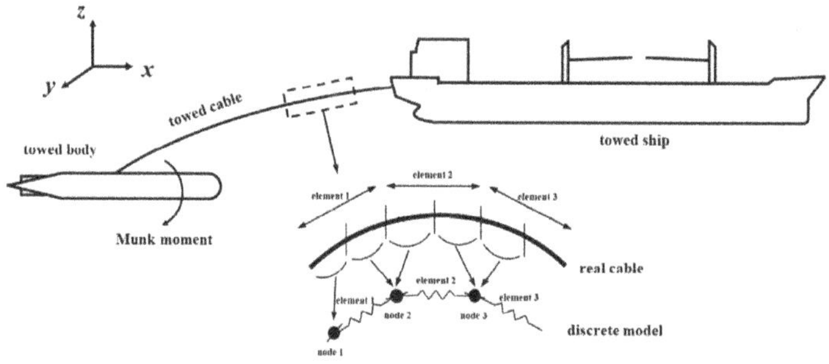

Figure 4. Schematic diagram of the ocean towing cable-array system model [9].

Moreover, significant progress has been made in utilizing numerical simulation methods to investigate the maneuvering characteristics of Escort Tugs, which play a pivotal role in marine engineering towing systems. Particularly crucial when assisting large vessels in entering and exiting ports, executing rescue missions, and towing large offshore platforms, the maneuvering performance of Escort Tugs directly impacts the safety and efficiency of towing operations. Piaggio employed the Reynolds-Averaged Navier–Stokes (RANS) approach within Computational Fluid Dynamics (CFD), combined with OpenFOAM software, to conduct numerical simulations of the maneuvering characteristics of Escort Tugs [10,11]. The research focused on analyzing the hydrodynamic properties of the tug's hull and skeg (a rudder-like appendage) and how these properties influence the tug's maneuvering performance. The study encompassed pure drift, pure yaw, and combined drift–yaw conditions, and by altering the size and shape of the skeg, the mechanisms through which it affects hydrodynamic properties and maneuvering performance were analyzed. Mauro proposes an improved method based on a nonlinear thrust distribution algorithm, including thruster interactions, and adds lateral balance modeling of the tug to assess the stability of the tug during escort operations, providing a flexible method for initial prediction of the tug's escort capability [12,13].

These studies not only provide essential theoretical foundations for the design of Escort Tugs, but also offer scientific guidance for the selection and operation of tugs in towing systems. The maneuvering performance of Escort Tugs directly affects the dynamic response and stability of towing systems. By optimizing the hydrodynamic characteristics of tugs, the overall performance of towing systems can be enhanced, minimizing tension fluctuations in towlines and reducing potential safety hazards during towing operations. These studies offer new perspectives for the design and operation of towing systems, especially in complex marine environments, where the maneuvering performance of Escort Tugs is crucial to the safety and efficiency of towing systems.

By integrating AQWA and STAR-CCM+ along with various empirical formulas, Chen conducted an in-depth study on the wet towing operation of semi-submerged floating wind turbines [14]. As shown in Figure 5, the numerical model of the floating wind turbine (FOWT) towing system was constructed in AQWA to investigate the effects of different towing speeds and environmental loads on the hydrodynamic characteristics and power of the FOWT towing system. By comparing the results from different standards and those from STAR-CCM+, a refined empirical formula for calculating the towing resistance of FOWT was determined and validated using the results from AQWA. This study not only optimized the calculation method for towing resistance, but also proposed the optimal correction schemes under different environmental conditions, providing an important reference for practical engineering applications.

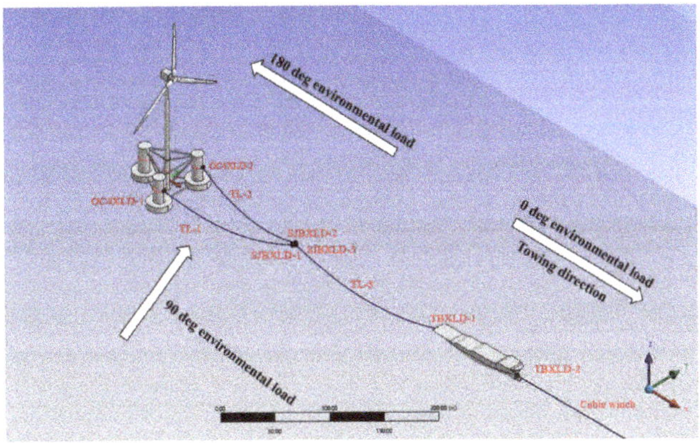

Figure 5. Numerical model of the towing system of the FOWT in various environmental load directions in AQWA [14].

The combined use of the strengths of these two software packages to solve towing problems is one of the important methods in the numerical simulation approach. AQWA is used to calculate the hydrodynamic coefficients of FPSO in the frequency domain, while OrcaFlex is used to simulate the wet towing transportation process in the time domain. By establishing a complex system model that includes a tugboat, towing cable, and FPSO, and considering environmental factors such as wind, waves, and currents, Huang can realize synchronous simulation of dynamic response and structural interaction of the entire wet towing system [15]. On this basis, by setting different parameters (such as towing speed, sea conditions, cable length, etc.), the impact of these parameters on wet towing transportation can be analyzed. It is possible to achieve synchronous simulation of the dynamic response and structural interaction of the entire wet towing system. On this basis, by setting different parameters, including acceleration, displacement, attitude, and structural interaction between FPSO and the towing cable, guidance can be provided for actual towing operations. In recent years, with the development of automation and intelligence of towing systems, more and more attention has been paid to the impact of environmental interference on towing safety. Zhang propose a framework for preventing collisions in ship towing operations in the presence of environmental disturbances [16]. In this study, the direction and magnitude of the towing force of the tugboat can be adjusted in real time through a multi-agent control strategy to ensure that the towing system can sail along the predetermined trajectory while maintaining a safe distance between the tugboat and the towed boat. This method not only improves the safety of towing operation, but also provides a new idea for the automatic control of towing systems.

2.2. Theoretical Model Method

When studying complex towing problems, transforming the challenges of actual engineering into mathematical problems through a series of reasonable assumptions and the neglect of interference terms is an extremely important research approach. By establishing mathematical models to describe the mechanical and kinematic responses of the tugboat–towline–towed body system, a crucial method of investigation is employed [17,18]. Zhu and Hu, taking the TLP (Tension Leg Platform) as the object of study, propose an analytical method for studying the wave resistance and maneuverability of multi-link towing systems based on Newton's laws and the general formulas for rigid body motion. According to the seakeeping and maneuverability models, the motion equation of the *i*th object can be expressed as [19]:

$$\underbrace{M_{RB_i}\dot{\nu}_i + C_{RB_i}(\nu_i)\nu_i}_{\text{rigid-body forces}} + \underbrace{A(0)_i\dot{\nu}_{s_i} + C_A(\nu_{s_i})\nu_{s_i} + N_n(\nu_{s_i})}_{\text{maneuvering terms}}$$
$$+ \underbrace{\sum_{j=1}\left(A_{ij}(\infty)\dot{\nu}_{f_j} + \int_0^t K_{ij}(t-\tau,U)\nu_{f_j}(\tau)d\tau\right)}_{\text{seakeeping terms}} + \underbrace{D_i\nu_i + D_{q_i}\nu_i|\nu_i|}_{\text{damping terms}} + \underbrace{g_i(\eta_i)}_{\text{hydrostatic forces}} = \tau_{m_i} + \tau_{w_i}, (i,j = 1 \rightarrow \text{anchor}, 2\ \text{platform}) \quad (1)$$

where v_s and v_f are the slowly and fast varying velocity components of v, referred as low-frequency velocity and wave-frequency velocity, respectively. $A(0)$ and $A(\infty)$ are the low-frequency and infinite-frequency added mass matrices, respectively, C_A denotes the hydrodynamic Coriolis–Centripetal matrix, $N_n(v_s)$ is the nonlinear damping force on the hull in the maneuvering model, $K(t, U)$ represents the speed-dependent retardation function at forward speed U, Dv and $Dqv\,|\,v\,|$ are damping forces excluded from the seakeeping and maneuvering models, $g(\eta)$ is the restoring force, τ_m is the towing force, and τ_w represents the wave excitation loads. The innovative aspect of this theoretical model lies in the use of a low-pass filter to separate the slow and fast motion components in maneuvering and seakeeping calculations, enabling the simultaneous resolution of maneuvering and seakeeping issues on the same timescale. The limitation of the model is that it only considers the effects of towing forces and wave loads, neglecting the impact of wind and current loads. The combined effects of wind, waves, and currents are a key and challenging factor threatening the safety of towing operations. One direction for future development of towing mathematical models is to fully consider the combined effects of wind, wave, and current loads.

The stability of the heading in ship towing systems is of paramount importance. In actual operations, both the tugboat and the towed vessel are subject to varying degrees of wind influence. Early research considered that the course stability of the towed barge is typically unstable under conditions ranging from crosswind to oblique wind. Subsequently, Kijima's theoretical modeling and linear analysis revealed that as the wind direction changed from headwind to tailwind, the course stability of the two towed vessels tended to be unstable [20,21]. In a recent representative study, Fitriadhy proposed a numerical model to analyze the course stability of tugboats under uniform and constant wind conditions. Under uniform and constant wind conditions, the stable and unstable towing conditions at different wind speeds and angles were identified using linearized motion equations. Nonlinear time domain simulations captured the dynamic interactions during the towing process and analyzed their stability. This research provided new perspectives and tools in the establishment and analysis of mathematical models, especially in considering the effects of wind and nonlinear dynamic responses, offering a more precise and comprehensive method for the stability analysis of towing systems [22,23]. Furthermore, Sinibaldi and Karasuno regarded the tugboat as a mass point with predetermined motion, while the towed vessel was considered a 4-degrees-of-freedom (surge/sway/yaw/roll) nonlinear dynamic system. The model took into account the influence of wind on the towed vessel and employed novel computational theories to solve for the maneuvering forces and moments at large drift angles, overcoming the deficiencies of traditional derivative models under such conditions [24,25]. In the study of towing systems, the use of bifurcation analysis to investigate the stability of towing systems is innovative because it can identify the stability and instability of the system under different parameters.

When conducting towing operations in broken ice areas, the presence of ice blocks of varying sizes, shapes, and thicknesses may pose additional challenges to towing operations. Therefore, reliable prediction of the ice area towing system is necessary, especially for large floating bodies that require multiple tugs to navigate cooperatively, as shown in Figure 6. Han established a 3-degrees-of-freedom (3-DOF) maneuvering model group (MMG) model to simulate the coupled motion of tugboat and towed structure, taking into account the surge, sway, and yaw motion responses of both the tugs and the towed structure [26]. It is noteworthy that Marnix and Matuttis proposed the mechanism of ice contact force on the towing system (Figure 7), including the Centroid Voronoi Tessellation (CVT) algorithm to generate a random broken ice field with polygonal-shaped ice blocks, and the use of the two-dimensional Discrete Element Method (DEM) to simulate the behavior of ice blocks during the towing process [27,28].

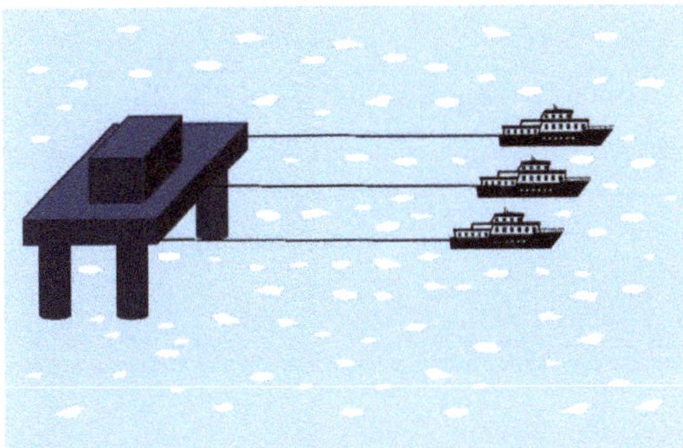

Figure 6. Ice region multi-tug towing system [26].

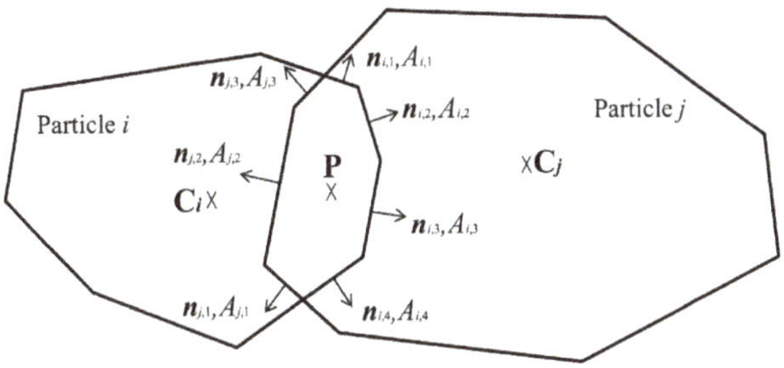

Figure 7. Polygonal particle contact model [26].

2.3. *Experimental Method*

Since the environmental loads in the towing process are extremely complex and involve multi-body coupled motions, it is difficult to accurately simulate the real working conditions by using only mathematical models or numerical simulation; therefore, many researchers obtain reliable data by using towing model experiments or full-scale towing experiments. In addition, the reliability of mathematical models and simulation models can be further verified by using experimental data [29–31].

2.3.1. Reduced-Scale Model Towing Experiment

Simulating ship towing in a water tank using ship models can yield relatively rich and accurate experimental data while keeping costs relatively controllable, thus it has been widely applied by researchers. As shown in Figure 8, Guo designed a novel marine towing experimental device, which differs from traditional towing experimental methods. The cable is constrained on the carriage frame through a connecting joint, and a motor-driven carriage simulates the motion speed and depth of the cable during towing to explore the dynamic response characteristics of the cable, as well as the response amplitude and frequency characteristics under different conditions. Research findings indicate that ship motion can lead to parametric excitation, causing an increase in the vibration amplitude of the cable [32]. A team from Seoul National University in South Korea conducted model tests in a towing tank to investigate the effects of cable types, the attitude of the towed object, and additional devices (such as fins, streamlined brackets, and spoilers) on towing stability. The study found that the type of cable has a significant impact on towing stability, with non-rotating steel cables being the most effective in suppressing yaw motion, and a lower center of gravity enhancing towing stability [33].

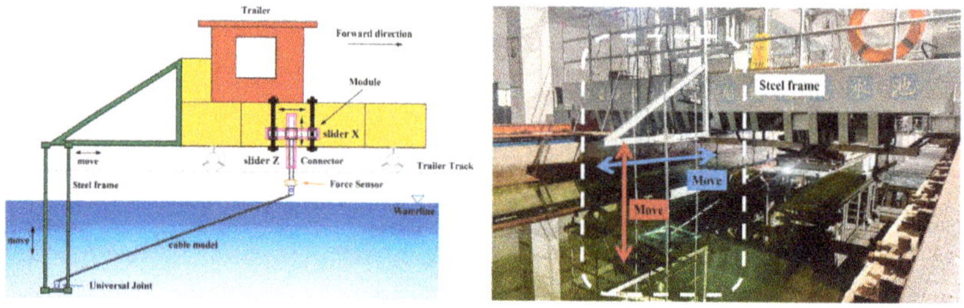

(a) (b)

Figure 8. Motor-driven cable towing experiment [32]. (**a**) Schematic diagram of the towing cable test; (**b**) detailed view of the cable suspension module.

Towing experiments can also be conducted using relatively simple equipment (Figure 9) [34]. In a circulating water channel (CWC) with a length of 4.0 m, a width of 1.8 m, and a depth of 1.2 m, the free fall of weights serves as the power source, and the sine movement of the guide rail simulates the directional changes during ship towing. By using the reflective devices on the FPSO model and a three-dimensional displacement meter, data on the ship's motion are extracted to study the towing stability and course stability of the FPSO under the influence of the tug's lateral motion. The innovation of this experimental scheme lies in the simulation of the impact of the tug's lateral motion on the motion response of the towed object with a relatively simple device, which was rarely considered in previous experiments. There are shortcomings: the tug's motion is simplified to sine motion, which deviates significantly from the actual situation; the self-elasticity and stiffness of the cable are neglected; and the effects of wind and waves are ignored. In addition, numerous researchers have conducted studies on towing problems from different perspectives using various types of water tanks [35,36].

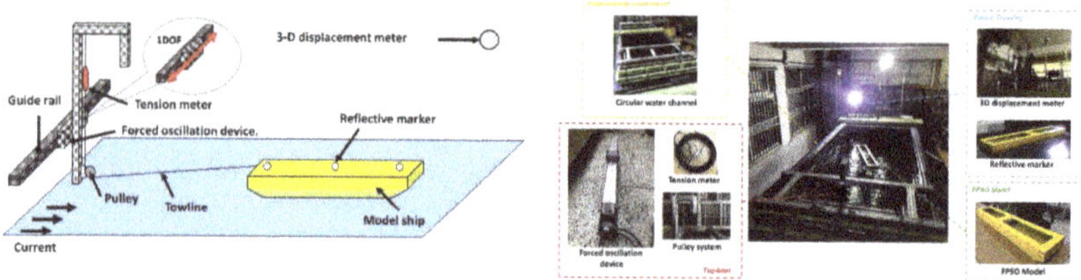

(a) (b)

Figure 9. Weight-gravity-driven ship towing experiment [34]. (**a**) Layout drawing of the towing test; (**b**) details of the test system components.

2.3.2. Full-Scale Towing Experiment

It is undeniable that, compared to towing experiments using ship models, data obtained from actual ship towing experiments possess greater reference value. The training vessel "SAE YU DAL" from the National Mokpo Maritime University, South Korea, conducted a sea towing experiment, with experimental data collected through load sensors and other measuring equipment. The results indicate that towing speed is the primary factor determining the towing force, and the ship's rolling and yawing movements have a significant impact on the towing force, which need to be considered in the total towing resistance [37]. The main advantage and innovation reside in conducting towing experiments with real vessels in open and real maritime environments, rendering the measured data highly reliable. The shortcomings of the experiment are that the dynamic motion of the ship's influence on the towing force was not fully accounted for, such as the dynamic changes in rolling and yawing; the versatility of the experimental vessel is uncertain, which may affect the extrapolation of the experimental results; and the experiment did not cover common factors affecting the towing force, such as ocean currents.

As shown in Table 1, scholars have used different methods to study the complex multi-object coupled motion problems involved in towing operations, among which, the numerical simulation method has the advantages of high efficiency, low cost, and high accuracy, and it can establish numerical models of ship towing, underwater equipment towing, and other numerical models according to the research needs, and set up a variety of environmental loading conditions to carry out the research, which can visually and accurately demonstrate the motion response of the towing system. In contrast, the theoretical model method is more abstract, and the research of the towing problem needs to rely on certain reasonable assumptions and is more difficult to study, but this method is an exploration of the mathematical nature of the motion problem of the towing system, and the progress of the research of this method can promote the overall enhancement of the level of research in this field. The advantage of the experimental method of researching this problem is that the difficulty is relatively small, and there is a certain degree of authenticity and reliability of the rich data at the same time, but the disadvantage is that it is costly, so future researchers should design the experimental method. Therefore, when designing the experimental program, future researchers need to ensure the accurate reduction of real working conditions and hydrodynamic characteristics of the ship, and also need to simplify the experimental equipment and reduce the cost as much as possible.

Table 1. Summary of solving methods of motion response problems involved in towing process.

Methods	Advantages	Disadvantages	Representative Articles
Numerical Simulation Method	High efficiency, low cost, high-precision simulation, comprehensive functions, and flexible parameter analysis.	It may not be able to fully and accurately reflect the dynamic response of the towing system in a complex marine environment.	The numerical simulation method considers the complex dynamic interactions under wave conditions through numerical models [6]. The hydrodynamic lift coefficient of the fins was predicted by using the vortex grid method [7]. The dynamic response of towed system with different Munk moment coefficients is simulated based on OrcaFlex [9].
Theoretical Model Method	Transform practical engineering challenges into mathematical problems, innovative analysis, and new perspectives of stability analysis.	Through reasonable assumption and ignoring interference terms, the model may have certain deviation from the actual situation. Not all factors are considered.	The theoretical model method considers the influence of wind and nonlinear dynamic response, and provides a more accurate and comprehensive method for the stability analysis of towed systems [22,23]. This model uses a novel computational theory to solve the control force and moment under large drift angle, overcoming the shortcomings of the traditional derivative model [24,25].
Experimental Method	The data are real and reliable, and the experimental equipment is simple.	The influence of hull dynamic motion on the drag force is not fully considered, and the generality of the experimental ship is uncertain, which may affect the extrapolation of the results. The cost of the experiment is relatively high.	Towing experiments are carried out with this method, and the towing stability and course stability of FPSO under the influence of tugboat lateral motion are studied by using reflector and three-dimensional displacement meter to extract ship motion data [34].

3. Dynamics Model and Mechanical Response of Cables

The towing cable, as a core component of the entire towing system, has a significant impact on the mechanical and kinematic responses of the tugboat and the towed object. Internationally, 20% of the total loss accidents involving large mobile installations during towing operations are caused by cable breakage, with the root cause of many accidents being the extreme and complex working conditions of the cables. The length of the cable typically ranges from 180 m to 300 m, and its motion and mechanical responses under the coupled loads of wind, waves, currents, and the combined tension of the tugboat and the towed object are extremely complex. It is particularly important to note that the instantaneous tension in the cable at the initial stage of the tug's start-up may reach 1.5 to 9 times that of the stable condition, which is highly likely to cause breakage accidents.

The corrosion resistance, wear resistance, and structural strength of the cable under dry and wet conditions are key factors affecting the safe conduct of towing operations. According to the specification requirements [38], the cable used for towing should be flexible galvanized steel wire ropes with a tensile strength of not less than 1.37 kN/mm^2. The length and minimum breaking load of the cable should be limited by the number of fittings, N. The number of fittings for the towing vessel is calculated as follows [38]:

$$N = \Delta^{\frac{2}{3}} + 2\left(Bh + S_{fun}\right) + \frac{A}{10} \tag{2}$$

where Δ is the displacement of the hull below the summer load waterline, t (tone); B is the width of the ship, m (meter); h is the effective height from the summer load waterline to the top of the topmost compartment, m (meter); S_{fun} is the effective positive projection area of the chimney, m^2 (square meter); A is the sum of the side projection areas of the hull and superstructure above the summer load waterline, as well as the deckhouse and chimneys with widths greater than $B/4$ on each floor within the length range L of the ship, m^2 (square meter).

3.1. Modeling of Cable Dynamics

During the towing operation, the cable is in a complex nonlinear motion state, and the hydrodynamic forces acting on the cable vary with time, thus numerical methods are relied upon for approximate solutions. There are three main approaches to studying the dynamic motion of cables: the lumped mass method, the finite element method, and the finite difference method. The fundamental idea behind these three cables hydrodynamic theoretical models is to establish a cable model by considering the material properties of the cable and the hydrodynamic loads, then to spatially and temporally discretize the cable motion equations, transforming them into a system of nonlinear equations for temporal integration [39]. In summary, the motion of the cable must satisfy the following two basic conditions:

(1) Newton kinetic condition: $F = ma$. In the formula, F is the force, m is the mass, kg, and a is the acceleration, m/s^2;

(2) Continuity conditions: for stretchable cables, the stress–strain relationship is used as the continuity condition, $T = f(\varepsilon)$; for non-stretchable cables, a kinematic continuity condition is required [40]:

$$\frac{\partial}{\partial s}\left(\frac{\partial r}{\partial t}\right) = \frac{\partial}{\partial t}\left(\frac{\partial r}{\partial s}\right) \tag{3}$$

where r is the position vector of a point on the cable, represented in vector form as $r = xi + yj + zk$, and x, y, and z are the coordinate components of the point in the fixed Cartesian coordinate system. s represents the length coordinate when the cable is not stretched, which is used to describe the position of a point on the cable. t is the unit tangent vector at a point on the cable, pointing in the direction of the cable's motion.

Building on the aforementioned two preconditions, the dynamic solution methods for cables established by scholars are summarized as follows:

3.1.1. Lumped Mass Method

The lumped mass method discretizes the cable into a series of elastic, frictionless springs connected by nodes, with the cable's mass, fluid resistance, inertial forces, gravity, and buoyancy all concentrated at the corresponding nodes. The kinematic relationships between nodes are considered as continuity conditions, with the influence of added mass also being included. The time differential vector is discretized to establish a system of differential equations for the motion of the cable. Walton [40] was a pioneer in using the lumped mass method to solve cable problems, and Polachek [41] built upon this by considering the stretchability of the cable, spatially discretizing the cable, and establishing a system of equations for cable motion, laying the foundation for the lumped mass method.

Since then, numerous researchers have improved, iterated, and applied the lumped mass method from different perspectives [42–47]. One of the latest applications of this method is in the study of the dynamic characteristics of hydrofoil control in underwater towing systems (Figure 10), in which Yang's modeling of the towing cable applied the concept of the lumped mass method (Figure 11) [48]. Innovatively, the Morison equation was combined with a CFD solver to determine the fluid resistance on each segment, with this combination capable of considering more complex interactions between the cable and fluid, including both tangential and normal forces. Mathematical modeling and experimental validation methods were used to explore issues such as hydrofoil shape optimization, speed control, and control methods.

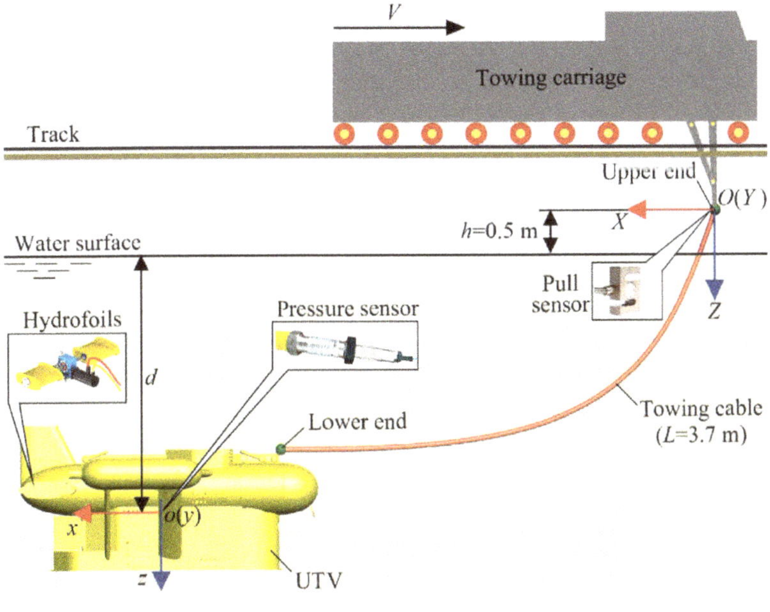

Figure 10. Schematic diagram of the towing test [48].

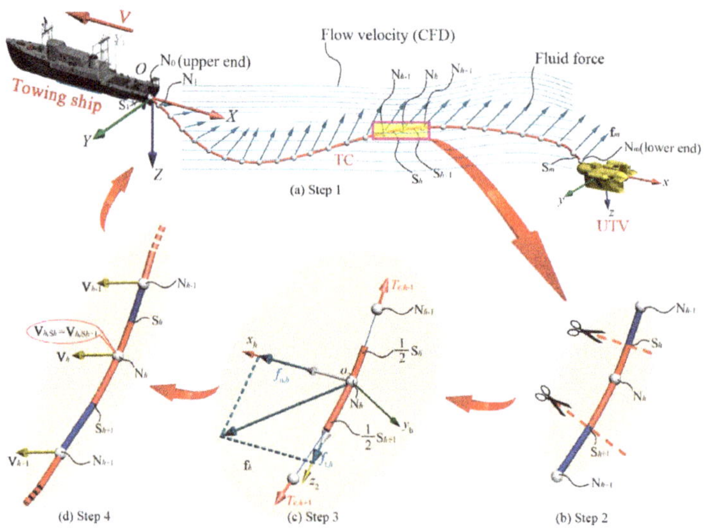

Figure 11. Modeling idea of cable centralized quality method [48].

The lumped mass method is not only used for towing cables in ship towing, but is also applied to the study of towed seismic cable system exploration. Since vibrations can affect signal transmission through vortex shedding, which in turn impacts exploration efficiency, it is of great significance to study the dynamic response characteristics of towed cables for vibration suppression and enhancing the reliability of marine seismic exploration systems. Guo et al. investigated the dynamic response characteristics of towed cable systems in marine seismic exploration, and they also employed the lumped mass method in their dynamic analysis model of the towed cables [49]. The study found that the cable weight and length have a minor impact on the towed cable configuration, mainly affecting the natural frequency and encounter velocity. Towing speed and the weight of the towing body significantly influence the towed cable configuration. The dynamic response of the towed cable exhibits characteristics of multi-mode participation, amplitude modulation, time-frequency sharing, and mode jumping.

3.1.2. Finite Difference Method

The finite difference method involves expanding the dynamic and continuity equations in the local coordinate system of the cable, establishing the governing equations for the cable's micro-elements, and using the Newton iteration method for numerical solution. In 1983, Ablow first applied the finite difference method to the study of cable motion [50], and subsequently, Milinazzo's revisions and expansions have led to a more efficient and stable form [51]. The advantage of this method is the use of implicit time and spatial integration, which allows it to remain stable when handling larger time steps; furthermore, it demands less computational effort and is suitable for simulating motions over large time scales. However, this theory is based on the assumption of linear material properties for the towing cable, and due to the limitations of the model's algorithm itself, it cannot simulate states of zero tension within the cable. The derivatives in the direction of cable length imply that the physical properties of the towing cable should be uniform and consistent, without large discontinuities. For non-uniform cables, improper handling may likely result in significant computational errors. The finite difference method has been widely applied, with numerous scholars using it to solve different types of cable motion problems [52–54]. For instance, when studying the vibration of cable tension during variable speed towing systems, reasonable simplifications are made based on the classical cable dynamics model, using the box format derived by Burgess as a foundation, and incorporating a slack factor within each time step to solve the discrete equations [55]:

$$\begin{aligned}&\left[M_{j+1}^{i+1}+M_{j}^{i+1}\right]\frac{Y_{j+1}^{i+1}-Y_{j}^{i+1}}{\Delta S_j}+\left[M_{j+1}^{i}+M_{j}^{i}\right]\frac{Y_{j+1}^{i}-Y_{j}^{i}}{\Delta S_j}=\\&\left[N_{j+1}^{i+1}+N_{j}^{i+1}\right]\frac{Y_{j+1}^{i+1}-Y_{j+1}^{i}}{\Delta t_k}+\left[N_{j}^{i+1}+N_{j}^{i}\right]\frac{Y_{j}^{i+1}-Y_{j}^{i}}{\Delta t_k}+Q_{j+1}^{i+1}+Q_{j}^{i+1}+Q_{j+1}^{i}+Q_{j}^{i}\end{aligned} \tag{4}$$

where M represents the mass coefficient matrix, which is used to describe the mass properties of the towed body; Y refers to the motion state vector of the towed body, encompassing the velocity and position information of the towed body in various directions; N denotes the time-related coefficient matrix, which is related to the coefficients that vary with time and affect the changes in the motion state of the towed body; S is the vector of unknowns, comprising unknowns such as the cable tension, cable motion velocity, and phase angle; and Q represents the source term vector, indicating the external forces acting on the towing system, such as fluid resistance.

3.1.3. Finite Element Method

The finite element method is based on Hamilton's principle of virtual work and derives the motion equations for cables. Similar to the lumped mass method, the finite element method discretizes the cable into a series of straight-line elements in space. Its advantage lies in overcoming the deficiency of the lumped mass method, which concentrates mass at discrete points, by enabling the integral calculation of cable elements along the entire length of the cable. Numerous scholars have utilized this method to conduct extensive research on slender structures such as cables or marine pipelines, significantly enhancing the precision and efficiency of marine engineering research [56–63]. During the towing process, the deformation of the towing cable is extremely complex and is influenced by various loads such as waves, wind, and the reaction forces from the vessel's structure. It is worth noting that the finite element method (FEM) excels in dealing with complex geometries, nonlinear problems, and fluid–structure interaction (FSI) issues. Therefore, the application of the finite element method in marine engineering towing systems has significant advantages. The convergence of finite element results is an important indicator for evaluating the accuracy of numerical simulations. The control of convergence mainly considers the following aspects: the rationality of mesh generation, the precision of iterative calculations, and numerical stability. For example, in the study of the dynamic stiffness characteristics of synthetic fiber cables, Zhang used the finite element software Abaqus to solve the motion response of synthetic fiber cables. During modeling, one-dimensional beam elements are employed, and after incorporating the dynamic stiffness empirical formula, the Jacobian matrix is [64]:

$$\frac{\partial\sigma}{\partial\varepsilon}=E=\left[\alpha+\beta L_m-\gamma L_a\right]M_{BL}/A \tag{5}$$

where E represents the element stiffness; A represents the cross-sectional area of the mooring cable; M_{BL} represents the minimum breaking load of the cable; L_m represents the average tension as a percentage of MBL; L_a represents the dynamic tension amplitude as a percentage of MBL; α, β, and γ are parameters related to the cable's structure and material; and $\frac{\partial\sigma}{\partial\varepsilon}$ represents the Jacobian matrix (stress–strain relationship). The method is traditionally suitable for solving static or quasi-static problems and is difficult to apply to dynamic equations that are inherently nonlinear in nature, such as those involving large motion accelerations and large deflections. Consequently, the finite element method is mostly applied to the static analysis of towed cables and is not suitable for describing the hydrodynamic performance of towed cables under unsteady conditions. In recent years, with the improvement of computer performance, the application scenarios of the finite element method have become more extensive, and the corresponding commercial software has become more abundant. It has been promoted and applied in various disciplines, including naval architecture and ocean engineering, civil engineering, and others.

As shown in Table 2, the following is a summary and refinement of the three dynamic solution methods for mooring cables mentioned above. It focuses on analyzing the applicable conditions, as well as the advantages and disadvantages of each method, and provides an overview of representative studies and their contributions. In future research on the dynamic response of mooring cables, the most optimal solution method can be matched according to the research focus, in order to reduce computational effort, enhance accuracy, and achieve greater efficiency.

Table 2. Summary of Dynamic Solution Methods for Towing cables.

Methods	Advantages	Disadvantages	Representative Articles
Lumped Mass Method	Clear physical meaning, simple algorithm, and considerable applicability and extensibility.	The performance is poor when dealing with non-elastic problems, the computational load is large, and it is not easy to converge when dealing with violent motion of cables.	The lumped mass method was first applied to solve the dynamics of cable systems [40]. A relatively complete solution system for the concentrated mass method has been essentially established [41].
Finite Difference Method	It remains stable when dealing with large time steps, has a smaller computational load, and is suitable for simulating motions on a large time scale.	When zero tension occurs in the cable, the results become singular, causing the calculation to become unstable; significant errors are likely to occur when dealing with cables with non-uniform physical properties or abrupt changes.	The finite difference method was first applied to solve the dynamics of cable systems [49]. The latest representative research is the application of this method to study the vibration of cable tension during variable-speed towing operations [54].
Finite Element Method	It overcomes the shortcomings of the lumped mass method, breaks free from the constraints of the Courant condition, thus allowing for greater flexibility in the selection of spatial and temporal step sizes in the simulation of system motion.	The material properties of the cable are assumed to be linear, and due to the limitations of the model's own algorithms, it is unable to simulate the state where there is zero tension in the cable.	The most representative recent research is that this method has been used to address the motion response issues of synthetic fiber cables, with a focus on studying the impact of cable dynamic stiffness on its dynamic characteristics [64].

3.2. *Mechanical Properties of Composite Cable*

The rapid development of material science has greatly enriched the types of cables, and composite cables such as aramid fiber, polyester fiber, and polyester have been gradually recognized in the field of ship towing with the advantages of light weight, high strength, strong corrosion resistance, and excellent economy [65]. At the same time, the mechanical and geometric nonlinearities of composite cables have also become a new challenge for the promotion and application of composite cables, for which many scholars have carried out research from the perspectives of abrasion resistance, tensile strength, weaving methods, and material types.

Material diversity is one of the characteristics of composite fiber cables, which allows for the customization of fiber and matrix materials according to different application scenarios. The mechanism by which different materials used in cable making affect its mechanical properties has become an issue that warrants investigation. It is worth noting that Huang's study on the impact of cable material diversity (Figure 12) on the stability of towing systems is representative in this field [66]. The study selected cables made from polyester, aramid, and HMPE as the base materials and introduced parameters such as static stiffness Ks and coefficients α and β of dynamic stiffness to define the mechanical properties of different materials. Its innovation lies in proposing a numerical model that can calculate the dynamic stiffness of the cable based on its average load. It discussed the dynamic yaw and lateral offset performance of the tugboat under various wind speeds and investigated the influence of different types of fiber cables on the cable's properties and initial length. The results indicate that the choice of cable material significantly affects the steering stability of the towing system; the greater the stiffness of the cable, the faster the vessel's yaw and offset can decay to a stable state. The influence of cable length is also significant; the shorter the cable length, the smaller the induced offset and yaw angle motion.

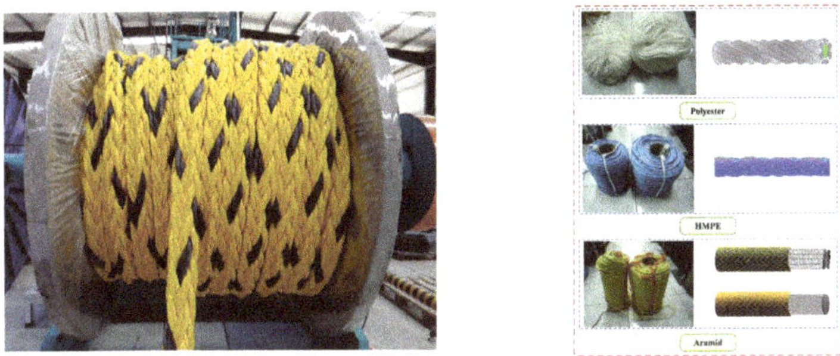

(a) (b)

Figure 12. Composite fiber cables [66]. (**a**) Composite material cables used in towing operations. (**b**) Three composite material cables and their structures.

The aforementioned conclusions have certain guiding significance for the application of composite cables in towing operations. Compared to traditional cables, aramid fibers and high-modulus polyethylene materials can achieve the same breaking load requirements at a smaller diameter, which can save storage space and greatly enhance the convenience of crew operations. It is foreseeable that composite material cables have a broad application prospect in future towing operations.

During the towing operation, the prolonged and repeated tension and friction can easily lead to fatigue fracture of the cable, which is undoubtedly a major cause of towing accidents. Due to the geometric nonlinearity and material nonlinearity of composite material cables, which differ from traditional steel wires, accurately studying their fatigue and fracture mechanisms is extremely challenging. As shown in Figure 13, researchers have proposed an empirical model for predicting the service life of braided HMPE synthetic cables under cyclic bending over sheaves (CBOS) loading. By combining experiments and numerical simulation, a life prediction model based on fracture patterns is proposed to predict and optimize the utilization of synthetic fiber cables in deepwater operations. MecTest software is used to test stainless steel wires and HMPE fibers of different diameters, simulating the wet underwater conditions, and testing the impact of fracture loads and repeated bending on material performance. Davies used linear regression analysis, residual strength measurement, and nonlinear regression analysis to evaluate the fatigue strength and service life of fiber ropes, as well as fracture behavior under different loads [67]. This research is of great significance for predicting the service life of synthetic fiber ropes used in deep-sea engineering applications.

Figure 13. 1000 kN cyclic tensile test and cyclic bending wear test [67].

In addition to material diversity, another characteristic of composite fiber cables is the diversity of their geometric structures. As shown in Figure 14, Li conducted systematic investigations on the dynamic stiffness and fatigue performance of HMPE fiber cables with varying diameters and braiding techniques [68]. Initially, based on full-scale experiments, the nonlinear dynamic stiffness of high-modulus polyethylene (HMPE) fiber cables under long-term cyclic loading was investigated, discussing the impact and evolution process of average load and load amplitude on axial stiffness. The experimental results showed that the dynamic stiffness of cables with different braiding structures and diameters varied slightly, primarily attributed to variations in braiding angles. At the same time, a newly derived empirical formula that considers braiding structure, diameter, and input load conditions was proven to have good reliability when compared with experimental results.

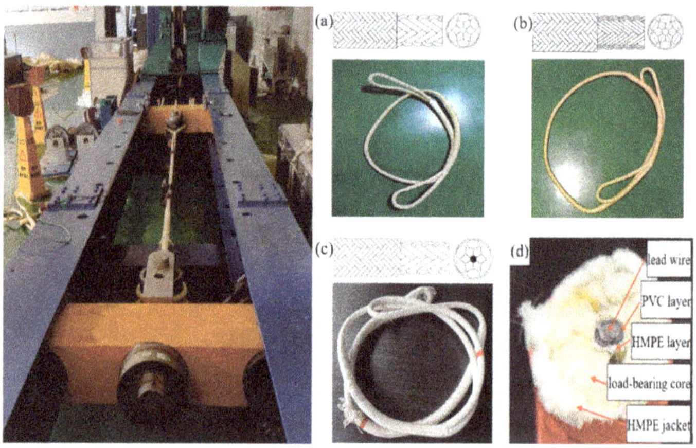

Figure 14. Fiber cables with different diameters and braided methods [69]. (**a**), (**b**) and (**c**) represent ropes with three different manufacturing methods, while (**d**) illustrates the functional layers of the rope cross-section.

In subsequent research, Lin studied the fatigue performance of high-modulus polyethylene (HMPE) fiber cables under dynamic loading through full-scale experiments and found that the stability of the internal structure has a significant impact on fatigue resistance [69]. A thermodynamics-based viscoelastic, plastic, creep model for loading cycles model was established to predict the fatigue life of HMPE fiber cables, taking into account the effects of internal friction and thermal effects on the long-term accuracy of the model. A new life estimation formula was proposed: $\log n(t) = \log n(NP) + \alpha t - \beta t$. The coefficients were solved using the method of least squares, considering the impact of cable length changes on fatigue life, and adjusting coefficients for cables of different materials and sizes to improve prediction accuracy.

4. Mechanical Response Analysis of Towed Objects

In towing operations, the towed objects not only include conventional vessels, but also offshore operating platforms, caissons, jacket structures, and the like. Research on the mechanical responses of the overall or local components of the aforementioned towed objects under towing conditions is becoming a new hot topic. Scientists and researchers primarily model the local or entire towed object, define the displacements and loads generated during the towing process within the model, and conduct numerical analysis. The existing research subjects are generally categorized as jack-up platforms with floating pontoons, Spar platforms, and other large components.

4.1. Offshore Platforms

4.1.1. Mat-Supported Jack-Up Platform

During the towing process, the lifting legs, reaching tens of meters in height, stand above the main hull and are subjected to the forces of wind, inclined gravity, and inertial forces due to motion. This results in significant bending moments at the root of the legs. Whether the root of the legs and key components such as the leg-supporting frame possess adequate structural strength is crucial to the safety of the entire platform. To address these issues, Teng attempted to calculate the dynamic wave pressure using potential flow theory and define three sets of design waves [70]. Based on the finite element model, the stress distribution of a mat-supported jack-up platform in the towing state can be analyzed, taking into account the connection methods of the mat, legs, and main hull, as well as their load characteristics. Research has found that the structural response of the mat under oblique waves is a limiting factor for the feasibility of platform towing; appropriately lowering the position of the mat while ensuring the platform's stability can significantly improve the platform's towing strength; and the most critical area is the region between the mat and the main hull.

4.1.2. Spar Platform

As resource exploration and development gradually move towards deep waters, based on the research of mat-supported jack-up platforms, many researchers have conducted systematic studies on the mechanical properties of Spar platforms under towing conditions. One of the representative studies is Abraham's use of Ansys software to build detailed finite element models of the Spar hull, supporting framework, and transportation ship using Ansys software. The WAMIT (Version 5.4) software was used to calculate the acceleration, hydrodynamic pressure, and forces acting on the ship–Spar system, assessing the fatigue damage of the Spar hull during towing, the vibration of trusses, risers, and tension members, as well as the impact and immersion of the hull's overhanging parts by waves. The study also investigated the structural strength, deflection, acceleration, and pressure of the Front Runner hull during towing, as well as fatigue damage and wind-induced vibration [71]. The systematic research path in this field, from the performance of Truss Spar structures under different environmental conditions, to the detailed design of wave plates leading to fatigue analysis of key connection points, and finally to the overall structural analysis under wet towing conditions, provides comprehensive technical support for the design and safe operation of Truss Spars [72–74]. The hydrodynamic performance and structural strength issues of Truss Spar platforms during loading, dry towing, floating unloading, and wet towing processes are very complex. The reference research approach is to establish the structural finite element model of the Truss Spar platform and analyze the structural strength during loading; use three-dimensional potential flow theory and design wave methods to assess the hydrodynamic performance of the semi-submersible ship towing the Spar platform, and calculate the structural strength caused by waves; propose a load solution and structural strength analysis plan for the double-hull phase during the floating unloading process, and solve the motion equations in the frequency and time domains based on three-dimensional potential flow theory; solve the wave loads and static loads on the Spar platform during the wet towing process, and propose an overall strength analysis plan for the platform considering the combined effects of wind, waves, and towing forces [75].

4.2. Other Large Components

Offshore substations are a crucial component of offshore wind farms, which are typically constructed, equipped, and commissioned onshore in factories before being transported as a whole to the target sea area via towing operations. The structural strength of the topside module of the offshore substation during towing is key to the safe transportation of the entire assembly. Lin designed the topside module using finite element analysis software based on the calculation method for heave acceleration applicable to the towing transportation of the topside module. By adjusting the ballast to ensure the center of gravity aligns with the barge, and using VBA programming to extract node masses and calculate loads under different conditions, the goal of optimizing the topside structure design is ultimately achieved [76]. When transporting jacket structures or large modules, Lu can utilize the same dry towing method as offshore substations. Structural strength finite element analysis is conducted on the towing/launching barge design scheme for jacket structures using Ansys software, and corresponding improvement suggestions and safe operating environment requirements are proposed for its structure [77]. The static stability and seakeeping performance of submerged floating tunnel modules (SFTMs) during wet towing operations have been investigated by Lee et al. [78]. Their study provides insights into the motion response and stability of large underwater structures under towing conditions, which is relevant to the transportation of other large offshore components. Additionally, the flexibility of the barge exerts a certain degree of impact on the structural strength of the module. Based on the design wave method in the frequency domain, Liang conducted three-dimensional fully coupled numerical simulation and calculation analysis of the transportation of a certain large module, establishing an overall structural model and hydrodynamic model. The research results indicate that analyzing the module and barge as a single system, taking into account the elastic influence of the barge structure, can more realistically reflect the interaction between the module and barge, enhancing the accuracy and reliability of transportation analysis [79]. The dynamic responses of multi-float systems during wet towing have been investigated by Cao et al. [80]. Their study provides insights into the motion characteristics and cable tensions of multi-float systems under various sea conditions. The findings highlight the importance of considering the interactions between multiple floating modules and their connectors during towing operations.

As shown in Table 3, the current analysis of the mechanical response of the towed objects primarily employs finite element modeling methods to simulate the environmental conditions of the towing operation and their impact on the overall and local aspects of the towed objects. The main subjects of research are mat-supported jack-up platform, Spar platforms, and other large components. The main shortcomings in the current research are the lack of model experimental methods to study such issues, the absence of exploration in local structural optimization, and the deficiency in optimization of towing-related components such as mooring posts, guideposts, bridle cables, and towing triangles. As the core target of towing operations, studying the mechanical properties of the towed objects to ensure structural safety is extremely important and contributes to enhancing the safety of towing operations.

Table 3. Summary of the research status of the mechanical response analysis of towed objects.

Towed Objects	Research Content	Research Method	Representative Articles
Mat-Supported Jack-Up Platform	The large bending moment generated at the root of the pile leg and the pile-fixing frame during towing due to the effects of wind force, tilting gravity, and inertial force of motion.	Wave dynamic pressure is calculated by potential flow theory and stress distribution is analyzed by finite element model.	Ref. [69] attempted to calculate the dynamic wave pressure using potential flow theory and to define three sets of design waves.
Spar Platform	The mechanical properties of Spar platform under towing operation were systematically studied.	The finite element model was established by Ansys software and the hydrodynamic performance was calculated by WAMIT software.	Ref. [70] investigated the structural strength, deflection, acceleration, and pressure of the Front Runner hull during towing, as well as fatigue damage and wind-induced vibration. Refs. [71–73] provided comprehensive technical support for the design and safe operation of Truss Spars. Ref. [74] proposed an overall strength analysis plan for the platform considering the combined effects of wind, waves, and towing forces.
Other Large Components	The structural strength of the block in the offshore booster station during towing is the key to the safe transportation of the whole assembly.	Finite element analysis software was used to design the upper block, and VBA programming was used to calculate the load.	Ref. [75] used VBA programming to extract node masses and calculated loads under different conditions. Ref. [76] conducted a finite element analysis on the structural strength of the jacket structure towing/launching barge design, and proposed corresponding improvement suggestions and safe operating environment requirements for its structure. Ref. [77] indicated analyzing the module and barge as a single system.

5. Challenges and Opportunities

Towing technology is widely used in the fields of marine accident rescue, port entry and exit of large ships, and transportation of large offshore platforms, but its research and development also face many challenges and opportunities. First of all, the towing system is a complex multi-body coupling system, and the motion response among the tug, the towed object, and the cable is interrelated, which is affected by the complex marine environmental loads such as wind, wave, and current, which makes it extremely difficult to accurately simulate and predict the dynamic response of the towing system. For example, in the starting stage of a tugboat, the large vibration of heave and pitch will affect the tension of the cable and the motion of the towed object, increasing the complexity of the system control. In addition, the study of mechanical properties of cable is also facing challenges, especially the mechanical nonlinearity and geometric nonlinearity of composite cable, which bring new research topics to its application in towing. However, with the continuous development of numerical simulation techniques, such as computational fluid dynamics (CFD) and finite element analysis (FEA), researchers are able to more accurately simulate the dynamic interaction effects of towing systems under different conditions, providing a powerful tool to solve the above challenges. The light weight, high strength, and corrosion resistance of composite cables also bring new opportunities for the development of towing technology, which is expected to play an important role in future towing operations. At the same time, the continuous deepening of interdisciplinary research provides a broad space for the innovation and development of towing technology. By integrating the theories and methods of ship and ocean engineering, mechanics, materials science, and other disciplines, complex problems in towing systems can be solved more comprehensively, and towage technology can be promoted to a more efficient, safer, and more environmentally friendly direction.

Combined with the current research status, the possible future development directions and research ideas for towing technology are proposed:

(1) The research on cable mechanical response in the initial stage of towing was emphasized. During towing operation, the cable tension in the initial acceleration stage is much greater than that in the stable stage, especially the sudden tension phenomenon at the beginning moment, which may lead to cable break or reduce the fatigue life of the cable to a certain extent. However, the traditional fatigue test and empirical formula usually neglect the impact of this phenomenon; a towing experiment and numerical model method can be used to study the change law of cable tension in the initial stage of towing, so as to carry out targeted fatigue experiments on cables and optimize the fatigue theory of towing cables.

(2) The formation towing experiment was carried out. At present, the test of simulated towing operation is mainly single towing, which has the advantages of being a relatively mature test, relatively simple operation, low cost, and little difficulty. The actual single towing operation must meet the following conditions: smaller tonnage of the towed vessel, closer towing distance, better operating sea conditions, and so on. However, with the gradual development of large ships, towing usually requires multiple tugs to work together to complete the operation. Therefore, it is of practical significance to try to carry out the towing test with the participation of multiple tugs and large tonnage of the towed ship model in the experimental study of the towing problem [81–83].

(3) The experimental and numerical simulation methods are employed to conduct an in-depth study on composite cables. The geometric nonlinearity and material nonlinearity of composite cables are the key points and difficulties in the research. To address these, a series of tests, including tension, fatigue, wear, and corrosion, are conducted to obtain relevant data. Furthermore, a refined finite element model of the cable is established and its accuracy is verified. This allows for the simulation of real-world towing conditions using finite element methods, providing guidance for the application of composite cables in towing operations.

(4) Structural optimization of components related to towing operations. Towing operation involves a range of relevant structural components on the tugboat and the towed objects. Especially when large vessels are being towed, the concentrated loads generated by the cables are significant, which places high demands on the structural strength of the components. To mitigate the effects of stress concentration, plastic deformation, and other phenomena generated by the transient load during the towing operation, the structural design of the relevant components can be optimized to improve the safety of the towing operation.

(5) Towing technology is advancing towards unmanned and automated operations. Traditional maritime towing operations face challenges such as safety risks, high costs, and strong operational dependencies. The development of unmanned ships and multi-agent technologies offers the potential to address these issues. Currently, research on coordinated towing by unmanned vessels is limited, primarily focusing on trajectory

control of large vessels by multiple tugs. These studies often simplify the role of tugs and towing ropes, without considering their dynamic responses. With the progress of intelligent ship technology, research has begun to focus on multi-agent systems, treating tugs as independent entities with dynamic responses. However, most of these studies are based on linear models, which may not accurately reflect the nonlinear characteristics of ships and could affect towing accuracy. Meanwhile, the computational complexity of multi-ship coordinated control algorithms may limit their real-time application.

6. Conclusions

The Ocean Engineering Towing System is widely applied in various fields, including the entry and exit of large vessels into ports, the rescue of distressed ships, and the transportation of large offshore platforms. The motion response and structural strength of the towing cable and the towed object are crucial factors affecting the safety of towing operations. The slender and flexible nature of the towline results in highly complex motion. Sudden changes in instantaneous tension during towing operations may lead to the breakage or damage of the towing cable. Compared with self-propelled vessels, the motion response of the towed object, influenced by both environmental loads and towline tension, is more variable, and the structural loading is more complex. To this end, this paper summarizes previous research and points out its deficiencies and areas that have not been fully explored or delved into, following a logical sequence from the whole (towing system) to the local component (cables, towed objects). First of all, for the research on the motion response of the towing system, this paper analyzes and reviews the classification of the numerical simulation method, theoretical model method, and experimental method, and summarizes the advantages and disadvantages of the different methods. Then, the development and latest research of three methods commonly used in the study of cable dynamics model (concentrated mass method, finite difference method, and finite element method) are described, and the necessity of studying the wear resistance, tensile strength, structure, and material type of composite cables is discussed. The towed object is a very important part of the towing system, the current research is biased to the dynamic research, and the analysis of the structure and mechanical response is less. This paper summarizes the current research on mechanical response analysis of the towed object, and proposes the direction of in-depth research on the optimization of key structural topology and the analysis of structural mechanical response under complex sea conditions.

Author Contributions: Conceptualization, K.Q. and J.Z. (Jianing Zhang); methodology, K.Q. and J.Z. (Jianing Zhang); validation, L.Z.; investigation, K.Q.; resources, K.Q. and R.G.; writing—original draft preparation, K.Q., J.Z. (Jinlong Zhang) and L.Z.; writing—review and editing, K.Q., J.Z. (Jinlong Zhang) and R.G.; supervision, J.Z. (Jianing Zhang); funding acquisition, J.Z. (Jianing Zhang). All authors have read and agreed to the published version of the manuscript.

Funding: This research was funded by the Foundation of CHINA COSCO SHIPPING GROUP, (Grant No. 2023-2-Z001-02-01), and the Dalian City's Public Recruitment Project for Key Technical Challenges (Grant No. 2023JB11SN006).

Institutional Review Board Statement: Not applicable.

Informed Consent Statement: Not applicable.

Data Availability Statement: Data will be made available upon request.

Conflicts of Interest: The authors declare no conflicts of interest.

References

1. "XinFu 102" Ultra-Large Container Ship Entering the Yangtze River Towing Project. Available online: https://www.ctdsb.net/c1716_202212/1596784.html (accessed on 6 December 2024).
2. IMO. Collision Between BW Seine and Barge SOL 2315 Towed by Tug SOL 1009 at Singapore Strait TSS on 5 February 2023.2024-05-20 [2024-10-4]. Available online: https://gisis.imo.org/Public/Shared/Public/Download.aspx?FileType=MCIR_InvestigationDocument&Reference=C1000552-SGP-R01 (accessed on 6 December 2024).
3. U.S. Navy. *U.S. Navy Towing Manual*; U.S. Government Printing Office: Washington, DC, USA, 2002.
4. Aydın, Ç.; Ünal, U.; Karabulut, U.; Sarıöz, K. Practical computational procedures for predicting steering and braking forces of escort tugs. *Ocean Eng.* **2018**, *166*, 159–171. [CrossRef]
5. Yang, X.; Wu, J.; Xu, S. Study on dynamic response of underwater towed system in ship propeller wakes using a new hydrodynamic model. *Ocean Eng.* **2022**, *265*, 112599. [CrossRef]
6. Yu, Y.; Woodward, M. Towed-body system dynamics and interaction. *Ocean Eng.* **2023**, *271*, 113719. [CrossRef]
7. Li, B.; Huang, W.; Liang, H. An efficient method to assess effect of fin on the course stability of towing system. *Ocean Eng.* **2020**, *217*, 108005. [CrossRef]
8. Figari, M.; Martinelli, L.; Piaggio, B.; Enoizi, L.; Viviani, M.; Villa, D. An All-Round Design-to-Simulation Approach of a New Z-Drive Escort Tug Class. *J. Mar. Sci. Eng.* **2020**, *142*, 031107. [CrossRef]
9. Zhang, D.; Zhao, B.; Zhu, K. Hydrodynamic Response of Ocean-Towed Cable-Array System under Different Munk Moment Coefficients. *Sustainability* **2022**, *14*, 1932. [CrossRef]
10. Piaggio, B.; Villa, D.; Viviani, M. Numerical analysis of escort tug manoeuvrability characteristics. *Appl. Ocean Res.* **2020**, *97*, 102075. [CrossRef]
11. Piaggio, B.; Villa, D.; Viviani, M.; Figari, M. Numerical analysis of escort tug manoeuvrability characteristics—Part II: The skeg effect. *Appl. Ocean Res.* **2020**, *100*, 102199. [CrossRef]
12. Mauro, F. A flexible method for the initial prediction of tugs escort capability. *Ocean Eng.* **2022**, *246*, 110585. [CrossRef]
13. Mauro, F. Thrusters modelling for escort Tug capability predictions. *Ocean Eng.* **2021**, *229*, 108967. [CrossRef]
14. Chen, M.; Chen, Y.; Li, T.; Tang, Y.; Ye, J.; Zhou, H.; Ouyang, M.; Zhang, X.; Shi, W.; Sun, X. Analysis of the wet-towing operation of a semi-submersible floating wind turbine using a single tugboat. *Ocean Eng.* **2024**, *299*, 117354. [CrossRef]
15. Huang, Y.; He, Y.; Liu, Y.; Qiu, M. Study of the wet-towing transportation of an FPSO with elastomeric-supporting topside modules. *Ocean Eng.* **2024**, *293*, 116602. [CrossRef]
16. Zhang, L.; Du, Z.; Valdez Banda, O.; Goerlandt, F.; Du, L.; Li, X. Collision prevention of ship towing operation under environmental disturbance. *Ocean Eng.* **2022**, *266*, 112870. [CrossRef]
17. Hu, Z.; Mou, J.; Chen, L.; Jia, X.; Chen, P. Model predictive control framework for green and smart tug escorting. *Ocean Eng.* **2023**, *280*, 114539. [CrossRef]
18. Jiang, X.; Zhu, Y.; Wang, L.; Wang, Y. Cooperative control for automatic towing operation by multiple DP tugboats with fully unknown model parameters. *J. Mar. Sci. Eng.* **2023**. [CrossRef]
19. Zhu, H.; Hu, C. A unified seakeeping and maneuvering analysis of multiple linked towing system with triangular Bodies. *Ocean Eng.* **2021**, *222*, 108577. [CrossRef]
20. Kijima, K.; Wada, Y. Course stability of towed vessel with wind effect. *J. Soc. Nav. Archit. Jpn.* **1983**, *1983*, 117–126. [CrossRef]
21. Kijima, K.; Varyani, K. Wind effect on course stability of two towed vessels. *J. Soc. Nav. Archit. Jpn.* **1985**, *1985*, 137–148. [CrossRef]
22. Fitriadhy, A.; Yasukawa, H. Course stability of a ship towing system. *Ship Technol. Res.* **2011**, *58*, 4–23. [CrossRef]
23. Fitriadhy, A.; Yasukawa, H.; Koh, K.K. Course stability of a ship towing system in wind. *Ocean Eng.* **2013**, *64*, 135–145.1. [CrossRef]
24. Sinibaldi, M.; Bulian, G. Towing simulation in wind through a nonlinear 4-DOF model: Bifurcation analysis and occurrence of fishtailing. *Ocean Eng.* **2014**, *88*, 366–392. [CrossRef]
25. Karasuno, K.; Okano, S.; Miyoshi, J.; Maekawa, K. Predictions of ship's hull hydrodynamic forces and maneuvering motions at slow speed based on a component-type mathematical model. *Proc. MARSIM2003 Kanazawa Jpn.* **2003**, 25–28.
26. Han, Y.; Zhu, X.; Zhou, L. Numerical simulation of multi-tug towing of a gravity-based structure in broken sea ice. *Ocean Eng.* **2022**, *261*, 112070. [CrossRef]
27. Marnix, V.D.B.; Raed, L.; Sveinung, L. An implicit time-stepping scheme and an improved contact model for ice-structure interaction simulations. *Cold Reg. Sci. Technol.* **2018**, *155*, 193–213.
28. Matuttis, H.G.; Chen, J.F. *Understanding the Discrete Element Method: Simulation of Non-Spherical Particles for Granular and Multi-Body Systems*; John Wiley & Sons Inc.: Hoboken, NJ, USA, 2014.
29. Kim, D.; Kim, J.; Kim, J.; Im, N. Development of ship collision avoidance system and sea trial test for autonomous ship. *Ocean Eng.* **2022**, *266*, 113120. [CrossRef]
30. Piaggio, B.; Viviani, M.; Martelli, M.; Figari, M. Z-Drive Escort Tug manoeuvrability model and simulation. *Ocean Eng.* **2019**, *191*, 106461. [CrossRef]

31. Piaggio, B.; Viviani, M.; Martelli, M.; Figari, M. Z-Drive Escort Tug manoeuvrability model and simulation, Part II: A full-scale validation. *Ocean Eng.* **2022**, *259*, 111881. [CrossRef]
32. Guo, L.; Yuan, Y.; Duan, Z.; Tang, W.; Xue, H. Experimental investigation on nonlinear dynamic response of towing cable under vessel motion. *Ocean Eng.* **2022**, *266*, 113170. [CrossRef]
33. Park, J.; Rhee, S.; Im, J.B.; Ji, B.; Lee, S. Experimental study on the towing stability of a towed underwater object. *Int. J. Nav. Arch. Ocean* **2023**, *15*, 100539. [CrossRef]
34. Park, S.H.; Lee, S.J.; Lee, S. Experimental investigation of towing- and course-stability of a FPSO towed by a tug-boat with lateral motion. *Int. J. Nav. Arch. Ocean* **2021**, *13*, 12–23. [CrossRef]
35. Sarraf, S.; Abbaspour, M.; Dolatshahi, K.M.; Sarraf, S.; Sani, M. Hydrodynamic interaction of a squat submarine towed by a marine vessel: Experimental investigation. *Ocean Eng.* **2023**, *271*, 113736. [CrossRef]
36. Wu, J.; Yang, X.; Xu, S.; Han, X. Numerical investigation on underwater towed system dynamics using a novel hydrodynamic model. *Ocean Eng.* **2022**, *247*, 110632. [CrossRef]
37. Nam, T.K.; Jung, C.H.; Kim, J.M.; Choi, H. A Study on the Calculation of Towing Force for the Disabled Ship and Its Experiments. *J. Nav. Port. Res.* **2014**, *38*, 463–470.
38. CCS. *Rules for Classification of Sea-Going Steel Ships*; China Classification Society (CCS): Beijing, China, 2024.
39. Yuan, Z.; Jin, L.; Tian, H.; Lu, Y. Comments on the Research of Hydrodynamic and Control Technology of Underwater Towed System. *Sci. Technol. Eng.* **2013**, *13*, 408–415, 420.
40. Walton, T.S.; Polachek, H. Calculation of transient motion of submerged cables. *Math. Comput.* **1960**, *14*, 27–46. [CrossRef]
41. Polachek, H.; Dawson, C. Transient motion of an elastic cable immersed in a fluid. *Math. Comput.* **1963**, *17*, 60–63. [CrossRef]
42. Delmer, T.N.; Stephens, T.C.; Coe, J.M. Numerical simulation of towed cables. *Ocean Eng.* **1983**, *10*, 119–132. [CrossRef]
43. Huang, S. Dynamic analysis of three-dimensional marine cables. *Ocean Eng.* **1994**, *21*, 587–605. [CrossRef]
44. Thomas, D.O. A numerical Investigation of Time Integration Schemes Applied to the Dynamic Solution of Mooring Lines. Ph.D. Thesis, The University of Newcastle Upon Tyne, Newcastle, UK, 1993.
45. Du, X.; Zhang, X. Influence of Towed Cable on Maneuverability of Underwater Vehicle. *Acta Armamentarii.* **2019**, *40*, 1476–1484. (In Chinese)
46. Wang, C.; Liu, Z.; Wen, W.; Zhang, Z.; Zhao, M. Numerical analysis of underwater towing cable under complex sea state. *Ocean Eng.* **2020**, *38*, 135–142. (In Chinese)
47. Zhu, Z.; Yin, Y.; Shen, H. Analysis on the dynamics properties of new type deep-water tau-wire mooring lines. *J. Ship. Mech.* **2019**, *23*, 180–189. (In Chinese)
48. Yang, X.; Sheng, S.; Wu, J.; Yue, W. Numerical analysis of dynamic behaviors of underwater towed system with hydrofoil manipulations. *Ocean Eng.* **2024**, *310*, 118791. [CrossRef]
49. Guo, L.; Yuan, Y.; Tang, W.; Xue, H. A numerical investigation on quasi-static configuration and nonlinear dynamic response characteristics of marine towing cable. *Ocean Eng.* **2021**, *240*, 110007. [CrossRef]
50. Ablow, C.M.; Schechter, S. Numerical simulation of undersea cable dynamics. *Ocean Eng.* **1983**, *10*, 443–457. [CrossRef]
51. Milinazzo, F.; Wilkie, M.; Latchman, S.A. An efficient algorithm for simulating the dynamics of towed cable systems. *Ocean Eng.* **1987**, *14*, 513–526. [CrossRef]
52. Howell, C.T. Investigation of the Dynamics of Low-Tension Cables. *Mass. Inst. Technol.* **1992**, *21*, 546–548.
53. Burgess, J.J. Bending stiffness in a simulation of undersea cable deployment. *Int. J. Offshore Eng.* **1993**.
54. Tjavaras, A.A. Dynamics of Highly Extensible Cables. Ph.D. Thesis, Massachusetts Institute of Technology, Cambridge, MA, USA, 1996.
55. Wang, Z. Towed Cable System Vibration Responses to Varying Towed Velocity. *Nav. Arch. Ocean Eng.* **2019**, *35*, 1–7. (In Chinese)
56. Ruan, W.; Qi, K.; Chen, M.; Liu, K.; Nie, Q. Finite element simulation of flexible pipe laying process by reel approach. *Ocean Eng.* **2023**, *41*, 127–139. (In Chinese)
57. Ruan, W.; Qi, K.; Sun, B.; Gao, X.; Li, J.; Han, X. Multi-scale coupling numerical modeling of metallic strip flexible pipe during reel-lay process. *Ocean Eng.* **2023**, *283*, 115209. [CrossRef]
58. Ruan, W.; Nie, Q.; Han, X.; Li, J.; Bai, Y.; Fu, X. Accurate modeling and safety evaluation of dented pipeline with internal pressure based on reverse modeling technique. *Ships Offshore Struc.* **2024**, 1–11. [CrossRef]
59. Leonard, J.W.; Nath, J.H. Comparison of finite element and lumped parameter methods for oceanic cables. *Eng. Struct.* **1981**, *3*, 153–167. [CrossRef]
60. Chatzopoulou, G.; Karamanos, S.A.; Varelis, G.E. Finite Element Analysis of Cyclically-Loaded Steel Pipes During Deep Water Reeling Installation. *Ocean Eng.* **2016**, *124*, 113–124. [CrossRef]
61. Liang, B.; Zhang, J.; Liu, X. Nonlinear Hysteretic Modeling and Parameter Identification of Steel strip winding reinforced composite pipe. *Ocean Eng.* **2021**, *235*, 108704. [CrossRef]
62. Paulling, J.R.; Webster, W.C. A Consistent Large-Amplitude Analysis of the Coupled Response of a TLP and Tendon System. In Proceedings of the Fifth International Conference on Offshore Mechanics and Arctic Engineering, Tokyo, Japan, 13–18 April 1986; Volume 3, pp. 126–133.

63. Garrett, D.L. Dynamic analysis of slender rods. *J. Energy Resour. Technol.* **1982**, *104*, 302–306. [CrossRef]
64. Zhang, R.; Fan, Z.; Chen, L.; Zhao, T. Dynamic Stiffness Characteristics of Synthetic Fiber Mooring Line based on Abaqus. *China Offshore Platf.* **2022**, *37*, 8–15. (In Chinese)
65. Song, B.; Jiang, R.; Li, T. Application and Development Status of Synthetic Fiber Cables for Marine Engineering at Home and Abroad. *China Text. Lead.* **2021**, 76–79. (In Chinese)
66. Huang, W.; Li, B. An investigation on material diversity of synthetic fiber ropes in the course stability of towing under wind. *Ocean Eng.* **2023**, *279*, 114410. [CrossRef]
67. Davies, P.; François, M.; Lacotte, N.; Vu, T.D. An empirical model to predict the lifetime of braided HMPE handling ropes under cyclic bend over sheave (CBOS) loading. *Ocean Eng.* **2015**, *97*, 74–81. [CrossRef]
68. Li, G.; Li, W.; Lin, S.; Li, H.; Sun, Y. Dynamic stiffness of braided HMPE ropes under long-term cyclic loads: A full-scale experimental investigation. *Ocean Eng.* **2021**, *230*, 109076. [CrossRef]
69. Lin, S.; Li, G.; Li, W.; Ye, H.; Li, H.; Pan, R.; Sun, Y. Experimental measurement for dynamic tension fatigue characteristics of HMPE fibre ropes. *Appl. Ocean Res.* **2022**, *119*, 103021. [CrossRef]
70. Teng, X.; Gu, Y. Strength Analysis of the Mat Supported Jack-Up Platform in Transit Condition. *J. Shanghai Jiao Tong Univ.* **2000**, *34*, 1723–1727. (In Chinese)
71. Abraham, M.A.; Nolte, J.; Peng, B.; Rains, W. Front Runner Spar Transportation Analysis—A Rigorous Approach. In Proceedings of the Offshore Technology Conference, Houston, TX, USA, 2–5 May 2005.
72. Lu, R.; Wang, J.; Erdal, E. *Time Domain Strength and Fatigue Analysis of Truss Spar Heave Plate*; Technip Offshore, Inc.: Houston, TX, USA, 2003.
73. Luo, Y.; Lu, R.; Wang, J.; Berg, S. Time-Domain Fatigue Analysis for Critical Connections of Truss Spar. In Proceedings of the 11th International Offshore and Polar Engineering Conference, Stavanger, Norway, 17–22 June 2001.
74. Wang, J.; Lu, R.; Lu, N. *Truss Spar Strength and Fatigue Analysis for Wet Tow*; Floating Systems Engineering Department; Technip Offshore, Inc.: Houston, TX, USA, 2003.
75. Yang, Y. Research on the Hydrodynamic Performance and Structural Strength of Truss Spar during the Load-out, Transportation and Float-off Processes. Master's Thesis, Shanghai Jiao Tong University, Shanghai, China, 2012. (In Chinese).
76. Lin, H.; He, X.; Zhang, L.; Zheng, B. Research on towing analysis for topside module of offshore substation. *Eng. J. Wuhan Univ.* **2018**, *51*, 34–37. (In Chinese)
77. Lu, C.; Lin, Y.; Ji, Z.; Qian, H. Structure Strength Analysis of the Barge during Jacket Transporting and Launching. *Ship. Build. China* **2005**, *46*, 24–33. (In Chinese)
78. Lee, I.; Jin, C.; Kim, S.; Kim, M. On the Static Stability and Seakeeping Performance of a Submerged Floating Tunnel Module in Wet Tow. *J. Mar. Sci. Eng.* **2025**, *13*, 77. [CrossRef]
79. Liang, Y.; Liu, B.; Dong, X. Global Strength Analysis of a Large Topside Dry Tow Transportation. *Ship & Ocean Eng.* **2014**, *43*, 104–106. (In Chinese)
80. Cao, X.; Zhang, H.; Zhang, N. Research on dynamic responses of multi-float systems during wet towing. *Ocean Eng.* **2024**, *312*, 119226. [CrossRef]
81. Cai, J.; Chen, G.; Yin, J.; Ding, C.; Suo, Y.; Chen, J. A Review of Autonomous Berthing Technology for Ships. *J. Mar. Sci. Eng.* **2024**, *12*, 1137. [CrossRef]
82. Choi, J.; Jang, J.; Woo, J. A Review of Autonomous Tugboat Operations for Efficient and Safe Ship Berthing. *J. Mar. Sci. Eng.* **2023**, *11*, 1155. [CrossRef]
83. Liu, H.; Wang, A.; Han, B.; Li, T.; Wang, D.; Peng, Z. Safety-critical anti-disturbance control of tugs for collaborative berthing. *Ocean Eng.* **2024**, *312*, 118972. [CrossRef]

Journal of
*Marine Science
and Engineering*

MDPI

Review

A Review on the Recent Process of Lazy Wave Risers

Weidong Ruan [1,*], Chengcheng Zhou [1], Hongyu Yang [1], Zhi Wang [1], Bo Sun [1] and Yong Bai [2,*]

[1] Civil Engineering College, Zhejiang University of Technology, Hangzhou 310023, China; 201806080127@zjut.edu.cn (C.Z.); 13819124120@163.com (H.Y.); 221123060196@zjut.edu.cn (Z.W.); sunbo2017@zjut.edu.cn (B.S.)

[2] College of Civil Engineering and Architecture, Zhejiang University, Hangzhou 310058, China

* Correspondence: wdruan@zju.edu.cn (W.R.); baiyong@zju.edu.cn (Y.B.)

Abstract: Lazy wave risers (LWRs) are designed with equidistant buoyancy blocks attached in the lower half of the riser, allowing the riser to take on an arch shape under the buoyancy forces provided by buoyancy blocks. This arch configuration can provide flexibility to the LWR arrangement and effectively isolate the dynamic responses between the offshore floating structure and the riser's touchdown zone (TDZ). Its design and application aim to address the issues of dynamic response and fatigue damage that traditional steel catenary risers (SCRs) face in deep water and complex marine environments. Given that research on the LWRs in the field of ocean engineering is not sufficiently abundant, the structural characteristics, hydrodynamic loads, global responses, fatigue damage assessment, and structural optimization progress of LWRs are systematically reviewed in this paper to provide references for researchers in related fields. Among these topics, the global response of LWRs is the main point of this review. This section details the theoretical analysis and numerical modeling methods employed in the study of LWRs' global response, explores the research advancements in the vortex-induced vibration (VIV) related to LWRs, and discusses corresponding experimental studies. Finally, the installation, transfer, and repair processes of LWRs are investigated. Additionally, the importance of leveraging advanced technologies from other fields and combining them with current advanced algorithms is emphasized in efforts to assess fatigue damage and optimize the structures of LWRs, ultimately achieving complementary advantages.

Keywords: buoyancy block; fatigue damage; global response; LWR; VIV

1. Introduction

Deepwater oil and gas development has been recognized as a core growth area in the global petroleum industry. To enhance their core competitiveness, major oil companies and related research institutions in the world are increasingly investing financial and material resources in the pursuit of breakthroughs in deepwater technology. Therefore, deep-sea resource exploration and development technologies have become a research hotspot in international marine engineering [1]. As a crucial approach for transporting offshore oil and gas resources, subsea risers are employed to facilitate the transportation of oil/gas from the subsea wellhead to the upper floating structures at sea in actual offshore oil and gas extraction projects [2,3]. Subsea risers are not only the most complex and critical components in the subsea production system, but also the most vulnerable and susceptible marine structures.

When offshore oil and gas exploitation progresses from shallow water areas to deep water areas, traditional SCRs need to adapt to complex and harsh deep-sea environments. The SCRs are mainly subjected to high-temperature and high-pressure load inside and complex marine environmental load, seabed resistance, and severe coupling motions from the upper floating structure outside. These factors may lead to potential issues such as strength failure and fatigue damage in the risers [4–6]. Yang et al. point out that the intense floating production storage and offloading (FPSO) motion induced by harsh environments may result in significantly amplified dynamic responses in traditional SCRs [7]. Therefore, traditional SCRs may not be suitable for the FPSO system in deepwater. To reduce the failure probability of subsea risers, Petrobras conducted research on various configurations of steel catenary risers focusing on extreme load conditions, vortex-induced vibrations, and wave-induced fatigue. They found that the LWR configuration is more suitable for deepwater applications in terms of structural performance and economic viability. The concept of LWR configuration was first proposed by Karunakaran et al. [8].

Compared with traditional SCRs, LWRs have better compliance, lower weight, and lower installation costs, exhibiting significant advantages in the future deepwater oil and gas exploitation [4,9]. As illustrated in Figure 1, LWRs utilize abundant buoyancy blocks bundled at equal intervals along the lower half of the riser [10]. The buoyancy provided by these blocks causes the riser to adopt an arch shape, effectively reducing the top tension of deepwater riser. This design can alleviate the coupling effects of floating structure motions on the riser's TDZ, thereby enhancing the fatigue life of the riser in the TDZ [11]. Buoyancy blocks are usually made from lightweight materials, such as synthetic foam or thin-walled air tanks, which have lower specific gravities than the seawater. As a result, the effective gravity of the buoyancy block becomes negative underwater, providing an upward lifting force.

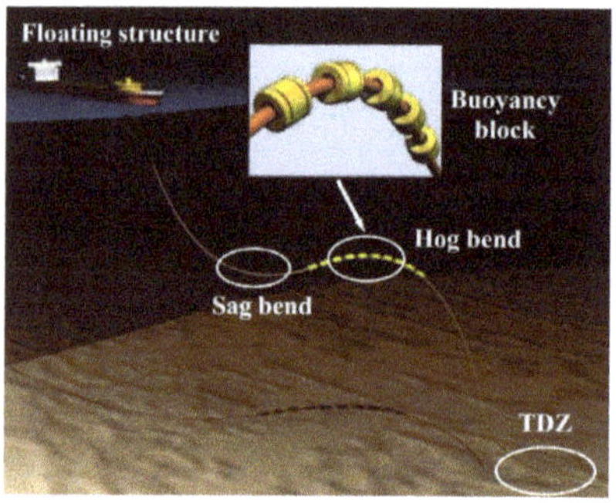

Figure 1. Deepwater LWR configuration [10].

On account of the introduction of buoyancy blocks, the mechanical properties of the riser segment bundled with equispaced buoyancy blocks are very different from those of the bare riser segment. Consequently, the LWRs are usually divided into three segments (suspended segment, buoyancy segment, and catenary segment, see Figure 2) for the global response study [12–17]. Meanwhile, some researchers divided the LWRs into four segments in order to distinguish the seabed resistance action [18–21]. As shown in Figure 2, there are four hazardous areas for the LWR: (1) hang-off location; (2) sag bend zone; (3) hog bend zone; (4) touchdown zone [16]. The hang-off location refers to the point where the riser is attached to the offshore production facility or platform. For steel lazy wave riser (SLWR), this connection is implemented by a flexible joint, which is treated as pinned [16,18,22,23]. For flexible LWR, the flexible riser is rigidly fixed to the floating structure with the bend stiffener constraint [22,24–26]. At this position, the riser is tension-controlled and able to withstand the significant dynamic loads applied. The hang-off angle should be designed within the range of 5 to 20 degrees to effectively prevent the risers from colliding with each other [27].

The sag bend zone refers to the downward bending section of the riser suspended segment due to the combined action of the riser self-weight, floating structure confinement, and lifting force of buoyancy segment. The requirements of fluid flow and structural strength in this zone should be guaranteed to prevent excessive bending stress throughout the design process. In addition, the sag bend zone should be designed as close to the seabed as possible, but the lowest point must avoid contact with the seabed [28]. For the hog bend zone, it exhibits an arch form in the buoyancy segment with the buoyancy effect of abundant buoyancy blocks. The capability of the LWR to absorb the dynamic movements is suggested to be associated with this arch form height [29]. It follows that the rational distribution of buoyancy blocks on the buoyancy segment is a key factor in forming the ideal "wave shape" [30,31]. The height difference between the hog bend peak and sag bend trough should be controlled within certain limits to guarantee the fluid transport environment in the riser. On this account, Ruan et al. [10] proposed the new concept of deepwater LWR configuration based on multiple waveform serial arrangement (i.e., multiple LWR), which can greatly reduce this height difference to improve the fluid transport environment in the riser.

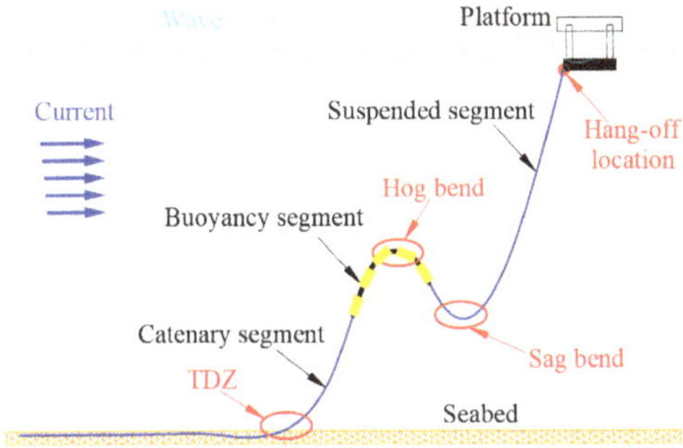

Figure 2. Schematic diagram of a LWR (adapted from [12]).

As illustrated in Figure 2, the TDZ of the riser is in the critical area where the catenary riser contacts with the seabed, and the touchdown point (TDP) is the point in the area where the catenary riser initially contacts with the seabed. Although the axial load of the riser in the TDZ is relatively small, the interaction between the riser flowline and seabed soil constantly occurs under the influence of the marine environmental loads and the floating body motion at the top, which makes this area a hot spot for riser fatigue analysis [32,33]. Ruan et al. [10] pointed out that compared to traditional catenary risers, a deepwater riser with lazy wave arrangement configuration can greatly attenuate the kinematic coupled response of the floating structure in the riser first sag bend section and effectively reduce the normal motion response of the LWR in the TDZ [10], which can effectively optimize the dynamic bending moment and antifatigue performance of the LWR in the TDZ. The results show that the maximum dynamic bending moments of the LWR in the TDZ decrease from 103.25 kN m to 64.86 kN m, with a decrease of up to 37.2%.

In recent years, scholars have conducted extensive and profound research on the global response of offshore risers, achieving remarkable outcomes. However, due to the complexities of deepwater environments and nonlinear boundary conditions, the analyses of offshore risers under complex deepwater conditions remain incomplete, particularly regarding the structural dynamic characteristics and motion responses of different deployment configurations. This study aims to summarize the primary achievements in the research on the global response of LWRs while identifying the current shortcomings and challenges. The review systematically organizes relevant research progress and emphasizes the contributions of various researchers in this field, to serve as a reference for future studies. In addition, key limitations in existing research, such as model accuracy, experimental validation, and multiple environmental factors, are presented. Furthermore, challenges in nonlinear behavior, fluid–dynamic interactions, and material property analysis are explored. This not only lays a foundation for a deeper understanding about the global response of LWRs, but also offers suggestions for subsequent research.

2. Hydrodynamic

The marine environment is a factor critical to the design and analysis of offshore risers and can cause multiple failure modes such as tensile failure, compression buckling, excessive bending, fatigue failure, and on-bottom instability.

2.1. Classification of Hydrodynamic Loads

The hydrodynamic loads caused by a marine environment can be divided into the following categories: (1) wave loads; (2) current loads. Waves are one of the most fundamental and significant hydrodynamic phenomena in the ocean. Wave loads include linear wave load, nonlinear wave load, and random wave load. Waves exert forces (such as drag force, inertial force, and buoyancy variations) on structures through the periodic motion of water particles. These forces have a significant impact on the motion and loading state of offshore riser systems. When calculating the wave forces on offshore risers, both inertial and viscous forces are important, which are often calculated by Morison equation [34]. Many experimental studies have provided the drag and inertia coefficients adopted in the Morison equation [35,36].

Currents are the steady water movement in the ocean and can be broadly categorized as tidal currents, wind-driven currents, ocean circulation, boundary currents including loop and eddy currents, and currents caused by internal waves and solitons. Currents are closely related to the density distribution and the flow into or out of an area, which is a function of the local topography and oceanography [34] (Figure 3a). For shallow water, since currents are frequently driven by tides, velocity–depth profiles can provide an adequate description; for deep water, both the velocity and direction of the currents change with the water depth, and are very complex to accurately describe (Figure 3b). The effects of currents on offshore riser systems are primarily manifested as constant drag forces and potential additional mass forces [6,29,37,38]. In addition, when ocean currents pass over offshore risers, a periodic vortex will be generated in the wake of the riser, and the periodic fluid force generated by the vortex falling off will stimulate the in-line flow (IF) and cross-flow (CF), and offshore risers will vibrate at the same time. This phenomenon is called vortex-induced vibration (VIV) due to ocean currents [6,29,37–39].

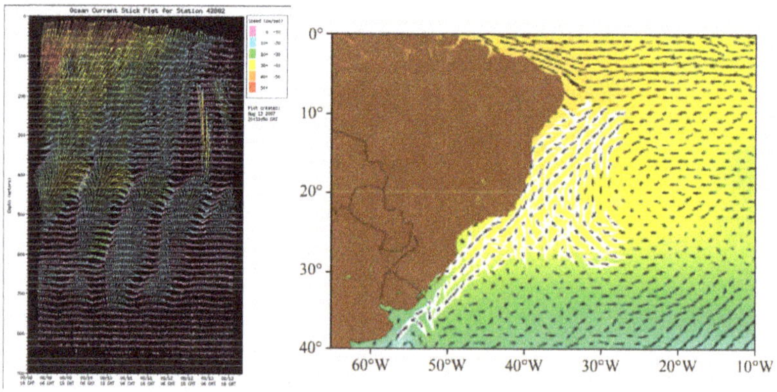

(a) (b)

Figure 3. Manifestation methods of ocean currents [34]. (**a**) Currents in the northern gulf of Mexico. (**b**) Brazil current as represented by the Mariano Global Surface Velocity Analysis.

Additionally, the waves and currents interaction (WCI) will further increase the complexity of the dynamic response of offshore riser systems [29,34]. Unlike the aforementioned, Yin et al. [27] divides the hydrodynamic loads acting on the LWRs into wave frequency loads, vortex-induced vibration due to currents, vortex-induced vibration due to vessel motion, loads due to internal flow, and low-frequency loads. The wave frequency loads can further be divided into wave loads acting on the riser and wave frequency vessel motion-induced vibration according to the position of wave action. For the latter, when the LWR is subjected to vessel motion (such as heave motion), the sag bend and hog bend will move relative to the surrounding water. Consequently, the relative motion and velocity may cause the riser to experience vibrations similar to the VIV in oscillating flows.

2.2. Solution of Hydrodynamic Loads

2.2.1. Wave Theory

In a marine environment, fluid behaves with viscous characteristics, and at the same time, there is surface tension on the fluid surface. These factors have a certain impact on the wave movement. Because the effects of fluid viscosity and surface tension are related to wave frequency, these effects are negligible in the wave frequency bands that are generally important for offshore engineering. Therefore, it is assumed that the fluid is an ideal fluid, and only the influence of gravity is considered, i.e., the fluid motion is irrotational [40]. Any wave can be determined by wave height (or wave amplitude), wavelength (or wave period), and water depth, and the movement of water particles can be determined according to different wave theories. The velocity potential $\Phi(x, y, z, t)$ for an ideal fluid should satisfy Laplace equation [41]:

$$\nabla^2\Phi(x,y,z,t) = \frac{\partial^2\Phi}{\partial x^2} + \frac{\partial^2\Phi}{\partial y^2} + \frac{\partial^2\Phi}{\partial z^2} = 0 \tag{1}$$

where x, y, and z denote the coordinates of any point in space; t is the time. In order to solve the ideal fluid velocity potential expression, the following boundary conditions can be determined:

(1) Wave surface condition ($z = \xi$): the free surface pressure must be equal to the atmospheric pressure (regardless of the surface tension). By applying the Bayesian equation to the free surface of the wave, it can be obtained:

$$\left.\frac{\partial\Phi}{\partial t} + \frac{1}{2}\left[\left(\frac{\partial\Phi}{\partial x}\right)^2 + \left(\frac{\partial\Phi}{\partial y}\right)^2 + \left(\frac{\partial\Phi}{\partial z}\right)^2\right]^2 + g\xi\right|_{z=\xi} = 0 \tag{2}$$

where g is the acceleration of gravity. Meanwhile, the kinematic boundary conditions are satisfied, that is, the water particles on the free surface always remain on the free surface:

$$\left.\frac{\partial\xi}{\partial t} + \frac{\partial\Phi}{\partial x}\frac{\partial\xi}{\partial x} + \frac{\partial\Phi}{\partial y}\frac{\partial\xi}{\partial y} - \frac{\partial\Phi}{\partial z}\right|_{z=\xi} = 0 \tag{3}$$

(2) Bottom boundary conditions ($z = -d$): the vertical velocity is zero at the horizontal seabed:

$$\left.\frac{\partial\Phi}{\partial z}\right|_{z=-d} = 0 \tag{4}$$

Since water particles cannot pass through the seabed boundary, they can only move along the tangential direction of the seabed boundary, that is, the normal velocity of water particle perpendicular to the seabed boundary is zero.

$$\left.\frac{\partial\Phi}{\partial n}\right|_{z=-d} = 0 \tag{5}$$

(3) Condition at infinity: $\Phi(x, y, z, t)$ and $\xi(x, y, z, t)$ remain bounded, sometimes even requiring that both Φ and ξ and their derivatives approach zero.

2.2.2. Current Theory

Since the periodicity of ocean currents is in the order of several months, relative to the water particle velocity induced by waves, it is relatively slow to change over time. Hence, it can be considered that the velocity and direction of ocean currents remain unchanged over time in a given sea state, that is, a current is regarded as a steady flow. Of course, the velocity and direction of ocean currents change at different depths. In practical engineering design, engineers generally adopt "current profile" once in a decade and in a hundred years to carry out engineering design for offshore risers [42,43]. If a detailed profile of the current velocity is lacking, the power function rule can be employed to simulate the current velocity v_c at the depth z [44]:

$$v_c = v_b + (v_s - v_b) \cdot \left[(z - z_b)/(z_s - z_b)\right]^{1/n} \tag{6}$$

where v_s and v_b represent the surface velocity and seabed velocity, respectively; z_b and z_s donate the surface vertical coordinate and seabed vertical coordinate, respectively, and n is the exponential coefficient. The influence of ocean currents on offshore structures, especially deepwater structures, is very large and its analysis will be more complicated.

2.2.3. Hydrodynamic Load Calculation

Since the diameter of the offshore riser is much less than the incident wavelength, the structure itself has little influence on the wave motion, and the wave effect on the offshore riser mainly contains viscous effect and added mass effect. The Morrison equation proposed by Morison et al. [45] is generally adopted to calculate the wave loads. Morison et al. [45] pointed out that the wave loads acting on the offshore riser per unit length in water can be regarded as the linear superposition of two kinds of hydrodynamic loads: (1) drag force, caused by the water particle velocity when the seawater flows through the cylinder; (2) inertial force, caused by the water particle acceleration. Because ocean currents are generally treated to be steady currents, the hydrodynamic load caused by ocean currents is mainly manifested as drag force. It is necessary to linearly superposition the water particle velocity caused by the wave and current first, and then to calculate the total drag force acting on the offshore riser.

Because of the catenary configuration of the LWR, the calculation of hydrodynamic loads on the inclined cylinder should be considered, which is similar to the calculation of hydrodynamic loads on the vertical cylinder. Considering that the water particle velocity and acceleration of the inclined cylinder are generally not in the same spatial plane, the corresponding water particle velocity and acceleration are generally described in vector form, so as to calculate the hydrodynamic loads acting on the inclined cylinder. The expression of Morison equation in vector form for the LWR is as follows [46]:

$$f = \rho_w \frac{\pi D^2}{4}\dot{v} + C_a \rho_w \frac{\pi D^2}{4}\left(\dot{v}_n - \ddot{r}_n\right) + \frac{1}{2} C_d \rho_w D \left(v_n - \dot{r}_n\right)\left|v_n - \dot{r}_n\right| + C_{a\,t} \rho_w \frac{\pi D^2}{4}\left(\dot{v}_t - \ddot{r}_t\right) + \frac{1}{2}\pi C_{d\,t} \rho_w D \left(v_t - \dot{r}_t\right)\left|v_t - \dot{r}_t\right| \tag{7}$$

where ρ_w and D represent the sea density and riser's external diameter, respectively; $\dot{r}$ and $\ddot{r}$ are the riser moving velocity and acceleration vectors, where the subscripts "n" and "t" represent the riser's normal direction and tangential direction, respectively; v and $\dot{v}$ denote the water particle velocity and acceleration vectors, where the meaning for the subscripts "n" and "t" are the same with riser velocity and acceleration vectors; C_d and C_a are the normal drag coefficient and added mass coefficient, meanwhile C_{dt} and C_{at} are the tangential drag coefficient and added mass coefficient.

3. Global Response of LWRs

Most researchers tend to carry out theoretical research and numerical simulation on the global response of LWRs because it is too time-consuming and costly to carry out experimental research on LWRs. The introduction of buoyancy blocks results in significant nonlinearity in the lazy wave configuration, and this configuration is notably affected by changes in the buoyancy blocks. Therefore, most scholars prefer to conduct large deformation nonlinear analyses and studies on the global response of LWRs [21,47,48]. OrcaFlex (version 11.4) is a widely applied dynamic simulation software in marine engineering and ship design, particularly for analyzing the dynamic behavior of marine structures, such as subsea pipelines, floating platforms, and offshore risers.

Based on the lumped mass method (LMM), this software allows for the setting of different boundary conditions, environmental conditions, and loading scenarios to adapt to various practical applications [9,13,49]. It also possesses powerful modeling capabilities that enable engineers to highly customize models based on specific requirements. Hu et al. [9] established OrcaFlex numerical models for the catenary riser and LWR to investigate how the response characteristics of offshore risers impact the stability and safety of the entire Offshore Compressed Air Energy Storage (OCAES) system (Figure 4). The results indicated that the structure of the LWR is more complex than the catenary riser; nevertheless, the former presents better response performance.

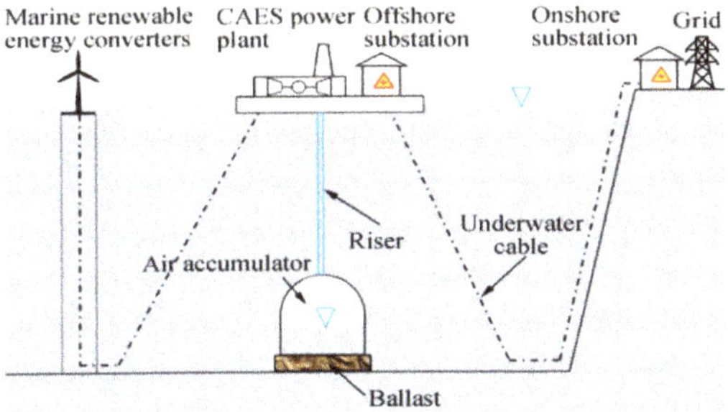

Figure 4. Typical configuration of an OCAES system [9].

3.1. *Mathematical Model and Software Tools*

The modeling of LWRs initially focused on two-dimensional (2D) models. With the improvement of computer computing power, there has been a gradual transition to three-dimensional (3D) models.

3.1.1. Two-Dimensional Model of LWRs

Early simulation software and insufficient computational power of computers hindered the ability to simulate 3D models of a LWR, leading to a significant computational burden. As a result, researchers primarily focused on developing 2D models for LWRs, accompanied by a series of simplifications and assumptions. For instance, the axial deformation and twisting behavior of the riser are often neglected [14,18,19]. Additionally, for the convenience of numerical modeling and mechanical analysis, the buoyancy segment of the LWR is simplified into a uniform pipe structure with constant inner and outer diameters based on the equivalence principles of hydrodynamic loads and buoyancy loads. Furthermore, during the global response analysis of the LWRs, it is usually necessary to consider the pipe–soil interaction. Therefore, the seabed in contact with the riser is generally assumed to be horizontal [15,17–19]. A such, most studies have treated the seabed as a horizontal and elastic foundation [13,14,18,50], typically modeled by the Winkler foundation model to simulate the normal seabed resistance loads.

In consideration of the boundary layer effects in the riser's TDZ and the buoyancy effects of the buoyancy segment, the LWR can be divided into several segments to carry out the analyses for 2D planar LWR models [13,14,18]. Compared to traditional small deformation beam theory or catenary theory, nonlinear large deformation beam theory has significant advantages in simulating rapid changes in the angle of the riser, making it more suitable for the mechanical analysis of deepwater risers with large deformations [51,52]. Wang et al. [18] employed nonlinear large deformation beam theory and small deformation beam theory to model the suspended segment (including the lower section, buoyancy section, and upper section) and the flowline segment, respectively (Figure 5). The effects of the pipe–soil interaction, the bending stiffness of the riser, and the large deformation characteristics of the suspended segment were accounted for in this model. Under given boundary conditions, the governing equations were derived according to the finite difference method, and the numerical solution was obtained. Equation (8) presents the governing equation for the flowline segment based on small deformation beam theory, while Equations (9) and (10) represent the governing equations for the suspended segment based on nonlinear large deformation beam theory.

$$EI_r\frac{d^4y_2}{dx^4} - T_2\frac{d^2y_2}{dx^2} + \mu y_2 + w_r = 0 \tag{8}$$

$$EI_r\frac{d^3\theta}{ds^3} - T\frac{d\theta}{ds} + w_r cos\theta = 0 \tag{9}$$

$$\frac{dT}{ds} = w_r sin\theta \tag{10}$$

where EIr denotes the bending stiffness of the bare riser; T is the riser's axial tension; T_2 represents the unknown constant axial tension at the TDP; w is the riser's apparent weight in the water per unit length. More detailed information about symbols can be seen in [18]. When calculating the mechanical analyses for the buoyancy section with Equations (9) and (10), the corresponding bending stiffness and apparent weight should be updated.

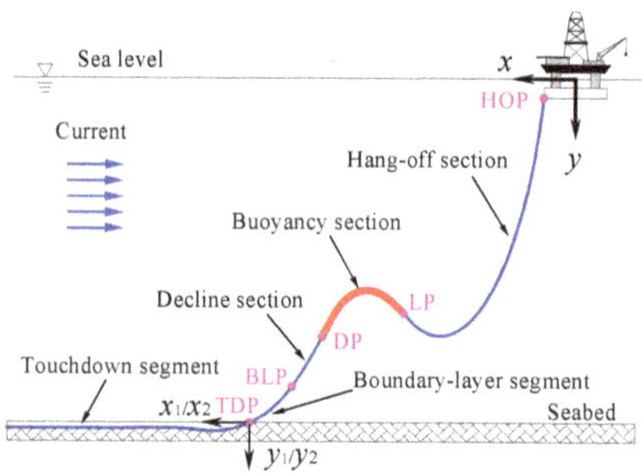

Figure 5. SLWR configuration [14].

However, this numerical model has a limitation issue, i.e., the length of the lower section is set as a known condition, which is hard to acquire through on-site measurements in actual engineering. Since the hang-off angle of the LWR can be obtained, some researchers took it as a known condition, instead of the length for the lower section. Then, Ruan et al. [14] proposed an efficient method to predict the accurate static stress range of deepwater SLWRs under the coupling effects of vessel slow drift motion and currents. The boundary layer effect in the riser's TDZ was well considered in this study, setting the riser segment close to the seabed as the individual boundary layer segment (Figure 5) and performing mechanical analysis according to small deformation beam theory. In contrast, catenary theory was adopted to calculate other sections of the suspended segment (including the hang-off section, buoyancy section, and decline section). This approach can effectively guarantee the continuity of the bending moment and shear force at the riser's TDP. It can be observed that the maximum axial stress range of the SCR occurs in the TDZ, reaching up to 112 MPa, which is 26% larger than that of the SLWR.

Unlike the solution of segmenting the LWR, Trapper [32] proposed a feasible numerical method for the LWR static configuration analysis. This method treated the entire riser as a continuous segment and modeled it according to nonlinear large deformation beam theory (Figure 6). Since the entire riser was considered as a continuous segment to conduct analysis, the need to solve the governing equations for each segment separately was eliminated, and the boundary continuity between each segment also did not need to be considered. The model can accommodate additional loads, constraints, or boundary conditions flexibly. Compared to segmenting the model, the continuous riser model can significantly reduce the complexity of the governing equations. Based on the principle of minimum potential energy, the configuration equation for the LWR in equilibrium is as follows:

$$U_{tot} = \frac{1}{2}\int_{L} EI\kappa(s)^2 ds + \frac{1}{2}\int_{L_s} k_s e(s)^2 ds - \int_{L} w_g y(s) ds - \int_{L_b} w_b y(s) ds - \int_{L} f_x x(s) ds - \int_{L} f_y y(s) ds - P_0 x(L) \quad (11)$$

where E and I represent the riser's Young's modulus and cross-sectional inertia moment, respectively; κ and $e(s)$ denote the riser's curvature and embedment into the seabed, respectively; L_s and L_b are the section length in contact with soil and buoyancy section length, respectively. The right terms in Equation (11) represent the bending elastic energy stored in a deformed riser, the elastic energy stored in the deformation of elastic seabed, the potential energy of the riser's submerged self-weight, and the potential energy of submerged self-weight for buoyancy blocks averaged over the section length, respectively.

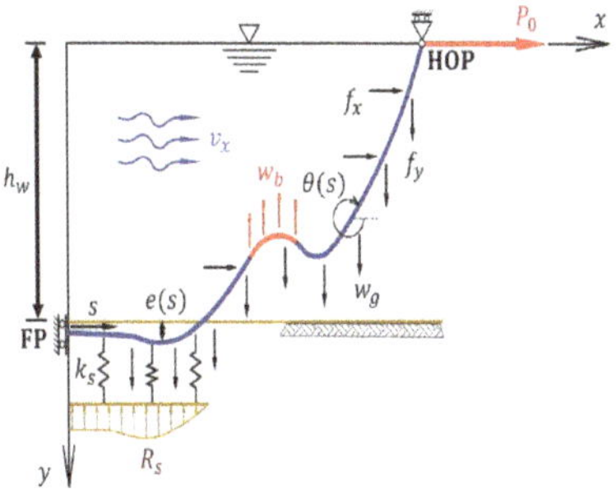

Figure 6. LWR scheme: geometry, loads, boundary conditions, and coordinate systems [32].

When investigating the global response of LWRs, the effect of the internal flow in the riser was generally ignored in numerical models. However, the internal flow can reduce the structural stiffness provided by the negative damping [53] and induce additional large riser displacements [54]. Therefore, Ruan et al. [12] established a dynamic model for the global response of LWRs under imposed top-end excitations and internal flow according to the elastic slender rod theory. The research found that the relative position and length of the buoyancy section can significantly impact the maximum bending moments in the sag bend zone, hog bend zone and TDZ, and the top tension at the hang-off point. However, this study did not specifically focus on the effects of the internal flow on the global dynamic response of the LWR. The elastic rod theory primarily contains a motion governing equation and a stretching constraint equation. By applying Galerkin's method for discretization and the tensor summation algorithm, a set of nonlinear ordinary differential equations in the time domain can be obtained:

$$\gamma_{ikm} M_{njm} \ddot{u}_{kj} - 2\rho_i A_i U \zeta_{ik} \dot{u}_{kn} + \alpha_{ikm} EI_m u_{kn} + \beta_{ikm} \overline{\lambda}_m u_{kn} = \mu_{im} q_{mn} + f_{in} \quad (12)$$

$$\frac{1}{2}\beta_{ikm} u_{in} u_{kn} - \frac{1}{2}\tau_m - \eta_{mn} \frac{\overline{\lambda}_n - (p_{on} A_o - p_{in} A_i) + \rho_i A_i U^2}{EA_n} = 0 \quad (13)$$

where i and k run from 1 to 4; j, m, and n run from 1 to 3. Detailed information about tensor expressions can be seen in [12].

Subsequently, Wang et al. [47] considered the effects of currents and internal flow and proposed a 2D nonlinear dynamic model for deepwater LWRs under normal and tangential top excitations based on Euler–Bernoulli beam theory and Keller Box finite difference approach. Through parameter analysis, it was found that the floating structure motion, internal fluid, and ocean current loads significantly affect the mechanical properties of LWRs. Jeong et al. [55] established a dynamic response analysis program for the LWRs accounting for environmental loads and slug flow. The dynamic responses of the LWRs under critical conditions with and without slug flow were analyzed and compared to provide a quantitative comparison of the effects of slug flow on the LWRs. As displayed in Figure 7, Cabrera-Miranda and Paik [56] modeled the LWR as an Euler–Bernoulli beam to investigate the nonlinear planar vibrations of LWRs excited by slug flow, in which the internal fluid was treated as a plug flow with time-space-varying mass in the form of a rectangular pulse train. The partial differential governing equations for the LWR planar motion in this model are as follows:

$$m\frac{\partial u}{\partial t} - (m+M)\frac{\partial \phi}{\partial t}v + \frac{MU}{1+e}(\frac{\partial U}{\partial s} - \Omega_3 v) = \frac{\partial T_e}{\partial s} - S_n\Omega_3 - (w_0 + Mg)\sin\phi - \frac{1}{2}\pi\rho D_0 C_{dt} u|u|\sqrt{1+e} \quad (14)$$

$$(m+m_a+M)\frac{\partial v}{\partial t} + m\frac{\partial \phi}{\partial t}u + \frac{MU}{1+e}(\frac{\partial v}{\partial s} + \Omega_3 U) = \frac{\partial S_n}{\partial s} + \Omega_3 T_e - (w_0 + Mg)\cos\phi - \frac{1}{2}\rho D_0 C_d v|v|\sqrt{1+e} \quad (15)$$

$$\frac{1}{EA_p} - \frac{\partial T_e}{\partial t} = \frac{\partial u}{\partial s} - \Omega_3 v \quad (16)$$

$$(1+e)\frac{\partial \phi}{\partial t} = \frac{\partial v}{\partial s} + \Omega_3 v \quad (17)$$

$$EI\frac{\partial \Omega_3}{\partial s} + S_n(1+e)^3 = 0 \quad (18)$$

$$\frac{\partial \phi}{\partial s} - \Omega_3 = 0 \quad (19)$$

where s denotes the local curvilinear Lagrangian coordinate related to the unstretched riser of length L; M represents the fluid mass in the riser. More detailed symbol explanations can be seen in [56].

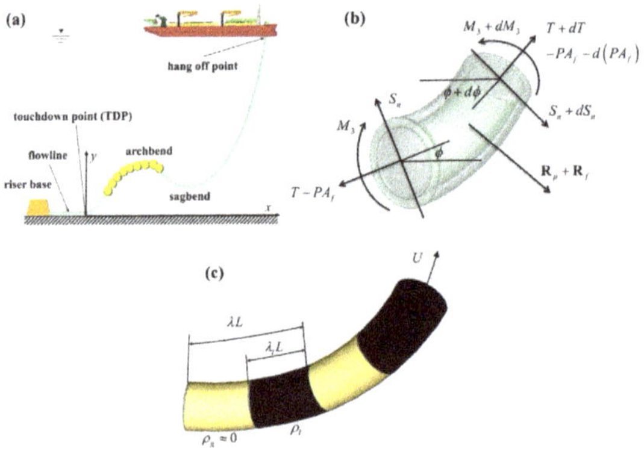

Figure 7. Riser subjected to two-phase flow-induced vibrations [56]. (**a**) sketch of steel lazy wave riser, (**b**) in-plane balance of forces and moments acting on a pipe conveying fluid after elongation and (**c**) internal fluid with time-space-varying mass in the form of a rectangular pulse train.

The global response analysis method for the LWRs has evolved from the initial small deformation beam theory and catenary theory to nonlinear large deformation beam theory, resulting in a mature theoretical system. Meanwhile, some researchers are still exploring new theoretical frameworks and progressively developing them. The theoretical system in other fields has been applied to the field of ocean engineering, and some achievements have been achieved, such as the absolute nodal coordinate formulation (ANCF) method [57]. The ANCF method is a dynamic analysis approach widely used in mechanical engineering, multi-body dynamics, and aerospace engineering in recent years, which can solve the problem of flexible bodies with large deformation and large rotation.

Furthermore, ANCF can alleviate inherent numerical problems associated with nonlinear dynamic analysis, such as non-physical oscillations and peak values. Therefore, the ANCF method is regarded as a promising theoretical approach for the dynamics analysis of flexible structures like risers and cables in offshore operations. Based on the ANCF and continuum mechanics theory, Zhang et al. [57] utilized the finite element method to establish a mechanical analysis model to investigate the static and dynamic responses of LWRs, extending the ANCF method to the field of marine engineering. The results obtained from the ANCF approach were larger than those by the LMM, and the maximum error reached about 6.65%. Wang et al. [58] applied the ANCF method to the simulation of flexible structures in marine operation and developed internal code MiNos using MATLAB program. Compared to OrcaFlex software, MiNos demonstrated a reduction in significant bending moment peaks due to the superior continuity of the ANCF elements.

3.1.2. Three-Dimensional Model of LWRs

Compared to 2D models of LWRs, 3D models can more accurately represent the shape and structure response of a riser, including its configuration, inclination angles, and connection points, thus providing a more comprehensive reflection of its actual design. Additionally, 3D models can account for loads from different directions, such as waves and currents, allowing for better simulation and analysis of a riser's dynamic response in a complex marine environment. Li et al. [48] employed the vector form intrinsic finite element (VFIFE) method to establish a physical model of the LWR and analyze its nonlinear behavior in 3D space. Differently from traditional mathematical models, the VFIFE method, based on vector mechanics theory and numerical computation, does not require the integrated stiffness matrix and the iterative solution for the governing equations, so it can effectively solve structural problems flexibly. As depicted in Figure 8, the initial riser is assumed to be an oblique straight line meshed into N segments by $N + 1$ particles. Meanwhile, the gravity, buoyancy forces, and hydrodynamic loads are loaded on the riser with ramp function to eliminate transient response and reduce the time to obtain steady-state response. Considering the effect of the structural damping, the governing equations for the particle can be expressed:

$$m_j \frac{d^2}{dt^2}\begin{bmatrix} x_j \\ y_j \\ z_j \end{bmatrix} = \begin{bmatrix} f_{jx}^{exi} \\ f_{jy}^{exi} \\ f_{jz}^{exi} \end{bmatrix} + \begin{bmatrix} f_{jx}^{int} \\ f_{jy}^{int} \\ f_{jz}^{int} \end{bmatrix} + \begin{bmatrix} f_{jx}^{dmp} \\ f_{jy}^{dmp} \\ f_{jz}^{dmp} \end{bmatrix} \tag{20}$$

$$\begin{bmatrix} I_{xx} & 0 & 0 \\ 0 & I_{yy} & 0 \\ 0 & 0 & I_{zz} \end{bmatrix} \frac{d^2}{dt^2}\begin{bmatrix} \theta_{jx} \\ \theta_{jy} \\ \theta_{jz} \end{bmatrix} = \begin{bmatrix} m_{jx}^{exi} \\ m_{jy}^{exi} \\ m_{jz}^{exi} \end{bmatrix} + \begin{bmatrix} m_{jx}^{int} \\ m_{jy}^{int} \\ m_{jz}^{int} \end{bmatrix} + \begin{bmatrix} m_{jx}^{dmp} \\ m_{jy}^{dmp} \\ m_{jz}^{dmp} \end{bmatrix} \tag{21}$$

where m_j denotes the mass of the particle j; the position vector $\mathbf{X}_j$ (x_j, y_j, z_j) in the particle motion formula and the direction vector $\boldsymbol{\theta}_j$ for the particle j are described by the global coordinates. More detailed symbol explanations can be seen in [48].

The above research adopts the uncoupled method to investigate the dynamic response of the riser system, which decouples the riser from the vessel movement and can reduce the dynamic top tension, TDZ stress, and fatigue damage [59–62]. This decoupled method does not consume excessive computational resources and is time-efficient. However, this method has certain limitations in the analysis of floating offshore systems: it separates the analysis of vessel motion from the load effects of mooring lines and risers into two steps, ignoring the interactions between them [63,64].

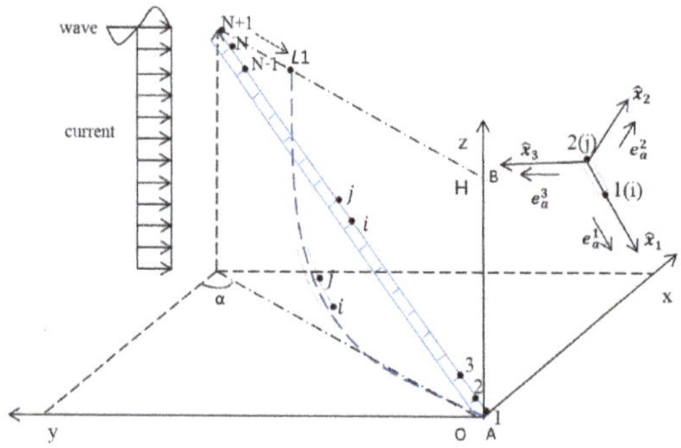

Figure 8. Riser model based on VFIFE method [48].

To comprehensively consider the interactions among various components in the system and improve the accuracy and reliability of structural analyses, coupled methods were developed. Ormberg and Larsen [63] conducted numerical simulations using both uncoupled and coupled methods, and they pointed out that the consistency of the model verification results with the coupled method was higher compared to the decoupled method. Then, Kim and Kim [15] developed the time-domain hull/mooring/riser fully coupled dynamic simulation program CHARM3D to carry out coupled numerical analyses. The transmitted forces from the mooring lines and risers to the platform are given as follows:

$$\widetilde{F}_P = \widetilde{K}(\widetilde{T}\widetilde{u}_P - \widetilde{u}_I) + \widetilde{C}(\widetilde{T}\dot{\widetilde{u}}_P - \dot{\widetilde{u}}_I) \tag{22}$$

Then, the hull response equation can be added into the riser/mooring equations in the time domain:

$$\left(\widetilde{M} + \widetilde{M}_a(\infty)\right)\ddot{\widetilde{u}}_p + \int_0^{\infty} \widetilde{R}(t-\tau)\dot{\widetilde{u}}_p \mathrm{d}\tau + \widetilde{K}_H\widetilde{u}_p = \widetilde{F}_D + \widetilde{F}^{(1)} + \widetilde{F}^{(2)} + \widetilde{F}_p + \widetilde{F}_w + \widetilde{F}_c + \widetilde{F}_{WD} \tag{23}$$

where $\widetilde{K}$ and $\widetilde{C}$ are the stiffness and damping matrices for mooring lines at the connection point; $\widetilde{M}$ and $\widetilde{M}_a$ are the mass and added mass matrix; $\widetilde{R}$ denotes the retardation function matrix. More detailed symbol explanations can be seen in [15].

By comparing the dynamic characteristics of the SCR and SLWR near TDZ, Kim and Kim [15] pointed out that the local dynamic buckling of the SCR near the TDZ will exacerbate the bending moment and bending stress responses in that area; in contrast, the LWR can effectively avoid local dynamic buckling near the TDZ due to its additional sag and hog bends, which can absorb the dynamic motion energy. In addition, the slender rod theory was employed in the CHARM3D program to carry out the static and dynamic analysis of mooring lines and risers. This theory was initially proposed by Nordgren [65] and Garrett [66]. Its major advantage lies in the fact that the motion governing equations are formulated directly in the global coordinate system and include all geometric nonlinearities, which can eliminate the cumbersome coordinate transformations between different coordinate systems and improve computing efficiency. Consequently, it has been widely applied in studies of static/dynamic response of marine slender rod structures [13,67–69]. Based on the initial slender rod theory, Paulling and Webster [70] improved and proposed a slender rod theory model with the assumption of small tensile deformations. To date, Texas A&M University has conducted extensive and systematic research on the slender rod theory and developed the fully coupled computation program CABLE 3D in Fortran to study the global coupled dynamic responses of floating production systems, anchor lines, and offshore risers [71–75].

Taking the effects of internal flow and seabed resistance into account, Cheng et al. [21] utilized 3D slender rod theory to investigate the dynamic performance of a LWR under vessel offsets and wave-current loads. This research indicated that when the LWR is subjected only to wave and current loads, internal flow parameters such as density, velocity, and top pressure have a significant impact on the LWR dynamic behavior. Based on a 3D modified slender rod theory, Li et al. [76] and Ruan et al. [77] employed an extended Korteweg-de Vries equation to simulate internal solitary waves (ISW) to study the nonlinear dynamic response of a LWR subjected to vessel motions, surface waves, ocean currents, and ISW. They discovered that ISW can induce significant oscillatory displacements in both the horizontal and vertical directions of the LWR, which notably increases the possibility of riser collision and shear damage. Yu et al. [78] established a self-developed program to describe the dynamic behavior of the coupled platform-mooring-riser system. The configurations of two types of risers (SCR and LWR) under the combined excitation of ISW and random waves were well compared. The results demonstrated that compared to SCRs, LWRs consistently exhibit greater motion responses with time differences due to complex configurations.

In addition, Yu et al. [78] established a fully coupled analysis model of floating platform-mooring-risers (SCRs and SLWRs) under the effects of ISWs and random waves. Rigid body dynamics theory and LMM were adopted to simulate the responses of the floating platform and riser/mooring lines, respectively, achieving the coupling effect between them through semi-coupled analysis. As shown in Figure 9, the LMM was employed to simulate the riser/mooring line configuration:

$$\boldsymbol{R}_k = \sum_{i=1}^{3} r_{i,k}\boldsymbol{n}_i, k = 1,2,\ldots,n \tag{24}$$

$$S_{k+1/2} = \boldsymbol{R}_{k+1} - \boldsymbol{R}_k = \sum_{i=1}^{3} (r_{i,k+1} - r_{i,k})\boldsymbol{n}_i, k = 1,2,\ldots,n-1 \tag{25}$$

$$l_{k+1/2} = |\boldsymbol{R}_{k+1} - \boldsymbol{R}_k| = \sqrt{\sum_{i=1}^{3} (r_{i,k+1} - r_{i,k})^2}, k = 1,2,\ldots,n-1 \tag{26}$$

$$\boldsymbol{t}_{k+(1/2)} = \frac{\boldsymbol{S}_{k+(1/2)}}{\boldsymbol{l}_{k+(1/2)}} = \frac{\sum_{i=1}^{3} (\boldsymbol{r}_{i,k+1} - \boldsymbol{r}_{i,k})\boldsymbol{n}_i}{\sqrt{\sum_{i=1}^{3} (\boldsymbol{r}_{i,k+1} - \boldsymbol{r}_{i,k})^2}}, \boldsymbol{k} = 1,2,\cdots,\boldsymbol{n}-1 \tag{27}$$

$$\boldsymbol{q}_k = \frac{\boldsymbol{R}_{k+1} - \boldsymbol{R}_{k-1}}{|\boldsymbol{R}_{k+1} - \boldsymbol{R}_{k-1}|} = \frac{\sum_{i=1}^{3} (\boldsymbol{r}_{i,k+1} - \boldsymbol{r}_{i,k-1})\boldsymbol{n}_i}{\sqrt{\sum_{i=1}^{3} (\boldsymbol{r}_{i,k+1} - \boldsymbol{r}_{i,k-1})^2}}, \boldsymbol{k} = 2,3,\cdots,\boldsymbol{n}-1 \tag{28}$$

where $\boldsymbol{R}_k$ denotes the position vector of the kth lumped mass point; the segment vector pointing from $\boldsymbol{R}_k$ to $\boldsymbol{R}_{k+1}$ is denoted as $S_{k+1/2}$, with its length defined as $l_{k+1/2}$; the tangent direction $\boldsymbol{q}_k$ at each node is approximated as the direction pointing between two adjacent nodes to calculate the hydrodynamic loads. More detailed symbol explanations can be seen in [78].

In summary, research on the global response of LWRs has gradually shifted from initial static response analyses to dynamic response analyses. Due to the complex dynamic loads (including wave loads, structural stiffness, and hydraulic resistance) acting on the LWRs, dynamic simulations are more effective and accurate compared to static simulations in reflecting real-world conditions. For this reason, design and validation techniques based on dynamic simulations have been actively researched in marine operations, which aligns with current and future research trends.

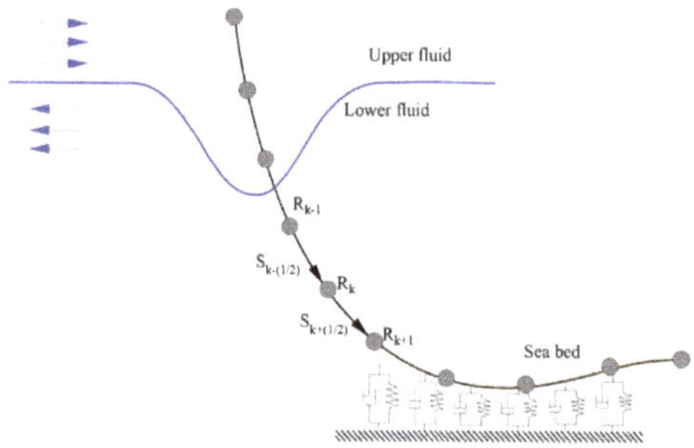

Figure 9. Discretization of riser/mooring based on the LMM [78].

3.2. Vortex-Induced Vibration

VIV could lead to the rapid accumulation of fatigue damage in risers and amplify the drag effects acting on the riser, which is induced by ocean currents and vessel motions. The occurrence and characteristics of the VIV are closely related to the Reynolds number. As the Reynolds number increases, the frequency and patterns of vortex shedding will change, which will affect the intensity and nature of the VIV. With low Reynolds numbers, vortex shedding usually presents a regular Karman vortex street pattern; however, with high Reynolds numbers, vortex shedding may become irregular, resulting in increased complexity and unpredictability of the VIV. Figure 10A showcases 3D vortex structures in the wake of a circular cylinder under different Reynolds number conditions [79]. These vortex structures include vortex loops and streamwise vortex pairs. With the increasing the Reynolds number, the vortex structures in the wake become more complex and unstable, exhibiting pronounced 3D characteristics [80]. Additionally, the flow field near the separation point on the cylinder varies across different Reynolds number regimes [81]. The period of vortex shedding is illustrated in Figure 10B. In this Figure, only the separatrices are shown, i.e., streamlines that leave or terminate at a saddle point.

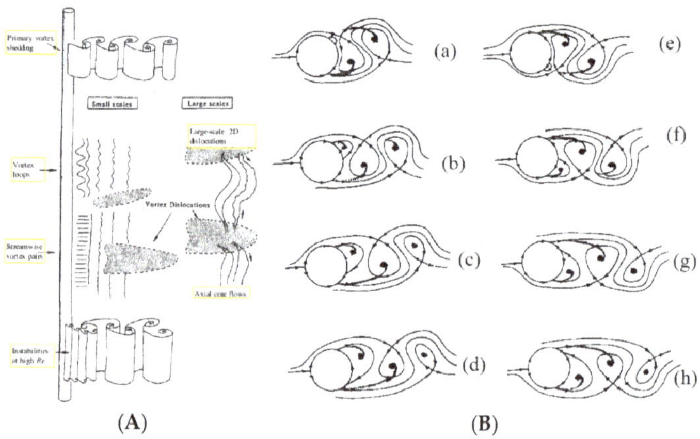

Figure 10. Vortex structures and vortex shedding period. (**A**) 3D features in the wake of a circular cylinder [79]. (**B**) Simplified model of vortex shedding period (**a**–**h**) [81].

In most cases, the time-varying and spatially varying ocean currents acting on the LWRs can be simplified to the steady current through a circular cross-section. When steady currents pass over offshore risers, periodic vortex will be generated in the wake of the riser, and offshore risers will vibrate at the same time, which is called VIV due to ocean currents [6,29,37–39]. However, rapidly time-varying currents and relative oscillations caused by vessel motions may lead to oscillatory currents. Under the action of vessel motions, the LWRs will be subjected to time-varying oscillatory currents, which may generate vorticity and periodic vortex-induced force, and finally cause the vibration of offshore risers. This fluid–structure interaction is called VIV due to vessel motion [82,83].

Research on the riser VIV primarily involves the application of computational fluid dynamics (CFD) for numerical simulations of the cylinder by both domestic and foreign scholars. However, studies on the VIV response for the combined structures with the riser and staggered buoyancy blocks are still relatively scarce. Jang and Kim [84] employed CFD simulation to numerically model the VIV response of the SLWR combined structure with staggered buoyancy blocks and riser and solved the structural equations of motion through modal analysis. As illustrated in Figure 11, Constantinides and Zhang [85] adopted a combination of CFD technology and semi-empirical software analysis to investigate the global response of a deepwater LWR with staggered buoyancy blocks.

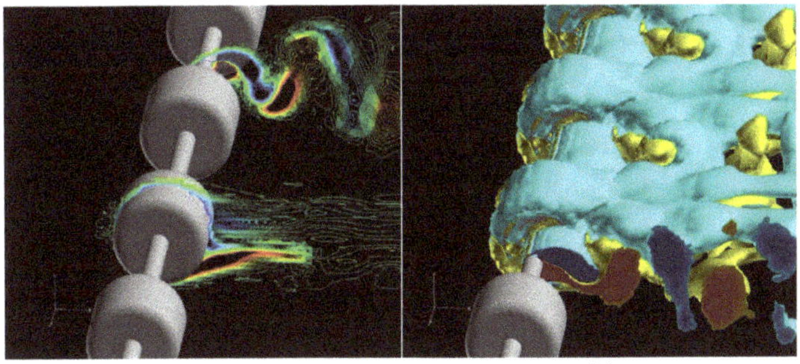

(**a**) (**b**)

Figure 11. CFD simulation of riser with staggered buoyancy blocks [85]. (**a**) Buoyancy module configuration illustrating flow over bare cylinder and buoyancy module. (**b**) Vortex structures over the buoyancy module area.

3.2.1. VIV Due to Vessel Motion

Due to the influence of vessel motion, the VIV response of the riser may become more complex. The vessel's roll and pitch motions would affect the formation and shedding patterns of vortices, subsequently altering the riser vibration characteristics [86]. Based on time-domain analysis method and the SIMA software workbench, Yin [82,83] simulated the VIV response of a full-scale LWR under the effect of vessel motion. This study constructed the total hydrodynamic loads in the CF direction:

$$F_{hydro,y} = \underbrace{\frac{1}{2}\rho D C_v |v_n| (j_3 \times v_n) \cos\phi_{exc}}_{vortex\ shedding} + \underbrace{\frac{1}{2}\rho D C_D |v_n| v_n}_{drag} + \underbrace{\frac{1}{4}\rho\pi C_M D^2 \dot{u}_n}_{water\ particle} \underbrace{- \frac{1}{4}\rho\pi (C_M - 1) D^2 \ddot{x}_n}_{cylinder\ added\ mass} \quad (29)$$

where the first term on the right side of Equation (29) represents the oscillating lift force due to vortex shedding, and the three last terms make up the Morison equation. More detailed symbol explanations can be seen in [82,83].

Research found that vessel motion will induce a normal relative velocity that varies along the LWR, with the largest value occurring at the lower part of the upper catenary. It was also pointed out that the VIV response of the LWR is dominated by the bare riser section. Additionally, Yin [82,83] highlighted the limitations of the time-domain models: since these time-domain models are semi-empirical, input parameters need to be obtained through experiments. Future improvements for the LWR's VIV due to vessel motion are needed in the following areas:

1. Rigid model (staggered buoyancy arrangement) test in oscillatory flow;
2. Computational fluid mechanics simulations to investigate 3D effects;

3. Experimental techniques to measure the displacement responses;
4. Higher-order components included in the time-domain model;
5. The consideration of in-line forces.

3.2.2. VIV Due to Current

Both the bare riser and buoyancy blocks in the LWR may experience the VIV under the influence of ocean currents. However, when the VIV occurs, the hydrodynamic characteristics of buoyancy blocks and their influence on the hydrodynamics of the bare riser section still require further investigation. A large number of studies indicated that the coverage of buoyancy blocks plays a crucial role in the competitive excitation between the bare riser section and buoyancy regions [37,87,88]. If the flexible cylinders with a staggered buoyancy system are linear, the frequency components in response will be the same as those in excitation forces; if the system is nonlinear, excitation frequency components will include those in the excitation forces and additional frequencies [87]. These additional frequencies arise from the nonlinear interaction of the cylinder's response at the excitation frequency components.

Subsequently, Rao et al. [37] investigated the VIV excitation characteristics of the buoyancy section and bare section of the LWR under five staggered buoyancy configurations. The research found that under almost all velocity conditions, the excitation response of the bare region predominates in pipes with a buoyancy block length-to-diameter ratio of 3. Furthermore, there were generally two excitation frequencies presented in the spectrum of risers with staggered buoyancy blocks, with the higher frequency related to the bare region and the lower frequency related to the buoyancy section. However, in addition to these two excitation frequencies, a third excitation frequency was observed (Figure 12), which is an unexpected combination of the two excitation frequencies and may arise from the nonlinear effects of hydrodynamic damping mechanisms.

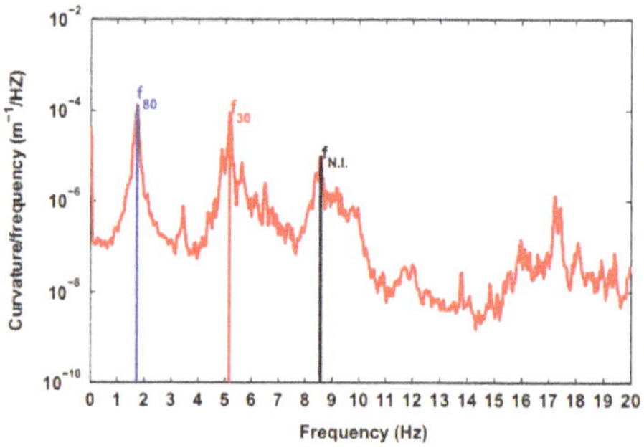

Figure 12. Schematic diagram of three frequencies [37].

To identify and compare the VIV responses in different regions, a frequency component identification method was proposed by Rao et al. [37]. The VIV excitations associated with 30 mm and 80 mm pipe diameters could be identified through the frequency components. The corresponding average VIV power can be computed numerically with Equations (30) and (31):

$$P_{30} = \int_L \frac{1}{T} \int_0^T c_{30} \dot{y}_{30}^2 dt dx \quad (30)$$

$$P_{80} = \int_L \frac{1}{T} \int_0^T c_{80} \dot{y}_{80}^2 dt dx \quad (31)$$

The reconstructed displacements y_{30} and y_{80}, and the damping c_{30} and c_{80} represent the corresponding values at fundamental excitation frequencies associated with 30 mm and 80 mm pipe diameters, respectively.

The forces induced on the risers by slug flows are intermittent, thus inducing excitations that may cause large oscillations on the suspended risers [89]. This internal flow-induced vibration can affect the VIV, and the two will interact, potentially causing resonance [90]. Bordalo and Morooka [89] established subroutines for internal flow forces to maintain a constant two-way dialog with the dynamic integrator of the pipeline movement simulator for solving the equations of motion for the riser (Figure 13). The research found that due to the influence of slug mass on the system's weight and inertia, the LWR oscillations induced by the VIV were altered in the presence of slugs. In the case of slug flow, intermittent mass distribution can lead to oscillatory motions in the suspended risers, thereby affecting the VIV response. Zhu et al. [91] investigated the VIV of a flexible LWR filled with various fluid media, including water and helium as typical examples of liquid and light gas, respectively. Similarly to [38], the VIV tests were conducted in the depth-averaged reduced velocity (Ur) range of 9.32–23.19 using non-intrusive optical measurement techniques with several high-speed cameras. The results indicated that when the internal fluid changed, both the excitation patterns and response amplitudes of the in-plane response were significantly adjusted, while the excitation patterns of the out-of-plane response showed no notable changes.

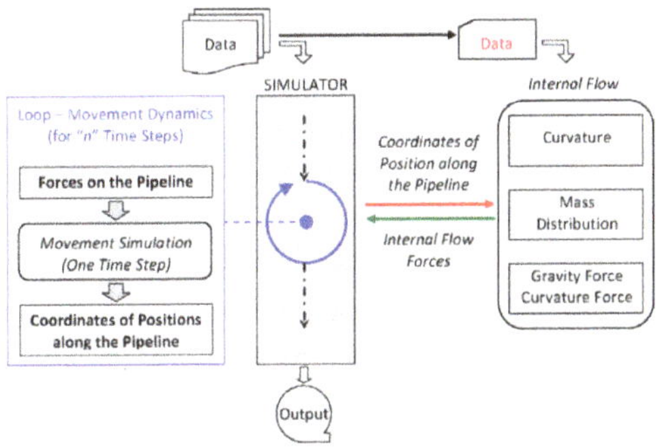

Figure 13. Interaction flowchart between pipeline simulator and internal flow subroutines [89].

3.3. *Experiment*

Due to the numerous assumptions commonly employed in numerical calculations, the results may be unreliable. To obtain accurate model results, the offshore engineering community tends to employ model tests to carry out research [92]. However, the limitations of existing laboratory facilities (such as water depth and flow velocity) make it impractical to perform full-scale LWR model testing at reasonable scales. Due to the large spatial dimensions of the offshore platform and its riser system, the model obtained by using the conventional scale ratio (1/50–1/100) still exceeds the scale range of the current testing facilities. At present, both physical model tests and numerical simulation have their own advantages and disadvantages, and the hybrid model test method which combines the advantages of both is the most widely applied and recognized solution for deepwater platform model testing. One of the key considerations in implementing hybrid model testing is to design an equivalent truncated system that matches the characteristics of the full-scale system, allowing it to serve as a substitute for the full-scale system in existing size models [93]. Currently, there are many studies on the global response tests of the SCRs [94], but there is little experimental research on the global response of the LWRs. Since fatigue damage estimations for risers within the industry tend to be conservative, the STREAM Joint Industry Project (JIP) was the first in the industry to compile and analyze full-scale field data from five risers (4 SCRs and 1 SLWR) to provide measurement data for modeling risers and advance fatigue damage research [95].

Research on the global response of LWRs has primarily focused on VIV experimental studies, such as the VIV response under different staggered buoyancy configurations [6,37,87,96]. In 2011, Shell Oil Company ran advanced experiments to investigate the effects of buoyancy block spacing on the VIV response at the Marintek Ocean Laboratory in Norway. The riser VIV model adopted a 38 m long riser (made of glass fiber) with an outer diameter of 0.03 m and a wall thickness of 1.5 mm, where the buoyancy block is 0.41 m in length and 0.08 m in diameter (Table 1) [6]. In this experiment, the ratio of the buoyancy diameter to the riser diameter was set as 2.67. This non-integer ratio was set to prevent the vortex shedding frequency of the bare riser from being a multiple of that of the buoyancy blocks, which is very critical in the LWR design. If the higher harmonics of the buoyancy block vortex shedding frequency are equal to the shedding frequency of the bare riser, this enhances the global VIV response of the LWR.

Table 1. The dimensions of the LWRs.

Organization	Outer Diameter	Wall Thickness	Buoyancy Block's Length	Buoyancy Block's Diameter
Shell Oil Company's model	0.03 m	0.0015 m	0.41 m	0.08 m
NDP's model	0.03 m	0.0015 m	0.15 m	0.15 m

Subsequently, to achieve an optimal design and arrangement of staggered buoyancy blocks that minimizes the VIV response, the Norwegian Deepwater Program (NDP) performed hydrodynamic testing on SLWR models subjected the VIV in the MARINTEK offshore basin [96]. The same type of fiberglass riser employed in Shell's VIV model experiments for the bare riser was selected in this program. A key difference from Shell's VIV model experiments was that the buoyancy block diameter was significantly larger than that of the bare riser. The buoyancy blocks were measured 0.15 m in length and 0.15 m in diameter in this program (Table 1).

The model was tested by towing the riser with different speeds to simulate uniform flow (Figure 14). Unlike the use of fins to suppress VIV in previous studies [34], strakes were utilized in this experiment to mitigate vortex shedding, thereby reducing vortex-induced vibrations. This program found that a large diameter ratio between the buoyancy blocks and bare riser may result in a substantially lower shedding frequency for the buoyancy blocks, potentially exciting much lower structural modes.

Figure 14. Test rig during testing—riser in water (adapted from [96]).

More and more new technologies have been applied in VIV experiments. Based on experimental models, Zhu et al. [91] conducted a study to investigate the VIV response of flexible LWRs filled with different fluid media. The experiment employed non-intrusive optical measurement techniques using a pair of high-speed cameras to simultaneously capture the IF and CF displacements with a sampling frequency of 100 Hz (Figure 15).

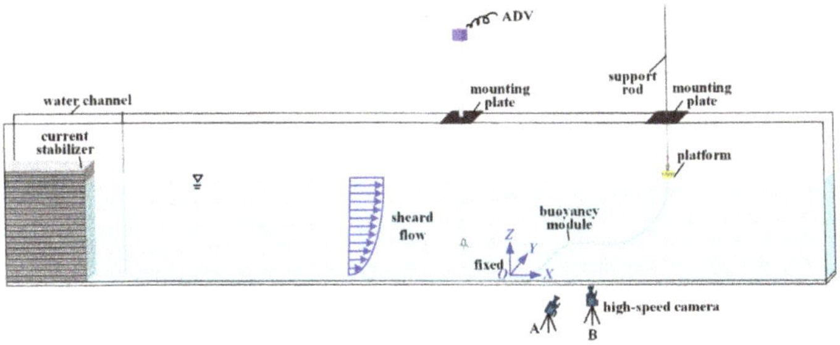

Figure 15. Schematic diagram of LWR's VIV test setup [91].

3.4. Abandonment and Recovery

In actual engineering projects, multiple offshore risers are typically installed on a floating platform. To improve construction progress and installation efficiency, a pre-lay approach is commonly adopted for the offshore laying and installation of marine risers. If the installation of the LWR is completed before the platform arrives, the LWR needs to be abandoned on the seabed and later recovered and reinstalled when the platform becomes available [97]. The LWR abandonment and recovery can be realized by controlling the tension and length of the A&R (Abandonment and Recovery) cable, as well as the position of the installation vessel. Many researchers have extensively studied the installation process of the SCRs [98–100], but there are fewer analyses on the installation process for LWRs, indicating a need for more comprehensive and in-depth research. There are numerous challenges that must be overcome during the abandonment, recovery, and tow transfer processes of LWR systems [101], which include determining the abandonment path for buoyancy blocks, applying tension to balance curve stability, ensuring the stability of the seabed's stagnation wave "hump", and applying stress at flexible joints, etc. Additionally, the recovery and transfer operations also face challenges related to gap monitoring between mooring lines and already installed risers.

According to the nonlinear large deformation beam theory, Wang et al. [102] divided the LWR transfer process into three stages (lowering stage, tension transfer stage, and recovery stage) to perform numerical analysis of the transfer process during the LWR installation. It was observed that the changes in the global riser configuration were very small compared to the local configurations near the pull-head and the TDP. In the same year, Wang et al. [97] pointed out that the LWR abandonment and recovery operations experienced a process with moving boundary conditions, making it difficult to determine the length of the riser suspended segment. Then, three different abandonment methods are employed for LWR installation operations: (1) maintain a constant top angle at the pull head during abandonment; (2) maintain a constant top tension at the pull head during abandonment; (3) maintain a constant configuration of the arch-shaped cross-section after the upper catenary portion of the SLWR contacts the seabed.

Given that hydrodynamic loads lead to continuous interaction between the riser and the seabed, the TDZ of the LWR is one of the fatigue hotspots [33]. Based on this, Wang et al. [50] developed an analytical model for the transfer process of the SLWR from the installation vessel to the production platform on an elastic seabed. This model facilitates the identification of an optimal installation path, ensuring the safety of the SLWR during the transfer process. Based on simulation-based design process, Jaewon et al. [23] utilized multi-body dynamics simulations to select effective parameters that influence the response and they leveraged open source PIDO (Process Integration and Design Optimization) software to develop a dynamic simulation model to identify the optimal shape of an SLWR during installation. As depicted in Figure 16, Gu et al. [103] proposed an efficient computational program based on the principle of minimum potential energy, which was successfully applied to investigate the static tension effects during the installation process of the SLWR.

By considering various vessel movement speeds and cable recovery rates to execute the recovery methods, the recovery operations for an SLWR were assessed.

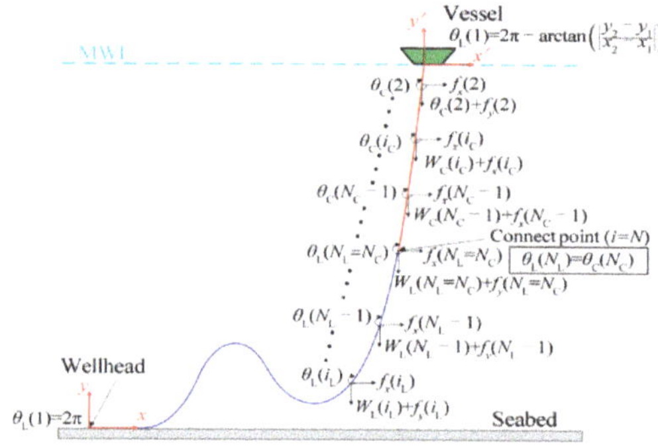

Figure 16. SLWR global coordinate system during abandonment and recovery [103].

4. Fatigue Damage of LWRs

The fatigue life of offshore risers is a critical aspect in the riser design process. Yang et al. [7] employed the finite element approach to carry out a dynamic stress analysis of the SLWR in the time domain, and systematically investigated the influence of hydrodynamic coefficients, structural damping, seabed stiffness, and floating structure motions on the fatigue damage of deepwater SLWR based on the S-N curve and Miner's linear cumulative damage rule. It was found that the motion characteristics of a floating structure have a significant impact on the fatigue damage of the SLWR. The research also indicated that increasing seabed stiffness by 15% reduces the minimum fatigue life in the TDZ from 158 years to 110 years from the base case. Based on Monte Carlo simulation with importance sampling (MCSIS), da Silva et al. [104] quantified the occurrence probability of failure modes and proposed a structural reliability fatigue solution to investigate the possibility of buoyancy length reduction in the LWR under fatigue failure modes (Figure 17). MCSIS is a numerical technique with variance reduction capability, and it is a method used to evaluate the probability of failure, requiring a relatively small number of samples (structural analyses) to obtain stable probability of failure (PF) results. This technique can be summarized by the following formula:

$$p_f = \frac{1}{N}\sum_{j=1}^{N}\left\{ I[G(x_j) \leq 0]\frac{f_X(x_j)}{h_X(x_j)} \right\} \tag{32}$$

where $G(\mathbf{X})$ represents the failure function; $f_X(x)$ represents the joint probability density function of all random variables $\mathbf{X}$; $h_X(x)$ denotes the so-called sampling distribution. Based on the Palmgren–Miner's summation rule for fatigue damage associated with S–N curves, the fatigue design criterion can be expressed as follows:

$$\frac{T_{Ref}K_D}{\sum\limits_{i=1}^{N} n_i(S_i)^m}\frac{\Delta}{SF} = \frac{T_{Ref}}{DT_{Ref}}\frac{\Delta}{SF} = \frac{FL}{SF} > T_{oper} \tag{33}$$

where K_D and m are the design S–N curve parameters; S_i denotes the stress cycles ranges identified in a reference period; Δ represents the allowable limit of accumulated fatigue damage; SF is a safety factor. In addition, the mean stress effect can be considered by means of the Gerber approach [105]. More detailed symbol explanations can be seen in [104].

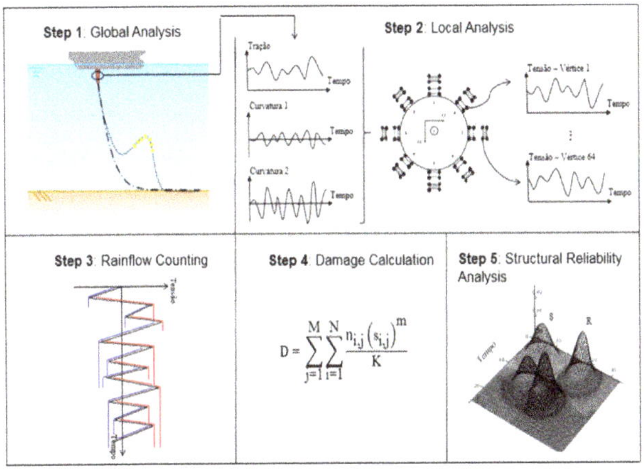

Figure 17. Fatigue calculation schematic sequence [104].

Hejazi et al. [106] employed extensive measured and hindcast metocean datasets in the fatigue design process of offshore risers. This research introduced a novel framework that leverages an "ANN-based technique" combined with the concept of a "representative (P50) year", which can make the direct application of large meteorological and oceanographic datasets estimate the riser fatigue life. Furthermore, the study proposed an innovative framework based on Monte Carlo methods to achieve more efficient long-term fatigue assessments of offshore risers. A fatigue monitoring approach for the SLWR based on a digital twin model was put forward by Lee et al. [107] to estimate the fatigue life of SLWRs in real-time. In this research, a machine learning algorithm was employed to estimate the 3D current distribution along the water column, which will be adopted as an additional load to refine the digital twin model. Based on a dual algorithm (DA), Chung et al. [108] put forward a novel structural health monitoring methodology for risers to accurately reproduce the movements and stresses along the whole lengths of the SCR and SLWR, with an error margin of approximately 5%, regardless of the riser's shape and material. This method not only can monitor the riser's condition, but also can assess the riser accumulated fatigue damage in real-time.

The buoyancy force provided by the buoyancy blocks makes the riser bulge and then produce a wave shape, which can effectively reduce the top tension of the riser and alleviate the coupling effect of the floating structure movement on the riser's TDZ, thus increasing the fatigue life of the area. In order to investigate the effect of buoyancy section number on the fatigue performance in the TDZ of the LWR, Ruan et al. [10] proposed a new concept of "multiple LWR", namely, LWR configuration based on multiple waveform serial arrangement (Figure 18).

Based on the improved slender rod theory, three riser configurations (catenary riser, LWR, multiple LWR) with different platform heave excitations and ocean currents were structured, and the corresponding dynamic response characteristics, excitation motion transfer paths, and fatigue damage hotspots were well compared. By constructing the lazy wave configuration, deepwater LWRs can attenuate the kinematic coupled response between the platform movement and riser's dynamic response in the TDZ, so as to optimize its anti-fatigue performance; compared to the LWR, multiple LWRs have some competitiveness in the anti-fatigue performance in the TDZ. Later, Ruan et al. [21] develop a systematic solution for the dynamic stress evaluation of flexible LWRs suffering from random wave and vessel motions, which can provide a good basis for fatigue analysis and damage assessment of LWRs.

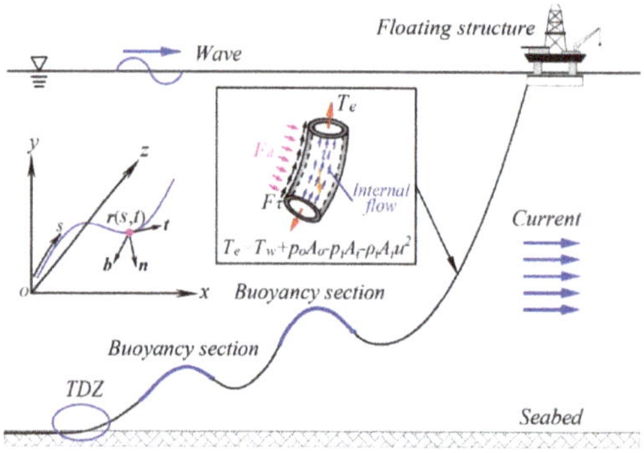

Figure 18. Schematic diagram of multiple LWR [10].

5. Structural Optimization of LWRs

The introduction of buoyancy blocks will increase the number of design parameters for the LWRs, making the optimization a pressing challenge. The unique structural form of LWRs also contributes to the configurational complexity, with different configurations varying significantly in terms of structural strength, mechanical performance, fatigue life, and economics. This brings significant uncertainty into the optimization design of LWRs. Given the numerous and interdependent design parameters, an increasing number of optimization algorithms have been applied in this field, achieving remarkable success. Based on genetic algorithm (GA), Tanaka et al. [30] took material and installation costs as the optimization goal to obtain an optimal LWR static configuration. Based on the given inner diameters and materials of the riser and buoyancy block, the optimizer was allowed to freely vary the number of riser segments, as well as the thickness, length, and buoyancy block diameter of each segment within user-defined ranges. Nevertheless, genetic algorithms have their limitations in this context. When dynamic and fatigue analyses are included, each evaluation of the objective function and constraints will take a longer time, potentially affecting the overall optimization efficiency. Furthermore, the optimized solution may be overly sensitive to small parameter changes, resulting in unrealistic configurations that become unfeasible due to manufacturing and installation tolerances and uncertainties.

De Pina et al. [31] employed a particle swarm optimization (PSO) algorithm to optimize the design of LWRs with the goal of minimizing cost, while considering requirements such as maximum equivalent stress, maximum total hanging angle, maximum tension, and minimum tension. This optimization process involved adjusting the length of each riser segment and the size of buoyancy materials. For specific optimization problems, the internal parameters in the PSO algorithm can significantly influence the optimization outcome. Aiming to identify the most cost-effective sizes, quantities, and positions of buoyancy blocks for lazy wave umbilical cables in installations, Yang et al. [109] proposed a multi-objective optimization method based on an approximation model for the dynamic installation of umbilical cables. The research took full account of the dynamic effects of environmental loads, resulting in safer and more accurate optimized results.

The optimization objective for LWRs is typically to achieve a cost-effective configuration while still being able to withstand extreme and fatigue loads, which leads to a balancing act between cost targets and multi-constraint engineering design problems. However, quantifying costs can be challenging due to factors such as material, installation, maintenance, and operation. Therefore, it is more appropriate to take objective functions based on structural criteria (e.g., strength, buckling, fatigue, and interference) during the design process. Ai et al. [17] achieved stress-based and fatigue-based objective optimizations in the distribution design of buoyancy blocks for the LWRs using genetic algorithm. It was noted that the definition of the objective function is not unique for critical design aspects. The distribution of buoyancy blocks is optimized to achieve an optimal configuration while minimizing stress, referred to as the stress objective. Therefore, the optimization problem can be defined as finding the solution vector x that minimizes the objective function f(x) within the given variable limits and constraints:

$$\begin{cases} \text{minimize Damage}(x) \text{ or Stress}(x)\ x \in R^n \\ x_i^l \le x_i \le x_i^u, i = 1, \ldots, 4 \\ \text{limit } \alpha, Z_{\max} \text{and} Z_{\min} \end{cases} \tag{34}$$

where x_i^l and x_i^u represent the lower and upper limits of the ith design variable.

Based on index optimization techniques, Ogbeifun et al. [110] came up with a 2D tabular optimization method for LWRs. This method can simplify the multi-dimensional problem into a 2D format, which can significantly reduce the computational resources required. In this approach, design variables are assembled in pairs, and indexes are assigned to the resulting design points for each combination, finally the optimal design points are tracked through index matching by data sorting and intersection operation techniques. This technology is utilized to optimize the LWR hanging length and hang-off tension, buoyancy section length, buoyancy module thickness, etc. Most studies employ standard GA or PSO to address problems with continuous design variables. However, discrete variables are involved in the design process for LWRs, such as the number of buoyancy blocks, which standard algorithms are not sufficient to handle. Therefore, improved algorithms are required to effectively deal with both continuous and discrete variables. Yu et al. [111] setup a self-developed program to calculate steel double LWR dynamics in a fully coupled platform/mooring/riser system, taking each segment length and buoyancy block parameter as design variables, and adopting the maximum static stress, horizontal motion range, and maximum dynamic stress standard deviation as optimization objectives. An improved PSO was proposed to handle both continuous and discrete design variables to automatically seek out the optimal solution. Furthermore, Feedforward Neural Networks (FNNs) were embedded into the optimization algorithm to replace time-consuming numerical calculations. The results showed that the FNNs provide high accuracy with an error of no more than 6.34%, and all determination coefficients are about 0.95.

6. Conclusions

In this paper, the research on LWRs over the past two decades is reviewed, outlining the unique structure of LWRs, hydrodynamic loads, global responses, fatigue damage, and the current state of research and achievements in the structural configuration optimization. In terms of global responses, the global static/dynamic response, vortex-induced vibration, installation, transfer, repair and other technical means, and research results are introduced from three aspects: numerical simulation, theoretical method, and model experiment.

The research status and advantages/disadvantages of the 2D mechanical model and 3D mechanical model for the LWRs are systematically summarized and compared. When conditions permit, 3D modeling should be the goal, and nonlinear large deformation theory should be employed. When computational resources are sufficient, fully coupled solutions should be adopted, avoiding uncoupled solutions as they can lead to low accuracy, especially in deep water scenarios. In model experiments, due to the inability to support full-scale models, it is essential to ensure that boundary conditions are satisfied when conducting truncated experiments to improve the accuracy of the results.

In addressing fatigue damage and the structural optimization of the LWRs, it is crucial to keep up with advancements and effectively utilize various advanced algorithms for analysis, ultimately achieving complementary advantages. The optimization of LWR configurations should incorporate target functions based on structural standards (such as strength, buckling, fatigue, and interference) during the design process. There are few studies on the LWR global response model test and the hydrodynamic test of the buoyancy section, so more numerous and deeper studies should be carried out to support the accurate calculation of the hydrodynamic loads in the buoyancy section of the LWR.

Author Contributions: Conceptualization, W.R. and Y.B.; methodology, W.R. and Y.B.; validation, Y.B.; investigation, C.Z. and B.S.; resources, W.R. and B.S.; writing—original draft preparation, W.R., C.Z., H.Y., and Z.W.; writing—review and editing, W.R., C.Z., H.Y., and Z.W.; supervision, B.S.; funding acquisition, W.R. All authors have read and agreed to the published version of the manuscript.

Funding: This research was funded by the Natural Science Foundation of Zhejiang Province (Grant No. LY23E090004), the Fundamental Research Funds for the Provincial Universities of Zhejiang (Grant No. RF-A2022006), the Ningbo Natural Science Foundation (Grant No. 2021J039), and the National Natural Science Foundation of China (Grant No. 51909236).

Institutional Review Board Statement: Not applicable.

Informed Consent Statement: Not applicable.

Data Availability Statement: Data will be made available upon request.

Conflicts of Interest: The authors declare no conflicts of interest.

References

1. Guo, X.; Fan, N.; Liu, Y.; Liu, X.; Wang, Z.; Xie, X.; Jia, Y. Deep seabed mining: Frontiers in engineering geology and environment. *Int. J. Coal Sci. Technol.* **2023**, *10*, 23. [CrossRef]
2. Bai, Q.; Bai, Y. 1—Introduction. In *Subsea Pipeline Design, Analysis, and Installation*; Bai, Q., Bai, Y., Eds.; Gulf Professional Publishing: Oxford, UK, 2014; pp. 3–21. [CrossRef]
3. Ruan, W.; Nie, Q.; Han, X.; Li, J.; Bai, Y.; Fu, X. Accurate modeling and safety evaluation of dented pipeline with internal pressure based on reverse modeling technique. *Ships Offshore Struct.* **2024**, 1–11. [CrossRef]
4. Yue, B.; Campbell, M.; Walters, D.; Thompson, H.; Raghavan, K. Improved SCR Design for Dynamic Vessel Applications. In Proceedings of the ASME 2010 29th International Conference on Ocean, Offshore and Arctic Engineering, Shanghai, China, 6–11 June 2010; pp. 495–504. [CrossRef]
5. Orimolade, A.P.; Karunakaran, D.; Meling, T.S. Steel Lazy Wave Risers From Turret Moored FPSO for Deepwater Harsh Environment. In Proceedings of the ASME 2015 34th International Conference on Ocean, Offshore and Arctic Engineering, Newfoundland, Canada, 31 May–5 June 2015. [CrossRef]
6. Jhingran, V.; Zhang, H.; Lie, H.; Braaten, H.; Vandiver, J.K. Buoyancy Spacing Implications for Fatigue Damage due to Vortex-Induced Vibrations on a Steel Lazy Wave Riser (SLWR). In Proceedings of the Offshore Technology Conference, Houston, TX, USA, 30 April–3 May 2012. [CrossRef]
7. Yang, H.; Li, H. Sensitivity analysis of fatigue life prediction for deepwater steel lazy wave catenary risers. *Sci. China Technol. Sci.* **2011**, *54*, 1881–1887. [CrossRef]
8. Karunakaran, D.; Nordsve, N.; Olufsen, A. An Efficient Metal Riser Configuration for Ship and Semi Based Production Systems. In Proceedings of the Sixth International Offshore and Polar Engineering, Los Angeles, CA, USA, 26–31 May 1996; pp. 156–162.
9. Hu, B.; Wang, Z.; Du, H.; Carriveau, R.; Ting, D.S.K.; Xiong, W. Response characteristics of flexible risers in offshore compressed air energy storage systems. *J. Marine. Sci. Appl.* **2019**, *18*, 353–365. [CrossRef]
10. Ruan, W.; Shi, J.; Sun, B.; Qi, K. Study on fatigue damage optimization mechanism of deepwater lazy wave risers based on multiple waveform serial arrangement. *Ocean Eng.* **2021**, *228*, 108926. [CrossRef]
11. Wu, M.; Huang, K. The comparison of various SCR configurations for bow turret moored FPSO in West Africa. In Proceedings of the 17th International Offshore and Polar Engineering Conference (ISOPE 2007), Lisbon, Portugal, 1–6 July 2007. Available online: https://onepetro.org/ISOPEIOPEC/proceedings-abstract/ISOPE07/All-ISOPE07/ISOPE-I-07-054/11041 (accessed on 24 August 2024).
12. Ruan, W.; Liu, S.; Li, Y.; Bei, Y.; Yuan, S. Nonlinear Dynamic Analysis of Deepwater Steel Lazy Wave Riser Subjected to Imposed Top-End Excitations. In Proceedings of the 35th ASME International Conference on Ocean, Offshore and Arctic Engineering, Busan, Republic of Korea, 19–24 June 2016. [CrossRef]
13. Ruan, W.; Bai, Y.; Cheng, P. Static analysis of deepwater lazy-wave umbilical on elastic seabed. *Ocean Eng.* **2014**, *91*, 73–83. [CrossRef]
14. Ruan, W.; Shang, Z.; Wu, J. Effective static stress range estimation for deepwater steel lazy-wave riser with vessel slow drift motion. *Ships Offshore Struct.* **2019**, *14*, 899–909. [CrossRef]
15. Kim, S.; Kim, M.H. Dynamic behaviors of conventional SCR and lazy-wave SCR for FPSOs in deepwater. *Ocean Eng.* **2015**, *106*, 396–414. [CrossRef]
16. Felisita, A.; Gudmestad, O.T.; Karunakaran, D.; Martinsen, L.O. Review of Steel lazy wave riser concepts for the North Sea. *J. Offshore Mech. Arct. Eng. Trans. ASME* **2017**, *139*, 15. [CrossRef]
17. Ai, S.M.; Xu, Y.; Kang, Z.; Yan, F. Performance comparison of stress-objective and fatigue-objective optimisation for steel lazy wave risers. *Ships Offshore Struct.* **2019**, *14*, 534–544. [CrossRef]
18. Wang, J.; Duan, M.; He, T.; Jing, C. Numerical solutions for nonlinear large deformation behaviour of deepwater steel lazy-wave riser. *Ships Offshore Struct.* **2014**, *9*, 655–668. [CrossRef]

19. Wang, J.; Duan, M. A nonlinear model for deepwater steel lazy-wave riser configuration with ocean current and internal flow. *Ocean Eng.* **2015**, *94*, 155–162. [CrossRef]
20. Yang, Z.; Yan, J.; Sævik, S.; Zhen, L.; Ye, N.; Chen, J.; Yue, Q. Multi-Objective Optimization Design of Flexible Risers Based on Bi-Scale Response Surface Models. In Proceedings of the ASME 2018 37th International Conference on Ocean, Offshore and Arctic Engineering. Volume 5: Pipelines, Risers, and Subsea Systems, Madrid, Spain, 17–22 June 2018. [CrossRef]
21. Cheng, Y.; Tang, L.Y.; Fan, T.H. Dynamic analysis of deepwater steel lazy wave riser with internal flow and seabed interaction using a nonlinear finite element method. *Ocean Eng.* **2020**, *209*, 107498. [CrossRef]
22. Ruan, W.; Nie, Q.L.; Lu, Y.T.; Chen, M.Y.; Liu, D.H.; Sun, B. A global-local approach for dynamic stress evaluation of lazy wave flexible risers subjected to random wave and vessel motion. *Ships Offshore Struct.* **2023**, *18*, 810–826. [CrossRef]
23. Oh, J.; Jung, D.; Kim, H.; Min, C.; Cho, S. A study on the simulation-based installation shape design method of steel lazy wave riser (SLWR) in ultra deepwater depth. *Ocean Eng.* **2020**, *197*, 106902. [CrossRef]
24. Ruan, W.; Bai, Y.; Yuan, S. Dynamic analysis of unbonded flexible pipe with bend stiffener constraint and bending hysteretic behavior. *Ocean Eng.* **2017**, *130*, 583–596. [CrossRef]
25. de Sousa, J.R.M.; Sousa, F.J.M.; Siqueira, M.Q.; Sagrilo, L.V.S.; de Lemos, C.A.D. A theoretical approach to predict the fatigue life of flexible pipes. *J. Appl. Math.* **2012**, *8*, 983819. [CrossRef]
26. Zhao, B. Fatigue Analysis of Flexible Riser-Effect of Mean Stress Correction Procedures. Master's Thesis, Institutt for Marin Teknikk, Trondheim, Norway, 2013. Available online: https://api.semanticscholar.org/CorpusID:107448295 (accessed on 24 August 2024).
27. Yin, D.; Lie, H.; Wu, J. Structural and hydrodynamic aspects of steel lazy wave riser in deepwater. *J. Offshore Mech. Arct. Eng. Trans. ASME* **2020**, *142*, 020801. [CrossRef]
28. Shanharan, R.; Anaturk, A.; Howells, H.; Lopes, M. Feasibility of Steel Lazy Wave Risers in the North Sea. In Proceeding of the MCD Deepwater Development, Addison-Wesley, Amsterdam, Netherlands, 3–5 April 2017. Available online: https://mcedd.com/wp-content/uploads/2017/Proceedings/14/MCEDD%202017%20Feasibility%20of%20Steel%20Lazy%20Wave%20Risers%20in%20the%20North%20Sea%20(Final).pdf (accessed on 24 August 2024).
29. Amaechi, C.V.; Wang, F.C.; Ye, J.Q. Investigation on Hydrodynamic Characteristics, Wave-Current Interaction and Sensitivity Analysis of Submarine Hoses Attached to a CALM Buoy. *J. Marine Sci. Eng.* **2022**, *10*, 120. [CrossRef]
30. Tanaka, R.L.; de Arruda Martins, C. A Genetic Algorithm Approach to Steel Riser Optimization. In Proceedings of the 25th International Conference on Offshore Mechanics and Arctic Engineering. Volume 1: Offshore Technology; Offshore Wind Energy; Ocean Research Technology; LNG Specialty Symposium, Hamburg, Germany, 4–9 June 2006; pp. 271–277. [CrossRef]
31. de Pina, A.A.; Albrecht, C.H.; de Lima, B.S.L.P.; Jacob, B.P. Tailoring the particle swarm optimization algorithm for the design of offshore oil production risers. *Optim. Eng.* **2011**, *12*, 215–235. [CrossRef]
32. Trapper, P.A. Feasible numerical analysis of steel lazy-wave riser. *Ocean Eng.* **2020**, *195*, 106643. [CrossRef]
33. Hou, Z.; Gaudin, C.; Sahdi, F.; Randolph, M. Centrifuge modelling of whole-life pipe-soil interaction in clay with different overconsolidation ratios. *Geotechnique* **2022**, *73*, 1056–1070. [CrossRef]
34. Ertekin, R.C.; Rodenbusch, G. Wave, current and wind loads. In *Springer Handbook of Ocean Engineering*; Dhanak, M.R., Xiros, N.I., Eds.; Springer International Publishing: Cham, Switzerland, 2016; pp. 787–818. [CrossRef]
35. Sarpkaya, T.S.; Isaacson, M.; Wehausen, J.V. Mechanics of wave forces on offshore structures. *J. Appl. Mech.* **1982**, *49*, 466–467. [CrossRef]
36. Sarpkaya, T.S. Wave Forces on Offshore Structures. Cambridge University Press: Cambridge, UK, 2010. [CrossRef]
37. Rao, Z.; Vandiver, J.K.; Jhingran, V. Vortex induced vibration excitation competition between bare and buoyant segments of flexible cylinders. *Ocean Eng.* **2015**, *94*, 186–198. [CrossRef]
38. Zhu, H.; Zhao, H.; Srinil, N. Experimental investigation on vortex-induced vibration and solid-structure impact of a near-bottom horizontal flexible pipeline in oblique shear flow. *J. Fluids Struct.* **2021**, *106*, 103356. [CrossRef]
39. El-Reedy, M.A. (Ed.) 4—Offshore structures design. In *Marine Structural Design Calculations*; Butterworth-Heinemann: Oxford, UK, 2015; pp. 85–187. [CrossRef]
40. Chanson, H. *Applied Hydrodynamics: An Introduction to Ideal and Real Fluid Flowsi*, 1st ed.; CRC Press: Boca Raton, FL, USA, 2009. [CrossRef]
41. Papillon, L.; Costello, R.; Ringwood, J.V. Boundary element and integral methods in potential flow theory: A review with a focus on wave energy applications. *J. Ocean. Eng. Mar. Energy* **2020**, *6*, 303–337. [CrossRef]
42. Xu, P.; Gong, S. Pipelay parametric investigation of pipeline dynamic behaviours for deepwater S-lay operation. *Ships Offshore Struct.* **2020**, *15*, 1141–1155. [CrossRef]
43. Xu, P.; Du, Z.; Huang, F.; Javanmardi, A. Numerical simulation of deepwater S-lay and J-lay pipeline using vector form intrinsic finite element method. *Ocean Eng.* **2021**, *234*, 109039. [CrossRef]
44. Orcina. *Orcaflex Manual*; Version 9.8a; Orcina Ltd.: Ulverton, Cumbria, UK, 2014. Available online: https://www.orcina.com/webhelp/OrcaFlex/Default.htm (accessed on 16 February 2020).
45. Morison, J.R.; Johnson, J.W.; Schaaf, S.A. The force exerted by surface waves on piles. *J. Pet. Technol.* **1950**, *2*, 149–154. [CrossRef]
46. Ruan, W.; Dai, W.; Wu, J. Study on motion transfer rule and extreme dynamic response of SCR's top-end heave excitation. *J. Mar. Eng. Technol.* **2021**, *20*, 186–199. [CrossRef]

47. Wang, J.; Duan, M.; He, R. A nonlinear dynamic model for 2D deepwater steel lazy-wave riser subjected to top-end imposed excitations. *Ships Offshore Struct.* **2018**, *13*, 330–342. [CrossRef]
48. Li, X.; Guo, X.; Guo, H. Vector form intrinsic finite element method for nonlinear analysis of three-dimensional marine risers. *Ocean Eng.* **2018**, *161*, 257–267. [CrossRef]
49. Hasanvand, E.; Edalat, P. Evaluation of the safe and failure zones of mooring and riser systems in a CALM oil terminal. *J. Mar. Sci. Appl.* **2021**, *20*, 751–766. [CrossRef]
50. Wang, Y.; Duan, M.; Gu, J. Analytical model for transfer process of deepwater steel lazy-wave riser on elastic seabed. *J. Mar. Sci. Technol.* **2019**, *24*, 123–133. [CrossRef]
51. Guarracino, F.; Mallardo, V. A refined analytical analysis of submerged pipelines in seabed laying. *Appl. Ocean Res.* **1999**, *21*, 281–293. [CrossRef]
52. Sparks, C.P. *Fundamentals of Marine Riser Mechanics: Basic Principles and Simplified Analyses*; Penn Well: Tulsa, OK, USA, 2007. Available online: https://books.google.com.hk/books?id=z0urYVzcFLUC (accessed on 28 August 2024).
53. Irani, M.B.; Modi, V.J.; Welt, F. Riser dynamics with internal flow and nutation damping. In Proceedings of the Sixth International Offshore Mechanics and Arctic Engineering Symposium, Houston, TX, USA, 1 March 1987; Volume 3, pp. 119–125.
54. Chucheepsakul, S.; Huang, T. Influence of Transported Mass On the Equilibrium Configuration of Risers. In Proceedings of the Fourth International Offshore and Polar Engineering Conference, Osaka, Japan, 10 April 1994; ISOPE-I-94-124. Available online: https://onepetro.org/ISOPEIOPEC/proceedings-abstract/ISOPE94/All-ISOPE94/26093 (accessed on 28 August 2024).
55. Jeong, H.; Jang, B.S.; Kim, J.D.; Park, G.; Choi, J.; Lee, D. A study on effects of slug flow on dynamic response and fatigue damage of risers. *Ocean Eng.* **2020**, *217*, 107965. [CrossRef]
56. Cabrera-Miranda, J.M.; Paik, J.K. Two-phase flow induced vibrations in a marine riser conveying a fluid with rectangular pulse train mass. *Ocean Eng.* **2019**, *174*, 71–83. [CrossRef]
57. Zhang, C.; Kang, Z.; Ma, G.; Xu, X. Mechanical modeling of deepwater flexible structures with large deformation based on absolute nodal coordinate formulation. *J. Mar. Sci. Technol.* **2019**, *24*, 1241–1255. [CrossRef]
58. Wang, C.; Liu, J.; Li, B.; Huang, W. The absolute nodal coordinate formulation in the analysis of offshore floating operations, Part II: Code validation and case study. *Ocean Eng.* **2023**, *281*, 114650. [CrossRef]
59. Jamaludin, F.E.; Koto, J. Catenary offset buoyant riser assembly for Malaysian deepwater. *Sci. Eng.* **2017**, *12*, 9–14. Available online: https://isomase.org/JSOse/Vol.12%20Dec%202017/12-2.pdf (accessed on 28 August 2024).
60. Karunakaran, D.N.; Baarholm, R. COBRA: An Uncoupled Riser System for Ultradeep Water in Harsh Environment. In Proceedings of the Offshore Technology Conference, Houston, TX, USA, 6–9 May 2013. OTC-23986-MS. [CrossRef]
61. Karunakaran, D.; Aasen, H.; Baarholm, R. New Un-coupled Deepwater Riser Concept for Harsh Environment–Catenary Offset Buoyant Riser Assembly (COBRA). In Proceedings of the Deepwater Offshore Technology Conference, New Orleans, LA, USA, 1 October 2011; pp. 11–13.
62. Nurwanto, T.; Karunakaran, D.; Franciss, R. COBRA Riser Concept for Ultra Deepwater Condition. In Proceedings of the ASME 2013 32nd International Conference on Ocean, Offshore and Arctic Engineering, Nantes, France, 9–14 June 2013. V04BT04A030. [CrossRef]
63. Ormberg, H.; Larsen, K. Coupled analysis of floater motion and mooring dynamics for a turret-moored ship. *Appl. Ocean. Res.* **1998**, *20*, 55–67. [CrossRef]
64. Ormberg, H.; Fylling, I.J.; Larsen, K.; Sødahl, N. Coupled Analysis of Vessel Motions and Mooring and Riser System Dynamics. In Proceedings of the International Conference on Offshore Mechanics and Arctic Engineering, Yokohama, Japan, 13–17 April 1997; pp. 91–100. Available online: https://jglobal.jst.go.jp/en/detail?JGLOBAL_ID=200902146948838040#:~:text=Coupled%20an (accessed on 2 September 2024).
65. Nordgren, R.P. On Computation of the motion of elastic rods. *J. Appl. Mech.* **1974**, *41*, 777–780. [CrossRef]
66. Garrett, D.L. Dynamic analysis of slender rods. *J. Energy Res. Technol.* **1982**, *104*, 302–306. [CrossRef]
67. Ma, W.; Lee, M.; Zou, J.; Huang, E.W. Deepwater Nonlinear Coupled Analysis Tool. In Proceedings of the Offshore Technology Conference, Houston, TX, USA, 1 May 2000. [CrossRef]
68. Low, Y.M.; Langley, R.S. Time and frequency domain coupled analysis of deepwater floating production systems. *Appl. Ocean Res.* **2006**, *28*, 371–385. [CrossRef]
69. Gu, H.; Guo, H.; Li, X.; Li, F.; Liu, Z.; Cui, P. Three-dimensional dynamic analysis of deep-water steel steep wave riser considering internal solitary wave. *J. Mar. Sci. Technolo.* **2022**, *27*, 452–466. [CrossRef]
70. Paulling, J.R.; Webster, W.C. A Consistent, Large-amplitude Analysis of the Coupled Response of a TLP and Tendon System. In Proceedings of the Fifth International Symposium on Offshore Mechanics and Arctic Engineering, Tokyo, Japan, 13–18 April 1986; American Society of Mechanical Engineers: New York, NY, USA; pp. 126–133. Available online: https://pascal-francis.inist.fr/vibad/index.php?action=getRecordDetail&idt=7963711 (accessed on 2 September 2024).
71. Ran, Z.H. Coupled Dynamic Analysis of Floating Structures in Waves and Currents. Ph.D. Dissertation, Texas A&M University, College Station, TX, USA, 2000. Available online: https://www.proquest.com/openview/65c0d6db18d616309960b1ba01b72e1e/1?pq-origsite=gscholar&cbl=18750&diss=y (accessed on 2 September 2024).
72. Kim, Y.B. Dynamic Analysis of Multiple-Body Floating Platforms Coupled with Mooring Lines and Risers. Ph.D. Dissertation, Texas A&M University, College Station, TX, USA, 2003. Available online: https://oaktrust.library.tamu.edu/server/api/core/bitstreams/2efba84c-2509-40bd-8624-97b9eae21ed3/content (accessed on 2 September 2024).

73. Ryu, S. Hull/Mooring/Riser Coupled Motion Simulations of Thruster-Assisted Moored Platforms. Ph.D. Dissertation, Texas A&M University, College Station, TX, USA, 2005. Available online: https://www.proquest.com/openview/b1cb14c54820f776e06256136c6151ea/1?pq-origsite=gscholar&cbl=18750&diss=y (accessed on 2 September 2024).
74. Tahar, A.; Kim, M.H. Hull/mooring/riser coupled dynamic analysis and sensitivity study of a tanker-based FPSO. *Appl. Ocean Res.* **2003**, *25*, 367–382. [CrossRef]
75. Tahar, A.; Kim, M.H. Coupled-dynamic analysis of floating structures with polyester mooring lines. *Ocean Eng.* **2008**, *35*, 1676–1685. [CrossRef]
76. Li, F.; Guo, H.; Gu, H.; Liu, Z.; Li, X. Deformation and stress analysis of the deepwater steel lazy wave riser subjected to internal solitary waves. *J. Ocean Univ. China.* **2023**, *22*, 377–392. [CrossRef]
77. Ruan, W.; Chen, M.; Nie, Q.; Xu, P.; Li, J.; Wang, X. Dynamic response of steel lazy wave riser considering the excitation of internal solitary wave and ocean currents. *Ocean Eng.* **2024**, *294*, 116708. [CrossRef]
78. Yu, Y.; Zhao, M.; Li, Z.; Yu, J.; Liu, C.; Xu, L. Comparative dynamic analysis of SCR and LWR in the coupled platform-mooring-riser system under internal solitary wave conditions. *Ocean Eng.* **2023**, *267*, 113215. [CrossRef]
79. Williamson, C.H. Vortex dynamics in the cylinder wake. *Annu. Rev. Fluid Mech.* **1996**, *28*, 477–539. [CrossRef]
80. Buresti, G. Vortex shedding from bluff bodies. In *Wind Effects on Buildings and Structures*; Riera, J.D., Davenport, A.G., Eds.; Balkema Publishers: Rotterdam, The Netherlands, 1998; pp. 61–95. Available online: https://www.researchgate.net/publication/321367112 (accessed on 4 September 2024).
81. Perry, A.E.; Chong, M.S.; Lim, T.T. The vortex-shedding process behind two-dimensional bluff bodies. *J. Fluid Mech.* **1982**, *116*, 77–90. [CrossRef]
82. Yin, D. Heave motion induced vortex-induced vibrations of a full-scale steel lazy wave riser. *ASME J. Offshore Mech. Arct. Eng. Trans.* **2022**, *144*, 041905. [CrossRef]
83. Yin, D. On Vessel Motion Induced Vortex-Induced Vibrations of a Steel Lazy Wave Riser. In Proceedings of the ASME 2021 40th International Conference on Ocean, Offshore and Arctic Engineering; Volume 4: Pipelines, Risers, and Subsea Systems, Online, 21–30 June 2021. [CrossRef]
84. Jang, H.; Kim, J.W. Numerical Investigation for Vortex-Induced Vibrations of Steel-Lazy-Wave-Risers: Part II—CFD Study on Long Flexible Riser. In Proceedings of the ASME 2019 38th International Conference on Ocean, Offshore and Arctic Engineering, Glasgow, Scotland, UK, 9–14 June 2019. [CrossRef]
85. Constantinides, Y.; Zhang, M. VIV Assessment of Deepwater Lazy-Wave Risers. In Proceedings of the ASME 2014 33rd International Conference on Ocean, Offshore and Arctic Engineering, San Francisco, CA, USA, 8–13 June 2014. [CrossRef]
86. Wang, J.; Fu, S.; Baarholm, R.; Wu, J.; Larsen, C.M. Out-of-plane vortex-induced vibration of a steel catenary riser caused by vessel motions. *Ocean Eng.* **2015**, *109*, 389–400. [CrossRef]
87. Rao, Z.; Vandiver, J.K.; Jhingran, V. VIV Excitation Competition Between Bare and Buoyant Segments of Flexible Cylinders. In Proceedings of the ASME 2013 32nd International Conference on Ocean, Offshore and Arctic Engineering, Nantes, France, 9–14 June 2013. [CrossRef]
88. Li, A.; Wu, B.; Fan, D. Vortex-induced vibration of risers with staggered buoyancy modules of small aspect ratio. *Appl. Ocean Res.* **2022**, *120*, 103014. [CrossRef]
89. Bordalo, S.N.; Morooka, C.K. Slug flow induced oscillations on subsea petroleum pipelines. *J. Pet. Sci. Eng.* **2018**, *165*, 535–549. [CrossRef]
90. Bossio, V.B.M.; Blanco, A.A.J.; Casanova, M.E.L. Numerical modeling of the dynamical interaction between slug flow and vortex induced vibration in horizontal submarine pipelines. *J. Offshore Mech. Arct. Eng.* **2014**, *136*, 041803. [CrossRef]
91. Zhu, H.; Liu, W.; Gao, Y.; Deng, K.R.; Zhou, T.M. Experimental investigation on the vortex-induced vibration response of a lazy-wave flexible riser filled with different fluid media. *Phys. Fluids* **2024**, *36*, 037106. [CrossRef]
92. Guo, L.; Pang, X.; Kuang, J.; Liu, S.; Nie, Y. Experimental study on coupling dynamics of a high dimensional nonlinear riser structure and floating platform by truncated equivalent method. *Expe. Tech.* **2024**, *48*, 523–535. [CrossRef]
93. Kim, M.H.; Koo, B.J.; Mercier, R.M.; Ward, E.G. Vessel/mooring/riser coupled dynamic analysis of a turret-moored FPSO compared with OTRC experiment. *Ocean Eng.* **2005**, *32*, 1780–1802. [CrossRef]
94. Bridge, C.; Howells, H.; Toy, N.; Parke, G.A.R.; Woods, R. Full-scale model tests of a steel catenary riser. *WIT Trans. Built Environ.* **2003**, *71*, 107–116. Available online: https://www.witpress.com/Secure/elibrary/papers/FSI03/FSI03011FU.pdf (accessed on 5 September 2024).
95. Deka, D.; Chandra, Y.; Campbell, M.; Santala, M.; Constantinides, Y.; Jin, J.; Darilmaz, I.; Nadathur, R.; Yiu, F. STREAM JIP–Insights into Steel Catenary Riser Response Using Measured Data. In Proceedings of the Offshore Technology Conference, Houston, TX, USA, 6–9 May 2019. Available online: https://onepetro.org/OTCONF/proceedings-abstract/19OTC/2-19OTC/D021S018R005/180663 (accessed on 5 September 2024).
96. Wu, J.; Lie, H.; Constantinides, Y.; Baarholm, R.J. NDP Riser VIV Model Test With Staggered Buoyancy Elements. In Proceedings of the ASME 2016 35th International Conference on Ocean, Offshore and Arctic Engineering. Volume 2: CFD and VIV, Busan, Republic of Korea, 19–24 June 2016. [CrossRef]
97. Wang, J.; Duan, M.; Luo, J. Mathematical model of steel lazy-wave riser abandonment and recovery in deepwater. *Mar. Struct.* **2015**, *41*, 127–153. [CrossRef]

98. Du, D.; Acevedo, V.A.; Hamadi, R.; Parikh, S. Static and dynamic mechanism of an SCR clamp system in deep water installation. In Proceedings of the 27th International Ocean and Polar Engineering Conference, San Francisco, CA, USA, 25–30 June 2017. Available online: https://onepetro.org/ISOPEIOPEC/proceedings-abstract/ISOPE17/All-ISOPE17/ISOPE-I-17-496/18122 (accessed on 5 September 2024).
99. Wang, H.; He, N.; Xu, F.; Cheng, Y. The Development and Installation of the SCR Pull-In System for Lingshui 17-2 Project. In Proceedings of the 32nd International Ocean and Polar Engineering Conference, Shanghai, China, 5 June 2022. Available online: https://onepetro.org/ISOPEIOPEC/proceedings-abstract/ISOPE22/All-ISOPE22/ISOPE-I-22-187/493786 (accessed on 5 September 2024).
100. Wever, R.; Hendriks, P. Deepwater SCR Installation with the HLV Thialf. In Proceedings of the ASME 2009 28th International Conference on Ocean, Offshore and Arctic Engineering. Volume 3: Pipeline and Riser Technology, Honolulu, HI, USA, 31 May 2009; pp. 787–795. [CrossRef]
101. Thomas, B.; Benirschke, A.; Sarkar, T. Parque das Conchas (BC-10) Steel Lazy Wave Riser Installation: Pre-Abandonment, Recovery and Transfer Challenges. In Proceedings of the Offshore Technology Conference 2010, Houston, TX, USA, 3–6 May 2010; p. OTC-20605-MS. [CrossRef]
102. Wang, J.; Duan, M.; Wang, Y.; Li, X.; Luo, J. A nonlinear mechanical model for deepwater steel lazy-wave riser transfer process during installation. *Appl. Ocean Res.* **2015**, *50*, 217–226. [CrossRef]
103. Gu, J.; Huang, J.; Gao, L.; Chen, L.; Jia, J.; Wang, S. Abandonment and Recovery Operation of Steel Lazy-Wave Riser in Deep-water by Controlled Vessel and Cable Velocity Rate. *China Ocean Eng.* **2023**, *37*, 29–41. [CrossRef]
104. da Silva, V.R.M.; Sagrilo, L.V.S.; Vignoles, M.A. Lazy-wave buoyancy length reduction based on fatigue reliability analysis. *ASME J. Offshore Mech. Arct. Eng. Trans.* **2018**, *140*, 031602. [CrossRef]
105. Almar-Næss, A. *Fatigue Handbook: Offshore Steel Structures*, 1st ed.; Tapir Academic Press: Trondheim, Norway, 1999.
106. Hejazi, R.; Grime, A.; Randolph, M.; Efthymiou, M. An efficient probabilistic framework for the long-term fatigue assessment of large diameter steel risers. *Appl. Ocean Res.* **2022**, *118*, 102941. [CrossRef]
107. Lee, Y.; Jin, C.; Kim, M.; Xu, W. Digital twin approach with minimal sensors for Riser's fatigue-damage estimation. *Int. J. Nav. Archit. Ocean Eng.* **2024**, *16*, 100603. [CrossRef]
108. Chung, W.; Jin, C.; Kim, M. Dual-algorithm hybrid method for riser structural health monitoring using the fewest sensors. *J. Mar. Sci. Eng.* **2022**, *10*, 1994. [CrossRef]
109. Yang, H.; Wang, A.; Li, H. Multi-objective optimization for deepwater dynamic umbilical installation analysis. *Sci. China Phys. Mech. Astron.* **2012**, *55*, 1445–1453. [CrossRef]
110. Ogbeifun, A.; Oterkus, S.; Race, J.; Naik, H.; Moorthy, D.; Bhowmik, S.; Ingram, J. A tabular optimisation technique for steel lazy wave riser. *IOP Conf. Ser. Mater. Sci. Eng.* **2021**, *1052*, 012022. [CrossRef]
111. Yu, Y.; Zhao, M.; Zhang, B.; Jin, Z.; Su, Y.; Tian, H.; Pang, H. Configuration design of a steel double lazy wave riser based on metamodel-assisted metaheuristic algorithms. *Appl. Ocean Res.* **2024**, *151*, 104159. [CrossRef]

MDPI AG
Grosspeteranlage 5
4052 Basel
Switzerland
Tel.: +41 61 683 77 34

Journal of Marine Science and Engineering Editorial Office
E-mail: jmse@mdpi.com
www.mdpi.com/journal/jmse

www.ingramcontent.com/pod-product-compliance
Lightning Source LLC
Chambersburg PA
CBHW041941260926
51850CB00018B/376
* 9 7 8 3 7 2 5 8 7 1 7 4 2 *